PHOTOIMMUNOLOGY

This book is dedicated to Annette, Hannah and Marie Krutmann, and to Dr Harry B. and Charlotte M. Elmets

PHOTOIMMUNOLOGY

EDITED BY

Jean Krutmann MD

Professor of Dermatology
Clinical and Experimental Photobiology
and Photoimmunology
Department of Dermatology
Heinrich-Heine-University
Düsseldorf, Germany

AND

Craig A. Elmets MD

Associate Professor
Skin Diseases Research Center
Department of Dermatology
Case Western Reserve University
University Hospitals of Cleveland
Cleveland, USA

FOREWORD BY

Margaret L. Kripke PhD

Blackwell
Science

Editorial Offices:
Osney Mead, Oxford OX2 0EL
25 John Street, London WC1N 2BL
23 Ainslie Place, Edinburgh EH3 6AJ
238 Main Street, Cambridge
Massachusetts 02142, USA
54 University Street, Carlton
Victoria 3053, Australia

Other Editorial Offices:
Arnette Blackwell SA
1, rue de Lille, 75007 Paris
France

Blackwell Wissenschafts-Verlag GmbH
Kurfürstendamm 57
10707 Berlin, Germany

Feldgasse 13, A-1238 Wien
Austria

First published 1995

Set by Excel Typesetters Company, Hong Kong
Printed and bound in Great Britain
at the University Press, Cambridge

DISTRIBUTORS

Marston Book Services Ltd
PO Box 87
Oxford OX2 0DT
(*Orders*: Tel: 01865 791155
Fax: 01865 791927
Telex: 837515)

North America
Blackwell Science, Inc.
238 Main Street
Cambridge, MA 02142
(*Orders*: Tel: 800 215-1000
617 876-7000
Fax: 617 492-5263)

Australia
Blackwell Science Pty Ltd
54 University Street
Carlton, Victoria 3053
(*Orders*: Tel: 03 347-0300
Fax: 03 349-3016)

A catalogue record for this title
is available from the British Library

ISBN 0-86542-826-3

Library of Congress
Cataloging-in-Publication Data

Photoimmunology / edited by Jean Krutmann,
Craig A. Elmets; foreword by Margaret L. Kripke.
p. cm.
Includes bibliographical references
and index.
ISBN 0-86542-826-3
1. Skin – Diseases – Immunological aspects.
2. Photobiology. 3. Skin – Cancer. 4. Photosensitization, Biological. I. Krutmann, Jean, 1959– . II. Elmets, Craig A.
[DNLM: 1. Skin Diseases – etiology.
2. Skin – immunology. 3. Skin – radiation effects.
4. Immune System – radiation effects.
5. Ultraviolet Rays – adverse effects.
WR 140 P5755 1995]
RL97.P46 1995
616.5′071 – dc20
DNLM/DLC
for Library of Congress 94-43729
CIP

Contents

List of contributors

David R. Bickers Columbia University College of Physicians and Surgeons, The Presbyterian Hospital in the City of New York, Columbia–Presbyterian Medical Center, 161 Fort Washington Avenue, New York, NY 10032, USA

Elisabeth Bohnert PhD, Department of Dermatology, Mannheim Medical School, University of Heidelberg, Theodor-Kutzer-Ufer, D-68167 Mannheim, Germany

Stephen M. Breathnach MD, PhD, FRCP, Immunodermatology Unit, St John's Institute of Dermatology, St Thomas' Hospital, Lambeth Palace Road, London SE1 7EH, UK

Eric Brown BS, Department of Immunology – Box 178, M.D. Anderson Cancer Center, 1515 Holcombe Boulevard, Houston, TX 77030, USA

Ponciano D. Cruz Jr MD, Associate Professor, Department of Dermatology, University of Texas, Southwestern Medical Center, 5323 Harry Hines Boulevard, Dallas, TX 75235-9069, USA

H.-Joachim Deeg MD, Fred Hutchinson Cancer Research Center, M31, 1124 Columbia Street, Seattle, WA 98104, USA

Cherrie Donawho PhD, Department of Immunology – Box 178, M.D. Anderson Cancer Center, 1515 Holcombe Boulevard, Houston, TX 77030, USA

Richard L. Edelson MD, Department of Dermatology, Photobiology Laboratory, Yale University, 333 Cedar Street, PO Box 3333, New Haven, CT 06510-8058, USA

Craig A. Elmets MD, Department of Dermatology, Case Western Reserve University, University Hospitals of Cleveland, 2074 Abington Road, Cleveland, OH 44106, USA

Francis P. Gasparro PhD, Department of Dermatology, Photobiology Laboratory, Yale University, 333 Cedar Street, PO Box 3333, New Haven, CT 06510-8058, USA

Günter Goerz MD, Professor of Dermatology, Department of Dermatology, Heinrich-Heine-University Düsseldorf, Moorenstrasse 5, D-40225 Düsseldorf, Germany

Stephan Grabbe MD, Zentrum für Dermatologie, Westfälische Wilhelms-Universität, Von-Esmarch-Strasse 56, D-48149 Münster, Germany

Richard D. Granstein MD, MGH–Harvard Cutaneous Biology Research Center, Massachusetts General Hospital East, Building 149, 13th Street, Charlestown, MA 02129, USA

Conrad Hauser MD, Department of Dermatology, University of Geneva Medical School, Geneva, Switzerland

John L.M. Hawk MD, FRCP, Photobiology Department, St John's Institute of Dermatology, St Thomas' Hospital, Lambeth Palace Road, London SE1 7EH, UK

Erhard Hölzle MD, Professor of Dermatology, Department of Dermatology, University of Hamburg, Martinistrasse 52, D-20251 Hamburg, Germany

Herbert Hönigsmann MD, Division of Special and Enviromental Dermatology, Department of Dermatology, University of Vienna, Währinger Gürtel 18–20, A-1090 Vienna, Austria

Amminikutty Jeevan PhD, Department of Immunology – Box 178, M.D. Anderson Cancer Center, 1515 Holcombe Boulevard, Houston, TX 77030, USA

Ernst G. Jung MD, Department of Dermatology, Mannheim Medical School, University of Heidelberg, Theodor-Kutzer-Ufer, D-68167 Mannheim, Germany

Irene E. Kochevar PhD, Wellman Laboratories of Photomedicine, Department of Dermatology, Harvard Medical School, Massachusetts General Hospital, Boston, MA 02114, USA

Margaret L. Kripke PhD, Department of Immunology – Box 178, M.D. Anderson Cancer Center, 1515 Holcombe Boulevard, Houston, TX 77030, USA

Jean Krutmann MD, Professor of Dermatology, Clinical and Experimental Photoimmunology and Photobiology, Department of Dermatology, Heinrich-Heine-University Düsseldorf, Moorenstrasse 5, D-40225 Düsseldorf, Germany

Thomas A. Luger MD, Department of Dermatology and Ludwig Boltzmann Institute for Cell Biology and Immunobiology of the Skin, University of Münster, Von-Esmarch-Strasse 56, D-48149 Münster, Germany

H. du P. Menagé BSc, MRCP, Photobiology Department, St John's Institute of Dermatology, St Thomas' Hospital, Lambeth Palace Road, London SE1 7EH, UK

Hans F. Merk MD, Professor of Dermatology, Department of Dermatology, Medical Unit of RWTH Aachen, Pauwelsstrasse 30, D-52057 Aachen, Germany

Hasan Mukhtar PhD, Skin Diseases Research Center, Department of Dermatology, Case Western Reserve University, University Hospitals of Cleveland, 2074 Abington Road, Cleveland, OH 44106, USA

David A. Norris MD, Professor of Dermatology, Department of Dermatology, University of Colorado Health Sciences Center, 4200 East Ninth Avenue, Denver, CO 80262, USA

Thomas Schwarz MD, Department of Dermatology and Ludwig Boltzmann Institute for Cell Biology and Immunobiology of the Skin, University of Münster, Von-Esmarch-Strasse 56, D-48149 Münster, Germany

Georg Stingl MD, Professor of Dermatology, Department of Dermatology, Division of Immunology, Allergy and Infectious Diseases, University of Vienna, Währinger Gürtel 18–20, A-1090 Vienna, Austria

J. Wayne Streilein MD, The Schepens Eye Research Institute, Department of Ophthalmology, Harvard Medical School, 70 Stamford Street, Boston, MA 02114, USA

Uwe Trefzer MD, Department of Dermatology, Humboldt-University Berlin (Charité), Schumannstrasse 20/21, D-10098 Berlin, Germany

Beatrix Volc-Platzer MD, Department of Dermatology, University of Vienna, Währinger Gürtel 18–20, A-1090 Vienna, Austria

Susan L. Walker MSc, Department of Photobiology, St John's Institute of Dermatology, United Medical and Dental Schools of Guy's and St Thomas' Hospitals, St Thomas' Campus, University of London, Lambeth Palace Road, London SE1 7EH, UK

Anthony R. Young PhD, Department of Photobiology, St John's Institute of Dermatology, United Medical and Dental Schools of Guy's and St Thomas' Hospitals, St Thomas' Campus, University of London, Lambeth Palace Road, London SE1 7EH, UK

Syed I.A. Zaidi PhD, Department of Dermatology, Case Western Reserve University, University Hospitals of Cleveland, 2074 Abington Road, Cleveland, OH 44106, USA

Foreword

The field of photoimmunology represents the intersection of a number of diverse fields, including dermatology, photobiology, immunology, and carcinogenesis. Its origins can be traced to studies of photosensitivity diseases, studies of sunburning, tanning and UV carcinogenesis, and studies on the immunology of the skin. In the 1970s interest in this subject was stimulated by the convergence of three independent areas of investigation. These were studies on the use of phototherapy for treatment of cutaneous diseases, the immunology of UV-induced skin cancers, and the function of epidermal Langerhans cells. Findings in these areas all pointed towards a common conclusion, namely that exposing the skin to UV light has profound, systemic effects on the immune system and can lead to suppression of certain immune responses.

The first wave of progress in photoimmunology during the 1970s and early 1980s centered around characterization of the effects of UV radiation on various immune responses. Interest in the field has been rekindled recently by exciting new advances in our understanding of the mechanisms of UV-induced immune suppression. This information is already being applied to studies of cutaneous diseases and used in attempts to prevent the induction of undesirable immune responses. It should also permit the development of new strategies for preventing or reversing the deleterious immunosuppressive effects of UV radiation.

This volume provides an overview of the principles of photobiology and immunology as they relate to the skin, and summarizes the effects of UV radiation on immunological processes. New information on the mechanisms by which UV radiation can interact with immune cells and modify immune function is also presented. The role of photoimmunology in the pathogenesis of skin cancer and photodermatoses, and its potential for involvement in infectious disease processes is reviewed. Finally, current and potential applications of UV radiation as a means of achieving specific immunosuppression and its use in the therapy of cutaneous diseases are discussed.

Thus within two decades a few isolated observations have coalesced to form the legitimate field of biomedical science called photoimmunology. The impending threat of global increases in solar UV radiation as a result of decreased stratospheric ozone have added a note of urgency to these studies and have no doubt contributed to the rapid rate of advancement of this field. Photoimmunology has already contributed greatly to our understanding of the immunophysiology of the skin and the normal regulation and functioning of the immune system; it promises to continue to improve our understanding of the pathogenesis of cutaneous diseases and to provide new avenues for therapy in the future.

Margaret L. Kripke
Houston, Texas

Preface

Man's relationship to the sun is precariously balanced between the benefits it provides in terms of life-sustaining heat, light and energy, and the repercussions engendered by injudicious overexposure. Within the biomedical sciences extraordinary attention has been given to the latter subject, and identification of the cellular, biochemical and molecular consequences of ultraviolet radiation exposure from the sun and other artificial light sources has been an issue of paramount concern. One consequence of this investigation has been the appreciation that ultraviolet radiation is a potent modulator of immunological function. Photoimmunology, which is the discipline of biomedical sciences that studies ultraviolet and other non-ionizing electromagnetic radiation effects on the immune system, brings together the interrelated fields of dermatology, immunology and photobiology.

Photoimmunology is only 20 years old, but has become one of the most rapidly advancing and expanding fields of biomedicine. A comprehensive review of the discipline has not appeared in over a decade. In that time significant progress has been made in filling the gaps between our knowledge of fundamental photobiology and immunology, and the resulting photoimmunological consequences. The application of molecular biological techniques has also contributed greatly to an understanding of the mechanisms that are operative in ultraviolet radiation-induced immunomodulation. A comprehensive treatise on the subject is thus long overdue.

In this book we have sought to develop a state-of-the-art study on the experimental and clinical aspects of photoimmunology. We have divided the book into four parts. Part I is designed to provide a basic understanding of photobiological and immunological principles. In Part II information regarding experimental aspects of photoimmunology is presented. Part III is devoted to clinical photoimmunology, with special emphasis on the relevance of photoimmunologic mechanisms to the pathogenesis of disease. Part IV will review the current status of photoimmunology as a therapeutic modality.

We hope that this book will prove to be a standard reference for photoimmunology, and thereby bring this increasingly important subspecialty closer to the medical, immunological and biological communities.

Jean Krutmann
Craig A. Elmets
Düsseldorf and *Cleveland, Ohio*

Abbreviations used in this book

ACLE	Acute cutaneous lupus erythematosus
AD	Atopic dermatitis
Ag	Antigen
APC	Antigen-presenting cell
ATPase	Adenosine triphosphatase
BFGF	Basic fibroblast growth factor
BMT	Bone marrow transplantation
CAD	Chronic actinic dermatitis
CCLE	Chronic cutaneous lupus erythematosus
CGRP	Calcitonin gene-related peptide
CHS	Contact hypersensitivity
CLA	Cutaneous leukocyte antigen
ConA	Concanavalin A
CTCL	Cutaneous T-cell lymphoma
CTL	Cytotoxic T lymphoma
DC	Dendritic cells
DNA	Deoxyribonucleic acid
DETC	Dendritic epidermal T cells
DNCB	Dinitrochlorobenzene
DNFB	Dinitrofluorobenzene
DPCP	Diphencyprone
DTH	Delayed-type hypersensitivity
ELAM-1	Endothelial cell lymphocyte adhesion molecule 1
EMR	Electromagnetic radiation
ETAF	Epidermal cell-derived thymocyte activating factor
FGF	Fibroblast growth factor
FITC	Fluorescein-isothiocyanate
GAT	L-Glutamic acid^{60}-L-alanine^{30}-L-tyrosine^{10}
GM-CSF	Granulocyte/macrophage colony-stimulating factor
GvHD	Graft-versus-host disease
HNK	Long-term cultured normal human keratinocytes
HPLC	High performance liquid chromatography
HSV	Herpes simplex virus
IARC	International Agency for Cancer Research
ICAM-1	Intercellular adhesion molecule 1
ICHB	Isolated congenital heart block
IFN	Interferon
IGF	Insulin-like growth factor
IL	Interleukin(s)
IL-1	Interleukin 1
IL-1R1	Type 1 interleukin 1 receptor
IL-6	Interleukin 6
KC	Keratinocyte(s)
KLH	Keyhole limpet hemocyanin
LC	Langerhans cell(s)
LFA-1	Lymphocyte function-associated antigen 1
LPS	Bacterial lipopolysaccharide
MAIDS	Murine acquired deficiency syndrome
M-CSF	Macrophage colony stimulating factor
MECLR	Mixed epidermal cell lymphocyte response
MED	Minimal erythemal dose
MGSA	Melanoma growth stimulating activity
MHC	Major histocompatibility complex
mICAM-1	Membrane-associated ICAM-1
MIP-1α	Macrophage inflammatory protein-1α
8-MOP	8-Methoxypsoralen
MSH	Melanocyte stimulating hormone
NGF	Nerve growth factor
NGFR	Nerve growth factor receptor

NLE	Neonatal lupus erythematosus
PBMC	Peripheral blood mononuclear cells
PC	Prohormone convertases
PCR	Polymerase chain reaction
PDGF	Platelet-derived growth factor
PDT	Photodynamic therapy
PE	Phycoerythrin
POMC	Proopiomelanocortin
PSS	Progressive systemic sclerosis
rh	Recombinant human
SALT	Skin-associated lymphoid tissues
SCF	Stem cell factor
SCI	Inhibitor of stem cell proliferation
SCLE	Subacute cutaneous lupus erythematosus
sICAM-1	Soluble ICAM-1
SIS	Skin immune system
SLE	Systemic lupus erythematosus
SPF	Sun protection factor
SSR	Solar simulated radiation
TAA	Tumor-associated antigens
TCR	T-cell antigen receptor
TGF	Transforming growth factor
T_h	T helper (cell)
T_s	T suppressor (cell)
TNCB	Trinitrochlorobenzene
TNF	Tumor necrosis factor
TNFα	Tumor necrosis factor-alpha
TNFR	Tumor necrosis factor receptor
TNP	Trinitrophenol
TSTA	Tumor-specific transplantation antigens
UCA	Urocanic acid
UV	Ultraviolet
UVA	Ultraviolet A
UVB	Ultraviolet B
UVBR	Ultraviolet B radiation
UVB-R	UVB resistant
UVB-S	UVB susceptible
UVC	Ultraviolet C
UVR	Ultraviolet radiation
VUV	Vacuum ultraviolet
XP	Xeroderma pigmentosum

PART 1
BASIC PHOTOIMMUNOLOGY

1: The skin immune response

Georg Stingl & Conrad Hauser

Introduction

All higher organisms require an intact immune system to detect and destroy invading microorganisms (bacteria, fungi, viruses and parasites) and to eliminate cells that have undergone malignant transformation. Anatomically, the first barrier to microbiologic invasion is the skin, a structure that was long considered only a passive barrier against that invasion. Over the last two decades, however, concepts of a previously unrecognized role for the skin have unfolded, a role in which bone marrow-derived dendritic leukocytes residing in the skin initiate and regulate the immune responses that protect it [111]. In combination with immunomodulatory resident (keratinocytes, mast cells, endothelial cells) and passenger (T lymphocytes recirculating between skin and regional lymph nodes) cells, these leukocytes have been termed SALT (skin-associated lymphoid tissues) and SIS (skin immune system) [14, 114].

In this review, we will describe and discuss the immunological functions of the cellular elements of SALT/SIS and, based on this information, will construct a hypothetical model of the molecular and cellular pathways operative in immune responses originating and terminating in the skin.

Antigen-presenting cells

Langerhans cells

Langerhans cells (LC) are dendritic cells of bone marrow derivation that reside mainly within stratified squamous epithelia but have also been shown to occur in mesenchymal tissues such as the dermis and lymph nodes. In the epidermis, they are usually located at a suprabasal level and constitute approximately 2–4% of all epidermal cells. LC visualization at the light microscopic level requires the use of appropriate histochemical and/or immunolabeling techniques (Fig. 1.1). Ultrastructurally, LC exhibit unique trilaminar cytoplasmic organelles (Birbeck granules) that allow their identification [111].

Phenotypic features of Langerhans cells

In the past few years, it has become clear that the phenotype of LC is determined by the quality and quantity of mediators present in their microenvironment. Resident and freshly isolated LC display non-specific esterase and adenosine triphosphatase (ATPase) activity, and carry surface-bound major histocompatibility complex (MHC) class I antigens as well as the panhemopoietic marker CD45. They are the only cells within the normal epidermis to express Fc IgG receptors type II (FcγRII/CD32) [95], Fc IgE receptors type I (FcεRI) [12, 128], C3bi receptors (CD11b/CD18), and MHC class II antigens [111]. CD1a, a non-polymorphic MHC class I-like molecule is also expressed by these cells and is currently the most reliable marker for human LC in the epidermis and in freshly isolated epidermal cells suspensions (Fig. 1.2) [111]. Very recent studies by Tang *et al.* [120] have shown that LC *in situ* express cadherins and that freshly isolated LC adhere to keratinocytes *in vitro* through E-cadherin. These authors concluded from their findings that expression of E-cadherins by LC promotes their persistence in the epidermis.

In epidermal cell cultures, LC acquire a pronounced dendritic configuration, and undergo profound changes in their marker repertoire. While

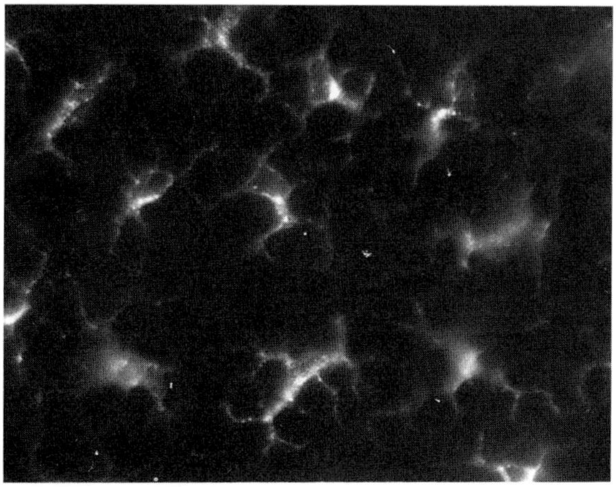

Fig. 1.1 Visualization of Langerhans cells in a sheet preparation of murine epidermis by means of a fluorochrome-labeled anti-MHC class II monoclonal antibody; ×640 (courtesy of Adelheid Elbe, PhD).

the expression of certain markers (FcγRII, ATPase activity, Birbeck granules, E-cadherin) continuously decreases, the surface density of other molecules greatly increases in culture: MHC class I and class II antigens [94, 98, 121, 130], RFD1 [94, 121], CD25 [94, 109], and several molecules which promote the binding of antigen-presenting cells to lymphocytes and, at the same time, provide an important signal for lymphocyte activation: CD40 [94], CD54 [94], CD58 [121] and B7/BB1 [60, 118]. Thus, LC in culture become essentially indistinguishable from certain MHC class II-bearing dendritic cells (DC) of lymphoid tissues which are extremely potent stimulators of both primed and naive T lymphocytes [110]. It is now clear that the keratinocyte-derived cytokine granulocyte/macrophage colony-stimulating factor (GM-CSF) is mainly responsible for the phenotypic and functional (see below) changes which LC undergo in epidermal cell cultures [47].

Functional capacities of Langerhans cells

Several groups of investigators have demonstrated that LC-containing epidermal cells can function as antigen-presenting cells, and as such can induce primary and secondary T-cell responses towards alloantigens, haptens, and soluble protein antigens including microbial antigens [111, 112, 129]. For a better understanding of this work, let us briefly reconsider a few basic principles of antigen presentation.

T cells recognize antigen via a clonally distributed polymorphic receptor molecule, the T-cell receptor (TCR), but they are unable to bind native antigen. Peptides derived from antigen and bound to a groove formed by the MHC molecules represent the physiological ligands of the TCR. Endogenously synthesized antigens such as tumor and viral antigens are degraded within the cytoplasm into peptides that are subsequently transported into an intracellular compartment that crosses the export pathway of nascent MHC class I molecules. Endogenous peptides bound to these MHC molecules are then transported to the cell surface for T-cell recognition. Extracellular antigens such as bacteria and proteins and immune complexes, as well as modified membrane proteins, are internalized into acidic intracellular compartments and degraded into peptides. Some of these peptides are protected from further degradation by binding to freshly synthesized MHC class II molecules that are transported through these compartments before appearing at the cell surface. The generation of peptide–MHC complexes is accomplished by specialized cells referred to as antigen-presenting cells, such as monocyte/macrophages, B cells, and cells of the DC lineage, including LC. These cells display (present) the peptide–MHC complex on their surfaces for recognition by T cells carrying specific receptors. The formation of a complex between TCR–MHC–peptide is required, but not sufficient, for the activation of the T cell (review in [35]). Additional signals, collectively referred to as second signals, are necessary for most T cell responses. In the past few years, second signals have started to be defined at the molecular level. The interactions that represent these second signals include the ligation of various cell surface receptors, such as B7/BB1, CD54 and CD58 by their respective counterreceptors on T cells and the release of soluble mediators such as IL-1 and IL-6 that bind to specific receptors on T cells. With regard to most T-cell responses, naive T cells have

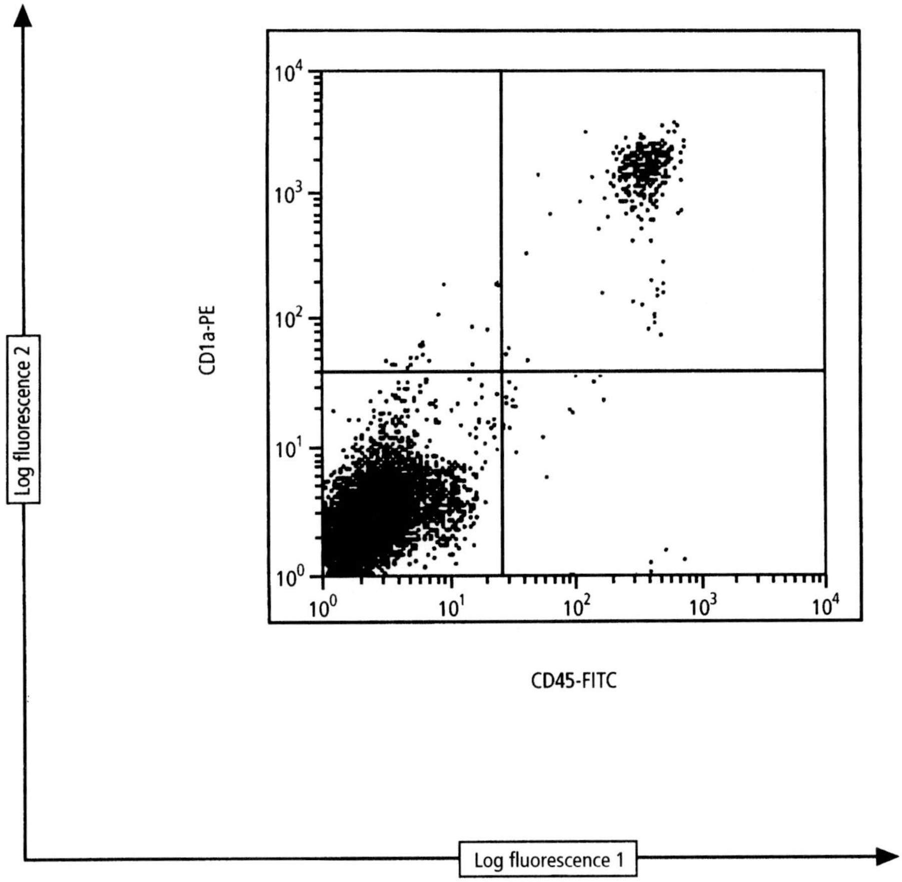

Fig. 1.2 Flow cytometric analysis of a human epidermal cell suspension using phycoerythrin (PE)-labeled anti-CD1a (ordinate) and fluorescein-isothiocyanate (FITC)-labeled anti-CD45 (abscissa) monoclonal antibodies. CD45/CD1a double-positive cells are located in the right upper quadrant and represent the LC population. CD45/CD1a double-negative cells are located in the left lower quadrant and mainly consist of keratinocytes. Single CD1a-positive cells (left upper quadrant) and single CD45-positive cells (right lower quadrant) occur very infrequently (courtesy of Dirk Strunk, MD).

more stringent requirements for second signals than previously activated cells (review in [71]).

The observation that certain cytokines apparently regulate the phenotype of LC, together with the development of purification techniques for both freshly isolated and cytokine-activated (cultured) LC, made it possible to address experimentally the basic mechanisms of LC antigen-presentation function. The results showed that freshly isolated LC are perfectly capable of processing large protein antigens, and of activating primed T cells, but are only poor stimulators of unprimed T cells [52, 93]. In contrast, cultured LC cannot efficiently process large protein antigens but are extremely effective in triggering naive, resting T cells [52, 93]. We have recently begun to understand the mechanisms underlying these functional differences between freshly isolated and cultured LC. It appears that the antigen-processing capacity of LC correlates with the level of biosynthesis of MHC class II molecules and with the expression of invariant chain [54, 87], and is dependent upon the presence of certain acidic organelles, such as endosomes and Birbeck granules [36, 113]. The possibility that transforming growth factor (TGF)-β, which is expressed by freshly isolated, but not cultured LC [39], keeps LC in an immature stage remains to be investigated. On the other hand, the GM-CSF/interleukin (IL)-1-induced acquisition of their immunostimulatory capacity for resting T cells [47] may be linked to an increased synthesis of: (i) adhesion molecules and other membrane moieties (e.g. β2 [103] and α4 [2] integrins, CD25 [94, 121], CD40 [94], CD54 [94, 103], CD58 [121], B7/BBI [60, 118] and, perhaps, the newly described ligand for CTLA-4 and CD28, B70/B7-2 [7, 31]); and/or (ii) cytokines, particularly IL-1β [25, 28, 49, 68, 82, 96] and IL-6 [96].

Ample evidence indicates that these *in vitro* findings are of relevance for the *in vivo* situation. Larsen *et al.* [61] documented that a few hours after skin transplantation, LC begin to enlarge and exhibit increased amounts of surface-bound MHC class II molecules. Subsequently a marked reduction in the number of epidermal LC occurs, concomitantly with the appearance of strongly Ia^+ cells in the dermis of the transplants. In keeping with this finding, Aiba *et al.* [1, 2] found that after application of a contact sensitizer, LC appear larger than normal, exhibit more intense anti-Ia and anti-α4 integrin staining, and are severalfold more potent in their T-cell stimulatory capacity than LC from non-treated or vehicle-treated animals. Concerning the role of cytokines in LC maturation and migration *in vivo*, Kaplan *et al.* [55] found that intradermal injection of GM-CSF results in LC recruitment, and Enk *et al.* [25] provided evidence for a critical role of IL-1β in the induction of contact hypersensitivity. They found that intradermal injection of IL-1β (but not IL-1α or TNFα) causes an enhancement of LC MHC class II expression, and that injection of an anti-IL-1β antibody into the skin prior to trinitrochlorobenzene (TNCB) treatment prevents sensitization to this allergen. Other investigators have identified antigen-bearing LC/DC in draining lymph nodes following the application of contact sensitizers [59, 66, 102], and have demonstrated that these cells can form clusters with lymphocytes [18], and can deliver signals resulting in antigen-specific T-cell activation [66]. On the basis of these findings, it appears likely that LC function involves two components: (i) LC bind and process antigen in the skin; and (ii) then acquire the capacity to emigrate and to trigger resting T cells in the lymphoid organs.

With the exception of the recent finding that cytokine-activated LC can activate allogeneic $CD8^+$ T lymphocytes in the absence of $CD4^+$ helper cells [23], our knowledge of LC-induced T-cell activation stems from studies of the interaction of LC and $CD4^+$ T cells*. Results have shown that hapten-modified LC can induce proliferation and IL-2 production in naive resting $CD4^+$ T cells [41]. When these preactivated cells are cultured with IL-2, they produce IL-2, but not IL-4, upon antigenic restimulation *in vitro* and mediate contact sensitivity *in vivo* [40]. Repeated stimulation of these cells with hapten-modified LC results in a modulation of their cytokine secretion profile from the original Th1 to a Th0 and, finally, Th2 pattern. Such LC-activated Th2 cells can incude IgE synthesis [42] and directly mediate tissue inflammation [72]. These and other findings indicate that LC can promote the differentiation of postthymic $CD4^+$ T cells into subsets with distinct immune functions.

The observation by Simon *et al.* [104] that ultraviolet (UV) irradiation of LC abrogates their stimulatory capacity for Th1, but not Th2, clones leads to the exciting concept that it may be possible to influence the outcome of cutaneous immune reactions (e.g. contact hypersensitivity (CHS), atopic dermatitis) by therapeutic manipulation of the LC population. Based on the evidence presented, one can assume that the major *in vivo* function of LC is to provide sensitizing signals in the induction of immune responses against anti-

*Based on the original observation by Mosmann *et al.* [70] that long-term T-cell clones do not produce a random pattern of cytokines, it is now agreed that activation of resting T cells can result in the generation of different helper cell subsets: (i) T helper (Th)1 cells which preferentially produce IL-2, interferon (IFN)-γ and IL-12; (ii) Th2 cells which synthesize IL-4, IL-5, IL-10 and IL-13, but not IL-2 and IFN-γ; and (iii) Th0 cells which have an unrestricted cytokine secretion pattern (review in [91]).

gens introduced into the skin (e.g. contactants, microorganisms) or newly generated in the skin (e.g. tumor antigens). The validity of this assumption is supported by additional *in vivo* studies.

1 While the application of a contact sensitizer to a skin area with high LC density leads to the induction of CHS, the application of the same contactant to a skin area deficient in LC results in antigen-specific unresponsiveness [24, 122].

2 Antigen-bearing LC induce sensitization even when administered via routes (e.g. intravenously) that favor the induction of tolerance [86, 117].

3 Compared to transplants with high LC density (body skin), transplants devoid of LC (central portion of cornea), or subtotally depleted of LC (tape-stripped skin), enjoy a prolonged survival on MHC class II-disparate recipients [115, 116].

4 GM-CSF-activated LC-containing epidermal cells, but not LC-depleted epidermal cells, are capable of presenting tumor antigens for the generation of protective antitumor immunity *in vivo* [37].

Given the enormous immunostimulatory potency of LC and related DC, it can be expected that the elucidation of the exact molecular mechanism(s) of this function will enable us to influence the final outcome of an immune response (induction of immunity or tolerance) at very early stages. More rapid progress in this research area has been hampered by the relative paucity of these cells and by difficulties in isolating and purifying them from various tissues. Fortunately, recent developments in various laboratories now suggest that these limitations may be solved and that cytokines [16, 51, 53, 89] and/or retroviruses [78] can be used to propagate LC/DC *in vitro* and even to generate LC/DC lines.

Dermal dendritic cells

In the past few years, several investigators have described a population of dendritic cells in the human dermis which are primarily located in perivascular areas [17, 107]. Morphologically these so-called dermal dendrocytes have a folded nucleus and a highly ruffled, irregular surface. Their cytoplasm is relatively dark and contains the organelles needed for an active cellular metabolism, but is devoid of Birbeck granules [107]. Immunostaining *in situ* [17, 107] as well as immunolabeling studies on single cell suspensions [69] suggest that these cells exhibit ATPase and non-specific esterase activity and display the CD45$^+$, CD1a^{lo}, CD1b,c,$^+$, CD11b$^+$, CD11c^{hi}, CD32$^+$, HLA-DR$^+$, factor XIIIa$^+$, CD3$^-$, CD19$^-$, CD14$^-$, CD15$^-$ phenotype. At a functional level, these cells are potent stimulators of allogeneic T cells [69].

Very recently, Lenz *et al.* [61] isolated MHC class II^{hi}, CD11/CD18$^+$, CD54$^+$, B7/BB1$^+$, CD40$^+$, CD1a^{lo}, CD36$^+$, CD3$^-$, CD19$^-$, CD20$^-$, CD14$^-$, CD68$^-$ dendritic cells from 2–5 day human dermal explant cultures. These cells proved to be excellent stimulators of the allogeneic mixed leukocyte reaction and powerful accessory cells in the oxidative mitogenesis assay. While their relationship to dermal dendrocytes and/or to the epidermal LC population remains to be determined, all data so far available are compatible with the assumption that there exists – in analogy to resident LC within the epidermis – an indigenous population of dendritic leukocytes in the dermis which, upon cytokine activation, can develop into powerful stimulators of resting T cells.

Immunomodulatory resident cells

The contention that the regulatory effects of cytokines on phenotype and function of dendritic antigen-presenting cells *in vitro* is of great *in vivo* relevance gains support from immunomorphological studies documenting a close proximity between dendritic antigen-presenting cells and resident skin cells capable of producing soluble mediators of the inflammatory and immune response.

Dermal dendritic cells are members of the 'dermal microvascular unit' and can conceivably receive signals from the other cellular components of this unit, i.e. endothelial cells, pericytes, mast cells and T cells [106]. Within the epidermis, the Langerhans cell enjoys an even closer association with its symbionts. Recently, Hosoi *et al.* [50] have elegantly shown that calcitonin gene-related peptide (CGRP)-containing nerves frequently impinge on LC of the human epidermis. The importance of this finding can be deduced from the additional observation that the neuropeptide CGRP substantially inhibits LC antigen presentation function. While LC may also have occasional encounters with melanocytes and Merkel cells, their most important epidermal partner is unquestionably the keratinocyte.

Keratinocytes are capable of producing and secreting growth factors and hormones as well as various mediators of the inflammatory reaction and the immune response, such as eicosanoids and cytokines (reviews in [58, 65]). Keratinocyte-derived cytokines include IL-1, IL-6, IL-7 [48], IL-8 and other chemokines, IL-10 [27], IL-12 (T. Schwarz, personal communication), GM-CSF, and tumor necrosis factor α (TNFα), as well as some of the factors regulating the growth of certain epithelial and/or mesenchymal cells, e.g. stem cell factor (G. Schuler, personal communication), TGFα and -β, platelet-derived growth factor (PDGF), and basic fibroblast growth factor (bFGF) (review in [65]). It should be kept in mind that – with the notable exceptions of IL-1, IL-7 and TGFβ – most of these mediators are not constitutively produced by keratinocytes or other skin cells, but only after the delivery of certain noxious stimuli (hypoxia, trauma, ultraviolet radiation, certain chemicals, etc.) [28, 56, 65]. Application of contact allergens, for example, leads to an upregulation in the transcription of various cytokine genes in keratinocytes including those (IL-1, IL-10, GM-CSF, TNFα) which affect LC viability, phenotype and function [26, 47, 57]. One must therefore assume that the induction of CHS or, more generally, of an immune response to an epicutaneously applied antigen, does not only require antigen/allergen uptake by LC, but also hapten-induced alterations in cytokine production by keratinocytes. The actions of these and other mediators on LC via specific receptor sites [4, 92] will then trigger a series of events finally resulting in the emigration and immunological maturation of these cells. The reader should not get the impression that the keratinocyte–LC interaction is a unidirectional process. Evidence exists that freshly isolated LC produce the macrophage inflammatory protein-1α (MIP-1α) [49, 68], a heparin-binding cytokine which is identical to a recently identified inhibitor of hemopoietic stem cell proliferation (SCI) [38]. Parkinson *et al.* [79] have recently shown that SCI/MIP-1α inhibits the proliferation of epidermal keratinocytes *in vitro*, and concluded that this factor plays an important regulatory role not only in hemopoiesis, but also keratopoiesis. It will be interesting to see whether the numerical reduction of LC in psoriasis [83] is also reflected in a decreased epidermal production of SCI/MIP-1α. If this were the case, one could speculate that a deficient SCI/MIP-1α production by LC of psoriatic patients may be causally linked to keratinocyte hyperproliferation characteristic of this disease.

Evidence is accumulating that keratinocytes, in addition to their secretory capacities, can functionally interact with lymphocytes in an antigen-specific fashion. The observation that keratinocytes can be effectively lysed by $CD8^+$ cytotoxic T cells in an MHC class I-restricted fashion [119] may be of pathogenetic significance in herpes simplex virus infections, graft-versus-host disease, and other skin disorders with a lichenoid infiltrate and keratinocyte necrosis. While keratinocytes in culture and in normal skin do not constitutively express MHC class II molecules, the expression of these moieties may be induced by IFN-γ both *in vitro* [10, 126] and *in vivo* [73]. It has been amply documented that keratinocytes synthesize and express MHC class II antigens in many skin disorders with prominent lymphocytic infiltration [6, 127]. This phenomenon has been attributed to IFN-γ production by infiltrating T cells. Functionally, class II-bearing keratinocytes are capable of inducing proliferation in allogeneic $CD4^+$ cell lines, but not in resting T cells [33, 74]. Although unable to process exogenous protein antigen, class II^+ keratinocytes can present processing-independent peptide to a T-cell hybridoma [33]. More important, perhaps, was the observation that such class II-bearing keratinocytes induce non-responsiveness (tolerance) to subsequent antigen-specific stimulation in Th1 clones [8, 33]. The *in vivo* importance of this *in vitro* finding has been demonstrated by the observation that injection of hapten-modified, class II^+ keratinocytes into naive mice leads to a transient specific hyporesponsiveness to subsequent sensitization by hapten [34]. Because the same phenomenon has been shown to occur in models using antigen-presenting cells which cannot provide the so-called 'second signal' necessary for T-cell activation, one can assume that keratinocytes are unable to deliver auxiliary stimuli critical to this event. In keeping with this reasoning is the recent finding by Gaspari *et al.* [32] that transfection of the B7/BB1 gene into MHC class II-bearing keratinocytes allows them to function as stimulators of the primary allogeneic mixed leukocyte reaction.

Lymphocytes

The lymphocytes of mammalian skin belong almost exclusively to the T-cell lineage as defined by the surface expression of CD3-associated TCR.

To date two major species of TCR have been identified. TCR α/β heterodimers are expressed on most peripheral T cells and recognize the nominal antigen in conjunction with MHC-encoded antigens [3]. The more recently identified TCR γ/δ heterodimers are present on early fetal thymocytes, on a minor fraction of adult thymocytes and peripheral T cells, and, at least in the mouse system, on a varying percentage of lymphocytes populating epithelial tissues [88]. Although the occurrence of signal transduction via CD3-TCR γ/δ has been demonstrated under several experimental conditions, the physiological ligand of this type of TCR has yet to be identified. Some evidence exists that TCR γ/δ may be receptors for first-line defense (e.g. enterotoxins, heat shock proteins); that they function as adhesion molecules (homing receptors); that they recognize MHC or MHC-like molecules (MHC class I and class II, Qa and TL in mouse, CD1 in humans); that they recognize antigen in conjunction with MHC or MHC-like structures; and that they behave as antiself receptors, thereby regulating ontogeny and response of TCR α/β-bearing cells [29].

Studies in rodents

The mouse epidermis harbors a population of basally located cells that are $CD45^+$, Thy-1^+, asialo-$GM1^+$, $CD3^+$, $CD5^-$, $CD4^-$, $CD8^-$, MHC class II^-, and uniformly dendritic in shape. They were originally termed 'Thy-1^+ dendritic epidermal cells' [11, 123] and are now generally referred to as dendritic epidermal T cells (DETC) [108]. Remarkably, the TCR diversity of murine DETC and, to a lesser extent, of their rat counterparts [99] is very limited. The overwhelming majority of these cells carry TCRγ/δ heterodimers [21, 108] and, at least in the mouse system, they are encoded almost exclusively by $V\gamma3/J\gamma1/C\gamma$ and $V\delta1/D\delta2/J\delta1/C\delta$ genes rearranged in the canonical sequence [5, 46]. Essentially, the same TCR configuration is found on early fetal thymocytes [43]. Together with the observation that the continuous disappearance of $V\gamma3/V\delta1$ cells from the fetal thymus is accompanied by the appearance of the first $V\gamma3/V\delta1$ cells in the skin [43], this finding led to the theory that $V\gamma3/V\delta1$ fetal thymocytes are predecessors of $V\gamma3/V\delta1$ DETC. An alternative hypothesis states that DETC maturation from a TCR^- into a TCR^+ cell occurs in the skin itself. Both theories have received some experimental support [22, 44, 76, 80, 81, 101], but the relative contribution of the thymic versus the cutaneous microenvironment in the physiological maturation of DETC precursors into DETC has yet to be determined. Independent of this question, it is experimentally well established that, upon their arrival in the skin at the end of gestation, DETC (or their precursors) undergo profound mitotic activity [22, 80]. Based on the early observation that optimal DETC proliferation is achieved by the combined use of Concanavalin A (ConA) + IL-2 [75, 124], IL-2 was originally considered to be the relevant DETC-stimulating cytokine *in vivo*. The absence of IL-2 bioactivity in mouse skin cell suspensions, as well as the lack of polymerase chain reaction (PCR)-detectable IL-2 mRNA in normal murine skin [56], renders this possibility quite unlikely. It rather appears that IL-7, which is constitutively produced by murine keratinocytes [48, 56], is the relevant growth factor for DETC. This cytokine maintains the long-term survival of DETC lines and, together with the costimulatory mitogen ConA, induces proliferation in freshly isolated DETC and DETC lines [67]. Translating these *in vitro* findings to the *in vivo* situation, one could postulate that keratinocyte-derived IL-7: (i) secures the long-term survival of DETC in the murine epidermis; and (ii) in combination with an additional signal – perhaps delivered by stressed keratinocytes [45] – drives these cells into proliferation. It is quite conceivable that the latter event and its consequences (mediator production?) are important defense mechanisms protecting the skin from noxious pathogens.

Studies in humans

Although several investigators have searched for the human equivalent to murine DETC, no comparable T-cell population has so far been identified in human skin. Most T cells of human skin are round to polygonal in shape, reside in the dermis rather than in the epidermis, and predominantly express TCR α/β rather than TCR γ/δ [13, 14, 30] (Fig. 1.3). Intradermal T cells are clustered preferentially around postcapillary venules and around the appendages, and consist of both $CD4^+$ and $CD8^+$ cells [14]. Most of the intraepidermal T cells exhibit the $CD4^-$, $CD8^+$, $CD2^+$, $CD3^+$/TCR α/β^+ phenotype, and are located within the basal layer of the epidermis and the acrosyringial epithelium [30]. Their highest density is encountered within the plantar epidermis. The further observation by these authors that the vast majority of T cells of human skin are $CD45RO^+$ suggests that they are memory cells [30].

The derivation of these intraepidermal memory cells can be explained in two ways. First, they might be generated by LC-mediated antigenic stimulation of naive T cells residing indigenously within the epidermis. Alternatively, they could enter the skin/epidermis in an already sensitized state. Because the ontogenetic development of skin T cells has not yet been carefully studied in humans, no meaningful statement can currently be made about the validity of the former concept. In contrast, there exist some experimental data, mainly derived from studies with the monoclonal antibody HECA-452, which support the latter theory. This antibody reacts with 10–20% of peripheral blood T cells and with 85% of lymphocytes in inflammatory skin lesions [85]. HECA-452-reactive cells have the $CD45^{low}$, $CD44^{high}$ phenotype and thus qualify as memory cells [15]. Cell attachment experiments have shown that memory T cells, but not resting T cells, adhere to endothelial cell lymphocyte adhesion molecule 1 (ELAM-1, E-selectin)$^+$ cells [100] and that the HECA-452$^+$ T-cell subset is the predominant E-selectin-binding population among circulating lymphocytes [84]. It has further been observed

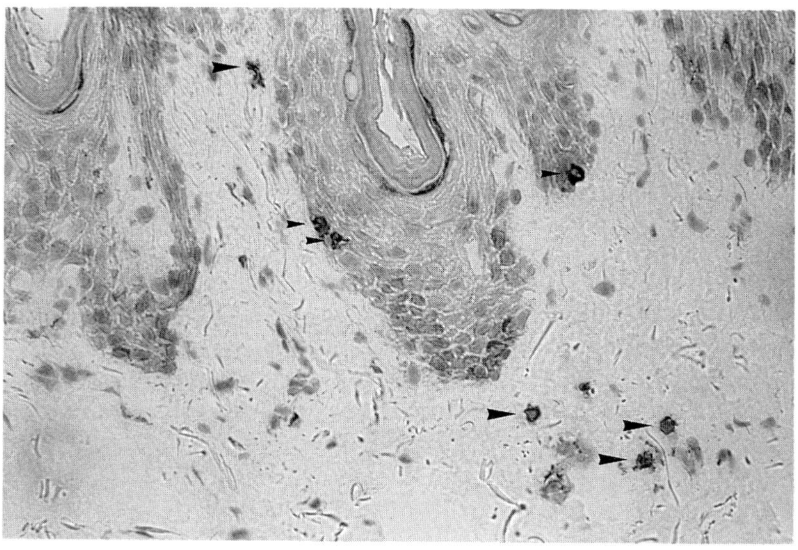

Fig. 1.3 Detection of epidermal (small arrowheads) and dermal (large arrowheads) T cells in a cryostat section of normal human skin using a monoclonal anti-CD3 antibody in an indirect immunoperoxidase technique; ×400 (courtesy of Ingrid Simonitsch, MD).

that E-selectin is preferentially expressed on endothelial cells of venules within inflammatory skin lesions and that these venules are associated with a predominantly HECA-452$^+$ T-cell infiltrate [84]. These data strongly suggest that the interaction between HECA-452-reactive molecules (termed CLA; cutaneous leukocyte antigen) and E-selectin mediates the adhesion of memory T cells to the endothelial cells of the dermal microvasculature and thus promotes the migration of memory T cells into the dermis. The further migration of these cells into the epidermis involves various cytokines, chemotactic factors, and adhesion molecules. Primarily on the basis of *in vitro* studies, one would assume that members of the IL-8 family (chemokines) [77, 97] are the most important keratinocyte-derived T-cell chemotactic signals. *In vitro*, IFN-γ and TNFα are capable of inducing chemokine production by keratinocytes [9] and presumably play an important role in T-cell recruitment to the epidermis. Additionally, these cytokines trigger the expression of intercellular adhesion molecule 1 (ICAM-1; CD54) by keratinocytes [9], rendering these cells capable of binding lymphocyte function-associated antigen 1 (LFA-1; CD11a/CD18) [20]. These data support the concept that the epidermotropic migration of dermal T cells is initiated by certain keratinocyte-derived cytokines (the production of which can also be induced or enhanced by secretory products of activated T cells themselves, e.g. IFN-γ) and that their attachment to and, finally, their entrance into, the epidermis can be mediated by the CD11a/CD18-CD54 interaction. In fact, the recent observations that: (i) transgenic mice that constitutively express CD54 in basal keratinocytes show exaggerated T-cell-mediated inflammatory responses (T.S. Kupper, personal communication); and (ii) intraepidermal lymphocytes in T-cell-mediated skin diseases are frequently found in close proximity to CD54$^+$ keratinocytes [63, 105, 125], provide additional evidence for the inflammation-enhancing activity of CD54 on keratinocytes. Returning to the question concerning the factors responsible for the expansion of these lymphocytes, it appears that, in analogy to the mouse system, IL-7 can act as a powerful costimulatory molecule for T cells of human skin. IL-7 is produced by human keratinocytes [48] and is a T-cell growth factor for *in vivo* primed antigen-specific T cells [64]. Thus, keratinocyte-derived IL-7 may well be of pathogenetic significance in a variety of inflammatory skin diseases. The additional findings: (i) that IL-7 can act as a growth factor for Sezary cells [19]; and (ii) that IL-7 transgenic mice often exhibit cutaneous lymphoproliferation and lymphomas [90], even suggest that IL-7 is an important progression factor in cutaneous T-cell lymphoma.

Conceptual view of the cutaneous immune response (Fig. 1.4)

On the basis of the material presented so far, we hope that the reader agrees with our contention that the skin is an immunological microcosmos containing all the cellular and molecular elements needed for initiation, regulation and expression of an immune response. The experimental data described and discussed above are best compatible with the SALT concept initially proposed by Streilein [114]. It appears that epicutaneous or intracutaneous application of an antigen evokes changes in several cell systems of the skin. Antigen-presenting cells (Langerhans cells, dermal dendritic cells) pick up the antigen, process it and reexpress it in the form of a peptide/MHC complex. Other skin cells such as keratinocytes, mast cells, endothelial cells and fibroblasts react to antigenic challenge with changes in their secretory pattern for soluble mediators some of which (e.g. GM-CSF, IL-1) may, in turn, affect the phenotype and function of antigen-presenting dendritic cells. As a consequence, antigen-laden dendritic cells will then leave their epidermal and/or dermal residence and will migrate via the dermal lymphatics (as veiled cells) to the paracortical areas of the draining lymph nodes.

At this site, they present (as lymphoid dendritic cells) the antigen in an immunologically relevant fashion to the surrounding resting T cells. In order to fulfill their task (i.e. to mediate or at least to initiate effector mechanisms against the noxious pathogens), T-cell blasts thus generated must find

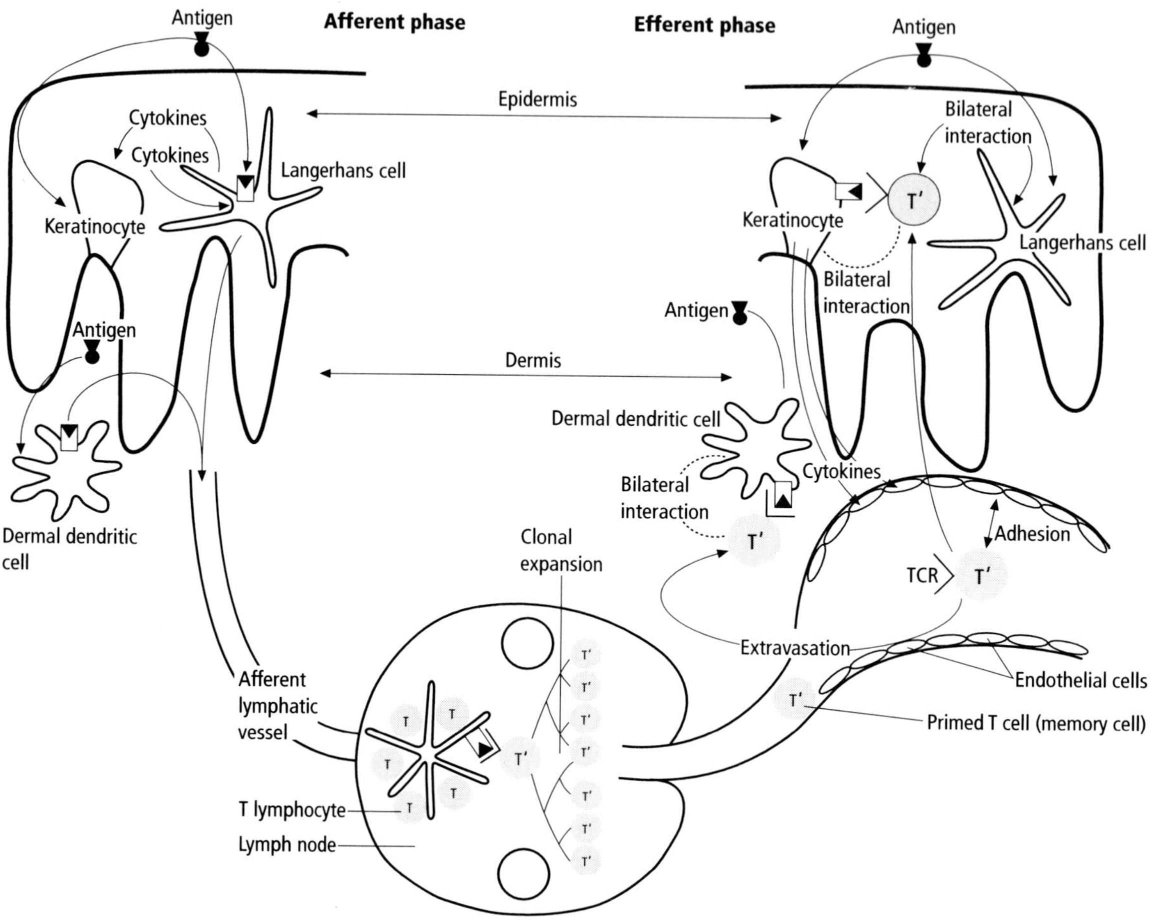

Fig. 1.4 Synoptic view of the mechanisms of initiation and expression of cutaneous immune responses (courtesy of Kai M. Müller, MD).

their way back to the skin and, ideally, should accumulate at the cutaneous sites harboring the antigen. The fact that they accomplish this task is due to the interaction of tissue-specific homing receptors on lymphocytes (e.g. cutaneous lymphocyte antigen) and corresponding addressins expressed on the surface of cytokine-activated endothelial cells of the dermal microvasculature of inflamed skin (e.g. E-selectin). Upon arrival in the skin and receipt of a renewed antigenic stimulus by professional antigen-presenting cells (LC, dermal dendritic cells) and of costimulatory signals by other cells (e.g. keratinocyte-derived IL-7), these sensitized T cells can undergo clonal expansion resulting optimally in the generation of protective effector mechanisms against the pathogen. Alternatively, sensitized T cells may encounter the antigen on the surface of non-professional antigen-presenting cells (e.g. MHC class II-bearing keratinocytes). In analogy to the *in vitro* situation, such an event could conceivably lead to clonal anergy and, thus, to the downregulation of a potentially protective (antitumor, antimicrobial) or harmful (contact hypersensitivity, autoreactive) immune response. The fact that many physicochemical agents – including powerful therapeutic tools used by dermatologists (e.g. UV radiation, corticosteroids, cyclosporin A, retinoids etc.) – exert diverse and profound effects on the various constituents of SALT, makes it mandatory for dermatologists to understand the basic principles of skin immunobiology, immunopharmacology and photoimmunology, and to carefully evaluate risks and benefits of a given treatment modality.

Acknowledgments

This work was supported, in part, by grants Nos P-8536 and S-6702 from the Austrian Science Foundation, Vienna, Austria, by a grant from the Sandoz Research Institute, Vienna, Austria, and by grants Nos 32.27159.89 and 31.30930.91 from the Swiss National Foundation for Scientific Research. We thank Mrs. Laura A. Stingl for helpful suggestions and Mrs. Päivi Vahteristo for her excellent help in the preparation of this manuscript.

References

1 Aiba, S., S.I. Katz: Phenotypic and functional characteristics of *in vivo*-activated Langerhans cells. *J. Immunol.* 145 (1990), 2791–2796.

2 Aiba, S., S. Nakagawa, H. Ozawa, K. Miyake, H. Yagita, H. Tagami: Up-regulation of α4 integrin on activated Langerhans cells: analysis of adhesion molecules on Langerhans cells relating to their migration from skin to draining lymph nodes. *J. Invest. Dermatol.* 100 (1993), 143–147.

3 Allison, J.P., L.L. Lanier: Structure, function, and serology of the T-cell antigen receptor complex. *Annu. Rev. Immunol.* 5 (1987), 503–540.

4 Arenberger, P., L. Kemény, R. Rupec, T. Bieber, T. Ruzicka: Langerhans cells of the human skin possess high-affinity 12(5)-hydroxyeicosa tetraenoic acid receptors. *Eur. J. Immunol.* 22 (1992), 2469–2472.

5 Asarnow, D.M., T. Goodman, L. LeFrancois, J.P. Allison: Distinct antigen receptor repertoires of two classes of murine epithelium-associated T cells. *Nature* 341 (1989), 60–62.

6 Auböck, J., N. Romani, G. Grubauer, P. Fritsch: HLA-DR expression on keratinocytes is a common feature of diseased skin. *Br. J. Dermatol.* 114 (1986), 465–472.

7 Azuma, M., D. Ito, H. Yagita, K. Okumura, J.H. Phillips, L.L. Lanier, C. Somoza: B70 is a second ligand for CTLA-4 and CD28. *Nature* 336 (1993), 76–79.

8 Bal, V., A. McIndoe, G. Denton, D. Hudson, G. Lombardi, J. Lamb, R. Lechler: Antigen presentation by keratinocytes induces tolerance in human T cells. *Eur. J. Immunol.* 20 (1990), 1893–1897.

9 Barker, J.N.W.N., V. Sarma, R.S. Mitra, V.M. Dixit, B.J. Nickoloff: Marked synergism between tumor necrosis factor-α and interferon-γ in regulation of keratinocyte-derived adhesion molecules and chemotactic factors. *J. Clin. Invest.* 85 (1990), 605–608.

10 Basham, T.Y., B.J. Nickoloff, T.C. Merigan, V.B. Morhenn: Recombinant gamma interferon induces HLA-DR expression on cultured human keratinocytes. *J. Invest. Dermatol.* 83 (1984), 88–90.

11 Bergstresser, P.R., R.E. Tigelaar, J.H. Dees, J.W. Streilein: Thy-1 antigen bearing dendritic cells populate murine epidermis. *J. Invest. Dermatol.* 81 (1983), 286–288.

12 Bieber, T., H. de la Salle, A. Wollenberg, J. Hakimi, R. Chizzonite, J. Ring, D. Hanau, C. de la Salle: Human epidermal Langerhans cells express the high affinity receptor for immunoglobulin E(FcεRI). *J. Exp. Med.* 175 (1992), 1285–1290.

13 Bos, J.D., M.B.M. Teunissen, I. Cairo, S.R. Krieg, M.L. Kapsenberg, J. Borst: T-cell receptor γδ bearing cells in normal human skin. *J. Invest. Dermatol.* 94 (1990), 37–42.

14 Bos, J.D., I. Zonneveld, P.K. Das, S.R. Krieg, C.M. van der Loos, M.L. Kapsenberg: The skin immune system (SIS): distribution and immunophenotype of lymphocyte subpopulations in normal human skin. *J. Invest. Dermatol.* 88 (1987), 569–573.

15 Budd, R.C., J.-C. Cerottini, C. Horvath, C. Bron, T. Pedrazzini, R.C. Howe, H.R. MacDonald: Distinction of virgin and memory T lymphocytes. Stable acquisition of the Pgp-1 glycoprotein concomitant with antigenic stimulation. *J. Immunol.* 138 (1987), 3120–3129.

16 Caux, C., C. Dezutter-Dambuyant, D. Schmitt, J. Banchereau: GM-CSF and TNF-α cooperate in the generation of dendritic Langerhans cells. *Nature* 360 (1992), 258–261.

17 Cerio, R., C.E.M. Griffiths, K.D. Cooper, B.J. Nickoloff, J.T. Headington: Characterization of factor XIIIa positive dermal dendritic cells in normal and inflamed skin. *Br. J. Dermatol.* 121 (1989), 421–431.

18 Cumberbatch, M., I. Illingworth, I. Kimber: Antigen-bearing dendritic cells in the draining lymph nodes of contact sensitized mice: cluster formation with lymphocytes. *Immunology* 74 (1991), 139–145.

19 Dalloul, A., L. Laroche, M. Bagot, M.D. Mossalayi, C. Fourcade, D.J. Thacker, D.E. Hogge, M. Merle-Béral, P. Debré, C. Schmitt: Interleukin-7 is a growth factor for Sezary lymphoma cells. *J. Clin. Invest.* 90 (1992), 1054–1060.

20 Dustin, M.L., K.H. Singer, D.I. Tuck, T.A. Springer:

Adhesion of T lymphoblasts to epidermal keratinocytes is regulated by interferon γ and is mediated by intercellular adhesion molecule 1 (ICAM-1). *J. Exp. Med.* 167 (1988), 1323–1340.

21 Elbe, A., O. Kilgus, T. Hünig, G. Stingl: T-cell receptor diversity in dendritic epidermal T cells in the rat. *J. Invest. Dermatol.* 102 (1994), 74–79.

22 Elbe, A., O. Kilgus, R. Strohal, E. Payer, S. Schreiber, G. Stingl: Fetal skin – a site of DETC development. *J. Immunol.* 149 (1992), 1694–1701.

23 Elbe, A., S. Schleischitz, D. Strunk, G. Stingl: Fetal skin-derived MHC class I^+, MHC class II^- dendritic cells stimulate MHC class I-restricted responses of unprimed $CD8^+$ T cells. *J. Immunol.* 153 (1994), 2878–2889.

24 Elmets, C.A., P.R. Bergstresser, R.E. Tigelaar, P.J. Wood: Analysis of the mechanism of unresponsiveness produced by haptens painted on skin exposed to low-dose ultraviolet irradiation. *J. Exp. Med.* 158 (1983), 781–794.

25 Enk, A.H., V.L. Angeloni, M.C. Udey, S.I. Katz: An essential role for Langerhans cell-derived IL-1β in the initiation of primary immune responses in skin. *J. Immunol.* 150 (1993), 3698–3704.

26 Enk, A.H., V.L. Angeloni, M.C. Udey, S.I. Katz: Inhibition of Langerhans cell antigen-presenting function by IL-10. A role for IL-10 in induction of tolerance. *J. Immunol.* 151 (1993), 2390–2398.

27 Enk, A.H., S.I. Katz: Identification and induction of keratinocyte-derived IL-10. *J. Immunol.* 149 (1992), 92–95.

28 Enk, A.H., S.I. Katz: Early molecular events in the induction phase of contact sensitivity. *Proc. Natl. Acad. Sci. U.S.A.* 89 (1992), 1398–1402.

29 Ferrick, D.A., P.S. Ohashi, V. Wallace, M.A. Schilham, T.W. Mak: Thymic ontogeny and selection of αβ and γδ T cells. *Immunol. Today* 10 (1989), 403–407.

30 Foster, C.A., H. Yokozeki, K. Rappersberger, F. Koning, B. Volc-Platzer, A. Rieger, J.E. Coligan, K. Wolff, G. Stingl: Human epidermal T cells predominantly belong to the lineage expressing α/β T cell receptor. *J. Exp. Med.* 171 (1990), 997–1013.

31 Freeman, G.J., J.G. Gribben, V.A. Boussiotis, J.W. Ng, V.A. Restivo, Jr, L.A. Lombard, G.S. Gray, L.M. Nadler: Cloning of B7-2: a CTLA-4 counter-receptor that costimulates human T cell proliferation. *Science* 262 (1993), 909–911.

32 Gaspari, A.A., B. Ferbel, Z. Chen, F. Razvi, R. Polakowska: Accessory and alloantigen-presenting cell functions of A 431 keratinocytes that stably express the B7 antigen. *Cell. Immunol.* 149 (1993), 291–302.

33 Gaspari, A.A., S.I. Katz: Induction and functional characterization of class II MHC (Ia) antigens on murine keratinocytes. *J. Immunol.* 140 (1988), 2956–2963.

34 Gaspari, A.A., S.I. Katz: Induction of *in vivo* hyporesponsiveness to contact allergens by hapten-modified Ia^+ keratinocytes. *J. Immunol.* 147 (1991), 4155–4161.

35 Germain, R.N., D.H. Margulies: The biochemistry and cell biology of antigen processing and presentation. *Annu. Rev. Immunol.* 11 (1993), 403–450.

36 Girolomoni, G., P.D. Cruz, Jr, P.R. Bergstresser: Internalization and acidification of surface HLA-DR molecules by epidermal Langerhans cells: a paradigm for antigen processing. *J. Invest. Dermatol.* 94 (1990), 753–760.

37 Grabbe, S., S. Bruvers, R.L. Gallo, T.L. Knisely, R. Nazareno, R.D. Granstein: Tumor antigen presentation by murine epidermal cells. *J. Immunol.* 146 (1991), 3656–3661.

38 Graham, G.J., E.G. Wright, R. Hewick, S.D. Wolpe, N.M. Wilkie, D. Donaldson, S. Lorimore, I.B. Pragnell: Identification and characterisation of an inhibitor of haemopoietic stem cell proliferation. *Nature* 344 (1990), 442–444.

39 Gruschwitz, M.S., O.P. Hornstein: Expression of transforming growth factor type beta on human epidermal dendritic cells. *J. Invest. Dermatol.* 99 (1992), 114–116.

40 Hauser, C.: Cultured epidermal Langerhans cells activate effector T cells for contact sensitivity. *J. Invest. Dermatol.* 95 (1990), 436–440.

41 Hauser, C., S.I. Katz: Activation and expansion of hapten- and protein-specific T helper cells from nonsensitized mice. *Proc. Natl Acad. Sci. U.S.A.* 85 (1988), 5626–5628.

42 Hauser, C., C.M. Snapper, J. Ohara, W.E. Paul, S.I. Katz: T helper cells grown with hapten-modified cultured Langerhans cells produce interleukin 4 and stimulate IgE production by B cells. *Eur. J. Immunol.* 19 (1989), 245–251.

43 Havran, W.L., J.P. Allison: Developmentally ordered appearance of thymocytes expressing different T-cell antigen receptors. *Nature* 335 (1988), 443–445.

44 Havran, W.L., J.P. Allison: Origin of $Thy\text{-}1^+$ dendritic epidermal cells of adult mice from fetal thymic precursors. *Nature* 344 (1990), 68–70.

45 Havran, W.L., Y.H. Chien, J.P. Allison: Recognition of self antigens by skin-derived T cells with invariant γδ antigen receptors. *Science* 252 (1991), 1430–1432.

46 Havran, W.L., S. Grell, G. Duwe, J. Kimura, A. Wilson, A.M. Kruisbeek, R.L. O'Brien, W. Born, R.E. Tigelaar, J.P. Allison: Limited diversity of T-cell receptor γ-chain expression of murine $Thy\text{-}1^+$ dendritic epidermal cells revealed by Vγ3-specific monoclonal antibody. *Proc. Natl Acad. Sci. U.S.A.* 86 (1989), 4185–4189.

47 Heufler, C., F. Koch, G. Schuler: Granulocyte/macrophage colony-stimulating factor and interleukin 1 mediate the maturation of murine epidermal Langerhans cells into potent immunostimulatory

dendritic cells. *J. Exp. Med.* 167 (1986), 700–705.

48 Heufler, C., G. Topar, A. Grasseger, U. Stanzl, F. Koch, N. Romani, A.E. Namen, G. Schuler: Interleukin 7 is produced by murine and human keratinocyts. *J. Exp. Med.* 178 (1993), 1109–1114.

49 Heufler, C., G. Topar, F. Koch, B. Trockenbacher, E. Kämpgen, N. Romani, G. Schuler: Cytokine gene expression in murine epidermal cell suspensions: interleukin 1β and macrophage inflammatory protein 1α are selectively expressed in Langerhans cells but are differently regulated in culture. *J. Exp. Med.* 176 (1992), 1221–1226.

50 Hosoi, J., G.F. Murphy, C.L. Egan, E.A. Lerner, S. Grabbe, A. Asahina, R.D. Granstein: Regulation of Langerhans cell function by nerves containing calcitonin gene-related peptide. *Nature* 363 (1993), 159–163.

51 Inaba, K., M. Inaba, N. Romani, H. Aya, M. Deguchi, S. Ikehara, S. Maramatsu, R.M. Steinman: Generation of large numbers of dendritic cells from mouse bone marrow cultures supplemented with granulocyte/macrophage colony-stimulating factor. *J. Exp. Med.* 176 (1992), 1693–1702.

52 Inaba, K., G. Schuler, M.D. Witmer, J. Valinsky, B. Atassi, R.M. Steinman: Immunologic properties of purified epidermal Langerhans cells. Distinct requirements for stimulation of unprimed and sensitized T lymphocytes. *J. Exp. Med.* 164 (1986), 605–613.

53 Inaba, K., R.M. Steinman, M. Witmer-Pack, H. Aya, M. Inaba, T. Sudo, S. Wolpe, G. Schuler: Identification of proliferating dendritic cell precursors in mouse blood. *J. Exp. Med.* 175 (1992), 1157–1167.

54 Kämpgen, E., N. Koch, F. Koch, P. Stöger, C. Heufler, G. Schuler, N. Romani: Class II major histocompatibility complex molecules of murine dendritic cells: synthesis, sialylation of invariant chain, and antigen processing capacity are down-regulated upon culture. *Proc. Natl Acad. Sci. U.S.A.* 88 (1991), 3014–3018.

55 Kaplan, G., G. Walsh, L.S. Guido, P. Meyn, R.A. Burkhardt, R.M. Abalos, J. Barker, P.A. Frindt, T.T. Fajardo, R. Celona, Z.A. Cohn: Novel responses of human skin to intradermal recombinant granulocyte/macrophage colony-stimulating factor: Langerhans cell recruitment, keratinocyte growth, and enhanced wound healing. *J. Exp. Med.* 175 (1992), 1717–1728.

56 Kilgus, O., E. Payer, S. Schreiber, A. Elbe, R. Strohal, G. Stingl: *In vivo* cytokine expression in normal and perturbed murine skin-analysis by competitive quantitative polymerase chain reaction. *J. Invest. Dermatol.* 100 (1993), 674–680.

57 Koch, F., C. Heufler, E. Kämpgen, D. Schneeweiß, G. Böck, G. Schuler: Tumor necrosis factor α maintains the viability of murine epidermal Langerhans cells in culture, but in contrast to granulocyte/macrophage colony-stimulating factor, without inducing their maturation. *J. Exp. Med.* 171 (1990), 159–171.

58 Kragballe, K., G.J. Fisher, J.J. Voorhees: Eicosanoids in inflammatory and immunologic skin disorders. In: *Skin Immune System (SIS), part III,* (1990), pp. 243–255. Ed.: J.D. Bos, CRC Press, Inc.; Boca Raton, FL.

59 Kripke, M.L., C.G. Munn, A. Jeevan, J.M. Tang, C. Bucana: Evidence that cutaneous antigen-presenting cells migrate to regional lymph nodes during contact sensitization. *J. Immunol.* 145 (1990), 2833–2838.

60 Larsen, C.P., S.C. Ritchie, T.C. Pearson, P.S. Linsley, R.P. Lowry: Functional expression of the costimulatory molecule, B7/BB1, on murine dendritic cell populations. *J. Exp. Med.* 176 (1992), 1215–1220.

61 Larsen, C.P., R.M. Steinman, M. Witmer-Pack, D.F. Hankins, P.J. Morris, J.M. Austyn: Migration and maturation of Langerhans cells in skin transplants and explants. *J. Exp. Med.* 172 (1990), 1483–1493.

62 Lenz, A., M. Heine, G. Schuler, N. Romani: Human and murine dermis contain dendritic cells: isolation by means of a novel method and phenotypic and functional characterization. *J. Clin. Invest.* 92 (1993), 2587–2596.

63 Lisby, S., E. Ralfkiaer, R. Rothlein, G.L. Vejlsgaard: Intercellular adhesion molecule-1 (ICAM-1) expression correlated to inflammation. *Br. J. Dermatol.* 120 (1989), 479–484.

64 Londei, M., A. Verhoef, C. Hawrylowicz, J. Groves, P. De Berardinis, M. Feldmann: Interleukin 7 is a growth factor for mature human T cells. *Eur. J. Immunol.* 20 (1990), 425–428.

65 Luger, T.A., T. Schwarz (Eds.): *Epidermal Growth Factors and Cytokines,* (1994); Marcel Dekker, Inc., New York.

66 Macatonia, S.E., S.C. Knight, A.J. Edwards, S. Griffiths, P. Fryer: Localization of antigen on lymph node dendritic cells after exposure to the contact sensitizer fluorescein isothiocyanate. Functional and morphological studies. *J. Exp. Med.* 166 (1987), 1654–1667.

67 Matsue, H., P.R. Bergstresser, A. Takashima: Keratinocyte-derived IL-7 serves as a growth factor for dendritic epidermal T cells in mice. *J. Immunol.* 151 (1993), 6012–6019.

68 Matsue, H., P.D. Cruz, Jr, P.R. Bergstresser, A. Takashima: Langerhans cells are the major source of mRNA for IL-1β and MIP-1α among unstimulated mouse epidermal cells. *J. Invest. Dermatol.* 99 (1992), 537–541.

69 Meunier, L., A. Gonzales-Ramos, K.D. Cooper: Heterogeneous populations of class II MHC$^+$ cells in human dermal cell suspensions. Identification of a small subset responsible for potent dermal antigen-presenting cell activity with features analogous to

Langerhans cells. *J. Immunol.* 151 (1993), 4067–4080.

70 Mosmann, T.R., H. Cherwinski, M.W. Bond, M.A. Giedlin, R.L. Coffman: Two types of murine helper T cell clone. I. Definition according to profiles of lymphokine activities and secreted proteins. *J. Immunol.* 136 (1986), 2348–2357.

71 Mueller, D.L., M.K. Jenkins, R.H. Schwartz: Clonal expansion versus functional clonal inactivation: a costimulatory signalling pathway determines the outcome of T cell antigen receptor occupancy. *Annu. Rev. Immunol.* 7 (1989), 445–480.

72 Müller, K.M., F. Jaunin, I. Masouyé, J.-H. Saurat, C. Hauser: Th2 cells mediate IL-4 dependent local tissue inflammation. *J. Immunol.* 150 (1993), 5576–5584.

73 Nathan, C.F., G. Kaplan, W.R. Levis, A. Nusrat, M.D. Witmer, S.A. Sherwin, C.K. Job, F.R.C. Path, C.R. Horowitz, R.M. Steinman, Z.A. Cohn: Local and systemic effects of intradermal recombinant interferon-γ in patients with lepromatous leprosy. *N. Engl. J. Med.* 315 (1986), 6–15.

74 Niederwieser, D., J. Auböck, J. Troppmair, M. Herold, G. Schuler, G. Boeck, J. Lotz, P. Fritsch, C. Huber: IFN-mediated induction of MHC antigen expression on human keratinocytes and its influence on *in vitro* alloimmune responses. *J. Immunol.* 140 (1988), 2556–2564.

75 Nixon-Fulton, J.L., P.R. Bergstresser, R.E. Tigelaar: Thy-1^+ epidermal cells proliferate in response to Concanavalin A and interleukin 2. *J. Immunol.* 136 (1986), 2776–2786.

76 Ogimoto, M., G. Matsuzaki, Y. Yoshikai, Y. Tauchi, K. Nomoto: Appearance of TCR-$\alpha\beta^+$ $CD4^-$ $CD8^-$ skin intraepithelial lymphocytes in radiation bone marrow chimeras. *J. Immunol.* 151 (1993), 3000–3006.

77 Oppenheim, J.J., C.O.C. Zachariae, N. Mukaida, K. Matsushima: Properties of the novel proinflammatory supergene 'intercrine' cytokine family. *Annu. Rev. Immunol.* 9 (1991), 617–648.

78 Paglia, P., G. Girolomoni, F. Robbiati, F. Granucci, P. Ricciardi-Castagnoli: Immortalized dendritic cell line fully competent in antigen presentation initiates primary T cell responses *in vivo*. *J. Exp. Med.* 178 (1993), 1893–1901.

79 Parkinson, E.K., G.J. Graham, P. Daubersies, J.E. Burns, C. Heufler, M. Plumb, G. Schuler, I.B. Pragnell: Hemopoietic stem cell inhibitor (SCI/MIP-1α) also inhibits clonogenic epidermal keratinocyte proliferation. *J. Invest. Dermatol.* 101 (1993), 113–117.

80 Payer, E., A. Elbe, G. Stingl: Circulating $CD3^+$/TCR V$\gamma 3^+$ fetal murine thymocytes home to the skin and give rise to proliferating dendritic epidermal T cells. *J. Immunol.* 146 (1991), 2536–2543.

81 Payer, E., R. Strohal, R. Kutil, A. Elbe, G. Stingl: Demonstration of a $CD3^+$ lymphocyte subset in the epidermis of athymic nude mice. Evidence for T cell receptor diversity. *J. Immunol.* 149 (1992), 413–420.

82 Peguet-Navarro, J., C. Dalbiez-Gauthier, C. Dezutter-Dambuyant, D. Schmitt: Dissection of human Langerhans cells' allostimulatory function: the need for an activation step for full development of accessory function. *Eur. J. Immunol.* 23 (1993), 376–382.

83 Petzelbauer, P., G. Stingl, K. Wolff, B. Volc-Platzer: Cyclosporin A suppresses ICAM-1 expression by papillary endothelium in healing psoriatic plaques. *J. Invest. Dermatol.* 96 (1991), 362–369.

84 Picker, L.J., T.K. Kishimoto, C.W. Smith, R.A. Warnock, E.C. Butcher: ELAM-1 is an adhesion molecule for skin-homing T cells. *Nature* 349 (1991), 796–799.

85 Picker, L.J., S.A. Michie, L.S. Rott, E.C. Butcher: A unique phenotype of skin-associated lymphocytes in humans. Preferential expression of the HECA-452 epitope by benign and malignant T cells at cutaneous sites. *Am. J. Pathol.* 136 (1990), 1053–1068.

86 Ptak, W., D.W. Ruzycka, P.W. Askenase, R.K. Gershon: Role of antigen-presenting cells in the development and persistence of contact hypersensitivity. *J. Exp. Med.* 151 (1980), 362–375.

87 Puré, E., K. Inaba, M.T. Crowley, L. Tardelli, M.D. Witmer-Pack, G. Ruberti, G. Fathman, R.M. Steinman: Antigen processing by epidermal Langerhans cells correlates with the level of biosynthesis of major histocompatibility complex class II molecules and expression of invariant chain. *J. Exp. Med.* 172 (1990), 1459–1469.

88 Raulet D.H.: The structure, function, and molecular genetics of the γ/δ T cell receptor. *Annu. Rev. Immunol.* 7 (1989), 175–207.

89 Reid, C.D.L., A. Stackpoole, A. Meager, J. Tiperkae: Interactions of tumor necrosis factor with granulocyte-macrophage colony-stimulating factor and other cytokines in the regulation of dendritic cell growth *in vitro* from early bipotent $CD34^+$ progenitors in human bone marrow. *J. Immunol.* 149 (1992), 2681–2688.

90 Rich, B.E., J. Campos-Torres, R.I. Tepper, R.W. Moreadith, P. Leder: Cutaneous lymphoproliferation and lymophomas in interleukin 7 transgenic mice. *J. Exp. Med.* 177 (1993), 305–316.

91 Röcken, M., K.M. Müller, C. Hauser: Induction of T helper cell subsets. *Springer Semin. Immunopathol.* 13 (1992), 289–302.

92 Romani, N., C. Heufler, F. Koch, G. Topar, E. Kämpgen, G. Schuler: Cytokines and Langerhans cells. In: *Epidermal Growth Factors and Cytokines*, (1994), pp. 345–363. Eds.: T.A. Luger, T. Schwarz, Marcel Dekker, Inc., New York.

93 Romani, N., S. Koide, M. Crowley, M. Witmer-Pack, A.M. Livingstone, C.G. Fathman, K. Inaba, R.M. Steinman: Presentation of exogenous protein antigens by dendritic cells to T cell clones. Intact

protein is presented best by immature, epidermal Langerhans cells. *J. Exp. Med.* 169 (1989), 1169–1178.

94 Romani, N., A. Lenz, H. Glassel, H. Stössel, U. Stanzl, O. Majdic, P. Fritsch, G. Schuler: Cultured human Langerhans cells resemble lymphoid dendritic cells in phenotype and function. *J. Invest. Dermatol.* 93 (1989), 600–609.

95 Schmitt, D.A., D. Hanau, T. Bieber, C. Dezutter-Dambuyant, D. Schmitt, M. Fabre, G. Pauly, J.-P. Cazenave: Human epidermal Langerhans cells express only the 40-kilodalton Fcγ receptor (IcRII). *J. Immunol.* 144 (1990), 4284–4290.

96 Schreiber, S., O. Kilgus, E. Payer, R. Kutil, A. Elbe, C. Müller, G. Stingl: Cytokine pattern of Langerhans cells isolated from murine epidermal cell cultures. *J. Immunol.* 149 (1992), 3524–3534.

97 Schröder, J.-M., M. Sticherling, P. Smid, E. Christophers: Interleukin 8 and structurally related cytokines. In: *Epidermal Growth Factors and Cytokines*, (1994), pp. 89–112. Eds.: T.A. Luger, T. Schwarz, Marcel Dekker, Inc., New York.

98 Schuler, G., R.M. Steinman: Murine epidermal Langerhans cells mature into potent immunostimulatory dendritic cells *in vitro*. *J. Exp. Med.* 161 (1985), 526–546.

99 Shimada, S., N. Shibagaki, T. Tanaka, Y. Kubota, Y. Hashimoto, K. Tamaki: Identification of an equivalent to murine Thy-1^+ dendritic epidermal cells in the rat epidermis. *J. Dermatol. Sci.* 3 (1992), 68–71.

100 Shimizu, Y., S. Shaw, N. Graber, T.V. Gopal, K.J. Horgan, G.A. Van Seventer, W. Newman: Activation-independent binding of human memory T cells to adhesion molecule ELAM-1. *Nature* 349 (1991), 799–802.

101 Shiohara, T., N. Moriya, H. Hayakawa, K. Arahari, H. Yagita, M. Nagashima, H. Ishikawa: Bone marrow-derived dendritic epidermal T cells express T cell receptor-αβ/CD3 and CD8. Evidence for their extrathymic maturation. *J. Immunol.* 150 (1993), 4323–4330.

102 Silberberg-Sinakin, I., G.J. Thorbecke, R.L. Baer, S.A. Rosenthal, V. Berezowsky: Antigen-bearing Langerhans cells in skin, dermal lymphatics and in lymph nodes. *Cell. Immunol.* 25 (1976), 137–151.

103 Simon, J.C., P.D. Cruz, Jr, R.E. Tigelaar, R.D. Sontheimer, P.R. Bergstresser: Adhesion molecules CD11a, CD18, and ICAM-1 on human epidermal Langerhans cells serve a functional role in the activation of alloreactive T cells. *J. Invest. Dermatol.* 96 (1991), 148–151.

104 Simon, J.C., R.E. Tigelaar, P.R. Bergstresser, D. Edelbaum, P.C. Cruz, Jr: Ultraviolet B radiation converts Langerhans cells from immunogenic to tolerogenic antigen-presenting cells. Induction of specific clonal anergy in $CD4^+$ T helper 1 cells. *J. Immunol.* 146 (1991), 485–491.

105 Singer, K.H., D.T. Tuck, H.A. Sampson, R.P. Hall: Epidermal keratinocytes express the adhesion molecule intercellular adhesion molecule-1 in inflammatory dermatoses. *J. Invest. Dermatol.* 92 (1989), 746–750.

106 Sontheimer R.D.: Perivascular dendritic macrophages as immunobiological constituents of the human dermal microvascular unit. *J. Invest. Dermatol.* 93 (1989), 96S–101S.

107 Sontheimer, R.D., T. Matsubara, L.L. Seelig, Jr: A macrophage phenotype for a constitutive, class II antigen-expressing, human dermal perivascular dendritic cell. *J. Invest. Dermatol.* 93 (1989), 154–159.

108 Steiner, G., F. Koning, A. Elbe, E. Tschachler, W.M. Yokoyama, E.M. Shevach, G. Stingl, J.E. Coligan: Characterization of T cell receptors on resident murine dendritic epidermal T cells. *Eur. J. Immunol.* 18 (1988), 1323–1328.

109 Steiner, G., E. Tschachler, M. Tani, T.R. Malek, E.M. Shevach, W. Holter, W. Knapp, K. Wolff, G. Stingl: Interleukin 2 receptors on cultured murine epidermal Langerhans cells. *J. Immunol.* 137 (1986), 155–159.

110 Steinman, R.M.: The dendritic cell system and its role in immunogenicity. *Annu. Rev. Immunol.* 9 (1991), 271–296.

111 Stingl, G., C. Hauser, K. Wolff: The epidermis: an immunologic microenvironment. In: *Dermatology in General Medicine*, fourth edition, (1993), pp. 172–197. Eds.: T.B. Fitzpatrick, A.Z. Eisen, K. Wolff, I.M. Freedberg, K.F. Austen, McGraw-Hill, Inc., New York.

112 Stingl, G., E.M. Shevach: Langerhans cells as antigen-presenting cells. In: *Epidermal Langerhans Cells*, (1991), pp. 159–190. Ed.: G. Schuler, CRC Press, Boca Raton, FL.

113 Stössel, H., F. Koch, E. Kämpgen, P. Stöger, A. Lenz, C. Heufler, N. Romani, G. Schuler: Disappearance of certain acidic organelles (endosomes and Langerhans cell granules) accompanies loss of antigen processing capacity upon culture of epidermal Langerhans cells. *J. Exp. Med.* 172 (1990), 1471–1482.

114 Streilein, J.W.: Skin-associated lymphoid tissues (SALT): origins and functions. *J. Invest. Dermatol.* 80 (1983), 12S–16S.

115 Streilein, J.W., L.W. Lonsberry, P.R. Bergstresser: Depletion of epidermal Langerhans cells and Ia immunogenicity from tape-stripped mouse skin. *J. Exp. Med.* 155 (1982), 863–871.

116 Streilein, J.W., G.B. Toews, P.R. Bergstresser: Corneal allografts fail to express Ia antigens. *Nature* 282 (1979), 326–327.

117 Sullivan, S., P.R. Bergstresser, R.E. Tigelaar, J.W. Streilein: Induction and regulation of contact hypersensitivity by resident bone marrow-derived,

dendritic epidermal cells: Langerhans cells and Thy-1^{+} epidermal cells. *J. Immunol.* 137 (1986), 2460–2467.

118 Symington, F.W., W. Brady, P.S. Linsley: Expression and function of B7 on human epidermal Langerhans cells. *J. Immunol.* 150 (1993), 1286–1295.

119 Symington, F.W., E.B. Santos: Recognition of keratinocytes by cytotoxic T cells specific for conventional HLA-class-I alloantigen. *J. Invest. Dermatol.* 95 (1990), 224–228.

120 Tang, A., M. Amagai, L.G. Granger, J.R. Stanley, M.C. Udey: Adhesion of epidermal Langerhans cells to keratinocytes mediated by E-cadherin. *Nature* 361 (1993), 82–85.

121 Teunissen, M.B.M., J. Wormmeester, S.R. Krieg, P.J. Peters, I.M.C. Vogels, M.L. Kapsenberg, J.D. Bos: Human epidermal Langerhans cells undergo profound morphologic and phenotypical changes during *in vitro* culture. *J. Invest. Dermatol.* 94 (1990), 166–173.

122 Toews, G.B., P.R. Bergstresser, J.W. Streilein: Epidermal Langerhans cell density determines whether contact hypersensitivity or unresponsiveness follows skin painting with DNFB. *J. Immunol.* 124 (1980), 445–453.

123 Tschachler, E., G. Schuler, J. Hutterer, H. Leibl, K. Wolff, G. Stingl: Expression of Thy-1 antigen by murine epidermal cells. *J. Invest. Dermatol.* 81 (1983), 282–285.

124 Tschachler, E., G. Steiner, H. Yamada, A. Elbe, K. Wolff, G. Stingl: Dendritic epidermal T cells: activation requirements and phenotypic characterization of proliferating cells. *J. Invest. Dermatol.* 92 (1989), 763–768.

125 Vejlsgaard, G.L., E. Ralfkiaer, C. Avnstorp, M. Czajkowski, S.D. Marlin, R. Rothlein: Kinetics and characterization of intercellular adhesion molecule-1 (ICAM-1) expression on keratinocytes in various inflammatory skin lesions and malignant cutaneous lymphomas. *J. Am. Acad. Dermatol.* 20 (1989), 782–790.

126 Volc-Platzer, B., H. Leibl, T. Luger, G. Zahn, G. Stingl: Human epidermal cells synthesize HLA-DR alloantigens *in vitro* upon stimulation with γ-interferon. *J. Invest. Dermatol.* 85 (1985), 16–19.

127 Volc-Platzer, B., O. Majdic, W. Knapp, K. Wolff, W. Hinterberger, K. Lechner, G. Stingl: Evidence of HLA-DR antigen biosynthesis by human keratinocytes in disease. *J. Exp. Med.* 159 (1984), 1784–1789.

128 Wang, B., A. Rieger, O. Kilgus, K. Ochiai, D. Maurer, D. Födinger, J.-P. Kinet, G. Stingl: Epidermal Langerhans cells from normal human skin bind monomeric IgE via FcεRI. *J. Exp. Med.* 175 (1992), 1353–1365.

129 Will, A., C. Blank, M. Röllinghoff, H. Moll: Murine epidermal Langerhans cells are potent stimulators of an antigen-specific T cell response to *Leishmania major*, the cause of cutaneous leishmaniasis. *Eur. J. Immunol.* 22 (1992), 1341–1347.

130 Witmer-Pack, M.D., J. Valinsky, W. Oliver, R.M. Steinman: Quantification of surface antigens on cultured murine epidermal Langerhans cells: rapid and selective increase in the level of surface MHC products. *J. Invest. Dermatol.* 90 (1988), 387–394.

2: Primary processes in photobiology and photosensitization

Irene E. Kochevar

Introduction

Alterations in immune functions that are induced by ultraviolet (UV) or visible radiation are initiated by absorption of light energy by molecules. The light energy causes these molecules, called chromophores, to transform into new molecules called photoproducts. Cells respond to photoproduct molecules in several ways. Some photoproducts are removed by repair mechanisms, others induce signal transduction pathways and still others are toxic to cells. The observable response in skin may occur within minutes of light exposure (i.e., urticaria) or may take days to be expressed (i.e., inhibition of contact hypersensitivity), or much longer periods (i.e., tumors). The processes from light absorption to expression of a biological response are the focus of studies in photobiology. Cutaneous photobiology also includes responses such as sunburn, tanning and photoaging.

This chapter will address the primary molecular steps in skin responses to UV and visible radiation, namely the absorption of light by molecules and the formation of photoproduct molecules. In addition the primary processes in photosensitivity will be discussed. The biochemical and immunological mechanisms for each specific response to UV and visible radiation will be considered in chapters devoted to these topics.

Ultraviolet and visible radiation

The wavelengths of radiation that initiate photobiological responses are in the UV and visible wavelength ranges of the electromagnetic spectrum. The spectrum of radiation from the sun that reaches the Earth's surface includes wavelengths above 290 nm in the UV and the entire visible spectrum. The sun is the major source of radiation that initiates photobiological responses in humans. Artificial sources (i.e., sun lamps) also emit UV radiation and are used to produce photobiological responses experimentally. The UV and visible spectral regions are in the center of the electromagnetic spectrum (Table 2.1) which includes shorter wavelength radiation (X-rays, γ-rays) and longer wavelength radiation (microwaves, radiowaves).

Electromagnetic radiation

A short introduction to electromagnetic radiation (EMR) will be given to help clarify the relationship between light energy and wavelength, and to discuss the concepts involved in light absorption by molecules in tissue. EMR is a form of energy that can be described both as a continuous wave (wave description) or as a series of packets of energy called photons (particle description). Each description is used to explain certain aspects of the interaction of EMR with matter. Absorption of EMR by molecules is best described using the particle description. A molecule takes up all the energy in each particle or photon during the process of absorption.

In the wave description, EMR is thought of as a wave with regular oscillations of electric and magnetic fields that are perpendicular to each other and are perpendicular to the direction of propagation. The number of oscillations per second is called the frequency of the radiation. The energy, E, of a photon of EMR is directly proportional to

Table 2.1 Wavelengths and energies associated with regions of the electromagnetic spectrum

Wavelength (nm)	Radiation region	Photon energy (J)
0.1	X-ray	2.0×10^{-15}
10	Soft X-ray	2.0×10^{-17}
100	Vacuum UV	2.0×10^{-18}
250	UVC	8.0×10^{-19}
300	UVB	6.6×10^{-19}
350	UVA	5.6×10^{-19}
400	Violet	5.0×10^{-19}
470	Blue	4.2×10^{-19}
530	Green	3.7×10^{-19}
600	Yellow	3.2×10^{-19}
700	Red	2.9×10^{-19}
1 000	Near infrared	1.9×10^{-19}
100 000	Far infrared	1.9×10^{-21}
$>10^6$	Microwaves and radiowaves	$<1.9 \times 10^{-22}$

the frequency, ν, of the oscillations of the waves (Equation 1).

$$E = h\nu \qquad (1)$$

In this equation, h is Planck's constant (6.63×10^{-34} J/s) and ν has the units of s^{-1}. The frequency of the radiation is related to the wavelength of the radiation, λ, because the product of ν and λ equals the speed of EMR in a vacuum, 3×10^8 m/s. Therefore, energy is related to wavelength by Equation 2:

$$E = hc/\lambda \qquad (2)$$

where E is the energy of the photon in joules, c is the speed of light in meters per second and λ is the wavelength in meters. This expression shows that the energy of a photon of EMR is inversely proportional to the wavelength of the light. From Equation 2, a photon of 300 nm radiation contains twice the energy of a 600 nm photon. This will become important later when the photoreactions produced by different wavelengths of UV and visible radiation are discussed. Equation 2 also shows the relationship between the wave description (wavelength) and the particle description (photons) of EMR.

Subdivisions of UV and visible radiation

The subdivisions of UV and visible ranges are generally described by their wavelengths rather than their frequencies. Within the UV (200–400 nm) and visible (400–760 nm) light ranges, the most useful unit for wavelength is the nanometer (nm). Ultraviolet and visible radiation ranges are important in photobiology because the photons have sufficient energy to cause photochemical reactions of biological chromophores. Radiation at shorter wavelengths causes ionization of tissue chromophores and is used for radiation therapy. A short wavelength limit for biologically important UV has been set at 200 nm because shorter wavelengths are not transmitted through air since they are absorbed by oxygen. Ultraviolet radiation below 200 nm can be transmitted through a vacuum, hence its name, vacuum ultraviolet (VUV). The division between ultraviolet and visible radiation is generally considered to be at 400 nm which is approximately the short wavelength limit of normal human vision.

The wavelength range of UV is usually divided into three regions: UVC, 200–290 nm; UVB, 290–320 nm; and UVA, 320–400 nm (Fig. 2.1). The border between UVC and UVB is 290 nm because ozone in the stratosphere absorbs all shorter wavelengths so that they do not reach the Earth's surface. The emission from germicidal bulbs is at 254 nm in the UVC range. This wavelength was used for many years to study the effects of UV on DNA and microorganisms. UVC radiation is strongly absorbed by nucleic acids and proteins and is highly cytotoxic and mutagenic.

The border between UVB and UVA is usually set at 320 nm. The UVB wavelength range is sometimes called the sunburn spectrum. Wavelengths above 320 nm are not absorbed strongly by protein and nucleic acids, and moderate doses of UVA do not cause erythema in normal skin in the absence of photosensitizing agents. UVA radiation is occasionally called black light. Although it is convenient to divide the UV wavelength range into UVA, UVB and UVC, it is important to remember that all the wavelengths within a particular band do not have equal effects because tissue chromo-

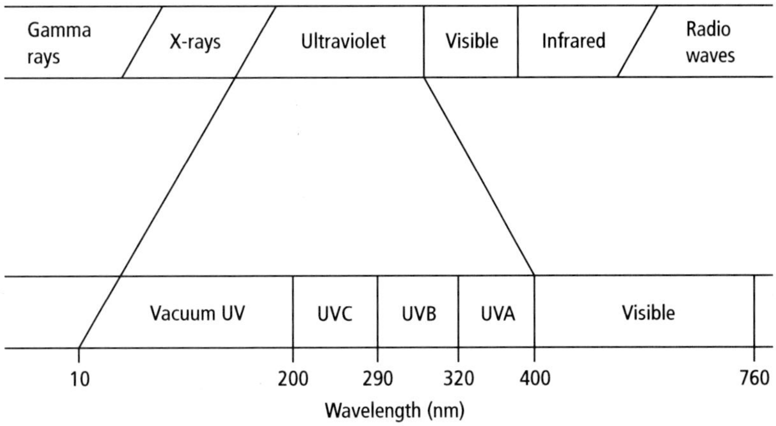

Fig. 2.1 Major regions in the electromagnetic spectrum with emphasis on the ultraviolet range.

phores do not absorb equally throughout a band. This is particularly apparent in the UVB range where the minimal erythema dose (MED) is about 100 times smaller at 290 nm than at 320 nm. Recently, a division of the UVA wavelength range has been proposed consisting of UVA II (320–340 nm) and UVA I (340–400 nm). This proposal was based on the 'UVB-like' effects of UVA II. Thus, the division of the UV range into bands is convenient but must not be taken to mean that the responses to the different ranges are absolutely different.

Absorption of light by molecules

The first step in a photobiological response is absorption of photons by endogenous or exogenous molecules in the tissue. The light absorbing molecules are called chromophores. When a photon is absorbed, the electromagnetic energy of the photon is transferred to the molecule and the photon disappears. The energy becomes part of the chemical energy of the molecule. The molecule is now in an excited state. It has extra energy and is capable of changing its chemical structure to form a new molecule that is called a photoproduct. This is a very important process, because chemical changes are the underlying cause for photobiological responses. The molecular and electronic structure of a chromophore determine which wavelengths of visible and UV radiation are absorbed by the chromophore.

Electronic states of molecules

The nuclei form the framework for the molecular structure of a compound. Electrons are found in defined volumes around this framework, and the defined volumes are called orbitals. Each orbital has a maximum capacity of two electrons. The energy associated with the molecule is set by the distribution of the electrons around the nuclei. According to quantum mechanics, only specific distributions of electrons are allowed. Each specific distribution is called an electronic state of the molecule (Fig. 2.2). Each electronic state has a certain molecular energy associated with it, and a series of electronic states is separated by the finite energy gaps. At normal body temperature almost all molecules are in the lowest electronic state, which is called the ground electronic state. Electronic states with higher energy are called excited electronic states, or simply excited states, and are designated with higher numbers as the energy increases. Excited state molecules can undergo photochemical reactions that the ground states molecule cannot.

The electronic states of molecules shown in Fig. 2.2 are designated singlet or triplet states. This designation derives from a property of electrons

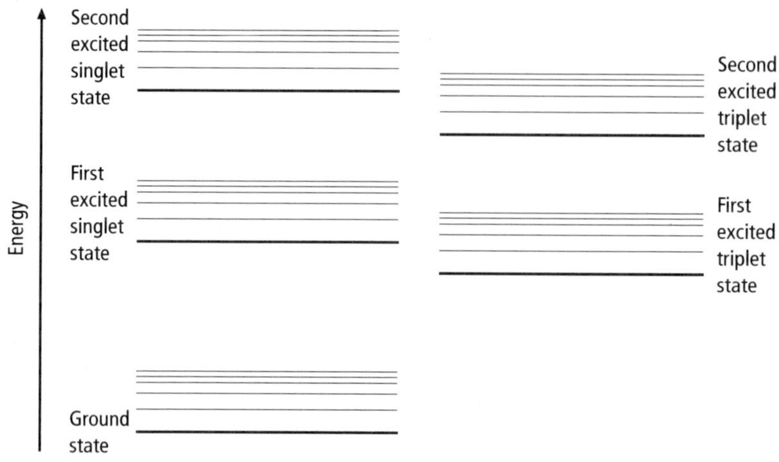

Fig. 2.2 Energy levels of electronic and vibrational states of a molecule. The ground state is a singlet state. Both excited singlet and excited triplet states are shown. Electronic states are denoted by thick lines, and vibrational states by thin lines.

called spin that can have a value of either plus 1/2 or minus 1/2. Only the spins on the two electrons with the highest energies in electronic states are important. When these electrons have the same spin value, both +1/2 or both −1/2, the electronic state is a singlet state. When the electrons have different spin values, the electronic state is a triplet state. Ground state molecules are almost always singlet states. Oxygen is a notable exception; the ground state of oxygen is a triplet state.

When a photon is absorbed by a ground state molecule, the values of the electron spins are not altered. Thus the excited state formed is a singlet excited state. The chemical reactions of singlet and triplet states often differ. When a molecule absorbs the energy of a photon, the electron distribution in the orbitals is changed. Only the wavelengths that have an energy equivalent to the energy gap between the ground state and an excited state can be absorbed. Only one photon is absorbed at a time.

Each chemical compound has a unique set of electronic energy states which is determined by the chemical structure of the molecule. Thus each molecule has a unique absorption spectrum. Remembering that longer wavelengths correspond to lower energies, it becomes apparent that compounds having a small energy gap between the ground and first excited states will absorb long visible wavelengths. In contrast, molecules having a large energy gap between the ground and the first excited single states will absorb shorter wavelengths in the UV portion of the spectrum.

It is often possible to estimate the wavelengths that a chromophore will absorb from the chemical structure of the compound. Compounds containing many conjugated double bonds (i.e., porphyrins) have small energy gaps between the ground and excited states, and absorb visible radiation. Molecules that contain few unsaturated rings or conjugated bonds have larger energy gaps between the ground state and excited singlet state, and consequently absorb higher energy UV photons.

In addition to electronic energy, molecules also possess vibrational and rotational energy due to the motions of the nuclei in the molecule. The energy associated with these vibrations and rotations is also quantized but the energy gaps between levels are much smaller than those between electronic energy levels. The energy of vibrations and rotations in a chemical structure combine with the electronic energy to give the total molecular energy. The energy of photons that are absorbed must match the energy gap between the total energy of the ground and excited states. The total energy is the sum of the electronic, vibrational and rotational energy. Because there are numerous vibrational and rotational levels, photons with a range of energies, not just a single energy, are absorbed.

Absorption spectrum

The probability that a photon will be absorbed, plotted against the wavelength, is called the absorption spectrum. Each molecule has a unique absorption spectrum. The absorption spectrum of an organic compound appears as a smooth curve (Fig. 2.3). Points along the curve represent the probability of absorption at that wavelength. Wavelengths having the highest probability of absorption are called the absorption maxima, λ_{max}, and are often used as identifying characteristics of a compound.

Absorption spectra of molecules are usually measured in a pure solution rather than in tissue where there are many interfering chromophores. However, more recently, techniques such as reflectance spectroscopy have been developed for measuring absorption spectra of chromophores in tissue.

Tissue contains many chromophores that absorb in the UV and visible ranges although not all of these molecules initiate photobiological responses after absorbing photons. Chromophores absorbing in tissue include DNA, aromatic amino acids, and 7-dehydrocholesterol. Aromatic amino acids and proteins absorb strongly but generally do not initiate photobiological responses. Absorption by protein in stratum corneum is an important protective factor against UV-induced effects on skin.

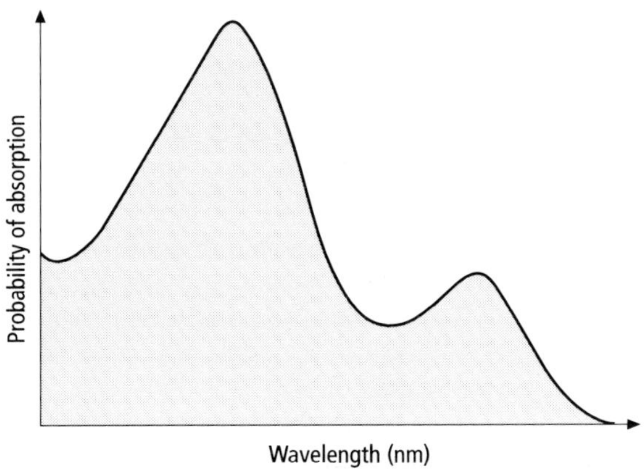

Fig. 2.3 Generalized absorption spectrum of a biological molecule. Each molecule has a distinct absorption spectrum.

7-Dehydrocholesterol contains conjugated double bonds and is converted to pre-vitamin D_3 upon absorbing a photon of UV radiation. Another photobiologic chromophore is urocanic acid. This molecule absorbs in the UV range and has been closely linked with several photoimmunological effects. The lipid components of tissue absorb only at short wavelengths and are not chromophores for cutaneous responses to biologically relevant wavelengths. Unsaturated fatty acid, sugars and glycosaminoglycans such as dermatan sulfate also absorb only in the UVC range.

Chromophores absorbing UVA radiation are present in skin at very low concentrations and include NADH and flavins. Chromophores for UVA-initiated photobiologic effects are not well defined. Photoimmunological effects of UVA radiation usually involve exogenous photosensitizers. The well-known photochemotherapeutic agent 8-methylpsoralen is excited with UVA radiation. Many chromophores absorbing visible light are based on a porphyrin structure, including hemoglobin, protoporphyrin and cytochromes. Hemoglobin has a strong absorption maximum in the blue portion of visible light around 410 nm, and has a lower absorption maxima in the 500–600 nm range. Absorption by hemoglobin is not responsible for photobiologic reponses of skin to visible radiations. Protoporphyrin, which has a similar absorption spectrum, is a chromophore for cutaneous photosensitivity.

Melanin in skin is a major factor in reducing UVB-induced responses such as sunburn and skin cancer. Bilirubin absorbs about 460 nm and is the chromophore for blue light therapy for neonatal jaundice. Beta-carotene absorbs about 450 to 470 nm but does not appear to be a chromophore for visible light effects in tissue.

Excited state molecules

After absorbing the energy of a photon, the molecule in its excited singlet state has a different

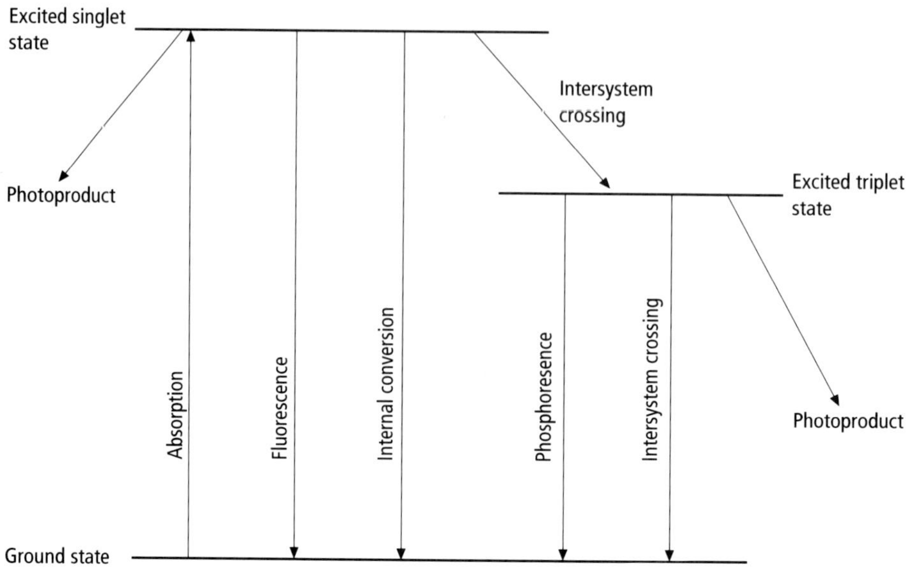

Fig. 2.4 Simplified energy level diagram for a molecule. When a molecule absorbs a photon it is transformed into an excited singlet state and now possesses the energy of the photon. The excited singlet state molecule can emit light (fluorescence), lose energy as heat (internal conversion), or convert into an excited triplet state molecule (intersystem crossing). The triplet state can emit light (phosphorescence) or return to the ground state. Both singlet and triplet excited state molecules can change their chemical structures and become photoproducts (photochemistry).

distribution of electrons and contains more energy than the ground state molecule. The processes that excited state molecules can undergo are shown in Fig. 2.4. Not all of these processes are equally likely to occur. The processes include emission of light (fluorescence), chemical reaction to photoproducts, intersystem crossing (interconversion between singlet and triplet states), and internal conversion in which the excited state returns to the ground state and the energy of the excited state is lost as heat. All of these processes occur with different probabilities for a particular excited state molecule. The probability that each process will occur depends on the specific chemical structure of the excited state molecule.

Excited singlet state molecules

The excited singlet state molecule that forms when the ground state molecule absorbs a photon typically exists for a few nanoseconds (10^{-9} s) and sometimes as briefly as a few picoseconds (10^{-12} s). Emission of fluorescence and internal conversion are photophysical processes that occur very rapidly and cause the excited state molecules to return to the ground state without changing their chemical structure. Fluorescence is the emission of the photon of light by the excited singlet state. The emission wavelengths from the excited singlet state are at longer wavelengths (that is, lower energy) than the absorption wavelengths of the molecule. The excited singlet state returns to the ground state by emitting a photon equal in energy to the gap between the excited state and the ground state energy levels. Each molecule has a unique fluorescence spectrum.

The excited singlet state can also return to the ground state by a process called internal conversion. In this process the excited singlet state molecule does not change its chemical structure. The energy is given off as heat to the solvent. Usually the amount of heat produced is not detectable because it is very small, and quickly dissipates into the tissue. The process of internal conversion is often described as a radiationless process. A laser is almost always required to produce large numbers of excited state molecules in a small tissue volume, the conditions required for a light-induced temperature increase. This effect is the basic principle for many laser therapies such as the treatment of port-wine stains. The chromophore for this laser therapy is hemoglobin since excited singlet states of hemoglobin return to the ground state by internal conversion.

The photochemical reactions of excited singlet state molecules generally involve only a rearrangement of the structure of the chromophore because the excited singlet states exist for a very short time. During their lifetime they do not have time to encounter other molecules in the cell before fluorescing, converting into a triplet state, or losing heat by internal conversion. For example, urocanic acid upon absorption of a UV photon undergoes a reaction from the singlet state which only involves rearrangement of the bonds between atoms in the molecule. Photochemical reactions are the most important processes of the excited state of molecules since the photoproduct initiates photobiological responses.

Conversion of a singlet excited state molecule into a triplet excited state molecule is called intersystem crossing. Triplet states are the source of many photochemical reactions in biological systems. Intersystem crossing is often the slowest photophysical process of excited singlet states because this conversion requires changing the sign of spin on an electron. This process is not allowed by quantum mechanics. Triplet states of a molecule have lower energies than the excited singlet states and possess different distributions of electrons. These factors cause the singlet and triplet states of a molecule to have different chemical reactivity. Some molecules undergo intersystem crossing very efficiently (i.e., protoporphyrin) whereas others, such as fluorescein, have very low yields of excited triplet states.

Excited triplet state molecules

Excited triplet state molecules exist for longer than excited singlet state molecules. The lifetime of an excited triplet state molecule may be a fraction of a second in the absence of oxygen. In biological systems the lifetime of an excited triplet state molecule is limited by its interaction with oxygen, which causes the excited triplet state molecule to convert into the ground state. During these interactions the energy of the triplet state is transferred to the ground state of the oxygen molecule in a process called energy transfer. Because of this interaction, triplet states in cells live about a few microseconds (10^{-6} s). While the excited triplet state molecule exists there is time for diffusion of molecules, and interactions of the excited triplet state molecule with other biomolecules in the cell. The longer lifetime of the triplet state molecule compared to the singlet state molecule is the reason that most photochemical reactions of chromophores with other biological molecules involve the triplet state.

Phosphorescence is the term for light emitted by an excited triplet state molecule. In this process the excited triplet state returns to the ground state. The spectrum of the light produced is a longer wavelength than the fluorescence of the same molecule because the energy gap between the triplet state and the ground state is smaller than that between the excited singlet state and the ground state. Phosphorescence is not usually strong enough to be seen in tissue at room temperature.

Light penetration into skin

The absorption of UV and visible radiation by chromophores in tissue is partially responsible for limiting light penetration into skin. The other major factor is scattering by particles and interfaces in the tissue. When UV and visible radiation strike the skin some of the photons are remitted, some are absorbed in various layers, and some are transmitted inward to successive layers of cells, until all the photons in the incident radiation are dissipated. Only a very small fraction of the absorbed

Table 2.2 Estimated optical penetration into very fair human skin

	Approximate depth of incident energy density decrease (μm)*		
Wavelength	37%	10%	1%
250	2	4.5	9
280	1.5	3.5	7
300	6	14	28
350	60	140	260
400	90	210	240
450	150	350	700
500	230	530	1100
600	550	1300	2500
700	750	1700	3400
800	1200	2700	5400†

* Calculated values (by permission of RR Anderson, from: Tissue optics and photoimmunology. In: *Photoimmunology*, ed. JA Parrish, ML Kripke, WL Morison, New York, Plenum Medical, 1983, pp, 61–76).
† Subcutaneous.

radiation is emitted as fluorescence. UV radiation less than 320 nm is mostly absorbed by proteins, DNA and other components of epidermal cells. About 5% of the incident light is reflected by surfaces of the stratum corneum. Scattering includes any process that deflects the path of optical radiation. Melanin which absorbs relatively uniformly over the visible wavelength, and is normally only present in the epidermis, acts largely as a density filter to diminish dermal remittance. Table 2.2 shows the light penetration of various wavelengths into skin deduced from calculations. Measurements have been made on skin separated into dermis and epidermis *in vitro* since *in vivo* measurements are not available. The photobiological responses of tissue are determined in part by the penetration of specific wavelengths into tissue.

Action spectra

An action spectrum is a way to describe the efficiency of each wavelength for producing a photobiological response. It is plotted like an absorption spectrum, effectiveness versus wavelength. Photobiologists use action spectra for identifying the chromophore in skin likely to be responsible for a particular response. When wavelengths in a certain range are particularly effective, fewer photons are needed. Peaks in the action spectrum correspond to the most effective wavelengths. Ideally, the action spectrum for a response corresponds to the absorption spectrum of the chromophore.

Action spectra for photobiologic responses in skin such as erythema, mutagenesis, immune system alterations, and photosensitivity have been reported. An action spectrum may differ from the absorption spectrum of the chromophore because absorbers in layers of tissue overlying the chromophore may filter out specific wavelengths. For example, the action spectrum for erythema in porphyria patients corresponds to the absorption spectrum of porphyrins showing maxima at about 400 nm, and between 500 and 600 nm [10].

To measure an action spectrum the tissue is exposed to varying doses of radiation at several wavelengths. The relative effectiveness of each wavelength is obtained either from the dose–response relationship or from the minimal effective dose, and is plotted versus wavelength. The effectiveness of each wavelength should be expressed relative to the number of photons delivered per area, rather than the amount of energy (J) delivered. This is necessary because the extent of a photobiological response varies with the number of photons delivered, and the number of photons per joule at each wavelength is not constant. A simple way to calculate the number of photons at a given wavelength is the following relationship:

$$\text{Number of photons in one joule} = 5 \times 10^{15} \times \text{wavelength in nanometers}$$

It may not be necessary to know a complete action spectrum for some purposes. Simply knowing the wavelength range, that is UVA or UVB, may be sufficient. These kinds of studies can be performed using a simple filtered lamp and a set of filters that cut off certain portions of the spectrum.

Quantum yield

The concept of a quantum yield is used to describe the efficiency of photophysical and photochemical processes. The quantum yield is the probability that a certain process such as fluorescence or photoproduct formation will occur when a photon is absorbed by a specific chromophore. This concept is based on the fact that each excited state molecule has a choice of processes (internal conversion, fluorescence, etc.) that it may undergo during its lifetime. These processes are in competition; each excited state molecule can undergo only one photochemical or photophysical process. The sum of all the probabilities (quantum yields) equals unity. The molecular and electronic structure of the excited state and the availability of another molecule with which to react determine the relative probabilities for each process to occur.

Reciprocity

According to the concept of the quantum yield, the number of molecules undergoing a photochemical or photophysical process (photoproduct formation, internal conversion, etc.) is proportional only to the number of photons absorbed. An extension of this concept is that the yield of a photoproduct is not dependent on the rate at which the photons are delivered but is only dependent on the total number of photons delivered to the sample. This concept is called reciprocity and can be tested experimentally by delivering the same number of photons over a varying time period. Reciprocity is a valid concept for most photobiological responses. For example the same UVB dose is required to produce erythema whether the exposure time is for seconds or several hours [15]. Reciprocity may not hold when the exposure time is long compared to the repair time in tissue. Recent work with high peak power lasers in photochemical systems also indicates that reciprocity is not maintained when the laser power is greater than about 10^5 W/cm^2. Research using high peak power lasers is just beginning to be reported in photosensitized biochemical and biological systems.

Photochemistry of endogenous chromophores

Chromophores in tissue are converted to photoproducts by several different photochemical reactions. Two of these reactions, photoaddition and photoisomerization, account for many of the light-induced changes in endogenous chromophores.

Photoaddition reactions of endogenous chromophores

The reaction of an excited state molecule with a ground state molecule to form a covalently linked photoproduct is called a photoaddition reaction. The product of this reaction is called a photoadduct. Photoaddition reactions on one strand of DNA, involving an excited state pyrimidine base with an adjacent pyrimidine base, are important for many of the effects of UV radiation in skin. In some cases, the carbon–carbon double bond of one pyrimidine ring reacts with the carbon–carbon double bond of the adjacent pyrimidine to form a cycloadduct. The product is called a cyclobutylpyrimidine dimer (Equation 3).

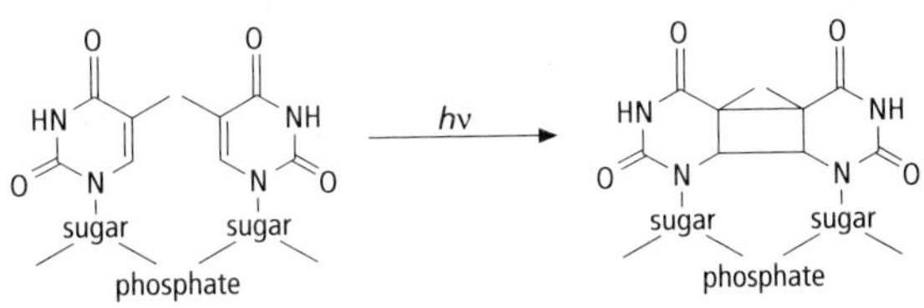

(3)

In a similar reaction, the carbon–carbon double bond of a pyrimidine adds to the carbon–nitrogen double bond of an adjacent cytosine to form an adduct that rearranges to a product in which the rings are linked by a single covalent bond (Equation 4). This type of adduct is called a (6-4) pyrimidine–pyrimidone photoproduct. Both of these types of photoaddition products are responsible for the cytotoxicity and mutagenicity of UV radiation.

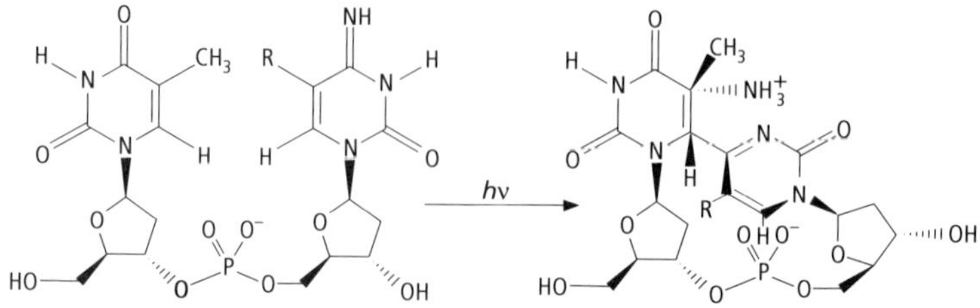

(4)

Photoisomerization reactions of endogenous chromophores

Photoisomerization reactions occur in molecules containing at least one carbon–carbon double bond that is not in a ring structure. The prefixes *cis* and *trans* designate the two arrangements of substituents that are attached to the carbon–carbon double bond, and the two structures are called *cis* and *trans* isomers. Photoisomerization converts one isomer into the other. *Cis* and *trans* forms of the molecule interconvert as shown for *trans*-urocanic acid (Equation 5). This process occurs when urocanic acid is in its excited singlet state. The formation of *cis*-urocanic acid by UVB exposure of skin is believed to be a primary event for certain effects of UV radiation on immune responses.

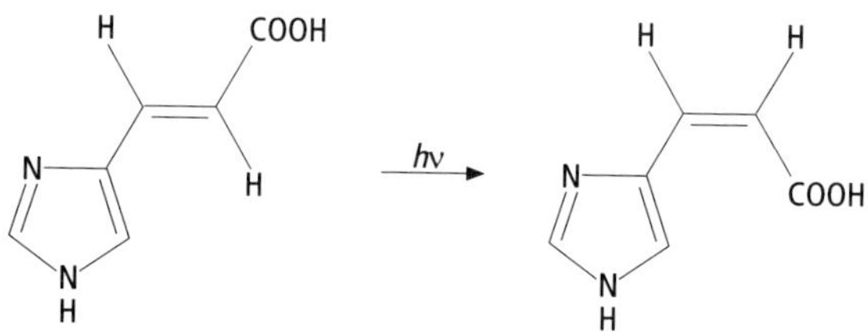

trans-urocanic acid — *cis*-urocanic acid

(5)

Other examples of photoisomerization are the *cis* to *trans* isomerization of bilirubin in skin during blue light phototherapy for neonatal hyperbilirubinemia and of retinal in rhodopsin which initiates a chain of reactions resulting in vision. The change in molecular shape that occurs when a molecule changes between *cis* and *trans* forms alters the ability of the molecule to bind to sites in tissue or, in the case of rhodopsin, alters membrane functions. Thus photoisomerization reactions provide a mechanism for transducing a UVB-induced chemical change into a biological response.

Photochemistry of photosensitizers

Exogenous photosensitizers render skin abnormally sensitive to wavelengths of light absorbed by the chromophore. The photosensitizers absorb UV or visible radiation and initiate chemical reactions with normal cellular components. Photosensitization is especially important in the UVA and visible wavelength ranges since there are few endogenous chromophores absorbing in these regions that undergo efficient photochemical reactions. Occasionally, a photosensitizer is an endogenous molecule. For example part of the damage produced in DNA by UVA radiation is believed to be initiated by excited states of endogenous chromophores such as NADH. In other cases the endogenous photosensitizer may be a molecule normally present in skin but present at a much higher concentration in photosensitive skin. The best example is certain porphyrias where protoporphyrin IX is the chromophore which causes photosensitivity in patients. Photosensitized alterations in immune function have not been studied as intensively as immune modulation initiated by photochemistry of endogenous UVB-absorbing chromophores. Photoimmunology related to psoralen photosensitization and photodynamic therapy will be discussed in Chapters 19 and 20 respectively.

The mechanisms for most photosensitization reactions fall into two categories, photoaddition and photosensitized oxidation.

Photoaddition reactions of photosensitizers

In photoaddition reactions of photosensitizers an excited state of the photosensitizer covalently binds to a cellular molecule. An important example is the reaction of the photosensitizing drug 8-methoxypsoralen (8-MOP) with pyrimidine bases in DNA. The 8-MOP molecule absorbs wavelengths in the UVA and UVB ranges, although UVA radiation is usually used because 8-MOP absorbs more strongly than DNA in this range. Because the excited states of 8-MOP exist for only very short times, the 8-MOP molecule must be adjacent to the pyrimidine base when it absorbs a photon in order for photoaddition to occur. A portion of the 8-MOP molecules in cells is reversibly intercalated into double-stranded DNA thus placing them in position for photoaddition reactions.

Theoretically, cycloadducts can be formed by reaction with a pyrimidine of either the double bond in the furan (five-membered) ring or the pyran (six-membered) ring of the psoralen molecule. Most evidence indicates that the predominant, if not exclusive, adduct is with the furan ring (Equation 6). The cycloadduct that is formed has an absorption spectrum in the UVB and UVA ranges.

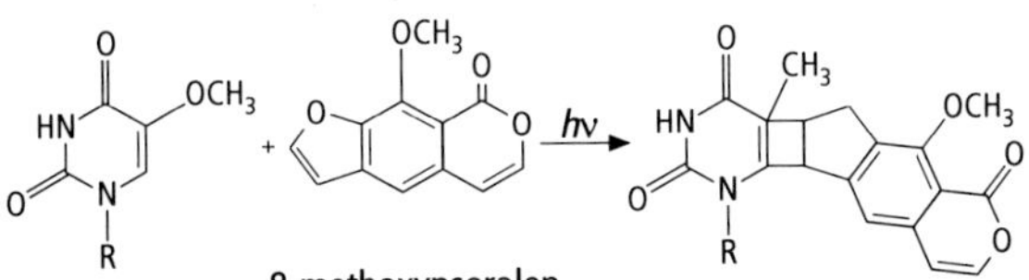

(6)

Adducts of 8-MOP with pyrimidine bases in DNA have been detected in human skin after topical application of the photosensitizer and exposure to UVA radiation [18]. The cycloadduct absorbs UV radiation and may covalently link to another pyrimidine in the opposite strand of DNA to form a crosslink between two DNA chains.

These photoaddition products with DNA are believed to be responsible for the phototoxicity of 8-MOP in human skin, and in the cytotoxic and mutagenic properties in psoralens in cell culture [14]. More recently, studies have shown that 8-MOP forms adducts with lipids in cell membranes, at least *in vitro* [4]. These photoproducts may also be involved in some of the phototoxicity responses to 8-methoxypsoralen. Other studies indicate that photosensitization by 8-MOP of DNA on the surface of effector T cells has been reported to render them immunogenic [16]. The mechanisms for immune modulation initiated by 8-MOP are discussed in Chapter 20.

Photoaddition reactions have also been hypothesized to be involved in formation of antigens involved in photoallergic responses. An excited state or free radical that is formed upon light absorption by the photoallergic drug or chemical is hypothesized to react with a protein in skin to form an antigen.

In vitro experiments indicate that many drugs that cause photoallergy also produce free radicals upon UV irradiation and in some cases light-induced covalent linkage of drugs to proteins has been demonstrated [11].

Photosensitized oxidation reactions

Many photosensitized responses to drugs or dyes result in oxidation of cellular molecules such as unsaturated lipids and proteins in membranes. These photosensitized oxidation reactions typically involve reactive oxygen species and/or free radicals. The reactive oxygen species include singlet oxygen ($^1O_2^*$), superoxide anion ($O_2^{\bullet -}$) and hydroxyl radical ($HO^\bullet$). An important feature of photosensitized oxidation reactions is that the initial product is often unstable and leads to further chemical changes, and in some cases to chain reactions. As a result of chain reactions, absorption of one photon can cause changes in many cellular molecules.

Singlet oxygen

Singlet oxygen is believed to be involved in the photosensitization process occurring with many common photosensitizers such as the porphyrins and dyes used in photodynamic therapy. Singlet oxygen is the excited state of molecular oxygen with the lowest energy. It is formed by transfer of energy from an excited state photosensitizer dye molecule, usually the triplet state, to ground state oxygen (Equation 7).

$$^{3}DYE^{*} + O_2 \longrightarrow DYE + {}^{1}O_2^{*} \qquad (7)$$

Although most ground state molecules are singlet states, ground state oxygen is an exception; it is a triplet state. When a triplet excited state of a photosensitizer interacts with a dissolved oxygen molecule it often transfers its energy to the oxygen and becomes a ground state singlet molecule. In this process the oxygen gains energy and becomes an excited singlet state. The energy of a $^{1}O_2^{*}$ molecule is about 23 kcal/mol which is substantially lower than the triplet energy of most photosensitizers. Thus, transfer of energy to oxygen is energetically feasible from the triplet state of most photosensitizers. When the triplet state energy of a photosensitizer is approximately the same as that of $^{1}O_2^{*}$, reversible energy transfer is observed. This situation has been reported for dyes that absorb at approximately 850 nm [6]. This result indicates that 850 nm is approximately the limit for the absorption maximum for dyes that photosensitize by a mechanism involving $^{1}O_2^{*}$.

Singlet oxygen is highly reactive towards many biological molecules. It reacts with proteins, lipids, DNA and many molecules less prevalent in cells. In pure water the $^{1}O_2^{*}$ molecule exists for about 4 μs before losing its energy to the solvent as heat (internal conversion). In cells $^{1}O_2^{*}$ exists for a much shorter time because it reacts very rapidly with cellular molecules surrounding it. The lifetime of $^{1}O_2^{*}$ has been estimated to be 0.17–0.32 μs in mammalian cells [2]. One consequence of its short lifetime is that $^{1}O_2^{*}$ reacts with molecules very close to the site where it was generated. In 0.2 μs, the $^{1}O_2^{*}$ molecules can only diffuse about 0.040 μm. This is a short distance compared to cell dimensions. For example the radius of the nucleus in a mammalian cell is about 5 μm. Thus $^{1}O_2^{*}$ only damages the molecules very close to where it was generated by energy transfer from a photosensitizer. This means that the location of photosensitized damage depends critically on the location of the photosensitizer in the cell or tissue.

The amino acids that react most rapidly with $^{1}O_2^{*}$ are histidine, tryptophan and methionine. During the reactions with $^{1}O_2^{*}$ the imidazole ring of histidine is destroyed, tryptophan is converted into kynurenic acid derivatives, and the sulfur in methionine is oxidized. The products of these reactions are involved in crosslinking between protein molecules although the mechanisms have not been delineated.

The characteristic reaction of $^{1}O_2^{*}$ with unsaturated lipids is shown in Equation 8. The product is an allylic hydroperoxide which is unstable in the presence of metal ion catalysts and breaks down to free radicals that can initiate chain oxidation reactions.

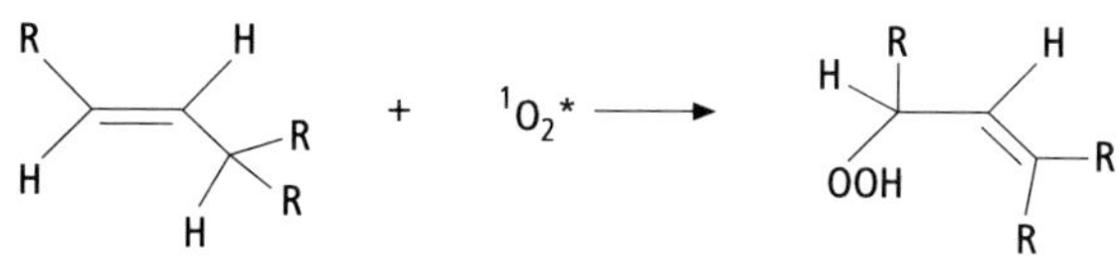

(8)

It is very difficult, and often impossible, to determine which reactive species is responsible for a photosensitized response in cells or tissue. When a photosensitized reaction is studied in solution, many techniques are available to assess the molecular mechanism. The complexity of cells precludes most of these techniques from producing unambiguous answers. For example $^{1}O_2^{*}$ emits phosphorescence at 1270 nm which can be detected in solution even though the quantum yield is very low (about 10^{-6}). The correlation of the quenching of this emission with the formation of an oxidized product in solution is good evidence that $^{1}O_2^{*}$ is involved in the photosensitized oxidation of a substrate. In cells, $^{1}O_2^{*}$ reacts very rapidly with proteins and other components so that its emission cannot be seen even with very sensitive instrumentation [8].

Other methods that are used to detect $^{1}O_2^{*}$ in solution are also much less informative in cells. For example in aqueous solution $^{1}O_2^{*}$ exists about 15 times longer in D_2O than in H_2O and therefore has a greater opportunity to cause oxidation of a substrate in the deuterated solvent. If photosensitized oxidation of a substrate is more effective by approximately a factor of 15 in D_2O than in H_2O the mechanism for the oxidation probably involves reaction with $^1O_2^*$. This technique is often not successful in cells even when $^1O_2^*$ is participating in a photosensitized mechanism since the lifetime of $^1O_2^*$ in cells is limited by its reaction with cellular molecules rather than by deactivation by the solvent. The $^1O_2^*$ lifetime in D_2O is likely to be only slightly longer than that in H_2O in cells. Thus the enhancement factor may be only 2–3. Other factors may also interfere with this technique, such as the difference in apparent pH between H_2O and D_2O environments.

Quenchers of $^1O_2^*$ such as azide ion, DABCO and beta-carotene are used in solution to support the involvement of $^1O_2^*$ in a photosensitized oxidation mechanism. These quenchers are not entirely specific for $^1O_2^*$ but are often used as a first step in mechanistic analysis. In cells there are additional difficulties with these quenchers since they are not likely to be evenly distributed throughout the cell due to their hydrophilic or hydrophobic character. A quencher may also not localize in the same sites as the photosensitizer which is a problem since the $^1O_2^*$ is only found close to the photosensitizer.

A method that can be used to determine whether or not $^1O_2^*$ is formed by energy transfer from an excited state of a photosensitizer involves the formation of a specific oxidation product between $^1O_2^*$ and cholesterol [7]. Singlet oxygen reacts with cholesterol to form an allylic hydroperoxide by the reaction shown in Equation 8. Only one stereoisomer is formed and it can be detected by chromatography after chemical reduction. Other photo-oxidation mechanisms involving radicals also yield hydroperoxides with cholesterol but at least two isomers are formed (Equation 9). To test for formation of $^1O_2^*$ by this technique, radiolabeled cholesterol is incorporated into cells prior to photosensitization. The hydroperoxides formed during the irradiation are reduced, isolated and identified by chromatography.

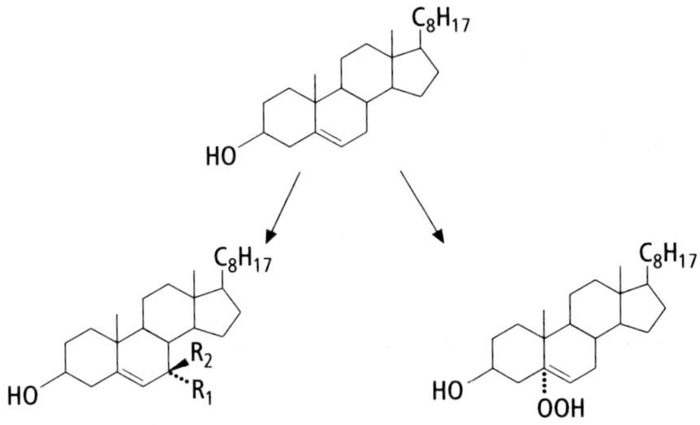

	R1	R2
7α-OOH	OOH	H
7β-OOH	H	OOH

5α-OOH

(9)

Superoxide anion

Superoxide anion, $O_2^{\bullet -}$, is molecular oxygen with an extra electron. It is usually formed by transfer of an electron from an excited state of a photosensitizer to ground state molecular oxygen (Equation 10). In addition to $O_2^{\bullet -}$, this reaction produces a cation radical of the photosensitizer, $PS^{\bullet +}$, which typically picks up an electron from oxidizable molecules in the cell to return to its ground state. Superoxide anion is also formed when an electron is transfered to a photosensitizer excited state by a donor molecule, D, and the $PS^{\bullet -}$ molecule transfers an electron to molecular oxygen (Equation 11). In aqueous solution at physiological

$$^3DYE^* + O_2 \longrightarrow DYE^{\bullet +} + O_2^{\bullet -} \qquad (10)$$

$$^*PS + D \longrightarrow PS^{\bullet -} + D^{\bullet +} \xrightarrow{O_2} PS + O_2^{\bullet -} \qquad (11)$$

pH, $O_2^{\bullet -}$ is in equilibrium with its protonated form, the hydroperoxyl radical, $HO_2^{\bullet}$. The hydroperoxyl radical is much more reactive towards biological molecules than is $O_2^{\bullet -}$, but is not as reactive as $^1O_2^*$ or hydroxyl radical, $HO^{\bullet}$. Because of its low reactivity $O_2^{\bullet -}$ can diffuse in cells to sites distant from the chromophore. Superoxide anion is destroyed by the enzyme superoxide dismutase, SOD, and in solution this enzyme can be used to determine whether $O_2^{\bullet -}$ is involved in a photosensitized oxidation reaction. In cells and tissue, quenching by SOD is not usually effective because this enzyme does not penetrate into cells. Recent studies have explored the use of 'SOD mimics' which are small molecules that destroy $O_2^{\bullet -}$ although they are not entirely specific for $O_2^{\bullet -}$ [12].

Although $O_2^{\bullet -}$ is not highly reactive towards most biomolecules, one of its breakdown products, hydroxyl radical ($HO^{\bullet}$), is very highly oxidizing. Hydroxyl radical is produced by the metal ion catalysed dismutation of $O_2^{\bullet -}$ as shown in Equations 12 and 13. These reactions convert $O_2^{\bullet -}$ and hydrogen peroxide to $HO^{\bullet}$ and hyroxide anion. Hydroxyl radical reacts with nearly diffusion-controlled rates with many organic molecules. The reactions of $HO^{\bullet}$ have been well studied because $HO^{\bullet}$ is produced by ionizing radiation in water such as occurs during radiation therapy. Hydroxy radical adds to carbon–carbon double bonds to form adducts such as hydrates of pyrimidine bases in cellular DNA. Other reactions of $HO^{\bullet}$ include abstracting hydrogen atoms and accepting electrons from organic molecules. Because of its high reactivity, $HO^{\bullet}$ in cells has a very short lifetime, approximately 2 ns. During its existence $HO^{\bullet}$ travels less than the thickness of a mammalian cell membrane (about 7 nm). It is difficult to quench $HO^{\bullet}$ in cells because the concentration of the quenchers such as mannitol must be as great as that of intracellular substances that react with $HO^{\bullet}$.

$$O_2^{\bar{}} + Fe^{3+} \longrightarrow O_2 + Fe^{2+} \qquad (12)$$

$$H_2O_2 + Fe^{2+} \longrightarrow HO\cdot + OH^- + Fe^{3+} \qquad (13)$$

Cellular sites and mechanisms of photosensitization

The subcellular site where photosensitized damage occurs is highly likely to be the site where the photosensitizer is localized because, as discussed above, most reactive species formed from excited states of photosensitizers only diffuse a short distance in cells. Subcellular sites that are often altered by photosensitization include the plasma membrane, mitochondria, lysozomes and DNA. Because of the interest in photodynamic therapy, there is a vast literature on the cellular sites of photosensitizers and their mechanisms of action; only a few examples will be discussed here.

Certain photosensitizers such as merocyanine 540 selectively localize in the plasma membranes of cells and others are found mainly in the plasma membrane early during incubation before they diffuse into other cellular membranes. Hydrophobic photosensitizers partition into the plasma membrane and other cellular membranes. Irradiation in the presence of photosensitizers causes oxidation of membrane lipids and proteins, and crosslinking between membrane proteins. These chemical changes cause alterations in membrane function such as decrease in membrane potential [17], increase in intracellular Ca^{2+} [3], and leakage of K^+ [5]. The relationships between these photosensitized plasma membrane changes and cell toxicity or altered immune cell function are currently being investigated.

Many hydrophobic photosensitizers that possess a delocalized positive charge selectively localize in mitochondria apparently because of the potential across the inner mitochondrial membrane [13]. Upon exposure to visible radiation in the presence of these photosensitizers, the mitochondria show damage to the cristae and are eventually disrupted. The detailed molecular mechanisms for all of these photosensitizers are not known, but at least for certain cyanine dyes photosensitized inhibition of complex I in the electron transport chain occurs [1].

Certain classes of photosensitizing dyes localize primarily in lysozomes because after entering the acidic milieu inside this organelle they become protonated [9]. The mechanism for cytotoxicity photosensitized by these dyes may involve disruption of the lysozomes which releases hydrolytic enzymes into the cytoplasm.

Photosensitized damage to DNA has been postulated to be the mechanism of psoralen phototoxicity as discussed above. Other photosensitizers that intercalate into DNA and cause damage such as DNA strand breaks include methylene blue and acridine. These photosensitizers may react directly, usually by electron transfer, with the nucleotides or may generate $^1O_2{}^*$ which reacts with the guanine bases in DNA.

Summary

The effects of UV and visible radiation on the immune system are initiated by absorption of photons by specific chromophores. The excited states that are produced undergo unimolecular reactions such as photoisomerization, or bimolecular reactions such as photoaddition or energy transfer to oxygen. Many of the photochemical reactions likely to be involved in photoimmunology have been characterized. Identifying the biochemical and cell biological mechanisms that connect the initial photochemistry to the ultimate changes in immune function is a stimulating and challenging goal for researchers in photoimmunology.

References

1 Ara, G., J.R. Aprille, C.D. Malis, S.B. Kane, L. Cincotta, J. Foley, J.V. Bonventre, A.R. Oseroff: Mechanisms of mitochondrial photosensitization by the cationic dye, *N,N*-bis(2-ethyl-1,3-dioxylene) kryptocyanine (EDKC): preferential inactivation of complex I in the electron transport chain. *Cancer Research* 47 (1987), 6580–6585.

2 Baker, A., J.R. Kanofsky: Quenching of singlet oxygen by biomolecules from L1210 leukemia cells. *Photochem. Photobiol.* 55 (1992), 523–528.

3 Ben-Hur, E., T.M.A.R. Dubbelman, J. Van Steveninck: Pthalocyanine-induced photodynamic changes of cytoplasmic free calcium in Chinese hamster cells. *Photochem. Photobiol.* 54 (1991), 163–166.

4 Cadet, J., P. Vigny, W.R. Midden: Photoreactions of furocoumarins with biomolecules. *J. Photochem. Photobiol. B: Biology* 6 (1990), 197–206.

5 Dubbelman, T.M.A.R., J. Van Steveninck: Photodynamic effects of hematoporphyrin-derivative on transmembrane transport systems of murine L929 fibroblasts. *Biochim. Biophys. Acta* 771 (1984), 201–207.

6 Firey, P.A., W.E. Ford, J.R. Sounik, M.E. Kenney, M.A.J. Rodgers: Silicon naphthalocyanine triplet state and oxygen: a reversible energy-transfer reaction. *J. Am. Chem. Soc.* 110 (1988), 7626–7630.

7 Girotti, A.W.: Photodynamic lipid peroxidation in biological systems. *Photochem. Photobiol.* 51 (1990), 497–509.

8 Gorman, A.A., M.A.J. Rodgers: Current perspectives of singlet oxygen detection in biological environments. *J. Photochem. Photobiol. B: Biology* 14 (1992), 159–176.

9 Lin, C.-W., J.R. Shulok, S.D. Kirley, L. Cincotta, J. Foley: Lysosomal localization and mechanism of uptake of Nile Blue photosensitizers in tumor cells. *Cancer Res.* 51 (1991), 2710–2719.

10 Magnus, I.A.: The cutaneous porphyrins. *Seminars in Hematology* 5 (1968), 380–408.

11 Morison, W.L., I.E. Kochevar: Photoallergy. In: *Photoimmunology*, (1983), pp. 227–253. Eds.: J.A. Parrish, M.L. Kripke, W.L. Morison, Plenum, New York.

12 Ortel, B., R.W. Gange, T. Hasan: Investigations of a manganese-containing mimic of superoxide dismutase in riboflavin phototoxicity in human cells *in vitro*. *J. Photochem. Photobiol. B: Biology* 7 (1990), 261–276.

13 Oseroff, A.R., D. Ohuoha, G. Ara, D. McAuliffe, J. Foley, L. Cincotta: Intra-mitochondrial dyes allow selective *in vitro* photolysis of carcinoma cells. *Proc. Natl Acad. Sci. USA* 83 (1986), 9729–9733.

14 Papadopoulo, D., D. Averbeck: Genotoxic effects and DNA photoadducts induced in Chinese hamster V79 cells by 5-methoxypsoralen and 8-methoxypsoralen. *Mutat. Res.* 151 (1985), 281–291.

15 Park, Y.K., R.W. Gange, P.C. Levins, J.A. Parrish: Low and moderate irradiances of UVB and UVC irradiation are equally erythemogenic in human skin. *Photochem. Photobiol.* 40 (1984), 667–669.

16 Perez, M.I., Y. Yamane, L. John, F.P. Gasparro, R.L. Edelson: DNA associated with the cell membrane is involved in the inhibition of the skin rejection response induced by infusions of photodamaged alloreactive cells that mediate rejection of skin allograft. *Photochem. Photobiol.* 55 (1992), 839–849.

17 Specht, K.G., M.A.J. Rodgers: Plasma membrane depolarization and calcium influx during cell injury by photodynamic action. *Biochim. Biophys. Acta* 1070 (1991), 60–68.

18 Yang, X.Y., F.P. Gasparro, V.A. DeLeo: 8-Methoxypsoralen-DNA adducts in patients treated with 8-methoxypsoralen and ultraviolet A light. *J. Invest. Dermatol.* 92 (1989), 59–63

3: Photobiology of ultraviolet radiation-induced DNA damage

Ernst G. Jung & Elisabeth Bohnert

Photochemical background

Electromagnetic radiation, of which the visible portion is called 'light', is capable of interacting with molecules in the following ways:

1 transmission (direction of propagation is not affected)

2 reflection, reemission, refraction, dispersion, interference (direction of propagation is changed)

3 absorption (no further propagation, extinction).

The first two possibilities are described by the laws of optics. Absorption has been characterized by the laws of quantum physics, which are of major importance for the fields of photochemistry and photobiology. In order to understand energy absorption by molecules and the resulting primary and secondary photochemical events, the following laws have been formulated:

Law of Gotthus–Draper. This states that only light absorbed by a molecule can produce photochemical change in the molecule. This indicates that radiation in spectral regions in which no absorption occurs fails to induce photochemical changes. On the other hand, the reversal of this law is not true, since photons may be absorbed without causing photochemical effects, since the energy may produce luminescence or may be transferred into caloric energy.

Law of Stark–Einstein. This states that the sum of the quantum yields for each individual photophysical and photochemical process must be unity. In other words, each photon that is absorbed causes the activation of one single molecule. This indicates that a primary photochemical reaction requires one quantum per molecule, or that 1 mol (molecular weight of a given substance) equals 6023×10^{23} quanta corresponding to Loschmidt's number. However, not every quantum that is absorbed will cause a photochemical reaction, and therefore the concept of quantum yield Φ has been introduced:

$$(\Phi) = \frac{\text{number of molecules of product formed}}{\text{number of quanta absorbed}}$$

In a complex system and during ineffective processes, the quantum yield may be as slow as 10^{-6}, whereas a photochemical chain reaction may produce a very high quantum yield of up to 10^{6}.

Law of Reciprocity of Bunsen–Roscoe. This states that the photochemical effect of a given reaction depends on the product of intensity and time of exposure rather than the single value of one of these determinants. This indicates that only the number of quanta is important for the photochemical reaction, independent of the irradiation time. With regard to biological systems, however, the law of reciprocity is only of limited relevance since a variety of repair mechanisms may occur during longer irradiation periods. The photobiological effect of a given dose (energy fluence) is thus smaller for longer irradiation periods compared to shorter ones (protrahation effect).

Law of Lambert–Beer. This law represents the first quantitative approach to the study of light absorption. It says that the fraction of incident, parallel light absorbed by a medium is porportional to the thickness of the medium traversed and is independent of the incident intensity of light.

$$\text{extinction } E = \log. \frac{I_0}{I}$$

with I_0 being the intensity of the incident light and I being the optical path in centimeters.

The extinction of diluted solutions is proportional to the concentration of the substance dissolved. This principle is used for example for photometrical and spectrophotometrical measurements. It is not valid for complex systems such as human skin [9].

Photobiology of desoxyribonucleic acid (DNA)

The principal components of DNA are purine and pyrimidine bases, which are capable of absorbing electromagnetic radiation of 230–300 nm. This wavelength's range is therefore responsible for the main effects of electromagnetic radiation on DNA. Radiation of longer wavelengths, however, may contribute to DNA damage as well if other chromophores are present which are capable of transferring energy to DNA.

Ultraviolet irradiation of both isolated and cellular DNA molecules was found to generate a large variety of photoproducts, of which pyrimidine dimers were the most frequent. In experiments in which the DNA components pyrimidine and purine were exposed in aqueous solution or within an ice matrix to irradiation of 254 nm, the generation of cyclobutyl pyrimidine dimers, and of 6-hydroxy-dihydropyrimidine, as well as other pyrimidine photoadducts was observed [42]. Cyclobutane dimers are known to exist in four different configurations, which differ in the spatial assembly of their hydrogen atoms and methyl groups. Dimerization of two neighboring bases is responsible for the formation of *cis*-syn cyclobutane dimers within one strand of DNA (Fig. 3.1). If thymiylyl (3′-5′) thymidin (TpT) in aqueous solution is irradiated with UVC (240–290 nm) radiation, both the generation of cyclobutane dimers and of pyrimidin–(6-4) pyrimidon photoproducts (TpT_4) may be observed [16, 35, 50]. The latter photoproduct may be spliced by alkalic elution of DNA [26] and may be transformed via photolysis into a Dewar isomer (Fig. 3.2) [45]. Dimer formation leads to conformational changes of the DNA double helix (Fig. 3.3). The relative ratios of the formation of cyclobutane dimers versus (6-4) photoproducts may vary to quite an extent and are mainly a function of the (A + T) to (G + C) ratio of a given DNA, the amount of irradiation applied, and the wavelength [32, 33].

Studies employing DNA isolated from fungi and bacteria indicate that irradiation at 254 nm in a dose range of 2–7 J/m² leads to the generation of the cyclobutane dimers CC, CT, and TT in a ratio of 1:1:2 [48]. In certain DNA regions (6-4) photoadducts are more frequent at neighboring

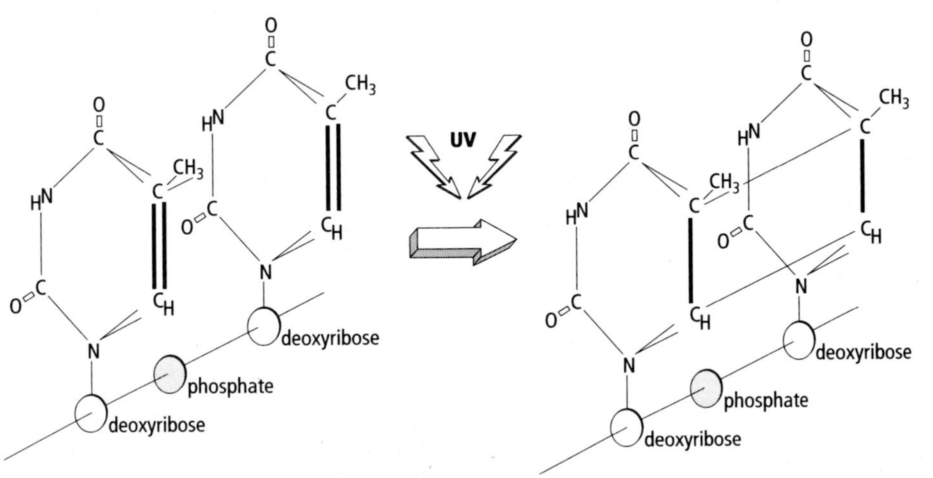

Fig. 3.1 UV-induced *cis*-syn-cyclobutan dimerization of adjacent thymidines within a DNA strand.

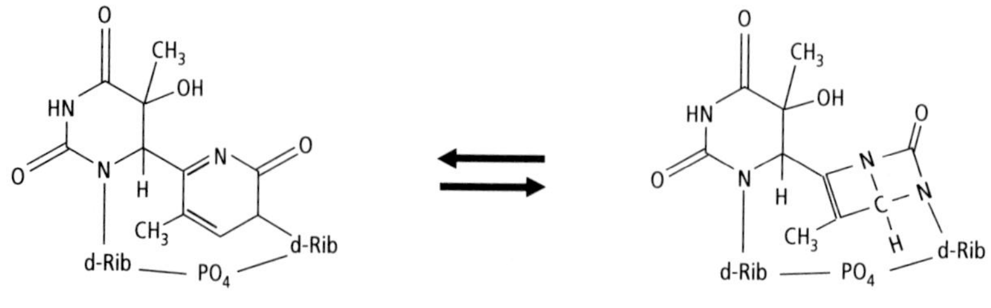

Fig. 3.2 (6-4) Photoproducts: pyrimidin–(6-4) pyrimidon (TpT_4) and the Dewar isomer (TpT_3).

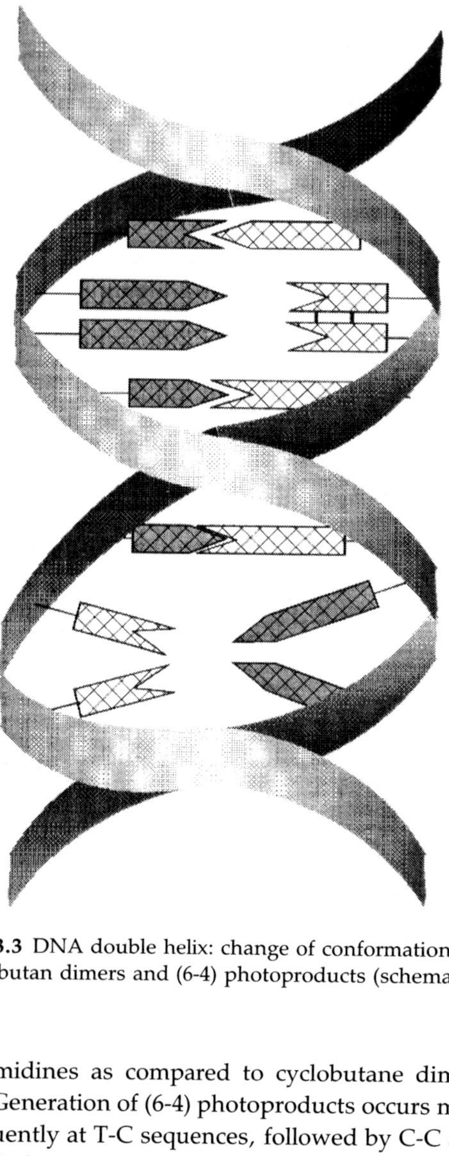

Fig. 3.3 DNA double helix: change of conformation by cyclobutan dimers and (6-4) photoproducts (schematic).

pyrimidines as compared to cyclobutane dimers [5]. Generation of (6-4) photoproducts occurs most frequently at T-C sequences, followed by C-C and finally by TT locuses [26, 40]. Generation of (6-4) photoproducts in DNA molecules does not occur at C-T sequences. Pyrimidine dimers may also be generated within the skin through irradiation with ultraviolet radiation of longer wavelengths (UVA; >340 nm).

Cyclobutane dimers are acid resistant and thus may be detected chromatographically after acid hydrolysis. The development of monoclonal antibodies specifically directed against cyclobutane dimers or (6-4) photoproducts first made it possible to detect both types of DNA damage independently of each other [27, 29, 31].

It is also possible to detect cyclobutane dimers by employing endonucleases, which may be isolated from *Micrococcus luteus* or from T4 bacteriophages, and which are capable of cleaving normal DNA into single nucleic acids, whereas dimer molecules remain unaffected [12, 34].

Formation of pyrimidine (6-4) pyrimidone is associated with a switch of the absorption spectrum into the red. The (6-4) photoproduct exhibits an absorption maximum at 310 nm, and exposure to radiation of the identical wavelength is capable of photolysing these products.

Within the UVB and UVC range, DNA itself is the chromophore and, as a consequence, absorption of UV radiation by DNA may even cause splicing of the DNA strand. Radiation of 254 nm produces strand breaks less frequently compared to pyrimidine dimers. With longer wavelengths, however, the frequency of strand breaks will increase, and at 365 nm the ratio of dimers to strand breaks is 1:1, whereas at 365–500 nm, this ratio decreases [11, 37]. At these higher wavelengths, intracellular chromophores such as porphyrines, riboflavines, NADH, and others, are capable of absorbing radiation energy, which will then cause strand breaks via energy transfer into DNA.

Biological consequences of UV radiation-induced DNA damage: repair reactions

DNA sequences are highly conserved throughout many generations. Under normal conditions, spontaneous mutations are rare events and occur at a frequency of one to 10^9 base pair replications [1]. In order to achieve this low mutation level, a number of mechanisms are required which allow a given cell to repair the DNA damage that is permanently generated by environmental factors.

Excision repair

Excision repair is the most important mechanism for the repair of UV radiation-induced dimer photoproducts within cellular DNA. Experimental proof for the existence of excision repair was first provided by studies employing different strains of *E. coli* [14]. Excision repair represents a highly complex process, which includes several enzymatically regulated steps. Since this repair mechanism occurs without additional irradiation with ultraviolet radiation or visible light, it was also called 'dark repair'. During dark repair, UV-induced pyrimidine dimers, together with adjacent nucleotides, are replaced by newly synthesized, intact nucleotides of the identical DNA strand, which are synthesized by using the undamaged complementary DNA strand as a template. This form of DNA synthesis, which occurs during G_1 and G_2 phases of the cell cycle, is called unscheduled DNA synthesis, as opposed to scheduled DNA synthesis, which is the semiconservative replication of DNA during S-phase. Excision repair consists of several consecutive steps. Each step is mediated by a specific enzymatic activity. A specific endonuclease first incises the DNA strand near the dimer (incision), and consecutively another enzyme, which exhibits endo- and exonuclease activity, excises the dimer containing oligonucleotide (excision). A helicase will subsequently separate the excised oligonucleotide from the complementary DNA strand [47] and at the same time a new oligonucleotide is synthesized by a polymerase. The newly synthesized material is eventually ligated with the parental DNA strand. This repair process requires several hours and, during a sunburn reaction, is known to precede the apparance of a clinically visible UV-induced erythema. Time kinetic studies indicate that after a quick onset, the reaction will slow down and become asymptotic. Since it is possible to discriminate between pyrimidine dimers and (6-4) photoadducts, Mitchell *et al.* have analysed the repair kinetics for these two photoproducts separately [27–31]. Within human cells, (6-4) photoproducts were removed by 75% 4 h after irradiation, whereas 17 h were required in order to achieve a 70% reduction in the number of cyclobutane dimers [28, 29]. These studies, which are based on experiments employing DNA damage specific monoclonal antibodies, are in agreement with earlier studies in which DNA damage was assessed chromatographically [36] or via T4-endonuclease [30].

Excision repair may be measured by several methods. One way is to determine the time kinetics of reduction in the number of dimers within DNA isolated from UV-irradiated cells as an indirect measure of excison repair capacity [43]. In addition, it is possible to isolate polynucleotides (containing up to 70 nucleotides), after they have been excised. This method allows a direct assessment of the number of incisions, which reflects the number of dimers or the activity of the endonucleases. Freeman *et al.* determined the number of UV radiation-induced single strand breaks by employing alkalic gel electrophoresis [10], whereas Hincks and Coulombe used an alkalic filtration technique in order to separate DNA fragments from high molecular DNA [14].

Unscheduled DNA synthesis may also be measured as the amount of newly synthesized DNA by employing radioactively labeled, DNA specific nucleotides such as 3H-thymidine. The amount of radioactivity incorporated into DNA may be measured by means of a liquid scintillation counter or autoradiographically.

Recent studies indicate that not all cyclobutane

dimers within a given cell are equally accessible for endonucleases. This has led to a differentiation between 'core DNA' versus 'transcriptionally active DNA', and to the introduction of the concept that DNA repair occurs in a domain-oriented fashion [7, 15, 20]. It has been demonstrated, for example, that DNA damage within domains which encode for vital functions is preferentially repaired [3, 21].

Photoreactivation

Cyclobutane dimers as well as (6-4) photoadducts may be repaired by an enzyme-dependent, light-mediated mechanism. Photolysis of (6-4) photoproducts by radiation of 310 nm has already been referred to [45]. Under normal circumstances, photorepair is observed if UV irradiation is followed by subsequent exposure of cells to radiation in the range of 300–600 nm, which will result in the activation of the photoreactivating enzyme.

Photoreactivation of pyrimidine dimers has been found to occur in a large variety of microorganisms, but also in certain mammalian cells, and maybe even in human fibroblasts [44]. There is also some evidence that photorepair may be observed *in vivo* in human skin [19].

A photoreactivating enzyme has been isolated from yeast and shown to be capable of repairing dimers in fibroblasts isolated from xeroderma pigmentosum patients [38]. Photolyases have been isolated and purified in quantities which allowed a detailed analysis of the mechanisms underlying DNA repair via photoreactivation [39].

Wood [53] successfully removed more than 95% of cyclobutane dimers present in UV-irradiated DNA by using DNA-photolyase which had been purified from *E. coli*. Photolyase treatment led to a 20–40% decrease of excision repair, indicating that the major percentage of dimers consisted of (6-4) photoadducts, which may be rapidly repaired in mammalian cells. The observation that enzymatic photoreactivation of cyclobutane dimers is associated with a faster repair of (6-4) photoproducts may be explained by the possibility that excision repair associated enzymes are available in larger quantities for the removal of (6-4) photoproducts. Photoreactivation may offer one possible explanation for the observation that following broad band UV radiation exposure less DNA damage is caused compared to damage induced by monochromatic radiation [52].

Postreplication repair

In bacterial cells it has been shown that during DNA replication, the polymerase leaves gaps opposite the dimer. The damaged region will be replicated, and afterwards, a DNA recombinational mechanism yields a new configuration of the strands, leading to opposition of the dimer containing strands by a DNA region containing the correct information for its replacement. This configuration can now be repaired by an excison type mechanism, and also the gapped DNA is opposed by complete information which will be filled in by polymerization, and finally sealed by ligation.

For human cells this repair mechanism may be of importance for the repair induced by ionizing radiation. In addition, there is a class of xeroderma pigmentosum patients, called XP-variants, in which postreplication repair is thought to be defective. Accordingly, these patients exhibit all symptoms of xeroderma pigmentosum patients, although they are able to undergo normal excision repair.

Xeroderma pigmentosum: a biological model for UV-induced DNA damage and defective repair

Information concerning the various DNA repair mechanisms has been derived to quite some extent from studies employing cells which were isolated from patients suffering from the hereditary disease xeroderma pigmentosum. Xeroderma pigmentosum patients belong to eight different genetically defined subgroups, seven of which are characterized by a defect in excision repair. Hybridization of normal fibroblasts with fibroblasts derived from a patient with xeroderma pigmentosum will lead to compensation of the molecular defect in excision repair present in XP fibroblasts, indicating that the two nulcei of the heterokaryotic cell are able to repair DNA damage successfully [13]. These fusion experiments have allowed the definition of seven complementation groups (A to G), which are characterized by distinct defects in excision repair [22, 24, 51]. In addition to defects in excision repair, some of these complementation groups were found to be associated with other genetic defects. For example, most patients of complementation group A exhibit neurological abnormalities; six patients of complementation group D also have trichothiodystrophie (TTD) and eight xeroderma pigmentosum patients (three XP-B; one XP-C; two XP-D and two XP-G) also have Cockayne syndrome [18, 49].

All fibroblast strains isolated from XP patients are characterized by a decreased colony forming ability following UV radiation exposure as compared to normal cells [23]. In addition, after UVB and UVC radiation exposure, increased sister chromatid exchange rates may be observed within their chromosomes [2].

The main defect of xeroderma pigmentosum cells is the lack of, or the decreased activity of, UV-specific endonucleases, which are required for dimer recognition and incision of the DNA strand [6, 22, 41]. In contrast, XP-variant cells exhibit no, or only a moderate, decrease in excison repair activity, but are thought to have a defect in post-replication repair [19, 25].

Recent progress in molecular biological techniques has greatly contributed to the understanding of the defects present in DNA damage repair deficient cells. By employing microinjection or electroporation, proteins and plasmids may be introduced into viable cells, which may help to elucidate or compensate the defect present in these cells [17, 46]. For example, these techniques have been successfully used to generate UV-resistant, XP-revertant cells, which are capable of repairing (6-4) photoadducts but which fail to repair cyclobutane dimers. Another example is the introduction of a gene isolated from bacteriophage T4 into xeroderma pigmentosum cells, which enabled these cells to specifically repair cyclobutane dimers [7, 8, 9].

References

1 Alberts, B., D. Bray, J. Lewis, M. Raft, K. Roberts, J.D. Watson: *Molecular Biology of the Cell*, (1983); Garland Publishing, New York.

2 Bohnert, E.: Hypermutabilität bei Xeroderma pigmentosum (XP). *Akt. Dermatol.* 17 (1991), 97–99.

3 Bohr, V.A., D.S. Okumoto, P.C. Hanawalt: Survival of UV-irradiated mammalian cells correlated with efficient repair in an essential gene. *Proc. Natl Acad. Sci. USA* 83 (1986), 3830–3822.

4 Boyce, R.P., P. Howard-Flanders: Release of UV light-induced thymidine dimers from DNA in *E. coli* K 12. *Proc. Natl Acad. Sci. USA* 51 (1964), 293–300.

5 Brash, D.E., S. Seetharam, K.H. Kraemer, M.M. Seidman, A. Bredberg: Photoproduct frequency is not the major determinant of UV base substitution hot spots or cold spots in human cells. *Proc. Natl Acad. Sci. USA* 84 (1987), 3782–3786.

6 Cleaver, J.E.: Defective repair replication of DNA in xeroderma pigmentosum. *Nature (London)* 218 (1968), 652–656.

7 Cleaver, J.E.: DNA damage and repair in normal, xeroderma pigmentosum and XP revertant cells analyzed by gel electrophoresis: excision of cyclobutane dimers from the whole genome is not necessary for cell survival. *Carcinogenesis* 10 (1989), 1691–1696.

8 Cleaver, J.E., J.J. Wayne, W.C. Charles, D.L. Mitchell: Cyclobutane dimers and (6-4) photoproducts in human cells are mended with the same patch sizes. *Photochem. Photobiol* 54 (1991), 393–402.

9 Colicos, M.A., Y. Haj-Ahmad, K. Valerie, E.E. Henderson, A.J. Rainbow: Construction of a recombinant adenovirus containing the denV gene from bacteriophage T4 can partially restore the DNA repair deficiency in xeroderma pigmentosum fibroblasts. *Carcinogenesis* 12 (1991), 249–255.

10 Freeman, S.E., A.D. Blackett, D.C. Monteleone, R.B. Setlow, B.M. Sutherland: Quantitation of radiation-, chemical-, or enzyme-induced single strand breaks in nonradioactive DNA by alkaline gel electrophoresis: application to pyrimidine dimers. *Anal. Biochem.* 158 (1986), 119–129.

11 Freeman, S.E., R.W. Gange, J.C. Sutherland, E.A. Matzinger, B.M. Sutherland: Production of pyrimidine dimers in DNA of human skin exposed *in situ* to UVA radiation. *J. Invest. Dermatol.* 88 (1987), 430–433.

12 Friedberg, E.C., J.J. King: Endonucleolytic cleavage of UV-irradiated DNA controlled by the V+ gene in phage T4. *Biochem. Biophys. Res. Comm.* 37 (1969), 646–651.

13 Giannelli, F., P.H. Croll, S.A. Lewin: DNA repair synthesis in human heterokaryons formed by normal and UV-sensitive fibroblasts. *Exp. Cell Res.* 78 (1973), 175–185.

14 Hincks, J.R., R.A. Coulombe: Rapid detection of DNA-interstrand and DNA-protein cross-links in mammalian cells by gravity-flow alkaline elution. *Environ. Mol. Mutagen.* 13 (1989), 211–217.
15 Jensen, K.A., M.J. Smerdon: DNA repair within nucleosome cores of UV-irradiated human cells. *Biochemistry* 29 (1990), 4773–4782.
16 Johns, H.E., M.L. Pearson, J.C. LeBlanc, C.W. Helleiner: The ultraviolet photochemistry of thymidyly-(3′-5′) thymidine. *J. Mol. Biol.* 9 (1964), 503–524.
17 DeJonge, A.J.R., W. Vermeulen, W. Keijzer, J.H.J. Hoeijma-Kers, D. Bootsma: Microinjection of *Micrococcus luteus* UV-endonuclease restores UV-induced unscheduled DNA synthesis in cells of 9 xeroderma pigmentosum complementation groups. *Mutat. Res.* 150 (1985), 99–105.
18 Jung, E.G.: Photocarcinogenesis in the skin. *J. Dermatol.* 18 (1991), 1–10.
19 Jung, E.G., E. Bohnert: Lichtbiologie der Haut. In: *Jadasohn, Handbuch der Haut- und Geschlechtskrankheiten*, vol. I/4, (1979), pp. 459–540. Ed.: A. Marchionini, Springer, Berlin.
20 Kantor, G.J., L.S. Barsalou, P.C. Hanawalt: Selective repair of specific chromatin domains in UV-irradiated cells from xeroderma pigmentosum group C. *Mutat. Res.* 235 (1990), 171–180.
21 Keyse, S.M., R.M. Tyrrell: Rapidly occurring DNA excision repair events determine the biological expression of UV induced damage in human cells. *Carcinogenesis* 8 (1987), 1251–1256.
22 Kleijer, W.J.: *Defective DNA repair in human skin disease xeroderma pigmentosum*, Thesis (1973); Brouder-Offset BV, Rotterdam.
23 Kraemer, K.H., A.D. Andrews, S.F. Barrett, J.H. Robbins: Colony-forming ability of ultraviolet-irradiated xeroderma pigmentosum fibroblasts from different DNA repair complementation groups. *Biochem. Biophys. Acta* 442 (1976), 147–153.
24 Kraemer, K.H., H.G. Coon, R.A. Petinga, S.F. Barrett, A.E. Rahe, J.H. Robbins: Genetic heterogeneity in xeroderma pigmentosum: complementation groups and their relationship to DNA repair rates. *Proc. Natl Acad. Sci. USA* 72 (1975), 59–63.
25 Lehmann, A.R., S. Kirk-Bell, C.F. Arlett, M.C. Paterson, P.H.M. Lohman, E.A. de Weerd-Kastelein, D. Bootsma: Xeroderma pigmentosum cells with normal levels of excision repair have a defect in DNA synthesis after UV-irradiation *Proc. Natl Acad. Sci. USA* 72 (1975), 219–223.
26 Lippke, J.A., L.K. Gordon, D.E. Brash, W.A. Haseltine: Distribution of UV light-induced damage in a defined sequence of human DNA: Detection of alkaline-sensitive lesions at pyrimidine nucleoside–cytidine sequences. *Proc. Natl Acad. Sci. USA* 78 (1981), 3388–3392.
27 Mitchell, D.L., J.P. Allison, R.S. Nairns: Immunoprecipitation of pyrimidine (6-4) pyrimidone photoproducts and cyclobutane pyrimidine dimers in UV-irradiated DNA. *Radiat. Res.* 123 (1990), 299–303.
28 Mitchell, D.L., D.E. Brash, R.S. Nairn: Rapid repair kinetics of pyrimidine (6-4) photoproducts in human cells are due to excision rather than conformational change. *Nucl. Acids Res.* 18 (1990), 963–971.
29 Mitchell, D.L., C.A. Haipek, J.M. Clarkson: (6-4) Photoproducts are removed from the DNA of UV-irradiated mammalian cells more efficiently than cyclobutane pyrimidine dimers. *Mutat. Res.* 143 (1985), 109–112.
30 Mitchell, D.L., R.S. Nairn, J.A. Alvillar, J.M. Clarkson: Loss of thymine dimers from mammalian cell DNA: the kinetics for antibody-binding sites is not the same as that for T4 endonuclease V sites. *Biochem. Biophys. Acta.* 697 (1982), 270–277.
31 Mitchell D.L., B.S. Rosenstein: The use of specific radioimmunoassays to determine the action spectrum for the photolysis of (6-4) photoproducts. *Photochem. Photobiol.* 45 (1987), 781–786.
32 Patrick, M.H.: Studies on thymine-derived UV photoproducts in DNA. I. Formation and biological role of pyrimidine adducts in DNA. *Photochem. Photobiol.* 25 (1977), 357–372.
33 Patrick, M.H., R.O. Rahn: Photochemistry of DNA and polynucleotides: photoproducts. In: *Photochemistry and Photobiology of Nucleic Acids*, vol. II, (1976), pp. 35–95. Ed.: S.Y. Wang, Academic Press, New York.
34 Peak, M.J., J.G. Peak, B.A. Carnes: Induction of direct and indirect single-strand breaks in human cell DNA by far- and near-ultraviolet radiations: Actions spectrum and mechanisms. *Photochem. Photobiol.* 45 (1987), 381–387.
35 Pearson, M.L., F.P. Ottensmeyer, H.E. Johns: Properties of an unusual photoproduct of UV-irradiated thymindylyl-thymidine. *Photochem. Photobiol.* 4 (1965), 739–747.
36 Regan, J.D., J.E. Trosko, W.L. Carrier: Evidence for excision of ultraviolet-induced pyrimidine dimers from the DNA of human cells *in vitro*. *Biophys. J.* 8 (1968), 319–325.
37 Rosenstein, B.S., J.M. Ducore: Induction of DNA strand breaks in normal human fibroblasts exposed to monochromatic ultraviolet and visible wavelengths in the 240–546 nm range. *Photochem. Photobiol.* 39 (1983), 51–55.
38 Roza, L., W. Vermeulen, J.B. Bergen-Henegouwen, A.P. Eker, N.G. Jaspers, P.H. Lohman, J.H. Hoeijmakers: Effects of microinjected photoreactivating enzyme on thymine dimer removal and DNA repair synthesis in normal human and xeroderma pigmentosum fibroblasts. *Cancer Res.* 50 (1990), 1905–1910.

39 Sancar, G.B.: DNA photolyases: physical properties, action mechanism, and roles in dark repair. *Mutat. Res.* 236 (1990), 147–160.

40 Sancar, A., W.D. Rupp: A novel repair enzyme: uvrABC excision nuclease of *Escherichia coli* cuts a DNA strand on both sides of the damaged region. *Cell* 33 (1983), 249–260.

41 Setlow, R.B., J.D. Regan, J. German, W.L. Carrier: Evidence that xeroderma pigmentosum cells do not perform the first step in the repair of UV damage to their DNA. *Proc. Natl Acad. Sci. USA* 64 (1969), 1035–1041.

42 Smith, K.C., P.C. Hanawalt: *Molecular Photobiology*, 1969, Academic Press, New York.

43 Strauss, B.T., M. Searashi, M. Robbins: Repair of DNA, studied with a nuclease specific for UV-induced lesions. *Proc. Natl Acad. Sci. USA* 56 (1966), 932–939.

44 Sutherland, B.M.: Photoreactivating enzyme from human leukocytes. *Nature (London)* 248 (1974), 109–112.

45 Taylor, J.-S., M.P. Cohrs: DNA, light and Dewar pyrimidinones: the structure and biological significance of TpT3. *J. Am. Chem. Soc.* 109 (1987), 2834–2835.

46 Tsongalis, G.J., W.C. Lambert, M.W. Lambert: Correction of the ultraviolet light induced DNA-repair defect in xeroderma pigmentosum cells by electroporation of a normal human endonuclease. *Mutat. Res.* 244 (1990), 257–263.

47 Tuteja, N., R. Tuteja, K. Rahman, L.Y. Kang, I.A. Falasch: A DNA helicase from human cells. *Nucl. Acids Res.* 18 (1990), 6785–6792.

48 Unrau, P., R. Wheatcroft, B.J. Cox: Methods for the assay of UV light induced pyrimidine dimers in saccharomyces cerevisiae. *Biochim. Biophys. Acta.* 269 (1972), 311–317.

49 Venema, J., L.H. Mullenders, A.T. Natarajan, A.A. VanZeeland, L.V. Mayne: The genetic defect in Cockayne syndrome is associated with a defect in repair of UV-induced damage in transcriptionally active DNA. *Proc. Natl Acad. Sci. USA* 87 (1990), 4707–4711.

50 Wang, S.Y.: Pyrimidine bimolecular photoproducts. In: *Photochemistry and Photobiology in Nucleic Acids*, vol. I, (1976), pp. 326–356. Ed.: S.Y. Wang, Academic Press, New York.

51 DeWeerd-Kastelein, E.A., W. Keijzer, D. Bootsma: Genetic heterogeneity of xeroderma pigmentosum demonstrated by somatic cell hybridisation. *Nature New Biol.* 238 (1972), 80–83.

52 Wells, R.L., A. Han: Action spectra for killing and mutation of Chinese hamster cells exposed to mid- and near-ultraviolet monochromatic light. *Mutat. Res.* 129 (1984), 251–258.

53 Wood, R.D.: Repair of pyrimidine dimer ultraviolet light photoproducts by human cell extracts. *Biochem.* 28 (1989), 8287–8292.

4: Ultraviolet radiation and antigen presentation

Craig A. Elmets

Neutralization of exogenous pathogens that gain entrance into the internal environment, and identification and elimination of cells that may eventually develop into malignant neoplasms are activities performed by the immune system that are necessary for survival. T lymphocytes play a central role in this process, but are only effective at orchestrating such a response if they have become 'activated' to respond to antigen. The means by which this occurs, i.e. by which T cells become activated to respond to antigen, is termed antigen presentation. Antigen presentation has received intense scrutiny by investigators, not only because it will provide a better method for understanding basic immunological processes, but also because elucidation of the critical steps in the antigen presentation process holds promise for more efficient and effective mechanisms for immunization and tolerization.

Antigen presentation has been of considerable interest to photoimmunologists because of the early recognition that ultraviolet radiation interferes with the process [41]. More recent studies have investigated the role that UV-induced perturbations in antigen-presenting cell function play in the immunopathogenesis of skin cancer and allergic contact hypersensitivity [26, 75]. Moreover, intervention into the antigen presentation process by non-ionizing electromagnetic radiation, both with and without photosensitizers, is being investigated as a possible therapeutic modality for transplantation purposes to promote acceptance of skin, bone marrow and platelet allografts (see Chapter 21).

The purpose of this chapter will be to review basic aspects of the antigen presentation process, the methods for its evaluation, and the mechanisms by which ultraviolet radiation interferes with the process.

Antigen presentation

Although teleologically activation of T cells by antigens and mitogens would seem to be quite simple, in reality the process is quite complex, requiring an inductive stimulus (antigen or mitogen), specific subpopulations of T cells capable of recognizing the stimulus, and a second population of cells called antigen-presenting, or accessory, cells, whose function is to display the antigen or mitogen on the cell surface in such a manner that it can be recognized by the responding T-cell population [9, 76]. In other words, the addition of antigen to T cells without antigen-presenting cells does not result in T-cell activation (and in some instances even leads to clonal anergy, a form of tolerance, in the responding T cells).

The capacity to present antigen to T cells is restricted to a few cell types [9, 76]. Originally, this activity was thought to be confined to a small group of bone marrow-derived cells that included cells of monocyte/macrophage lineage, dendritic cells of the spleen and other tissues, B lymphocytes and Langerhans cells of the epidermis. However, over the years the concept of the antigen-presenting cell has broadened and now encompasses a variety of 'non-professional' antigen-presenting cell types, which can perform this function under special circumstances [20]. Examples include activated T lymphocytes, fibroblasts, endothelial cells, small bowel epithelial cells, and a number of other cell subpopulations. Several of these cell types have drawn the attention of photoimmunologists (Table 4.1).

Table 4.1 Antigen-presenting cell types

Antigen-presenting cell	Responding T-cell type
Epidermal Langerhans cell	T_h1 and T_h2 helper T cells
I-J$^+$ macrophage (mouse)	Suppressor T cells
CD36$^+$ (OKM5$^+$) macrophage	Suppressor T cells Suppressor inducer T cells
Thy-1$^+$ dendritic epidermal T cell	Suppressor T cells
Cells of monocyte/ macrophage lineage	T_h1 T cells

Langerhans cells are bone marrow-derived cells that migrate to the epidermis where they reside for long periods of time [28] (see also Chapter 1). Although they comprise only 2–4% of the entire epidermal cell population, they are the major antigen-presenting cells of the epidermis for T_h1 and T_h2 cells. Following contact with antigen they migrate to regional lymph nodes where they present antigen to relevant T lymphocyte populations. Based on a series of *in vitro* studies, current concepts of their role in the antigen presentation process suggest that, within the epidermis, they efficiently process antigens, but rapidly lose that function during their transit from skin to regional lymph nodes. However, by the time they have reached the regional lymph nodes they have undergone a series of adaptive changes that make them much more proficient at interacting with T cells, and thus much more effective antigen-presenting cells than they were in the epidermis.

I-J$^+$ macrophages represent a small subpopulation of antigen-presenting cells in murine skin [22]. These cells, so named because of their reactivity with anti-I-J$^+$ antibodies that identify T cells with suppressor activity and other antigen-presenting cells capable of activating suppressor T-cell populations, also express class II major histocompatibility determinants on their cell surface. These cells are thought to be the counterpart to epidermal Langerhans cells; rather than presenting antigen to helper T cells, I-J$^+$ macrophages preferentially activate suppressor T-cell populations. The human counterpart for the I-J$^+$ cell is not known although it is thought to be a population of macrophages that bear the CD36$^+$ (OKM5$^+$), HLA-DR$^+$, CD1a$^-$ immunophenotype.

CD36$^+$ (OKM5$^+$), HLA-DR$^+$, CD1a$^-$ cells are a subpopulation of macrophages that have been characterized primarily in humans [13]. Under normal conditions, they are not found in the epidermis and are present in only small proportions in the dermis. Following UV radiation exposure and a variety of their inflammatory insults, these cells migrate into the epidermis and their concentrations increase dramatically in the dermis. Like I-J$^+$ cells, they are responsible for activation of T-cell regulatory circuits. They are adept at activating CD8$^+$ suppressor/cytotoxic T cells and CD4$^+$ suppressor inducer T cells [5]. In addition to the I-J$^+$ cell described above, there appears to be a cell present in the skin of UV-irradiated mice (which may be similar to the I-J$^+$ cell) that also performs this function [11].

Cells of monocyte/macrophage lineage have been utilized primarily as a model to investigate the mechanism by which ultraviolet radiation modifies the signals provided by antigen-presenting cells for T-cell activation [35, 36, 37]. They are more plentiful than Langerhans cells and, like Langerhans cells, they are proficient at activating CD4$^+$ T cells. Their inactivation is also thought to be the mechanism by which ultraviolet radiation decreases the alloimmunizing capacity of peripheral blood and bone marrow, and thus is of practical importance as well. Moreover, because macrophages are present in substantial numbers in the dermis they also represent a possible target for the immunological effects of ultraviolet radiation exposure.

Thy-1$^+$ DETC represent a unique population of CD3$^+$, CD4$^-$, CD8$^-$ T cells that reside within the epidermis, and are part of a family of cells that are present in the blood and a number of other tissues [74]. Like Langerhans cells, they are bone marrow derived, are dendritic in appearance, and are present in nearly equivalent concentrations in most regions of the skin. They are distinguished from Langerhans cells in that thy-1$^+$ DETC bear $\gamma\delta$ T-cell receptors on their surface and are class II MHC negative. The precise function of thy 1^+ DETC is unknown, although a number of studies have provided evidence that thy-1^+ DETC act as antigen-presenting cells for suppressor T lymphocytes. Thy-1^+ DETC have not been detected in human epidermis; they have only been found in rodent epidermis.

Signals provided by antigen-presenting cells

Depending on the activating stimulus and on the type of responding T-cell population, at least four readily distinguishable signals are required for T-cell activation to occur (Table 4.2) [9, 76]. They include: (i) cell surface expression of MHC class II determinants (HLA-DR, -DQ and -DP in man; I-A and I-E in mice) [32]; (ii) an antigen-processing step in which the antigen is taken up by the antigen-presenting cell, degraded into immunogenic moieties within endosomes, and then combined and reexpressed on the cell surface with newly synthesized class II MHC molecules [9, 27]; (iii) production of various cytokines such as interleukin-1 and interleukin-6 [59]; and (iv) expression of various adhesion molecules such as ICAM-1, B7/BB-1 and LFA-3 [3, 67]. With respect to antigen presentation, adhesion molecules facilitate T-cell binding to antigen-presenting cells in a non-antigen specific manner and also communicate essential secondary signals to the responding T cell.

Table 4.2 Signals provided by antigen-presenting cells for T-cell activation

Class II MHC determinants
HLA-DR, -DQ and -DP (humans)
I-A and I-E (mice)
Antigen processing
Cytokines
Interleukin-1
Interleukin-6
Adhesion molecules
ICAM-1
B7
LFA-3

Models employed to assess UV effects on antigen presentation

Much of the interest in UV effects on antigen presentation is based on its role in the development of skin cancer. The complexity of working with tumor cells, whose immunogenic determinants are for the most part uncharacterized from a biochemical and molecular standpoint, has led to the use of alternative systems for photoimmunological studies on the antigen presentation process. Two of these model systems have been used extensively: (i) allergic contact hypersensitivity in mice, which is largely an *in vivo* system, and (ii) antigen- or mitogen-induced T-cell proliferation, which is an *in vitro* assay.

Allergic contact hypersensitivity

Allergic contact hypersensitivity is a popular *in vivo* method for evaluating photoimmunological effects on antigen presentation because the immune response to contact allergens bears many features in common with the immune response to UV-induced skin tumors [49]. In these assays hapten, applied to the skin, rapidly conjugates with cutaneous cellular proteins and membranes to form a complete antigen [6, 50]. Langerhans cells take up the complete antigen and migrate to regional lymph nodes where they present it to T cells. Activation and expansion of relevant T-cell clones results. The activated T cells then circulate to those sites where contact allergen has been applied where they produce an inflammatory response in an effort to eradicate those cells to which hapten has bound.

In mice, allergic contact hypersensitivity is assessed in the following manner. An immunizing dose of hapten is painted on the shaved abdominal skin of normal mice [50] or is conjugated directly to suspensions of antigen presenting cells *in vivo* or *in vitro*, which are then injected into naive syngeneic recipients. The degree to which sensitization has occurred can be determined by reapplying, several days later, a smaller dose of the hapten to the ears of the previously sensitized animals. The increment of ear swelling that occurs over subsequent days is used as a quantitative index of the contact hypersensitivity response.

Antigen- or mitogen-induced T-cell proliferation

Antigen presentation is frequently assessed *in vitro* in antigen- or mitogen-induced T-cell proliferation assays. In these systems, T lymphocytes are purified from peripheral blood of humans or from the spleen or lymph nodes of mice [35–37]. The T cells are then cultured for several days with antigen and with an untreated or UV-treated antigen-presenting cell population. Specific clones of T cells respond over the next few days by synthesizing new DNA and proliferating. T-cell incorporation of tritiated thymidine over the final hours of culture is used to quantify the proliferative response.

UVB radiation and antigen presentation

Although the first evidence to indicate that UV radiation might interfere with the process of antigen presentation came from studies in which it was observed that *in vitro* exposure of human peripheral blood to UVC radiation completely inhibited its ability to stimulate the proliferation of allogeneic T lymphocytes in the mixed lymphocyte reaction [41], a real interest in this effect from a photoimmunological standpoint came from studies from two different groups indicating that UVB radiation had an *in vivo* effect on antigen-presenting cells [26, 75]. Studies by Greene *et al.* [26] showed that splenic adherent cells from BALB/c mice that had been exposed *in vivo* to a total of 86 kJ/m^2 UVB radiation over 4 weeks developed a defect that crippled their capacity to present antigen to helper T cells. This defect was manifest as a decreased capacity to immunize animals to the hapten TNP when hapten-conjugated splenic adherent cells were injected subcutaneously into naive syngeneic recipients. Suppression of the contact hypersensitivity response was antigen specific and was associated with the generation of suppressor T lymphocytes. Of interest with respect to these studies was the fact that an antigen presenting defect could be demonstrated in the spleen, which was a site remote from that receiving UV radiation, suggesting that immunosuppressive mediators elaborated by the skin were able to suppress the antigen presenting capacity of splenic adherent cells.

Using much smaller doses, 200–400 J/m^2 UVB for four consecutive days, Toews *et al.* [75] were able to demonstrate that *in vivo* exposure to UVB radiation also suppressed the development of contact hypersensitivity, but in contrast to the studies of Greene *et al.* [26], suppression was only present at the UV-irradiated skin site. Treatment of animals in this manner not only inhibited the development of contact hypersensitivity, but also resulted in the development of long-lasting antigen-specific unresponsiveness that persisted long after the effects of UV radiation on the animal had disappeared. Subsequent studies showed that this form of unresponsiveness was also caused by antigen specific suppressor T lymphocytes that were CD4$^+$ and that inhibited the afferent limb of the immune response [18].

Because in the studies by Toews *et al.* [75] Langerhans cell concentrations and morphology, as assessed by histochemical staining for the Ca^{2+}/Mg^{2+} ATPase enzyme, were perturbed in UV-irradiated skin, the defect in the ability to contact sensitize animals was attributed to a deficiency in the capacity of epidermal Langerhans cells to present antigen to the relevant populations of T cells. Additional support for this hypothesis was provided by experiments showing that when UV-irradiated epidermal cell suspensions were used as antigen-presenting cells in antigen-induced T-cell proliferation assays, T-cell proliferation did not occur [68]. Moreover, when UV-irradiated epidermal cell suspensions containing Langerhans cells from murine skin were conjugated to TNP and were then used to immunize mice to that hapten, tolerance rather than sensitization developed [57].

While not to deny a critical role for UV-induced alterations in Langerhans cells in the *in vivo* photoimmunological effects of UVB, it is clear that suppression of contact hypersensitivity is not solely caused by its effects on Langerhans cells. Despite the fact that it is possible to impair Langerhans cell function with *in vivo* UVB radiation in all strains of mice, some strains do not develop a suppressed contact hypersensitivity response [69, 77]. Recent work would suggest that there are at least two populations of antigen-presenting cells in murine skin that are able to initiate contact hypersensitivity response. One of these is the Langerhans cell in the epidermis; the other is a dermal antigen-presenting cell, either a macrophage or a dermal dendritic cell [39]. Strain variation in the effects of UV radiation on the contact hypersensitivity response among mice might relate to differences in the number or functional capacity of their dermal antigen-presenting cells to initiate contact hypersensitivity responses.

Effects of UVB radiation on signals provided by antigen-presenting cells

Precisely how UV radiation perturbs the function of antigen-presenting cells is unclear. With respect to Langerhans cells, inhibition was initially attributed to a direct cytotoxic effect because concentrations of ATPase positive cells in the epidermis were diminished following *in vivo* exposure to UV radiation. However, with electron microscopy substantial numbers of ATPase positive Langerhans cells could still be demonstrated using a similar irradiation protocol [2], suggesting that the effect of UV radiation was to inhibit the activity of the ATPase enzyme used to demonstrate their presence, rather than to kill the cell altogether [30]. Moreover, studies by Bigby *et al.* have shown that the functional capacity of Langerhans cells that remain in UV-irradiated skin is impaired [8]. Similar observations with respect to viability have been made when human peripheral blood monocytes are exposed to UV radiation *in vitro* [36, 43]. Moreover, antimicrobial activity of this cell type is retained and is even enhanced at UV doses used to inhibit their antigen-presenting function [43].

The repeated observation that a decrease in the viability of UVB-irradiated cells is not a prerequisite for loss of antigen-presenting function has led to an investigation of the effect of UV radiation on the signals provided by accessory cells that are inhibited by UVB radiation. These studies have been accomplished largely using UV irradiation *in vitro*.

Interleukin-1 and other cytokines

A number of studies have demonstrated that UVB irradiation of cells of monocyte/macrophage lineage, in doses similar to those used to abrogate the presentation of antigen, also impairs their capacity to produce interleukin-1 [30, 51, 56]. However, re-addition of exogenous IL-1 to cultures containing UV-irradiated antigen-presenting cells has met with limited success in reconstituting the response [36]. These findings suggest that if UV-induced inhibition of IL-1 production is important to the loss of function of antigen-presenting cells, then it is not the only signal impaired by UV radiation.

In addition to soluble IL-1, macrophages also express a membrane form of the protein. Lange-Wantzin demonstrated that UV radiation inhibited membrane IL-1 on rat pulmonary macrophages [42]. Although these findings contrasted with those of Hertl *et al.* [29], who found no reduction in membrane IL-1 on human monocytes following UV radiation exposure, the discrepancy is in all probability due to technical differences in the isolation procedures for these two cell populations. In both human macrophages and in rat pulmonary macrophages, UV-induced downregulation of membrane IL-1 that is already present on the cell surface does not occur, suggesting that UVB radiation impairs the synthesis and/or movement of this molecule to the cell surface but does not reduce its expression once it is there.

When epidermal cells are exposed to UV radiation, an augmentation in the synthesis and release of IL-1 has been observed [38]. This has been attributed to keratinocytes which are proficient at producing IL-1, and which are known to augment their production of this cytokine following UV radiation exposure (see Chapter 5). Because of the limited number of purified cells that can be obtained, it has been impossible to determine whether UV radiation augments or inhibits Langerhans cell production of IL-1. Nevertheless, it is clear that substantial amounts of IL-1 are available to responding T cells when UV-irradiated epidermal cells are used as antigen-presenting cells in antigen-induced T-cell proliferation assays, making it unlikely that a deficiency in IL-1 production is responsible for the loss of epidermal antigen-presenting function.

MHC class II expression

Although there are some exceptions [4], most reports examining the effect of UV radiation on macrophage and monocyte expression of MHC class II determinants have shown that with UV doses sufficient to inhibit antigen-presenting function, no inhibition in class II MHC determinant expression occurs either immediately or with subsequent culture [29, 65]. B lymphoblastoid cell lines also do not modulate their expression of Class II MHC determinants in response to UV radiation [12, 27]. The most definitive studies in that regard have been performed by Aberer *et al.* using the Raji B-cell line [1]. They found that when radiolabeled detergent lysates of UVB-irradiated cells were immunoprecipitated, no reduction in the HLA-DR expression was observed. This corresponds well to flow cytometry findings using Epstein–Barr virus-transformed B lymphoblastoid cell lines [27].

The influence of *in vitro* UV irradiation on class II antigen expression in epidermal Langerhans cells is somewhat different than its effect on cells of macrophage/monocyte or B-cell lineage. While constitutive levels of class II antigen expression on human epidermal Langerhans cells are not perturbed by UVB radiation *in vivo*, the increase that normally occurs when they are put in culture does not take place [61]. This, in all likelihood, relates to decreased responsiveness to those agents that are responsible for augmenting class II antigen expression on this cell type.

With respect to *in vivo* UV radiation exposure, in human skin a depression in class II antigen positive Langerhans cells occurs, despite the fact that the overall numbers of HLA-DR positive cells remains relatively constant [12]. HLA-DR positive OKM5$^+$ macrophages that immigrate into UV-irradiated epidermis compensate for the loss of HLA-DR positive Langerhans cells. As will be discussed in detail below, although these immigrant macrophages express HLA-DR molecules, their antigen-presenting capacity differs significantly from Langerhans cells in that the former cell type, in contrast to Langerhans cells, provide a down-regulatory stimulus for T-cell responses.

Antigen processing

The effect of UVB radiation on antigen processing has only been evaluated indirectly. Studies by Hertl *et al.* [29] addressed this issue by using human B lymphoblastoid cell lines as antigen-presenting cells. They showed that ultraviolet radiation produced a dose-dependent reduction in the proliferative response of responding T-cell lines to the soluble antigen tetanus toxoid when it was added to cultures after the B lymphoblastoid cell lines were irradiated. However, if the B lymphoblastoid cell lines were incubated with tetanus toxoid for a period of time sufficient to allow antigen processing to take place, and were then UV-irradiated, the cells became resistant to the deleterious effects of UV radiation on antigen presentation. Similar effects were observed by Germain who found that proliferation of murine T cells to the polypeptide GAT could be restored if UV irradiation of murine macrophages followed rather than preceded pulsing with antigen [21]. In other studies, Shimada found that the inhibitory effects of UV radiation on the presentation of antigen to ovalbumin specific T-cell hybridomas could be reversed by substituting processed peptides of ovalbumin rather than the entire protein [62].

Adhesion molecules

A key feature of the antigen-presentation process is the clustering of T cells around the antigen-presenting cell. This occurs in an antigen non-specific manner. An investigation into the effects of UV radiation on antigen-presenting cell adhesion molecule expression was initiated when it was observed that the clustering of T cells around antigen-presenting cells was not present when UVB-irradiated monocytes were used in assays [36]. Although stimulus specific ligands on UV-irradiated monocytes remained intact, a marked downregulation in the expression of the adhesion molecule ICAM-1 was observed. ICAM-1 appeared to be unique in that regard since expression of several other cell surface membrane molecules including class II MHC determinants, FcγRI and LFA-3 were not inhibited.

In contrast to monocytes, UV radiation did not reduce ICAM-1 or other adhesion molecule expression on B lymphoblastoid cell lines, whose antigen-presenting function was resistant to the adverse effects of UVB radiation when a stimulus processing step was not required [29]. These findings lend support to the hypothesis that the differential sensitivity among accessory cells may be explained by different effects of UV radiation on ICAM-1 expression.

The issue of UVB radiation effects on adhesion molecule expression has also been examined in Langerhans cells. Freshly isolated Langerhans cells and Langerhans cells *in situ* express little or no ICAM-1 on their cell surface [16, 54, 72]. Following short-term culture, substantially greater amounts of this molecule are present. UVB radiation prevents this upregulation, but does not reduce its expression once it is already present on the cell surface [72, 73].

It is important to emphasize that a reduction in ICAM-1 may not be the sole adhesion molecule necessary for T-cell activation whose function is reduced by UVB exposure. In both human monocytes and in murine epidermal Langerhans cells, doses of UVB radiation that result in a complete reduction in antigen-presenting function are accompanied by only a partial inhibition in ICAM-1 expression [36, 72]. In this regard, recent studies have shown that IL-10 inhibits presentation of antigen to the T_h1 T helper cell subpopulation by inhibiting the expression of the B7/BB-1 adhesion molecule on the surface of antigen-presenting cells [17]. Ultraviolet radiation is a known stimulus for IL-10 secretion [52]. Thus, it is tempting to speculate that UV-induced production of IL-10 results in inhibition of B7 on antigen-presenting cells and thereby contributes to deficiencies in antigen-presenting cell function.

Differential effects of UVB radiation on different antigen-presenting cell populations

Depending on the cell population being studied and the type of *in vitro* assay being examined, substantial variation in the effect of UVB radiation on the antigen-presentation process has been observed [29, 78]. In situations in which antigen processing is required for effective presentation of antigen, antigen-presenting cells have been almost uniformly sensitive to UV radiation, providing considerable evidence that UV radiation interferes with the processing of antigen. In assays in which antigen processing is not required – for example, in T-cell responses to anti-CD3 antibodies, to Staphylococcal superantigens and to oxidative mitogens – or in situations in which antigen processing precedes UV radiation exposure, considerable heterogeneity among different cell types in the accessory cell response to UV radiation has been observed. While monocytes are susceptible to UV doses *in vitro* that range from 100 to 200 J/m^2 UVB, activated B lymphocytes (either Epstein–Barr transformed or activated with conditioned human lymphocyte medium) and dendritic cells require eight to ten times more UV radiation to inhibit their function. Freshly prepared Langerhans cells are also susceptible to UV radiation exposure in these assays, but their function is preserved if they have been placed in culture prior to UVB exposure [73]. These observations are consistent with the hypothesis that, with the exception of the antigen-processing step, if accessory cells are allowed to express the relevant accessory signals they will become resistant to the adverse effects of UV radiation on their function.

UV-irradiated accessory cells may also lose their ability to present antigen to some populations of T cells but retain it for others. For example, Simon *et al.* have shown that UV-irradiated Langerhans cells are unable to present the antigen KLH to T_h1 T-cell clones but are able to stimulate T_h2 T-cell clones to proliferate to the same antigen [63]. These findings are consistent with the hypothesis that the two different T-cell populations require different signals from antigen-presenting cells for their activation and that UV radiation interferes with those for T_h1, but not for T_h2, cell activation. The findings are also of interest because one function of T_h2 cells is to activate suppressor T lymphocytes, suggesting that differential activation of T-cell subpopulations by UV-irradiated Langerhans cells might in part be responsible for the imbalance between helper and suppressor T-cell populations in UV-irradiated individuals.

UVB radiation effects on premitotic events in T-cell activation

Most studies examining the effects of UVB-irradiated accessory cells on the T-cell activation process have utilized T-cell proliferation as an endpoint, and have shown that for UVB-sensitive antigen-presenting cells, T cells do not incorporate tritiated thymidine when they are used in *in vitro* assays. It should be noted, however, that T-cell proliferation is the culmination of a series of events, all of which must take place before transition from the resting to the fully activated state occurs. Investigation into the stages of the T-cell activation process that were deficient when UV-irradiated accessory cells were employed for these studies have shown that there is relatively little effect on the increase in free intracellular calcium in T cells, which is one of the earliest steps in the T-cell activation process [37]. T-cell upregulation in the expression of interleukin-2 receptors also occurs, albeit at reduced levels [13, 50]. IL-2 production by responding T cells is markedly reduced when UVB-irradiated antigen-presenting cells are employed [37, 78]. The doses necessary to achieve complete inhibition of IL-2 production are substantially less than that required to reduce tritiated thymidine incorporation to background levels, suggesting that IL-2 production is the most sensitive step in the process [37].

It should be emphasized that treatment of T cells with UVB-irradiated antigen-presenting cells is not without its consequences. When T cells that have been incubated with antigen and UV-irradiated Langerhans cells are harvested and restimulated with the same antigen and normal, untreated antigen-presenting cells, their inability to respond to antigen persists [64]. The T cells that respond in this manner have been described as being in a state of clonal anergy. The biological relevance of this to the phenomenon is unknown, but it has been postulated that similar mechanisms are operative when autoreactive T-cell clones are rendered inactive *in vivo* [58].

Direct versus indirect effects of UVB radiation on antigen-presenting cells

When antigen-presenting cells are exposed to UV radiation *in vitro*, their function is inhibited primarily by the direct actions of this form of radiant energy. However, following whole body *in vivo* UV radiation exposure, a variety of soluble mediators are produced that also operate to inhibit the function of antigen-presenting cells. This is obvious from the fact that antigen-presenting cell function is deficient in tissues such as the spleen that are remote from the site of UV radiation exposure – sites where the direct effects of UVB radiation could not possibly be operative [26]. At skin sites treated by UVB radiation, there is evidence to indicate that both direct and indirect effects of UVB radiation cooperate to inhibit the function of antigen-presenting cells. Using murine contact hypersensitivity as a model, Chung *et al.* were able to demonstrate a deficiency in the antigen-presenting function of skin that had been exposed *in vivo* to UVB radiation [10]. If an area of skin within the UVB irradiation site was shielded, its antigen-presenting function was still decreased. Pretreatment of animals with the cyclooxygenase inhibitor indomethacin reversed the inhibitory effect in the skin that had been shielded, but not in the unshielded skin site. The results suggest that there are two distinct mechanisms by which UVB radiation alters the antigen-presenting function of the skin – one of these is a direct effect on the antigen-presenting cell population and the other is through the production of soluble mediators that act indirectly on antigen-presenting cells.

Identification of the mediators induced by UV radiation that inhibit the presentation of antigen indirectly has been the source of considerable research interest, and, as a result, several potential candidates have been identified. As mentioned above, the fact that UVB-induced inhibition of contact hypersensitivity could be reversed by indomethacin has suggested that prostaglandins are involved in the process [10, 31]. Additional support for this concept comes from the observation that UVB exposure increases the production of prostaglandin E_2 [15, 48], and the fact that PGE_2 is a known inhibitor of signals required for antigen-presenting cell function in other systems [66]. It is of interest that IL-1 which is a cytokine whose synthesis is upregulated by UVB exposure [19, 25, 38] induces the synthesis of PGE_2. Evidence has been presented that IL-1 is an intermediary between ultraviolet radiation and the synthesis of PGE_2 [53].

Another cytokine that has been proposed to inhibit antigen-presenting cell function following UVB exposure is tumor necrosis factor-alpha (TNFα; see Chapter 5). *In vivo* and *in vitro* studies have shown that synthesis of this agent is upregulated following UVB exposure [34]. Precisely how this agent inhibits the presentation of antigen is unknown although there is evidence that the addition of TNFα to antigen-induced T-cell proliferation assays suppresses the response [7]. TNFα may also inhibit the migration of Langerhans cells to regional lymph nodes and thereby disrupt the activation of T cells in this manner as well [33].

There are also studies to suggest that *cis*-urocanic acid is a mediator of the indirect effects of UV radiation on the antigen-presentation process [46]. Urocanic acid is normally found as a *trans*-isomer in the stratum corneum [44]. Following UVB exposure, this agent undergoes isomerization from the *trans* to the *cis* form of the molecule. Intravenous injection of *cis*-urocanic acid into mice suppresses the induction of contact hypersensitivity [40] and the delayed type hypersensitivity response to the *Herpes simplex* virus type I [55], and also inhibits the antigen-presenting function of splenic dendritic cells [45]. Exactly how an antigen-presenting cell defect in the spleen contributes to a defect in the induction of cell-mediated responses in the skin remains to be determined. Administration of anti-TNFα antibodies reverses the immunosuppressive effect of *cis*-urocanic acid, suggesting that this cytokine is an intermediary in the immunosuppression caused by this agent [40].

Recent studies by Rivas and Ullrich have shown that UVB irradiation of keratinocyte cultures augments the production of interleukin-10 [52]. This is of interest because IL-10 production is known to favor activation of the T_h1 rather than the T_h2 subpopulation of T cells. This situation is similar to that seen following UVB exposure in which activation of T_h1 cells is deficient and T_h2 cells, one function of which is to activate T suppressor cells, is unperturbed [63].

Finally, a number of less well-characterized cytokines have been identified as being produced by UVB radiation and inhibiting the antigen-presentation process [60, 71]. These include one which appears to inhibit the activities of interleukin-1 (IL-1) [60].

UVB radiation and antigen presentation for suppressor T cells

An important part of the UV-induced alterations in the immune response is the generation of suppressor T lymphocytes [18]. Suppressor T cells also require antigen-presenting cells in order to become activated, and there has been considerable interest in identifying the antigen-presenting cell populations responsible for this activity. Within murine epidermis, at least three cell populations – $I\text{-}J^+$ macrophages, thy-1^+ DETC, and UVB-irradiated Langerhans cells – have this capability. For each cell type, activation of suppressor T-cell circuits has been shown to be resistant to UVB exposure.

The $I\text{-}J^+$ macrophage was the first cell type within the epidermis to be identified as having the capacity to present antigen to suppressor T lymphocytes [22, 24]. When azobenzene arsonate-derived UV-irradiated epidermal cells were injected subcutaneously into mice prior to their immunization with the same hapten, not only was the development of delayed-type hypersensitivity to that antigen prevented, but afferent acting suppressor T lymphocytes were generated that could adoptively transfer unresponsiveness to naive recipient mice. Activation of suppressor cells could be inhibited by removal of I-J$^+$ cells from the epidermal cell population. Further characterization of this cell population has demonstrated that the cells, in addition to being I-J$^+$ and UV resistant, are I-A$^+$, thy-1$^-$, adherent cells that are cyclophosphamide sensitive [23]. In humans, the CD36$^+$ (OKM5$^+$), HLA-DR$^+$, CD1a$^+$ macrophage is thought to perform an analogous function. It has been shown that these cells migrate into the epidermis following UV radiation *in vivo* and present antigen to CD8$^+$ suppressor/cytotoxic T lymphocytes and to CD4$^+$ suppressor/inducer T cells [5, 12]. This cell population has, furthermore, been implicated in the downregulation of DNCB contact hypersensitivity that occurs when that hapten is applied to UV-irradiated skin [14].

There is a substantial body of investigative evidence to indicate that thy-1$^+$ DETC are also responsible for activation of suppressor T cells following UV radiation exposure. The functional activity of this cell population was demonstrated by Sullivan *et al.* [70] when they showed in mice that intravenous administration of thy-1$^+$ DETC that had been derived with the contact sensitizer TNBS was an efficient method of inducing antigen-specific unresponsiveness and that such unresponsiveness was mediated by suppressor cells. It has been shown that the draining lymph nodes of UVB-irradiated mice contain a population of hapten-conjugated thy-1$^+$ cells, which when adoptively transferred into naive recipients results in the induction of suppressor T lymphocytes [47].

As was mentioned above, UVB-irradiated Langerhans cells represent a third population of antigen-presenting cells proficient at initiating suppressor T-cell circuits. This has been shown in experiments in which UVB-irradiated Langerhans cells were compared for their ability to present antigen to T_h1 and T_h2 lymphocyte clones in *in vitro* T-cell proliferation assays [63, 64]. T_h1 lymphocytes represent the population of T helper cells that mediate delayed-type hypersensitivity reactions, whereas T_h2 lymphocytes are responsible for inducing suppressor T-cell populations. Unirradiated Langerhans cells are proficient at presenting antigen to both types of T helper cells. When exposed to UVB radiation *in vitro* and then conjugated with hapten, Langerhans cells lose their ability to present antigen to T_h1 cells, but retain their ability to present antigen to T_h2 cells. Moreover, ineffective presentation of antigen by UV-irradiated Langerhans cells to T_h1 cells renders those cells anergic. Based on the capacity of UV radiation to downregulate the expression of ICAM-1 on the surface of antigen-presenting cells, and the critical role that this molecule plays in the antigen-presentation process, it has been hypothesized that T_h1 and T_h2 cells differ in the signals required for antigen presentation. T_h1 cells must have ICAM-1 on their surface for activation and UV radiation interferes with its expression. In contrast, ICAM-1 is not necessary for T_h2 activation and therefore UV radiation does not interfere with the antigen-presenting signals used to activate that cell population. Guided by these observations, experiments have been performed in which murine Langerhans cells were exposed *in vitro* to UV radiation and then conjugated with hapten. When those cells were given intravenously to naive recipient mice, antigen-specific suppressor T cells were generated, resulting in the generation of antigen-specific suppressor T lymphocytes.

Acknowledgment

This work was supported by NIH grants AR32593, AR39750 and CA43703.

References

1 Aberer, W., H. Leibl: Effect of UVB radiation on the biosynthesis of HLA-DR antigens. *Arch. Dermatol. Res.* 279 (1987), 321–326.

2 Aberer, W., G. Schuler, G. Stingl, H. Honigsmann, K. Wolff: Ultraviolet light depletes surface markers of Langerhans cells. *J. Invest. Dermatol.* 76 (1981), 202–210.

3 Altmann, D.M., N. Hogg, J. Trowsdale, D. Wilkinson: Cotransfection of ICAM-1 and HLA-DR reconstitutes human antigen presenting cell function in mouse L cells. *Nature* 338 (1989), 512–514.

4 Aprile, J.A., H.J. Deeg, T. Castner, L. Miller, R.

Storb: Impaired expression of class-II antigens on canine dendritic cells following ultraviolet irradiation. *Exp. Hematol.* 15 (1987), 452A.

5 Baadsgaard, O., D.A. Fox, K.D. Cooper: Human epidermal cells from ultraviolet light-exposed skin preferentially activate autoreactive $CD4^+2H4^+$ suppressor-inducer lymphocytes and $CD8^+$ suppressor/cytotoxic T lymphocytes. *J. Immunol.* 140 (1988), 1738–1744.

6 Bergstresser, P.R.: Immune mechanisms in contact allergic dermatitis. *Dermatol Clinics* 8 (1990), 3–11.

7 Beutler, B., A. Cerami: The biology of cachectin/TNF – a primary mediator of the host response. *Annu. Rev. Immunol.* 7 (1989), 625–655.

8 Bigby, M., R. Vargas, M.S. Sy: Production of hapten-specific T cell hybridomas and their use to study the effect of ultraviolet B irradiation on the development of contact hypersensitivity. *J. Immunol.* 143 (1989), 3867–3872.

9 Brodsky, F.M., L.E. Guagliardi: The cell biology of antigen processing and presentation. *Annu. Rev. Immunol.* 9 (1991), 707–744.

10 Chung, H.T., D.K. Burnham, B. Robertson, L.K. Roberts, R.A. Daynes: Involvement of prostaglandins in the immune alterations caused by the exposure of mice to ultraviolet radiation. *J. Immunol.* 137 (1986), 2478–2484.

11 Cooper, K.D., N. Duraiswamy, C. Hammerberg, E. Allen, C. Kimbrough-Green, W. Dillon, D. Thomas: Neutrophils, differentiated macrophages, and monocyte/macrophage antigen presenting cells infiltrate murine epidermis after UV injury. *J. Invest. Dermatol.* 101 (1993), 155–163.

12 Cooper, K.D., P. Fox, G. Neises, S.I. Katz: Effects of ultraviolet radiation on human epidermal cell alloantigen presentation: initial depression of Langerhans cell-dependent function is followed by the appearance of $T6^-Dr^+$ cells that enhance epidermal alloantigen presentation. *J. Immunol.* 134 (1985), 129–137.

13 Cooper, K.D., P. Fox, G. Neises, S.I. Katz: Antigen-presenting macrophages appear in human epidermis after ultraviolet radiation. *J. Invest. Dermatol.* 86 (1986), 363–370.

14 Cooper, K.D., L. Oberhelman, T.A. Hamilton, O. Baadsgaard, M. Terhune, G. LeVee, T. Anderson, H. Koren: UV exposure reduces immunization rates and promotes tolerance to epicutaneous antigens in humans: relationship to dose, $CD1a^-DR^+$ epidermal macrophage induction, and Langerhans cell depletion. *Proc. Natl Acad. Sci. USA* 89 (1992), 8497–8501.

15 DeLeo, V.A., D. Hanson, I.B. Weinstein, L.C. Harber: Ultraviolet radiation stimulates the release of arachidonic acid from mammalian cells in culture. *Photochem. Photobiol.* 41 (1985), 51–56.

16 De Panfilis, G., G.C. Manara, C. Ferrari, C. Torresani: Adhesion molecules on the plasma membrane of epidermal cells. II. The intercellular adhesion molecule-1 is constitutively present on the cell surface of human resting Langerhans cells. *J. Invest. Dermatol.* 94 (1990), 317–321.

17 Ding, L., P.S. Linsley, L.Y. Huang, R.N. Germain, E.M. Shevach: IL-10 inhibits macrophage costimulatory activity by selectively inhibiting the up-regulation of B7 expression. *J. Immunol.* 151 (1993), 1224–1234.

18 Elmets, C.A., P.R. Bergstresser, R.E. Tigelaar, P.J. Wood, J.W. Streilein: Analysis of the mechanism of unresponsiveness produced by haptens painted on skin exposed to low dose ultraviolet radiation. *J. Exp. Med.* 158 (1983), 781–794.

19 Gahring, L., M. Baltz, M.B. Pepys, R.A. Daynes: Effects of ultraviolet radiation on production of epidermal cell thymocyte activating factor/interleukin 1 *in vivo* and *in vitro*. *Proc. Natl Acad. Sci. USA* 81 (1984), 1198–1202.

20 Geppert, T.D., P.E. Lipsky: Antigen presentation by cells that are not of bone marrow origin. *Regional Immunol.* 2 (1989), 60–71.

21 Germain, R.N.: Accessory cell stimulation of T cell proliferation requires active antigen processing, Ia-restricted antigen presentation, and a separate non-specific 2nd signal. *J. Immunol.* 127 (1981), 1964–1966.

22 Granstein, R.D.: Epidermal I-J bearing cells are responsible for transferable suppressor cell generation after immunization of mice with ultraviolet radiation-treated epidermal cells. *J. Invest. Dermatol.* 84 (1985), 206–209.

23 Granstein, R.D., M. Askari, D. Whitaker, G.F. Murphy: Epidermal cells in activation of suppressor lymphocytes: further characterization. *J. Immunol.* 138 (1987), 4055–4062.

24 Granstein, R.D., A. Lowy, M.I. Greene: Epidermal antigen-presenting cells in activation of suppression. Identification of a new functional type of ultraviolet radiation resistant epidermal cell. *J. Immunol.* 132 (1984), 563–565.

25 Granstein, R.D., D.N. Sauder: Whole-body exposure to ultraviolet radiation results in increased serum interleukin-1 activity in humans. *Lymphokine Res.* 6 (1987), 187–193.

26 Greene, M.I., M.S. Sy, M.L. Kripke, B. Benacerraf: Impairment of antigen-presenting function by ultraviolet radiation. *Proc. Natl. Acad. Sci. USA* 76 (1979), 6591–6595.

27 Harding, C.V., F. Leyva-Cobian, E.R. Unanue: Mechanisms of antigen processing. *Immunol. Rev.* 106 (1988), 77–92.

28 Hauser, C., A. Elbe, G. Stingl: The Langerhans cell. In: *Physiology, Biochemistry, and Molecular Biology of the Skin*, vol. II, (1991), pp. 1144–1163. Ed.: L. Goldsmith, Oxford University Press, New York.

29 Hertl, M., D.R. Kaplan, J.D. Fayen, J.R. Panuska, J.J. Ellner, C.A. Elmets: The accessory function of B

lymphocytes is resistant to the adverse effects of UV radiation. *Eur. J. Immunol.* 21 (1991), 291–297.

30 Jakway, J.P., E.M. Shevach: Stimulation of T-cell activation by UV-treated, antigen-pulsed macrophages: evidence for a requirement for antigen processing and interleukin 1 secretion. *Cell. Immunol.* 80 (1983), 151–162.

31 Jun, B.D., L.K. Roberts, B.H. Cho, B. Robertson, R.A. Daynes: Parallel recovery of epidermal antigen-presenting cell activity and contact hypersensitivity responses in mice exposed to ultraviolet irradiation: role of a prostaglandin-dependent mechanism. *J. Invest. Dermatol.* 90 (1988), 311–316.

32 Kappes, D., J.L. Strominger: Human class II major histocompatibility complex genes and proteins. *Annu. Rev. Biochem.* 57 (1988), 991–1028.

33 Kimber, I., M. Cumberbatch: Stimulation of Langerhans cell migration by tumor necrosis factor α (TNF-α). *J. Invest. Dermatol.* 99 (1992), 48S–50S.

34 Kock A., T. Schwarz, R. Kirnbauer, A. Urbanski, A. Perry, J.C. Ansel, T.A. Luger: Human keratinocytes are a source for tumor necrosis factor α: evidence for synthesis and release upon stimulation with endotoxin or ultraviolet light. *J. Exp. Med.* 172 (1990), 1609–1614.

35 Krutmann J., M. Athar, I.U. Khan, D. Mendel, P.M. Guyre, H. Mukhtar, C.A. Elmets: Functional inhibition of the high affinity Fc receptor (FcγRI) on human monocytes by porphyrin photosensitization is highly specific and mediated by the generation of superoxide radicals. *J. Biol. Chem.* 264 (1989), 11407–11413.

36 Krutmann, J., I.U. Khan, F. Zhang, E.A. Rich, R.S. Wallis, J.J. Ellner, C.A. Elmets: The cell membrane is a major locus for UVB-induced inhibition of monocyte accessory function. *J. Clin. Invest.* 85 (1990), 1529–1536.

37 Krutmann, J., Z. Toossi, G.M. Kammer, J.J. Ellner, C.A. Elmets: Inhibition of human monocyte accessory function by ultraviolet B radiation has differential effects on premitotic events in OKT3-induced T cell activation. *J. Invest. Dermatol.* 94 (1990), 204–209.

38 Kupper, T.S., A.O. Chua, P. Flood, J. McGuire, U.Gubler: Interleukin 1 gene expression in cultured keratinocytes is augmented by ultraviolet irradiation. *J. Clin. Invest.* 80 (1987), 430–436.

39 Kurimoto, I., M. Arana, J.W. Streilein: Where in the skin does UVB radiation act to impair the induction of contact hypersensitivity? *J. Invest. Dermatol.* 100 (1993), 513.

40 Kurimoto, I., J.W. Streilein: Cis-urocanic acid suppression of hypersensitivity induction is mediated via tumor necrosis factor-α. *J. Immunol.* 148 (1992), 3072–3078.

41 Lafferty, K.J., I.S. Misko, M.A. Cooley: Allogeneic stimulation modulates the *in vitro* response of T cells to transplantation antigen. *Nature* 249 (1974), 275–276.

42 Lange-Wantzin, G., R. Rothlein, J. Kahn, R.B. Faanes: Effect of UV irradiation on expression of membrane IL 1 by rat macrophages. *J. Immunol.* 138 (1987), 3803–3807.

43 Mirando, W., H. Shiratsuchi, K. Tubesing, H. Toba, J.J. Ellner, C.A. Elmets: *In vitro* exposure of human monocytes to ultraviolet B radiation augments killing of Mycobacterium avium intracellulare. *J. Clin. Invest.* 89 (1992), 1282–1287.

44 Morrison, H., R.M. Diebel: Photochemistry and photobiology of urocanic acid. *Photochem. Photobiol.* 43 (1986), 663–665.

45 Nonan, F.P., E.C. DeFabo, H. Morrison: Cis-urocanic acid, a product formed by ultraviolet B irradiation of the skin, initiates an antigen presentation defect in splenic dendritic cells *in vivo*. *J. Invest. Dermatol.* 90 (1988), 92–99.

46 Norval, M., T. Simpson, J.A. Ross: Urocanic acid and immunosuppression. *Photochem. Photobiol.* 50 (1989), 267–275.

47 Okamoto, H., M.L. Kripke: Effector and suppressor circuits of the immune response are activated *in vivo* by different mechanisms. *Proc. Natl. Acad. Sci. USA* 84 (1987), 3841–3845.

48 Pentland, A.P., M. Mahoney, S.C. Jacobs, M.J. Holtzman: Enhanced prostaglandin synthesis after ultraviolet injury is mediated by endogenous histamine stimulation. *J. Clin. Invest.* 86 (1990), 566–574.

49 Perry, L.L., M.I. Greene: T cell subset interactions in the regulation of syngeneic tumor immunity. *Fed. Proc.* 40 (1981), 39–44.

50 Phanuphak, P., J.W. Moorhead, H.N. Claman: Tolerance and contact sensitivity to DNFB in mice. *J. Immunol.* 112 (1974), 115–123.

51 Rich, E., C.A. Elmets, H. Fujiwara, R.S. Wallis, J.J. Ellner: Deleterious effect of ultraviolet-B radiation on accessory function of human blood adherent mononuclear cells. *Clin. Exp. Immunol.* 70 (1987), 116–126.

52 Rivas, J.M., S.E. Ullrich: Systemic suppression of delayed-type hypersensitivity by supernatants from UV-irradiated keratinocytes: an essential role for keratinocyte-derived IL-10. *J. Immunol.* 149 (1992), 3865–3871.

53 Robertson, B.L., L. Gahring, R. Newton, R. Daynes: *In vivo* administration of interleukin 1 to normal mice depresses their capacity to elicit contact hypersensitivity responses: prostaglandins are involved in this modification of immune function. *J. Invest. Dermatol.* 88 (1987), 380–387.

54 Romani, N., A. Lenz, H. Gassel, H. Stössel, H. Stanzl, O. Majdic, P. Fritsch, G. Schuler: Cultured human Langerhans cells resemble lymphoid dendritic cells in phenotype and function. *J. Invest. Dermatol.* 93 (1989), 600–609.

55 Ross, J., S.E.M. Howie, M. Norval, J. Maingay, T. Simpson: UV irradiation of urocanic acid suppresses the delayed type hypersensitivity response to *H. simplex* virus in mice. *J. Invest. Dermatol.* 87 (1986), 630–633.

56 Sauder, D.N., F.P. Noonan, E.C. DeFabo, S.I. Katz: Ultraviolet radiation inhibits alloantigen presentation by epidermal cells: partial reversal by the soluble epidermal cell product, epidermal cell-derived thymocyte-activating factor (ETAF). *J. Invest. Dermatol.* 80 (1983), 485–489.

57 Sauder, D.N., K. Tamaki, A.N. Moshell, H. Fujiwara, S.I. Katz: Induction of tolerance to topically applied TNCB using TNP-conjugated ultraviolet-irradiated epidermal cells. *J. Immunol.* 127 (1981), 261–265.

58 Schwartz, R.H.: A cell culture model for T lymphocyte clonal anergy. *Science* 248 (1990), 1349–1356.

59 Schwarz, T., T.A. Luger: Pharmacology of cytokines in the skin. In: *Pharmacology of the Skin*, (1992), pp. 283–313. Ed.: H. Mukhtar, CRC Press, Boca Raton, FL.

60 Schwarz, T., A. Urbanski, F. Gschnait, T.A. Luger: UV-irradiated epidermal cells produce a specific inhibitor of interleukin 1 activity. *J. Immunol.* 138 (1987), 1457–1463.

61 Shimada, S., S.W. Caughman, S.O. Sharrow, S. Stephany, S.I. Katz: Enhanced antigen-presenting capacity of cultured Langerhans cells is associated with markedly increased expression of Ia antigen. *J. Immunol.* 139 (1987), 2551–2555.

62 Shimada, S., Z. Kovac, R.H. Schwartz, S.I. Katz: Effect of ultraviolet radiation on antigen presentation. *Clin. Res.* 33 (1985), 531A.

63 Simon, J.C., P.D. Cruz, P.R. Bergstresser, R.E. Tigelaar: Low-dose UVB-irradiated Langerhans cells preferentially activate CD4 cells of the Th1 subset. *J. Immunol.* 145 (1990), 2087–2091.

64 Simon, J.C., R.E. Tigelaar, D. Edelbaum, P.R. Bergstresser, P.D. Cruz: UVB radiation converts Langerhans cells from immunogenic to tolerogenic antigen presenting cells. Induction of specific clonal anergy in $CD4^+$ T helper 1 cells. *J. Immunol.* 146 (1991), 485–491.

65 Slater, L.M., S. Murray, J. Liu, B. Hudelson: Dissimilar effects of ultraviolet light on HLA-D and HLA-DR antigens. *Tissue Antigens* 15 (1980), 431–435.

66 Snyder, D.S., D.I. Beller, E. Unanue: Prostaglandins modulate macrophage Ia expression. *Nature* 299 (1982), 163–165.

67 Springer, T.A.: Adhesion receptors of the immune system. *Nature* 346 (1990), 425–434.

68 Stingl, G., L.A. Gazze-Stingl, W. Aberer: Antigen presentation by murine epidermal Langerhans cells and its alteration by ultraviolet B light. *J. Immunol.* 127 (1981), 1707–1713.

69 Streilein, J.W., P.R. Bergstresser: Genetic basis of ultraviolet-B effects on contact hypersensitivity. *Immunogenetics* 27 (1988), 252–258.

70 Sullivan, S., P.R. Bergstresser, R.E. Tigelaar, J.W. Streilein: Induction and regulation of contact hypersensitivity by resident, bone marrow-derived dendritic epidermal cells: Langerhans cells and $Thy\text{-}1^+$ epidermal cells. *J. Immunol.* 137 (1986), 2460–2467.

71 Swartz, R.P.: Role of UVB-induced serum factor(s) in suppression of contact hypersensitivity in mice. *J. Invest. Dermatol.* 83 (1984), 305–307.

72 Tang, A., M.C. Udey: Inhibition of epidermal Langerhans cell function by low dose ultraviolet B radiation. Ultraviolet B radiation selectively modulates ICAM-1 (CD54) expression by murine Langerhans cells. *J. Immunol.* 146 (1991), 3347–3355.

73 Tang, A., M.C. Udey: Differential sensitivity of freshly isolated and cultured murine Langerhans cells to ultraviolet B radiation and chemical fixation. *Eur. J. Immunol.* 22 (1992), 581–586.

74 Tigelaar, R.E., J.M. Lewis, J.L. Nixon-Fulton, P.R. Bergstresser: Thy-1 antigen-bearing dendritic epidermal cells in mice. In: *Physiology, Biochemistry, and Molecular Biology of the Skin*, vol. II, (1991), pp. 1164–1187. Ed.: L. Goldsmith, Oxford University Press, New York.

75 Toews, G.B., P.R. Bergstresser, J.W. Streilein, S. Sullivan: Epidermal Langerhans cell density determines whether contact hypersensitivity or unresponsiveness follows skin painting with DNFB. *J. Immunol.* 124 (1980), 445–453.

76 Unanue, E.R., P.M. Allen: The basis for the immunoregulatory role of macrophages and other accessory cells. *Science* 236 (1987), 551–557.

77 Yoshikawa, T., J.W. Streilein: Genetic basis of the effects of ultraviolet light B on cultaneous immunity. Evidence that polymorphism at the *Tnfa* and *Lps* loci governs susceptibility. *Immunogenetics* 32 (1990), 398–405.

78 Young, J.W., J. Baggers, S.A. Soergel: High dose UV-B radiation alters human dendritic cell costimulatory activity but does not allow dendritic cells to tolerize T lymphocytes to alloantigen *in vitro*. *Blood* 81 (1993), 2987–2997.

5: Effects of UV light on cytokines and neuroendocrine hormones

Thomas A. Luger & Thomas Schwarz

Introduction

The concept of the epidermis being an active element of the immune system has been confirmed within the last decade [24]. Accordingly, constituents necessary for the initiation of an immune response are present in the epidermis including antigen-presenting dendritic Langerhans cells [162], cells belonging to the T-cell family [161] and cells exhibiting the capacity to release a variety of inflammatory and immunologic mediators including cytokines, growth factors, neuroendocrine hormones and eicosanoids [146]. Although keratinocytes appear to be the major source of epidermal cytokines there is evidence that Langerhans cells and melanocytes can also produce immunomodulating factors [47, 76, 141]. These initial observations were based on biologic as well as biochemistry data and confirmed by detection of cytokine specific mRNAs in epidermal cells [98]. Keratinocytes are known to produce interleukins, hematopoietic growth factors, tumor necrosis factor-α and growth factors [146].

There is ample evidence for a close relationship between the immune and the neuroendocrine system [13, 170]. For example immunosuppression, a characteristic feature of the stress syndrome, appears to be exerted via peripheral effectors such as glucocorticosteroids and catecholamines [67]. Furthermore, the hypothalamic corticotropin releasing hormone (CRF) which stimulates glucocorticoid and adrenocorticotropin (ACTH) secretion has also been found to induce the synthesis of several proopiomelanocortin (POMC)-derived peptide hormones [25]. POMC is synthesized as a large prohormone which via the action of recently identified specific prohormone convertases (PC) is cleaved into peptide hormones such as ACTH, melanocyte-stimulating hormones (αMSH, βMSH, γMSH), endorphins (EP) and lipotrophins (LP) [26]. In addition to their well-defined specific effects as hormones, POMC-derived peptides exert potent immunomodulating functions and interact strongly with the cytokine network [13]. On the other hand, cytokines were also found to be able to modulate the production and function of these peptide hormones [25, 137]. There is recent evidence that within the epidermis keratinocytes, melanocytes and Langerhans cells are able to produce POMC peptides [109, 139, 156].

The constitutive production of both cytokines and neuroendocrine hormones by keratinocytes *in vitro* and *in vivo* is quite low, but can be enhanced by injurious stimuli. UV light is one of these inducers [74, 146]. As exposure to UV light leads to local and systemic inflammatory reactions and alterations of the immune response, keratinocyte-derived mediators seem to play an important role in the pathogenesis of UV-induced reactions [83, 145]. Therefore this chapter will focus briefly on epidermal cell-derived cytokines and neuroendocrine hormones which are known to be upregulated by UV light. In addition, possible implications of these findings for the pathogenesis of UV-induced dermatoses will be discussed.

Interleukins

Interleukins (IL) comprise a heterogenous group of mediators mainly affecting immunocompetent and inflammatory cell functions [12]. Among the 15 different interleukins which have been characterized so far IL-1, IL-3, IL-6, IL-7, IL-8, IL-10,

IL-11, IL-12 and IL-13 are known to be produced by keratinocytes [98, 146].

Interleukin 1

Interleukin 1 (IL-1) is a multifunctional non-specific cytokine which plays an important role in the mediation of immunologic and inflammatory reactions. Two different subtypes of IL-1 exist, named α and β, which are encoded by separate genes located on chromosome two. Although the two IL-1 species express only 26% homology they share the same broad spectrum of biological activities and bind to common receptors [32, 41]. Additionally, two different classes of IL-1 receptors (IL-1R) sharing a homology of 28% and both belonging to the immunoglobulin superfamily have been identified [29, 40]. The IL-1R type 1 (IL-1R1) is found on T cells, fibroblasts, endothelial cells, synovial cells, chondrocytes, hepatocytes and keratinocytes [22, 154]. The IL-1R type 2 (IL-1R2) is expressed on B cells, neutrophils and bone marrow cells [103]. The two receptors differ in their binding affinities, signal transduction and cell surface expression. Whereas IL-1 bound to the IL-1R1 is rapidly internalized, the IL-1/IL-1R2 complex remains at the cell surface and is subsequently shed and degraded [40, 158]. Moreover, there is evidence that IL-1α and IL-1β bind to different regions on the same receptor and that IL-1α reacts mainly with the IL-1R1, while IL-1β does so with the IL-1R2 [40].

Recently, a soluble form of the IL-1R2 has been found which retains its IL-1 binding capacity and thus may function as an IL-1 inhibitor. Soluble IL-1 receptors have been detected in human serum, synovial cell exudates and supernatants derived from human peripheral blood mononuclear cells, B-cell lines and a keratinocyte cell line [39, 168]. In addition, an IL-1 receptor antagonist (IL-1RA) belonging to the IL-1 gene family was isolated and cloned from human monocytes. Due to its similar structure with IL-1, the IL-1RA competes with the binding of IL-1α and IL-1β to the receptor, but does not have IL-1 activity [44].

The wide range of biological activities of IL-1 includes immunomodulating effects such as thymocyte costimulation, the induction of cytokine production, IL-2 receptor expression and chemotaxis by T lymphocytes. IL-1 enhances the effects of B-lymphocyte stimulatory factors and stimulates mediator production and cytotoxic activity by macrophages. The influences of IL-1 on hematopoietic stem cells include costimulation of CSF activities [38, 41]. Moreover, IL-1 has proinflammatory effects such as induction of fever and of acute phase proteins, and causes an inflammatory infiltrate when injected into the skin [58]. Endothelial cells in response to IL-1 produce cytokines and express surface molecules such as intercellular adhesion molecule (ICAM-1) and endothelial cell leukocyte adhesion molecule (ELAM-1) which play a crucial role during endothelial cell–leukocyte interactions in vasculitis [41]. Recently, IL-1 was also identified as a cytokine involved in neuroendocrine immune interactions, since it stimulates the release of hypothalamic and pituitary gland hormones such as proopiomelanocortin (POMC) [38, 41, 137].

IL-1 was the first cytokine which could be isolated from keratinocytes and was originally named epidermal cell-derived thymocyte activating factor (ETAF) [97, 136]. Meanwhile it became clear that ETAF is identical with IL-1 and transcripts of both IL-1α and IL-1β were detected in keratinocytes [8, 86]. Whereas macrophages mainly produce IL-1β, the major active IL-1 species released by keratinocytes is IL-1α. This may be explained by the fact that keratinocytes are lacking a specific protease which is necessary to cleave the biologically inactive 31 kD IL-1β into the active 17 kD form [106]. The release of IL-1 by keratinocytes both *in vitro* and *in vivo* seems to be low, however, it can be significantly upregulated by stimuli including lipopolysaccharide (LPS), tumor promotors and other cytokines [146]. Keratinocytes also express IL-1R1 upon stimulation indicating that IL-1 will regulate function and growth of epidermal cells in an autocrine way [22]. In contrast to monocytes, keratinocytes only produce an intracellular form of the IL-1RA which lacks the leader sequence and therefore is not secreted [18, 44].

The primary targets for UV light are epidermal cells, suggesting that many UV-induced inflammatory and immunologic reactions are mediated by epidermal cell-derived cytokines. This idea was confirmed recently by the *in vitro* finding of an increase in both extracellular and intracellular IL-1 production of UVB-exposed murine and human epidermal cells. The UV-induced IL-1 activity was cycloheximide sensitive, suggesting that *de novo* synthesis of IL-1 rather than cell membrane leakage was responsible for the increased activity [5]. These observations were further confirmed by the detection of IL-1 in the serum of UV light-exposed but not normal animals [6, 51]. The UV-mediated release of IL-1 by murine epidermal cells was reduced by the addition of hydrocortisone, suggesting that this activity might account for the anti-inflammatory effect of glucocorticosteroids on the skin [88].

The induction of IL-1 by UV light was also confirmed at the transcriptional level showing significantly increased steady state levels of IL-1α as well as IL-1β mRNA in human keratinocytes and IL-1α mRNA in murine keratinocytes after UVB exposure [87]. The effect of UVB light on IL-1 synthesis appears to be dose dependent, since high dose UVB (3 kJ/m^2) inhibited murine keratinocyte IL-1α expression at 4, 8, 24 and 48 h post-UV, whereas a lower dose of UVB (1 kJ/m^2) inhibited IL-1 at 4 and 8 h, but induced IL-1 expression at 24 and 48 h post-UV [8]. The IL-1 inducing effect of UVB light appears not to be unique to keratinocytes, as similar findings have been reported with murine peritoneal adherent cells and with a murine macrophage cell line (P388D1) [7].

An increased release of IL-1-like activity was demonstrated at the site of UVB exposure by using an *in vivo* skin chamber method [112]. IL-1 levels peaked 1 h after irradiation and returned to baseline by 2 h. Further evidence of *in vivo* IL-1 induction upon UV irradiation is provided by the detection of enhanced IL-1 levels in the sera of UVB-treated rodents [5, 51]. Human volunteers given one minimal erythema dose of UVB light developed increased serum IL-1 activity peaking 1 to 4 h after exposure and returning to baseline by 8 h post-UV [59]. Recently, IL-1 plasma levels have been found to be elevated in psoriatics undergoing phototherapy [80]. Significant IL-1 levels were detected in blood samples collected 4 h after treatment, while samples obtained before 1 or 2 h irradiation were negative. Plasma IL-1 was neutralized by antibodies to recombinant human IL-1α and IL-1β. Since the predominant keratinocyte derived subspecies is IL-1α, this suggests that UV-induced IL-1 in the circulation may originate from both keratinocyte and non-keratinocyte sources [80].

UVB seems to cause a dual effect on keratinocyte IL-1R1 expression. Within 12 h after UVB exposure a significant downregulation of IL-1R protein and mRNA was reported. In contrast, 24 to 48 h after UVB treatment an increase in keratinocyte IL-1R expression may be observed. Whether the shedding of IL-1R from keratinocytes is influenced by UV light is not yet established [85, 121].

The combination of 8-methoxypsoralen (8-MOP) plus UVA light (PUVA, oral photochemotherapy) is a widely used effective therapy for various skin disorders including psoriasis and cutaneous T-cell lymphoma [62]. From animal studies it is well established that PUVA suppresses inflammatory and immune reactions and this is also supported by clinical observations [84]. In order to obtain more insight into the mode of action of PUVA, freshly isolated murine epidermal cells were exposed to various doses of 8-MOP (5–100 ng/ml) followed by UVA exposure (10 kJ/m^2). Supernatants harvested thereafter showed a clear reduction in IL-1 levels [174]. The suppression was dependent on the dose of 8-MOP and not due to the induction of IL-1 inhibitors as described for UVB treatment [147]. A similar effect of *in vitro* PUVA was observed by using peripheral blood mononuclear cells (PBMC) as a source for IL-1β, showing a significant reduction of IL-1β protein in the supernatants. Moreover, these data were confirmed by Northern blot analysis demonstrating reduced IL-1β specific transcripts in PUVA treated PBMC, while 8-MOP or UVA alone were without effect [115].

Since UV irradiation *in vitro* affects the production of IL-1 and IL-1RA, its *in vivo* role in photoaging seemed to be of interest. Accordingly, a significantly decreased expression of IL-1RA mRNA in keratinocytes from sun-exposed skin in comparison to sun-protected skin from the same individual was observed, suggesting a possible role of IL-1 in photoaging and perhaps photocarcinogenesis [53].

Interleukin 6

The pleotropic cytokine interleukin 6 (IL-6) is released by a variety of cells including normal keratinocytes, Langerhans cells, melanocytes and epidermoid carcinoma as well as melanoma cell lines [71, 77, 142]. Human epidermal cells in culture spontaneously release only small amounts of IL-6. Addition of stimuli such as IL-1, tumor promoters, bacterial endotoxins or mitogens significantly increases IL-6 production by keratinocytes [71]. Moreover, high affinity IL-6 receptors have been detected on a variety of different cells and cell lines including keratinocytes. The IL-6 receptor consists of an 80 kd glycoprotein which is a member of the immunoglobulin supergene family, and associates with another 130 kd membrane protein. Currently, it is believed that the 80 kd/130 kd complex is required for signal transduction [151, 187].

IL-6 enhances the production of acute phase proteins, inhibits the secretion of albumin, and is one of the endogenic pyrogenes. It may also function as a second signal for thymocyte and T-lymphocyte proliferation, stimulate the activity of cytotoxic T cells as well as natural killer (NK) cells, and costimulate hematopoietic stem cell differentiation. In addition, IL-6 induces B-cell differentiation, immunoglobulin production and the proliferation of certain transformed B-cell lines [187]. The role of IL-6 in tumor development is controversial since, depending on the tumor type, IL-6 may stimulate or inhibit differentiation and proliferation [151]. Accordingly, transfected murine melanoma cells expressing high amounts of IL-6 appear to exhibit reduced tumor growth [9]. Recently, increased amounts of IL-6 were detected in psoriatic epidermis and in the serum of these patients, suggesting that IL-6 may play a role during the pathogenesis of psoriasis [117].

The effect of UV light on the release of IL-6 by epidermal cells has been investigated in several studies. Supernatants derived from UVB-exposed normal human keratinocytes or keratinocyte cell lines contained significantly increased levels of IL-6 bioactivity. These findings were also confirmed by Western and Northern blot analysis [70]. Addition of different corticosteroid preparations resulted in prevention of the UVB-mediated induction of IL-6, suggesting that the therapeutic effect of corticosteroids may be partly due to this downregulation. In contrast to UVB, UVA1 in a dose range from 100 to 1500 kJ/m^2 did not affect IL-6 release by keratinocytes [70]. However, it seems to be premature to claim that UVA does not influence IL-6 production by keratinocytes at all, because there are speculations that this may depend on the type of cells and dose of UVA used (T. Schwarz, unpublished observations).

Induction of IL-6 by UV light also seems to happen *in vivo*. Severe sunburn reaction is associated with fever, chills and induction of an acute phase response, events which could principally be mediated via IL-6 [187]. Accordingly, a single total body UVB exposure eliciting a marked sunburn reaction resulted in a significant increase of IL-6 in the circulation peaking at 12 h post-UV. Plasma IL-6 activity correlated remarkably with the fever course followed by an increase of acute phase proteins such as C-reactive protein [181]. These data indicate that IL-6 probably released by keratinocytes after UV exposure may gain access to the circulation and via its pyrogenic as well as acute phase inducing effects could function as an important mediator of systemic sunburn reaction.

Within recent years, the particular mechanisms involved in photoaging have raised much interest. It is known that UVA reaches the reticular dermis, making fibroblasts an accessible target. In addition, it has been shown that UVA induces the release of collagenase by fibroblasts [126, 138]. Since IL-6 stimulates collagenase production by fibroblasts, the relation between the UV-induced collagenase increase and IL-6 was studied. Confluent fibroblast layers exposed to UVA1 light (50–350 kJ/m^2) released significantly enhanced levels of IL-6 which was also confirmed by Northern blot analysis. Incubation of irradiated cells with IL-6 antisense oligonucleotides resulted in a significant, but not complete, downregulation of collagenase. The failure of complete blocking may be due to the fact that UVA-induced release of other mediators such as IL-1 by fibroblasts is involved in the induction of collagenase as well [186].

As has already been reported for IL-1, the effect of PUVA on the release of IL-6 was studied. Treatment of PBMC by a combination of 8-MOP (10–1000 ng/ml) with UVA (20 kJ/m^2) resulted in a significant reduction of IL-6 protein in the supernatants and in a decreased expression of IL-6 specific transcripts in PUVA-treated PBMC [116].

Interleukin 8

Interleukin 8 is a member of the supergene family of proinflammatory and chemotactic cytokines which are characterized by the presence of four conserved cysteines and were recently named chemokines [143]. One subfamily contains members characterized by the presence of a single amino acid between cysteine 1 and 2 (C-X-C family α chemokines). Members of this family which are preferentially chemotactic for neutrophils are IL-8 and melanoma growth-stimulating activity/growth related gene product (MGSA/GRO) [100, 143]. The second group is characterized by the loss of the amino acid between the cysteines (C-C family β chemokines) and consists of mediators which are not chemotactic for neutrophils but attract other leukocytes, including monocyte chemotactic activating factor/monocyte chemotactic protein 1 (MCAF/MCP-1), 'RANTES' (regulated on activation, normal T expressed and secreted), murine macrophage inflammatory protein 1α (MIP-1α) and MIP-1β and their human equivalents LD 78 and PAT 744 which were identified recently [143].

High affinity receptors for IL-8 which belong to the group of guanine nucleotide binding proteins were detected on neutrophils as well as on myelocytic cell lines (HL 60, TPH 1) [143]. IL-8 is produced by many different cells including keratinocytes, melanocytes and Langerhans cells [17, 50, 143]. It stimulates chemotaxis and enzyme release of neutrophils, but has no effect on eosinophils or monocytes. In addition, IL-8 was found to be chemotactic for basophils, lymphocytes and keratinocytes. Increased amounts of IL-8 were detected in 'neutrophilic dermatoses' such as psoriasis and palmoplantar pustulosis [143].

MCP-1 which belongs to the C-C family is produced by monocytes, fibroblasts, endothelial cells and keratinocytes. In contrast to IL-8, keratinocytes were found to spontaneously express MCP-1 mRNA and to secrete the protein after stimulation with IFN-γ. MCP-1 is a specific chemotactic factor for monocytes and stimulates histamine release in basophils and mast cells [91]. Recently, it has been shown that MIP-1α, another member of the C-C family, which inhibits the proliferation of keratinocytes and stem cell differentiation, is produced by murine Langerhans cells [99]. Moreover, MIP-1α is chemotactic for memory T cells, whereas MIP-1β is a chemoattractant for naive T lymphocytes [143].

Recently, a protein designed as C-C CKR that acts as receptor for C-C chemokines has been cloned [114]. It binds human and murine MIP-1α, MCP-1 and RANTES with varying affinity. Moreover, C-C chemokines were shown to bind a C-C CKR-1 related gene product encoded by cytomegalovirus, suggesting that C-C chemokines may play a role in the defense against viral infections [114].

Based on its biological functions, IL-8 seems to play an important role during inflammatory reactions. The main sources of IL-8 are macrophages and fibroblasts, however, it has recently been shown that IL-8 can be produced by keratinocytes as well [90]. In addition, other members of the C-X-C family like MGSA/GRO are found in the epidermis [130, 143]. Because UV light appears to be a contributing factor in the pathogenesis of malignant melanoma the question of whether UV light affects the release of these peptides by keratinocytes appears obvious. Accordingly, IL-8 mRNA expression was significantly upregulated in a human keratinocyte cell line (COLO 16) upon treatment with UVB (50 J/m^2) [102]. Moreover, GRO specific transcripts were found to be increased in human keratinocytes within 0.5 h of UVB exposure (70 J/m^2) and remained elevated for up to 6 h after UV treatment [101]. Since IL-8 is supposed to be an important mediator in the pathogenesis of psoriasis, the effect of *in vitro* PUVA on the IL-8 release by PBMC was studied, demonstrating reduced expression of IL-8 both at the protein and the mRNA level after treatment of PBMC with 8-MOP (10–1000 ng/ml) followed by UVA exposure (20 kJ/m^2) [115].

Interleukin 10

Interleukin 10 (IL-10) was originally described as cytokine synthesis inhibitory factor (CSIF) [107]. However, it became clear that in addition to T cells, IL-10 is also produced by a variety of other cells including B cells and keratinocytes [46, 110]. An uncharacterized open reading frame (BRCF-1) in the EBV genome shows striking homology to IL-10. This appears to be interesting because IFN-γ inhibits the generation and outgrowth of EBV-transformed B cells. Since both IL-10 and BCRF-1 inhibit IFN-γ synthesis, EBV may manipulate the immune response against virally infected cells, consequently promoting survival of the virus [107]. IL-10 inhibits the effector functions of T_h1 cells suggesting that it may play a role in determining the class of immune response directed against the particular antigen in the downregulation of delayed hypersensitivity. The production of different cytokines such as IL-1, IL-2, TNFα, IFN-γ and GM-CSF is also inhibited by IL-10. In addition, it has been shown to costimulate thymocytes and mast cells and to inhibit the antigen-presenting function of macrophages and B lymphocytes [110].

IL-10 production in the skin is enhanced upon application of contact allergens while tolerogens or irritants have no effect [47]. UV light is known to inhibit inflammatory and immune reactions, and the selective nature of the UV-induced suppression is similar to the biological activity of IL-10. Accordingly, it could be demonstrated that UVB light upregulates IL-10 production by murine keratinocytes [132]. This was demonstrated both on the protein and at the mRNA level. Moreover, there is evidence that keratinocyte derived IL-10 is involved in systemic immunosuppression [131].

Tumor necrosis factor-α

Tumor necrosis and cachexia were found to be mediated by a macrophage-derived cytokine, named cachectin/tumor necrosis factor-α (TNFα). Another cytotoxic factor released by lymphocytes was initially named lymphotoxin and is now called TNFβ. TNFα and TNFβ are antigenically distinct proteins, but they have similar biologic activities and bind to common receptors [2]. Two different types of TNF receptors (TNFRI, p75 and TNFRII, p55) have been characterized and cloned in human cells. Both receptors bind TNFα and TNFβ with high affinity, but are differentially expressed on many cells. Their expression is regulated independently and they seem to be responsible for the mediation of different TNF activities. Both TNF receptors share a similar structure with the nerve growth factor receptor (NGFR), the putative human cytokine receptor CD40, the human CD27 and CD30 antigens, as well as the Fas antigen. However, despite a homology of 30%, no binding of TNF or NGF to the heterologous receptors has been demonstrated. Moreover, as is the case with other receptors, soluble TNFR which retained their binding capacity have been demonstrated in body fluids and cell culture supernatants [2, 105, 176].

TNFα was characterized as a macrophage-derived mediator responsible for hemorrhagic necrosis of tumors and cachexia in animals [15]. Subsequently it became clear that TNFα is produced by a variety of cells and has a broad spectrum of biological activities. TNFα like IL-1 is a pyrogenic cytokine and activates inflammatory cells such as neutrophils, eosinophils, macrophages and fibroblasts [38]. Endothelial cells in response to TNFα proliferate, produce cytokines and express adhesion molecules such as ICAM-1 [15]. Moreover, TNFα enhances MHC class I and II antigen expression on various cells and costimulates T and B lymphocytes as well as hematopoietic stem cells [15]. Within the epidermis TNFα appears to be relevant for several reasons. Keratinocytes which were found to express only TNFRII respond to TNFα, with increased expression of ICAM-1 possibly being of importance in T cell keratinocyte interactions [75, 175]. Furthermore, TNFα is one of the cytokines involved in Langerhans cell chemoattraction and maturation [69, 73]. During the elicitation of contact hypersensitivity and the occurrence of cutaneous lesions in graft versus host disease, TNFα also seems to be one of the crucial mediators [127, 128].

Like many other cytokines, TNFα is produced by a variety of cells including keratinocytes [75]. Unstimulated epidermal cells express low levels of TNFα mRNA and release minimal amounts of biologically active TNFα. Synthesis and release of TNFα is significantly upregulated by tumor promotors, endotoxin and also by UV light. Accordingly, UVB exposure of normal human keratinocytes and epidermoid carcinoma cell lines resulted in a significant upregulation of TNFα mRNA expression followed by an increased release of TNFα [75]. These findings parallel immunohistological and immunoelectron microscopic observations showing TNFα in epidermal cells and after UVB exposure in an even increased staining intensity [79, 122]. Further evidence for the *in vivo* release of TNFα after UV exposure comes from the detection of increased TNFα levels in the serum obtained from humans 12 and 24 h after a single total body UVB exposure [75].

Treatment with UV light results in depletion of epidermal Langerhans cells and in impairment of contact hypersensitivity (CHS) [1, 173]. Similar effects may be obtained by intradermal injection of TNFα or the immunosuppressive photoreceptor *cis*-urocanic acid [119, 190]. Since the effects of UVB and *cis*-urocanic acid are reversed by anti-TNFα, the impairment of CHS by UVB appears to be mediated by converting *trans*-urocanic acid to *cis*-urocanic acid which causes the local release of TNFα, altering the function of Langerhans cells [89]. There is more evidence that TNFα is involved in the suppression of CHS. It was recently demonstrated that in inbred strains of mice impairment of CHS by UVB is a polymorphic trait polygenically dictated by susceptibility alleles at the lps and tnfα loci [163]. Strains of mice homozygous both for the $tnf\alpha^b$ and for the lps^n allele are UVB susceptible. Accordingly, injection of TNFα prevented the induction of CHS, and administration of a TNFα antibody reversed UVB-induced suppression of CHS in these strains [189]. These findings are confirmed by the observation that the release of TNFα by keratinocytes can be induced by UVB light [75]. Moreover, it was recently discovered that hapten specific T cells of UVB-suceptible mice are unable to secrete IL-3 in response to hapten stimulation. Therefore, IL-3 appears to play a role in the activation of T lymphocytes by UVB-exposed Langerhans cells and may allow the differentiation of UVB-susceptible and UVB-resistant individuals in humans [164].

The pathomechanisms leading to the formation of apoptotic keratinocytes, i.e. sunburn cells (SC), present in UVB-exposed epidermis are not fully understood. Since keratinocyte production of TNFα and TNFRII expression are modulated by UV exposure it was tempting to speculate that TNFα might be involved in SC formation. Accordingly, injection of a TNFα antiserum into UV-irradiated murine skin resulted in a significant reduction of SC. This effect was dose dependent and maximally pronounced when anti-TNFα was injected immediately after UV treatment. Moreover, DNA fragmentation occurring during apoptosis was inhibited in skin extracts obtained from animals which were UVB and anti-TNFα treated [182]. Since TNFα alone was not able to induce SC these findings indicate that TNFα seems to be necessary, but not sufficient alone, for the generation of apoptotic keratinocytes.

Because of the pronounced effects of UV light on epidermal cells and the constitutive expression of TNFRII by keratinocytes it was also investigated whether keratinocyte TNFRII expression is subjected to regulation upon UVB exposure. Accordingly, UVB treatment decreased the expression of TNFRII on keratinocytes in a dose dependent manner with maximum inhibition occuring 6 h post-exposure. However, 12–24 h after UVB treatment TNFRII expression reoccurred, and at 48 h it exceeded that observed in non-irradiated cells. The biphasic modulation of keratinocyte TNFRII expression could be effectively blocked by adding neutralizing TNFα-antibodies to UVB-irradiated keratinocytes, suggesting a TNFα-mediated autocrine regulatory pathway. Although supernatants of human keratinocyte cultures were found to contain significant amounts of soluble TNFRII but no TNFRI, the production of soluble TNFRII was not affected by treatment with UVB or TNFα [176].

Because IL-1 and IL-6 have been demonstrated to be downregulated by treatment of the combination of 8-MOP plus UVA, the effect of PUVA on the release of TNFα has been studied. Treatment of PBMC with 8-MOP (10–1000 ng/ml) followed by UVA exposure (20 kJ/m^2) resulted in a significant reduction of TNFα protein and TNFα-specific transcripts [115].

Hematopoietic growth factors

Colony stimulating factors (CSF) which induce stem cell colony formation in soft agar cultures include multilineage CSFs such as IL-3 and granulocyte/macrophage-CSF (GM-CSF) as well as more specific mediators such as granulocyte-CSF (G-CSF) and macrophage-CSF (M-CSF). In addition, several other cytokines were identified as exhibiting some CSF activities including stem cell factor (SCF), IL-1, IL-6, IL-4 and IL-5 [104, 140].

IL-3 is a non-specific hematopoietic growth factor which in the murine system is identical with mast cell growth factor (MCGF). In addition to affecting stem cell differentiation, IL-3 is a growth factor for some myeloid cell lines and induces phagocytosis by macrophages [108]. The IL-3 receptor was identified on stem cells, monocytes and T cells. Like the IL-5R and the GM-CSFR the IL-3R consists of two distinct chains of which the β-chain is identical for all three cytokine receptors [169]. IL-3 is produced mainly by T lymphocytes, however, it was recently shown that keratinocytes and melanocytes can also produce IL-3 [35, 77, 94].

GM-CSF, like IL-3, is a multicolony stimulating factor which also stimulates the function of neutrophils, eosinophils and macrophages, and is a growth factor for myeloleukemic cells. Moreover, GM-CSF inhibits the migration of neutrophils, induces phagocytosis, eicosanoid production and antibody-dependent cytotoxicity (ADCC), and stimulates tumoricidal and antimicrobial activity as well as cytokine production of macrophages [33]. The GM-CSF receptor, which is similar to the IL-3 receptor, was detected on several myeloid and non-myeloid cells [169]. Within the epidermis keratinocytes are able to produce GM-CSF which was discovered to be one of the mediators responsible for the maturation of Langerhans cells [73, 85].

G-CSF mainly induces the formation of granulocyte colonies. In addition to its effect on stem cells, G-CSF also activates mature neutrophils, induces phagocytosis, chemotaxis, enzyme release and antibody dependent cytotoxicity (ADCC). G-CSF is produced by many different cells including keratinocytes [21]. The receptor for G-CSF was recently characterized on neutrophils [33, 118].

In contrast to the other colony stimulating factors, M-CSF is a disulfide bridge linked homodimer. The M-CSF receptor was found to be expressed on monocytes and to be identical with the product of the c-fms oncogene. M-CSF induces the proliferation and differentiation of macrophage progenitor cells. In addition, M-CSF stimulates the production of cytokines, tumor cytotoxicity and enzyme release in macrophages. M-CSF was also found to be produced by many different cells including keratinocytes [30, 33].

Stem cell factor (SCF) is a highly conserved hematopoietic growth factor which was identified as being a ligand of the c-kit gene product, and was therefore named c-kit ligand. In mice SCF is the product of the SI (steel) locus, whereas the W locus codes for the c-kit tyrosinase receptor and thus it was named steel factor [54, 185]. *In vitro*, SCF stimulates the proliferation of IL-3 dependent and mature murine mast cells and was therefore described as 'mast cell growth factor' (MGF) [178, 185]. SCF is essential for the development of hematopoietic, germ and pigment cell lineages. It synergizes with other hematopoietic cytokines such as GM-CSF, G-CSF, IL-3 and erythropoietin to stimulate progenitor cells of myeloid and erythroid lineages [54]. An important function of SCF is its melanocyte stimulating capacity. Accordingly, in patients with piebaldism, which is an autosomal dominant defect in melanocyte development characterized by patches of hypopigmented skin and hair, a mutant of the c-kit gene was detected [55]. Similarly, a mutation of the mouse c-kit gene is associated with depigmentation, a defective proliferation and differentiation of stem cells and germ cells [179].

UVB exposure of the murine keratinocyte cell line PAM 212 results in increases of IL-3- and GM-CSF-specific mRNA within 1 h of UV treatment [52]. However, only GM-CSF-specific bioactivity, as determined by antibody neutralization studies, was shown to be increased above baseline in cell supernatants. These effects were only observed when cells were cultured until subconfluency. Under confluent conditions a decrease in bioactivity was detected after UV exposure for both growth factors, and no change in mRNA levels was found. In addition to cell density, calcium and sodium levels during irradiation altered the release of IL-3 and GM-CSF. These data demonstrate that UV light influences the release of CSFs and suggests that the state of cell growth and conditions of membrane ion transport influence the mechanisms regulating the secretion of those factors. Moreover, it was shown that UVB-induced GM-CSF secretion by PAM 212 cells is mediated by an enhanced release of IL-1α, since IL-1α antibodies completely inhibited UV-mediated GM-CSF mRNA expression [120].

There is also evidence that CSF induction by UV light takes place *in vivo*. Albino hairless mice (SKH-1) were treated i.p. with a single dose of 5-fluorouracil sufficient to cause complete transient bone marrow suppression [19]. Following a daily UVB exposure, soluble extracts of UVB-irradiated skin stimulated proliferation of normal murine bone marrow cells, CSF activity present in normal murine skin was enhanced, and recovery of neutropenia was accelerated. Thus, one may conclude that UVB irradiation leads to the cutaneous release of CSFs in amounts sufficient to reverse acute myelosuppression.

Growth factors

Growth factors represent a heterogenous group of mediators that were originally regarded as exclusive regulators of cell growth [34, 135, 188]. The family includes epidermal growth factor, transforming growth factors (TGF), fibroblast growth factors (FGF), nerve growth factor (NGF), platelet-derived growth factor (PDGF) and insulin-like growth factors (IGF). Keratinocytes synthesize and release a variety of growth factors including TGFα and β, PDGF, basic FGF and NGF. TGFα mRNA is expressed in the stratified epidermis and psoriatic plaques, and TGFβ transcripts were found in normal and psoriatic epidermis [49, 57]. PDGF has been shown to be released by epidermal cells in response to TGFβ [36]. Moreover, bFGF was recently reported as being expressed in proliferating keratinocytes and functioning as a growth factor for melanocytes [63]. Murine keratinocytes were recently demonstrated to express NGF [177].

Recently, it became evident that growth factors appear to have several immunomodulating functions. TGFβ inhibits IL-1 receptor expression and blocks the activities of IL-1, IL-2 and CSFs [183]. PDGF has been recognized as a mitogen and chemoattractant for inflammatory cells and may stimulate MHC class II antigen expression [183]. During inflammatory reactions the growth promoting and chemotactic activities of growth factors such as FGF, PDGF and TGF on endothelial cells may also be of particular importance [134, 172].

Although keratinocytes are able to produce growth factors, the mechanisms involved in the regulation of the release are not well characterized. At present there is evidence that UV light can induce the release of bFGF, TGFα and nerve growth factor by keratinocytes [63, 113, 177]. Consequently, TGFα has been suggested as being involved in the generation of UVB-induced epidermal hyperplasia [113]. In addition, TGFα release by melanocytes and melanoma cell lines was found to be increased upon UVC (254 nm) irradiation [45]. The biological relevance of these findings remains to be determined however.

Other cytokines

When investigating the effect of UV light on the release of cytokines by epidermal cells most of the research focused on the above-mentioned mediators. However, there are also indications that other, in some cases not yet well-defined, factors are affected by UV light as well.

One of the most important biological effects of UV light represents the compromise to the immune system [16]. This immunosuppression may contribute at least partially to UV-induced carcinogenesis [82]. Immunologic alterations caused by UV irradiation can be divided into two forms: local and systemic immunosuppression. Local alterations are due to a direct interaction between UV light and immunocompetent cells such as Langerhans cells in the UV-exposed area [11]. Higher doses of UV light influence immune responses such as induction of contact hypersensitivity (CHS) at sites not exposed directly to UV light ('systemic' immunosuppression) [65]. How the events occurring at the irradiated site lead to an abnormal response to an antigen applied at a distant non-UV-exposed skin area remains unclear. The most obvious explanation involves soluble mediators, whereby keratinocytes are the most likely source of such putative inhibitors. Accordingly, it has been shown that intravenous injection of supernatants obtained from UV-exposed PAM 212 cells into naive mice five days before hapten application results in suppression of induction of CHS [148]. This inhibition was not found when applying conditioned medium of unirradiated cells, indicating that UV irradiation induces the release of soluble suppressive mediator(s) by keratinocytes. Upon HPLC gel filtration this factor revealed a molecular weight between 25 and 50 kD. For the first time, this study could at least partially explain the phenomenon of UV-induced systemic immunosuppression by showing that UV irradiation of epidermal cells *in vitro* causes the release of immunoregulatory factors that mimic some of the *in vivo* immunosuppressive effects caused by UV light [148].

Meanwhile, the concept of UV-exposed keratinocytes as a source of immunosuppressive factors has also been supported by other studies. Injection of supernatants from UVB-exposed PAM 212 cells suppressed the induction of delayed-type hypersensitivity (DTH) to alloantigen and trinitrophenyl-modified self antigens in syngeneic and allogeneic mice [68]. Surprisingly, in this study, application of the very same supernatants had no effect on the induction of CHS to trinitrochlorobenzene. On the other hand, treatment of keratinocytes with extremely low doses of UVA resulted in the release of a mediator that blocked CHS [68]. These findings indicate that more than one keratinocyte-derived suppressor factor is induced by UV light and that the UV spectrum dictates which factor is released. The DTH inhibitor was further characterized and appeared to be cycloheximide as well as trypsin sensitive, and to bind to Concanavalin A suggesting it to be a glycoprotein of approximately 68 kD [180].

In addition, supernatants obtained from PAM 212 or freshly isolated murine epidermal cells were found to suppress the proliferative response of keyhole limpet hemocyanin (KLH) specific Th1 clones in the presence of KLH coupled Langerhans cells [43]. Preincubation of Langerhans cells, but not of Th1 cells, with the supernatants from irradiated keratinocytes suppressed Th1 proliferation suggesting that Langerhans cells and not Th1 cells are the initial target of the suppressive activity.

Treatment of keratinocytes with 8-MOP plus UVA was shown to result in the secretion of a mediator depressing the DTH reaction to alloantigen [10]. The release was observed regardless of whether the psoralen formed monofunctional or bifunctional adducts with DNA and did not depend on phototoxicity. In addition, keratinocytes treated with 8-MOP (400 ng/ml) and lower doses of UVA (50–100 J/m^2) released a factor that inhibited CHS, but not DTH. Interestingly, the CHS inhibitor was induced by low UVA doses (100–500 J/m^2) alone as well [10]. This also supports the concept that more than one immunosuppressive factor is released by UV-treated keratinocytes.

There is also evidence for *in vivo* release of suppressor factors. Injection of plasma obtained from UVB-exposed mice into normal animals suppresses the ability of the recipient mice to generate CHS reactions [167]. This seems to be due to the induction of antigen-specific T suppressor cells. Whether the suppressor factor found in the serum of UV-treated mice is related to one of the keratinocyte-derived inhibitors remains to be determined [68]. Moreover, a UVB-inducible serum factor that suppresses allergic contact dermatitis has also been described in guinea-pigs [31].

Conditioned medium of UV-irradiated PAM 212 cells as well as that from freshly prepared murine epidermal cells were found to contain an inhibitory activity blocking particular functions of IL-1 [147]. This finding was unexpected since at this time it was well established that supernatants derived from UV-exposed epidermal cells contained high levels of IL-1 [5, 87]. Consequently this inhibitor, provisionally named EC-contra-IL-1, was only found after chromatographic removal of the high amounts of IL-1 present in the supernatants. As in all previous studies only crude supernatants were tested this inhibitor was always missed. Partially purified EC-contra-IL-1 blocked both natural and recombinant IL-1 in a dose dependent manner, and exhibited a molecular weight of approximately 40 kD and an isoelectric point of 8.8. In subsequent studies an IL-1 inhibitor with similar biochemical characteristics was found in the sera of total body UVB-exposed mice, suggesting that the production of EC-contra-IL-1 is obviously not only an *in vitro* phenomenon [149]. This circulating factor could not be detected in unirradiated animals; maximum levels appeared between 12 and 24 h after UV exposure. The biological *in vivo* functions of such inhibitors are not yet clear, however, they may be part of a physiological feedback mechanism downregulating IL-1 activity after excessive release of IL-1.

There is clear evidence that sunlight contributes to the etiology of melanoma [78]. Consequently, it was observed that UVB light activates epidermal cells to release factors that stimulate the *in vitro* and *in vivo* growth of syngeneic melanoma cells, but not of other tumors [184]. Possible candidates responsible for this activity include bFGF, melanoma growth stimulatory activity (MGSA/GRO) and IL-8, factors which have been detected in the epidermis and shown to induce melanoma cell growth [50, 63, 101, 144].

Neuroendocrine hormones

There is increasing evidence that neuroendocrine hormones such as POMC peptides may influence the function of epidermal cells. Accordingly, melanotropins were of particular interest because of their effects on the pigment system. Most of the investigations were focused on αMSH which was found to stimulate cutaneous pigment cells from various species via a cAMP-dependent mechanism [123]. Recently, receptors for αMSH and ACTH have been detected on the surface of normal human and murine melanocytes as well as on melanoma cells [42]. Subsequently the cDNAs for ACTH and MSH receptors were cloned and the transcription of MSH receptor gene was detected in normal human melanocytes [111]. In humans it still is controversial whether or not αMSH regulates pigmentary responses [27]. However, generalized skin hyperpigmentation has been reported upon αMSH hypersecretion by the pituitary [124], and humans injected with αMSH or a synthetic analogue of αMSH showed increased skin darkening [92, 93]. This is further supported by the finding that UV irradiation upregulates αMSH receptor expression on melanocytes [23]. Recent studies indicate that POMC peptides may also affect keratinocyte functions. Accordingly, MSH binding has been detected in cultured keratinocytes and addition of βMSH or γMSH to mouse skin grown in organic culture resulted in an increased proliferation rate of keratinocytes [20, 155].

In addition, POMC peptides have been demonstrated to modulate immune and inflammatory reactions [13, 25, 170]. Accordingly, αMSH blocks particular functions of IL-1, i.e. induction of thymocyte proliferation and PGE_2 release by fibroblasts [28]. The ability of either IL-1 or TNFα to cause fever, leukocytosis and enhancement of plasma acute phase proteins was inhibited by the simultaneous administration of αMSH [133]. Moreover, αMSH reversed the depressive influences of IL-1 or TNFα on the effector phase of contact hypersensitivity in animals [133]. Recently, αMSH also was found to downregulate the expression of MHC class I antigens on mononuclear cells but not to affect class II antigen or ICAM-1 expression [95]. In addition, αMSH also appears to be a stimulator of the activity of natural killer cells [74].

There is even more evidence for the immunomodulatory ability of POMC-derived peptides since it was shown that αMSH inhibits the production of

IFN-γ by peripheral blood mononuclear cells (PBMC). The suppressive effect was demonstrated both at the protein and the mRNA level [95]. Moreover, ACTH seems to be able to modulate IL-4- and anti-CD40-driven release of IgE by PBMC. ACTH at physiological concentrations increased IgE synthesis, whereas higher doses of ACTH strongly inhibited IgE synthesis by PBMC. αMSH also affected human IgE synthesis and was found to downregulate IFN-γ mRNA synthesis [3]. These findings indicate that POMC may play a role in the pathogenesis of allergic diseases and may represent the postulated neuroendocrine arm in diseases such as atopic dermatitis.

Recently, it has been demonstrated that the release of αMSH is not restricted to pituitary cells. αMSH production has been found in various other tissues including brain, gastrointestinal tract, gonads and PBMC [13, 25]. Moreover, there is increasing evidence for POMC expression in the skin. Accordingly, γMSH-like immunoreactivity has been found in human nerve fibers [66], αMSH was identified in human and rodent epidermal extracts [171], and POMC gene expression was demonstrated in murine epidermal cells [4]. Furthermore, supernatants derived from normal human keratinocytes and epidermoid carcinoma cell lines contained αMSH as well as ACTH, and POMC mRNA expression was detected in keratinocytes upon stimulation with tumor promoters of IL-1 [74]. In addition, there is evidence that human Langerhans cells express POMC-specific mRNA which is upregulated upon stimulation with PMA or LPS [109], and POMC peptides as well as POMC mRNA were also detected in human and murine melanocytes [155].

These findings raise new and important issues concerning cutaneous inflammation. Since αMSH can antagonize particular biological activities of the proinflammatory cytokine IL-1, it may play a crucial role in the downregulation of inflammatory reactions in the skin [133]. In order to obtain more insights into these mechanisms, the signals influencing the release of αMSH by keratinocytes need to be studied. In particular the influence of UV irradiation on the release of αMSH by keratinocytes appears to be of relevance, because keratinocytes are the main target of UV light, which is known to cause local as well as systemic immunosuppression. Recently it was found that epicutaneous application of αMSH suppresses both the sensitization and elicitation limbs of the cutaneous immune response to potent contact sensitizers [129]. UV light is known to suppress the induction of contact hypersensitivity [65], and thus the UV-induced production of αMSH by keratinocytes may be involved in mediating the suppressive effect on the epicutaneous sensitization response. This is in agreement with previous studies in which mediators released from keratinocytes inhibited contact hypersensitivity when injected i.v. into test animals [68, 148].

Since the present study demonstrates the production of αMSH by human keratinocytes it was of interest whether UV irradiation would affect POMC peptide production by epidermal cells. Treatment of keratinocytes or epidermoid carcinoma cell lines with UVB light resulted in a significant induction of POMC mRNA expression as well as αMSH production [74, 139]. Preliminary data also indicate that upon irradiation with UVA1, epidermal cell αMSH production was increased (J. Krutmann, unpublished observation). These findings suggest a role of keratinocyte-derived αMSH in UV-induced pigmentation. This is further supported by the finding that UV irradiation upregulates αMSH receptor expression on melanocytes [23].

Biological significance of UV-mediated cytokine and neurohormone production

The effect of UV light on the release of epidermal cytokines and neuroendocrine hormones has been extensively investigated. The majority of these studies used *in vitro* techniques, mostly exposing normal keratinocytes or keratinocyte cell lines to UV light and then measuring bioactivity or protein levels of the respective factor in the supernatants. In addition, most of these data have been confirmed on the mRNA level, demonstrating changes in the transcripts after UV exposure. However, no detailed studies have been performed addressing whether an increase in mRNA is due to real enhanced transcription or due to a change in mRNA stability. Moreover, it is not quite clear whether there are unique UV-responsive elements on the respective genes encoding for cytokines or neuroendocrine peptide hormones in epidermal cells.

In order to study the UVB signal transduction pathways leading to the upregulation of cytokines, the IL-6 gene which is known to be induced upon UVB irradiation was investigated. Since the IL-6 promoter contains a consensus sequence for the binding of NFkB the effect of NFkB-like binding activity was investigated in human epidermal cell lines [152]. NFkB activation in the cytoplasm has been shown to be due to dissociation of an inactive NFkB-IkB complex. Cytosolic extracts from UVB-exposed epidermal cells showed a UV dose dependent increase of NFkB binding. These observations suggest a dissociation of the NFkB-IkB complex by UVB irradiation which is mediated independently from DNA damage, indicating an additional signaling pathway for the early UVB response residing at the cell membrane or in the cytosol [153].

These type of *in vitro* studies raise the question of *in vivo* relevance. This has been addressed by the use of immunohistochemistry, demonstrating that the staining pattern with TNFα in UV-exposed skin is more pronounced than in unirradiated controls [122]. This, however, does not answer whether the cells stained with the antibodies are the source of the respective cytokine or whether the staining is due to mediators released by other cells, but bound to the receptors expressed on epidermal cells [22, 176]. Another proof for *in vivo* induction is determination of cytokine levels in suction blister fluid raised in UV-exposed skin by using the skin chamber method [112]. Stronger evidence for *in situ* production of various cytokines by keratinocytes upon UV exposure may be generated from *in situ*-hybridization studies. According to preliminary reports, TNFα transcripts seem to be expressed to a higher extent on keratinocytes in PUVA-treated skin (P. Petzelbauer, personal communication). Recently, transgenic mice bearing a chloramphenicol acetyltransferase (CAT) construct that reports the synthesis of TNF in a tissue-specific manner, expressed CAT as well as TNF in several extrathymopoietic tissues when injected with LPS. In the same model it was shown that *in vivo* UV exposure results in CAT expression within the epidermis, suggesting that UV induction of TNFα in keratinocytes occurs *in vivo* [14, 56].

Further evidence for the *in vivo* induction of cytokines by UV light is provided by detecting increased levels in the circulation after *in vivo* UV exposure. The first studies were performed in animals [6, 51]. Subsequently, enhanced IL-1, IL-6 and TNFα serum levels have been found in human volunteers after total body UVB exposure [59, 75, 80, 181]. The release of these mediators may be relevant in the pathogenesis of systemic sunburn reaction, which is characterized by fever, leukocytosis and induction of an acute phase response. IL-1, IL-6 and TNFα are non-specific inflammatory cytokines exhibiting activities which could play a role in this process. Involvement of these mediators in sunburn reaction can only be proven definitely by blocking experiments with antibodies. These studies, however, are problematic from the ethical point of view. A likely alternative to address that issue in the future will be the application of physiological cytokine antagonists, such as the recently cloned IL-1 receptor antagonist [44] or drugs which inhibit the secretion of cytokines, as was recently shown to be true for pentoxifylline blocking the release of TNFα by leukocytes [150, 166].

The cellular source of the cytokines detected as enhanced in the circulation upon UV exposure is not yet completely clear. Since keratinocytes are the main target for UVB light and exhibit the capacity to release these mediators, it is easy to speculate that the circulating factors are keratinocyte derived. However, definite proof for that speculation is lacking. This is also complicated by the fact that there is strong evidence for an *in vivo* network of interacting cytokines [12, 96]. Accordingly, cytokines usually have multiple overlapping activities, they may induce each other, interfere with the expression of their receptors as well as those of other mediators and thus can affect cell function in a synergistic, additive, or antagonistic way. For example, IL-1 can induce IL-6, thus it is possible that the enhanced IL-6 levels found in the circulation are not a direct effect of UV light, but are at least partially indirect via the release of IL-1 [96].

UVB light is known to induce the release of eicosanoids like prostaglandin E2 by keratinocytes [60]. The secretion of these substances plays an important role in the mediation of UVB-induced erythema, which can be suppressed effectively by the application of cyclooxygenase inhibitors [157]. There are also recent indications that prostaglandin release is not directly due to UV light, but may be mediated by UV-modulated cytokine secretion. TNFα and IL-1 are known to stimulate prostaglandin release [37, 81]. Recently it has been shown that culture of UVB-irradiated keratinocytes in the presence of a TNFα antibody results in a 50% reduction of prostglandin E2 production, indicating that endogenous TNFα substantially contributes to increased prostaglandin synthesis by UV-treated keratinocytes [61]. Failure of complete suppression of prostaglandin secretion by TNFα antibodies may be due to the fact that other UV-induced cytokines like IL-1 are involved as well [125].

UV irradiation *in vitro* has been demonstrated to induce the expression of POMC mRNA in keratinocytes resulting in an increased production of POMC-derived peptides [139]. Although this has not been proven for UVB light so far, there is some evidence that treatment of epidermal cells with UVA1 also results in an increased production of the POMC peptide αMSH (J. Krutmann, unpublished observation). However, the *in vivo* relevance of these findings has yet to be established. There are only preliminary observations showing increased plasma levels of αMSH following UVA exposure [64]. In contrast to the *in vitro* data demonstrating an induction of αMSH immunoreactivity upon UVB exposure *in vivo*, UVB appears not to alter skin immunoreactive αMSH or plasma αMSH concentrations. Similarly, PUVA treatment had no effect on *in vivo* αMSH production [159]. Because of the profound immunomodulating effects of POMC peptides however, it is tempting to speculate that their induction by UV light may contribute to UV-induced immunosuppression.

The great majority of studies addressing the implications of UV light on epidermal cytokine release has focused on the effects of UVB light. This may be due to the fact that UVB is the most relevant part of the solar spectrum in terms of biological activity. UVB exhibits potent proinflammatory properties, induces immunosuppression, and contributes to carcinogenesis [82, 83]. The fact that most of the studies focus on UVB, however, must not result in the conclusion that UVA has no effects, as is frequently claimed. There is growing evidence that UVA may also be involved in carcinogenesis and that it may play an important role in premature skin aging [72, 160, 165]. Therefore, the effects of UVA light on the release of cytokines need to be studied in the future. However, one has to be aware that the UVA range is much broader than the UVB part and that within the UVA spectrum distinct wavelengths exhibit different biological activities. There is no doubt that lower wavelengths between 320 and 340 nm differ from the long wave part (340–400 nm). Awareness of this fact is extremely important when comparing studies using different UVA sources.

There are a few indications that in contrast to previous expectations UVA has an effect on the release of cytokines. Fibroblasts seem to respond quite well to UVA1 with enhanced production of IL-1 and IL-6 [186]. Keratinocytes obviously release mediators with immunosuppressive properties upon very low doses of UVA light [10]. However, much more needs to be done to elucidate the relationship between UVA light and cytokine secretion within the skin.

Concluding remarks

According to current studies there is no doubt that many of the biological activities of UV light are due to the induced release of cytokines and neuroendocrine hormones by different cells. However, one has to be aware that there is strong evidence for a complicated network of interacting mediators [12, 48], thus it would be too simple to relate one activity of UV light to a particular mediator. Additional studies are necessary for the further understanding of physiologic and pathologic reactions within the skin. This will also increase the knowledge of the mechanisms of the biologic actions of UV light.

Acknowledgment

This work has been supported by grants from the Deutsche Forschungsgemeinschaft (443/1-1) and the Volkswagenwerke (AZ.: I/67-420).

References

1 Aberer, W., G. Stingl, L.A. Stingl-Gazze, K. Wolff: Langerhans cells as stimulator cells in the murine primary epidermal cell–lymphocyte reaction: alteration by UV-B irradiation. *J. Invest. Dermatol.* 79 (1982), 129–135.

2 Adolf, G.R., M. Grell, P. Scheurich: Tumor necrosis factor. In: *Epidermal Growth Factors and Cytokines*, (1993), p. 63. Eds.: T.A. Luger, T. Schwarz, Marcel Dekker Inc., New York.

3 Aebischer, I., M.R. Stämpfli, A. Zürcher, S. Miescher, A. Urwyler, B. Frey *et al.*: Neuropeptides are potent modulators of human *in vitro* immunoglobulin E synthesis. *Eur. J. Immunol.* 24 (1994), 1908.

4 Amornsiripanitch, S., J.J. Nordlund: Messenger RNA specific for proopiomelanocortin (POMC) identified in murine epidermal cells (Abstract). *J. Invest. Dermatol.* 92 (1989), 395.

5 Ansel, J.C., T.A. Luger, I. Green: The effect of *in vitro* and *in vivo* UV irradiation on the production of ETAF activity by human and murine keratinocytes. *J. Invest. Dermatol.* 81 (1983), 519–523.

6 Ansel, J.C., T.A. Luger, I. Green: Fever and increased serum IL-1 activity as a systemic manifestation of acute phototoxicity in New Zealand White rabbits. *J. Invest. Dermatol.* 89 (1987), 32–37.

7 Ansel, J.C., T.A. Luger, A. Kock, D. Hochstein, I. Green: The effect of *in vitro* UV irradiation on the production of IL 1 by murine macrophages and P388D1 cells. *J. Immunol.* 133 (1984), 1350–1355.

8 Ansel, J.C., T.A. Luger, D. Lowry, P. Perry, D.R. Roop, J.D. Mountz: The expression and modulation of IL-1 alpha in murine keratinocytes. *J. Immunol.* 140 (1988), 2274–2278.

9 Armstrong, C.A., N. Murray, M. Kennedy, S.V. Koppula, D. Tara, J.C. Ansel: Melanoma-derived interleukin 6 inhibits *in vivo* melanoma growth. *J. Invest. Dermatol.* 102 (1994), 278.

10 Aubin, F., M.L. Kripke, S.E. Ullrich: Activation of keratinocytes with psoralen plus UVA radiation induces the release of soluble factors that suppress delayed and contact hypersensitivity. *J. Invest. Dermatol.* 97 (1991), 995–1000.

11 Baadsgaard, O.: *In vivo* ultraviolet irradiation of human skin results in profound perturbation of the immune system. Relevance to ultraviolet-induced skin cancer. *Arch. Dermatol.* 127 (1991), 99–109.

12 Balkwill, F.R., F. Burke: The cytokine network. *Immunol. Today* 10 (1989), 299–304.

13 Bateman, A., A. Singh, T. Kral, S. Solomon: The immune–hypothalamic–pituitary–adrenal axis. *Endocr. Rev.* 10 (1989), 92–112.

14 Beutler, B., G.E. Grau: Tumor necrosis factor in the pathogenesis of infectious diseases. *Crit. Care Med.* 21 (1993), S243.

15 Beutler, B., A. Cerami: Cachectin (tumor necrosis factor), an endogenous mediator of shock and inflammatory response. In: *Immunophysiology, the Role of Cells and Cytokines in Immunity and Inflammation*, (1990), pp. 226–237. Eds.: J.J. Oppenheim, E.M. Shevach, Oxford University Press, New York.

16 Bickers, D.R., L.C. Harber, M. Kripke: Photoimmunology. In: *Photosensitivity Diseases, Principles of Diagnosis and Treatment*, (1989), pp. 46–57. Eds.: L.C. Harber, D.R. Bickers, B.C. Decker Inc., Toronto.

17 Bieber, T., M. Sticherling, T. Rupec, J.M. Schroder: Human epidermal Langerhans cells release the neutrophil activating peptide IL-8 (Abstract). *J. Invest. Dermatol.* 96 (1991), 1013.

18 Bigler, C.F., D.A. Norris, W.L. Weston, W.P. Arend: Interleukin-1 receptor antagonist production by human keratinocytes. *J. Invest. Dermatol.* 98 (1992), 38–44.

19 Birchall, N., C. Gamba, T.S. Kupper: Cutaneous UVB irradiation enhances recovery from bone marrow suppression (Abstract). *Clin. Res.* 36 (1988), 801.

20 Birchall, N., S.J. Orlow, T. Kupper, J. Pawelek: Interactions between ultraviolet light and interleukin-1 on MSH binding in both mouse melanoma and human squamous carcinoma cells. *Biochem. Biophys. Res. Commun.* 175 (1991), 839–845.

21 Birchall, N., L. Park, T. Kupper: IL-4 binds to specific cell surface receptors on human keratinocytes and induces colony stimulating factor release and HLA-DR expression (Abstract). *Clin. Res.* 36 (1988), 374.

22 Blanton, R.A., T.S. Kupper, J.K. McDougall, S. Dower: Regulation of interleukin 1 and its receptor in human keratinocytes. *Proc. Natl. Acad. Sci. USA* 86 (1989), 1273–1277.

23 Bolognia, J., M. Murray, J. Pawelek: UVB-induced melanogenesis may be mediated through the MSH-receptor system. *J. Invest. Dermatol.* 92 (1989), 651–656.

24 Bos, J., M.L. Kapsenberg: The skin immune system (SIS): its cellular constituents and their interactions. *Immunol. Today* 7 (1986), 235–240.

25 Brown, S.L., J.E. Blalock. Neuroendocrine immune

interactions. In: *Immunophysiology, the Role of Cells and Cytokines in Immunity and Inflammation*, (1990), pp. 306–319. Eds.: J.J. Oppenheim, E.M. Shevach, Oxford University Press, Oxford.
26 Burbach, J.P.H., V.M. Wiegant: Gene expression, biosynthesis and processing of proopiomelanocortin peptides and vasopressin. In: *Neuropeptides: Basics and Perspectives*, (1990), pp. 45–103. Ed.: D. de Wied, Elsevier, Amsterdam.
27 Burchill, S.A., J.M. Marks, A.J. Thody: Tyrosinase synthesis in different skin types and the effects of alpha-melanocyte-stimulating hormone and cyclic AMP. *J. Invest. Dermatol.* 95 (1990), 558–561.
28 Cannon, J.G., J.B. Tatro, S. Reichlin, C.A. Dinarello: Alpha melanocyte stimulating hormone inhibits immunostimulatory and inflammatory actions of interleukin 1. *J. Immunol.* 137 (1986), 2232–2236.
29 Chizzonite, R., T. Truitt, P.L. Kilian, A.S. Stern, P. Nunes, K.P. Parker, K.L. Kaffka, A.O. Chua, D.K. Lugg, U. Gubler: Two high affinity interleukin 1 receptors represent separate gene products. *Proc. Natl. Acad. Sci. USA* 86 (1989), 8029–8033.
30 Chodakewitz, J.A., J. Lacy, S.E. Edwards, N. Birchall, D.L. Coleman: Macrophage colony-stimulating factor production by murine and human keratinocytes. Enhancement by bacterial lipopolysaccharide. *J. Immunol.* 144 (1990), 2190–2196.
31 Christensen, O.B., M.B. Christensen, L.M. Wall: UVB serum factor: suppression of allergic contact dermatitis in guinea pigs. *Contact. Dermatitis.* 16 (1987), 208–211.
32 Cork, M.J., G.W. Duff: Interleukin one. In: *Epidermal Growth Factors and Cytokines*, (1993), p. 19. Eds.: T.A. Luger, T. Schwarz, Marcel Dekker Inc., New York.
33 Cosman, D.: Colony-stimulating factors *in vivo* and *in vitro* (News). *Immunol. Today* 9 (1988), 97–98.
34 Cross, M., M. Dexter: Growth factors in development, transformation and tumorigenesis. *Cell* 64 (1991), 271–280.
35 Dalloul, A.H., M. Arock, C. Fourcade, J.Y. Beranger, P. Jaffray, P. Debre, M.D. Mossalayi: Epidermal keratinocyte-derived basophil promoting activity. Role of interleukin-3 and soluble CD23. *J. Clin. Invest.* 90 (1992), 1242–1247.
36 Damm, D., G. Shipley, C. Hart, J.C. Ansel: The expression and modulation of PDGF in normal human keratinocytes. In: *Molecular and Cellular Biology of Cytokines*, (1990), pp. 111–115. Eds.: J.J. Oppenheim, M.C. Powanda, M.J. Kluger, C.A. Dinarello, Wiley-Liss, New York.
37 Dayer, J.M., B. Beutler, A. Cerami: Cachectin/tumor necrosis factor stimulates collagenase and prostaglandin E2 production by human synovial cells and dermal fibroblasts. *J. Exp. Med.* 162 (1985), 2163–2168.
38 Dinarello, C.A.: Interleukin-1 and its biologically related cytokines. *Adv. Immunol.* 44 (1989), 153–205.
39 Dinarello, C.A., R.C. Thompson: Blocking IL-1: interleukin 1 receptor antagonist *in vivo* and *in vitro*. *Immunol. Today* 12 (1991), 404–410.
40 Dower, S.K., E.E. Qwarnstrom, R.C. Page, R.A. Blanton, T.S. Kupper, E. Raines, R. Ross, J.E. Sims: Biology of the interleukin-1 receptor. *J. Invest. Dermatol.* 94 (1990), 68S–73S.
41 Durum, S.K., J.J. Oppenheim, R. Neta: Immunophysiologic role of interleukin-1. In: *Immunophysiology, the Role of Cells and Cytokines in Immunity and Inflammation*, (1990), pp. 210–225. Eds.: J.J. Oppenheim, E.M. Shevach, Oxford University Press, New York.
42 Eberle, A.N.: *The Melanotropins* (1988). Karger, Basel.
43 Eerkes, K.J., A. Takashima, P.R. Bergstresser, P.D. Cruz: UVB-irradiated keratinocytes (KL) secrete a factor(s) that suppresses the antigen presenting function of Langerhans cells (LC): differential effects on CD4+ Th1 vs Th2 proliferation (Abstract). *J. Invest. Dermatol.* 98 (1992), 557.
44 Eisenberg, S.P., R.J. Evans, W.P. Arend, E. Verderber, M.T. Brewer, C.H. Hannum, R.C. Thompson: Primary structure and functional expression from complementary DNA of a human interleukin-1 receptor antagonist. *Nature* 343 (1990), 341–346.
45 Ellem, K.A., M. Cullinan, K.C. Baumann, A. Dunstan: UVR induction of TGF alpha: a possible autocrine mechanism for the epidermal melanocytic response and for promotion of epidermal carcinogenesis. *Carcinogenesis* 9 (1988), 797–801.
46 Enk, A., S.I. Katz: Identification and induction of keratinocyte-derived IL-10. *J. Immunol.* 149 (1992), 92–95.
47 Enk, A.H., S.I. Katz: Early molecular events in the induction phase of contact sensitivity. *Proc. Natl. Acad. Sci. USA* 89 (1992), 1398–1402.
48 Feldmann, M.: Cytokine networks: do we understand them well enough to facilitate clinical benefits? *Eur. Cytokine. Netw.* 2 (1991), 5–9.
49 Fisher, G., J. Elder, J. Voorhees, J. Kawalki, C. Petersen, R. Derynck, L. Ellingsworth: Transforming growth factor β mRNA is expressed in normal and psoriatic epidermis and has similar biological effects on normal and psoriatic lesional cultured keratinocytes (Abstract). *Clin. Res.* 36 (1988), 645.
50 Forster, E., R. Kirnbauer, A. Urbanski, A. Kock, T.A. Luger, T. Schwarz: Human melanoma cells produce interleukin-8 which functions as an autocrine growth factor (Abstract). *Arch. Dermatol. Res.* 283 (1991), 26.
51 Gahring, L., M. Baltz, M.B. Pepys, R. Daynes: Effect of ultraviolet radiation on production of epidermal cell thymocyte-activating factor/interleukin 1 *in vivo* and *in vitro*. *Proc. Natl. Acad. Sci. USA* 81 (1984), 1198–1202.

52 Gallo, R.L., R. Staszewski, D.N. Sauder, T.L. Knisely, R.D. Granstein: Regulation of GM-CSF and IL-3 production from the murine keratinocyte cell line PAM 212 following exposure to ultraviolet radiation. *J. Invest. Dermatol.* 97 (1991), 203–209.

53 Garmyn, M., M. Yaar, N. Boileau, C. Backendorf, B. Gilchrest: Effect of aging and habitual sun exposure on the genetic response of cultured human keratinocytes to solar-simulated irradiation. *J. Invest. Dermatol.* 99 (1992), 743–748.

54 Geissler, E.N., M. Liao, J.D. Brook, F.H. Martin, K.M. Zsebo, D.E. Housman, S.J. Galli: Stem cell factor (SCF), a novel hematopoietic growth factor and ligand for c-kit tyrosine kinase receptor, maps on human chromosome 12 between 12q14.3 and 12qter. *Somat. Cell Mol. Genet.* 17 (1991), 207–214.

55 Giebel, L.B., R.A. Spritz: Mutation of the KIT (mast/stem cell growth factor receptor) protooncogene in human piebaldism. *Proc. Natl. Acad. Sci. USA* 88 (1991), 8696–8699.

56 Giroir, B.P., J.H. Johnson, T. Brown, G.L. Allen, B. Beutler: The tissue distribution of tumor necrosis factor biosynthesis during endotoxemia. *J. Clin. Invest.* 90 (1992), 693–698.

57 Gottlieb, A.B., C.K. Chang, D.N. Posnett, B. Fanelli, J.P. Tam: Detection of transforming growth factor alpha in normal, malignant, and hyperproliferative human keratinocytes. *J. Exp. Med.* 167 (1988), 670–675.

58 Granstein, R.D., R. Margolis, S.R. Mizel, D.N. Sauder: *In vivo* inflammatory activity of epidermal cell-derived thymocyte activating factor and recombinant interleukin-1 in the mouse. *J. Clin. Invest.* 77 (1986), 1020–1027.

59 Granstein, R.D., D.N. Sauder: Whole-body exposure to ultraviolet radiation results in increased serum interleukin-1 activity in humans. *Lymphokine Res.* 6 (1987), 187–193.

60 Greaves, M.W.: Ultraviolet erythema: causes and consequences. *Curr. Probl. Dermatol.* 15 (1986), 18–24.

61 Grewe, M., U. Trefzer, M. Brockhaus, A. Kapp, E. Schöpf, J. Krutmann: Analysis of human keratinocyte prostaglandin synthesis: endogenous TNFα is involved in ultraviolet radiation-induced PGE2 synthesis (Abstract). *Arch. Dermatol. Res.* 94 (1990), 43.

62 Gupta, A.K., T.F. Anderson: Psoralen photochemotherapy. *J. Am. Acad. Dermatol.* 17 (1987), 703–734.

63 Halaban, R., R. Langdon, N. Birchall, C. Cuono, A. Baird, G. Scott, G. Moellmann, J. McGuire: Paracrine stimulation of melanocytes by keratinocytes through basic fibroblast growth factor. *Ann. NY Acad. Sci.* 548 (1988), 180–190.

64 Holzmann, A., P. Altmeyer, L. Stohr, G.N. Chilf: Beeinflussung des Alpha-MSH durch UVA-Bestrahlung der Haut-ein Funktionstest. *Hautarzt.* 34 (1983), 294–297.

65 Jessup, J.M., N. Hanna, E. Palaszynski, M.L. Kripke: Mechanisms of depressed reactivity to dinitrochlorobenzene and ultraviolet-induced tumors during ultraviolet carcinogenesis in BALB/c mice. *Cell Immunol.* 38 (1978), 105–115.

66 Johansson, O., A. Ljungberg, S.W. Han, A. Vaalasti: Evidence for gamma melanocyte stimulating hormone containing nerves and neutrophilic granulocytes in the human skin by immunofluorescence. *J. Invest. Dermatol.* 96 (1991), 852–856.

67 Khansari, A., A.J. Murgo, R.E. Faith: Effects of stress on the immune system. *Immunol. Today* 11 (1990), 170–175.

68 Kim, T.Y., M.L. Kripke, S.E. Ullrich: Immunosuppression by factors released from UV-irradiated epidermal cells: selective effects on the generation of contact and delayed hypersensitivity after exposure to UVA or UVB radiation. *J. Invest. Dermatol.* 94 (1990), 26–32.

69 Kimber, I., M. Cumberbatch: Stimulation of Langerhans cell migration by tumor necrosis factor alpha (TNF alpha). *J. Invest. Dermatol.* 99 (1992), 48S–50S.

70 Kirnbauer, R., A. Kock, P. Neuner, E. Forster, J. Krutmann, A. Urbanski, E. Schauer, J.C. Ansel, T. Schwarz, T.A. Luger: Regulation of epidermal cell interleukin-6 production by UV light and corticosteroids. *J. Invest. Dermatol.* 96 (1991), 484–489.

71 Kirnbauer, R., A. Kock, T. Schwarz, A. Urbanski, J. Krutmann, W. Borth, D. Damm, G. Shipley, J.C. Ansel, T.A. Luger: IFN-beta 2, B cell differentiation factor 2, or hybridoma growth factor (IL-6) is expressed and released by human epidermal cells and epidermoid carcinoma cell lines. *J. Immunol.* 142 (1989), 1922–1928.

72 Kligman, L.H.: The ultraviolet-irradiated hairless mouse: a model for photoaging. *J. Am. Acad. Dermatol.* 21 (1989), 623–631.

73 Koch, F., C. Heufler, E. Kampgen, D. Schneeweiss, G. Bock, G. Schuler: Tumor necrosis factor alpha maintains the viability of murine epidermal Langerhans cells in culture, but in contrast to granulocyte/macrophage colony-stimulating factor, without inducing their functional maturation. *J. Exp. Med.* 171 (1990), 159–171.

74 Köck, A., E. Schauer, T. Schwarz, T.A. Luger: Human keratinocytes synthesize and release neuropeptides such as α-MSH and ACTH. In: *Cellular and Cytokine Networks in Tissue Immunity,* (1991), pp. 105–110. Eds.: M. Meltzer, A. Mantovani, John Wiley & Sons, Inc., New York.

75 Köck, A., T. Schwarz, R. Kirnbauer, A. Urbanski, P. Perry, J.C. Ansel, T.A. Luger: Human keratino-

cytes are a source for tumor necrosis factor alpha: evidence for synthesis and release upon stimulation with endotoxin or ultraviolet light. *J. Exp. Med.* 172 (1990), 1609–1614.

76 Köck, A., T. Schwarz, M. Micksche, T.A. Luger: Cytokines and human malignant melanoma. Immuno- and growth regulatory peptides in melanoma biology. In: *Cancer Treatment and Research*, vol. 5, (1991), p. 41. Ed.: L. Nathanson, Kluwer Academic, Norwell, MA.

77 Köck, A., T. Schwarz, A. Urbanski, Z. Peng, M. Vetterlein, M. Micksche *et al.*: Expression and release of interleukin-1 by different human melanoma cell lines. *J. Natl Cancer Inst.* 81 (1989), 36.

78 Koh, H.K., B.E. Kligler, R.A. Lew: Sunlight and cutaneous malignant melanoma: evidence for and against causation. *Photochem. Photobiol.* 51 (1990), 765–779.

79 Kolde, G., K. Schulze-Osthoff, H. Meyer, J. Knop: Immunohistological and immunoelectron microscopic identification of TNF alpha in normal human and murine epidermis. *Arch. Dermatol. Res.* 284 (1992), 154–158.

80 Konnikov, N., S.H. Pincus, C.A. Dinarello: Elevated plasma interleukin-1 levels in humans following ultraviolet light therapy for psoriasis. *J. Invest. Dermatol.* 92 (1989), 235–239.

81 Krane, S.M., J.M. Dayer, L.S. Simon, M.S. Byrne: Mononuclear cell-conditioned medium containing mononuclear cell factor (MCF), homologous with interleukin 1, stimulates collagen and fibronectin synthesis by adherent rheumatoid synovial cells: effects of prostaglandin E2 and indomethacin. *Coll. Relat. Res.* 5 (1985), 99–117.

82 Kripke, M.L.: Immunology and photocarcinogenesis. New light on an old problem. *J. Am. Acad. Dermatol.* 14 (1986), 149–155.

83 Kripke, M.L.: Photoimmunology. *Photochem. Photobiol.* 52 (1990), 919–924.

84 Kripke, M.L., W.L. Morison, J.A. Parrish: Systemic suppression of contact hypersensitivity in mice by psoralen plus UVA radiation (PUVA). *J. Invest. Dermatol.* 81 (1983), 87–92.

85 Kupper, T.: The role of epidermal cytokines. In: *Immunophysiology, the Role of Cells and Cytokines in Immunity and Inflammation*, (1990), pp. 285–305. Eds.: J.J. Oppenheim, E.M. Shevach, Oxford University Press, New York.

86 Kupper, T.S., D.W. Ballard, A.O. Chua, J.S. McGuire, P.M. Flood, M.C. Horowitz, R. Langdon, L. Lightfoot, U. Gubler: Human keratinocytes contain mRNA indistinguishable from monocyte interleukin 1 alpha and beta mRNA. Keratinocyte epidermal cell-derived thymocyte-activating factor is identical to interleukin 1. *J. Exp. Med.* 164 (1986), 2095–2100.

87 Kupper, T.S., A.O. Chua, P. Flood, J. McGuire, U. Gubler: Interleukin 1 gene expression in cultured human keratinocytes is augmented by ultraviolet irradiation. *J. Clin. Invest.* 80 (1987), 430–436.

88 Kupper, T.S., J. McGuire: Hydrocortisone reduces both constitutive and UV-elicited release of epidermal thymocyte activating factor (ETAF) by cultured keratinocytes. *J. Invest. Dermatol.* 87 (1986), 570–573.

89 Kurimoto, I., J.W. Streilein: Deleterious effects of cis-urocanic acid and UVB radiation on Langerhans cells and on induction of contact hypersensitivity are mediated by tumor necrosis factor alpha. *J. Invest. Dermatol.* 99 (1992), 69S–70S.

90 Larsen, C.G., A.O. Anderson, J.J. Oppenheim, K. Matsushima: Production of interleukin-8 by human dermal fibroblasts and keratinocytes in response to interleukin-1 or tumour necrosis factor. *Immunology* 68 (1989), 31–36.

91 Leonard, E.J., T. Yoshimura: Human monocyte chemoattractant protein-1 (MCP-1). *Immunol. Today* 11 (1990), 97–101.

92 Lerner, A.B., J.S. McGuire: The effect of alpha- and beta melanocyte stimulating hormone on the skin colour of man. *Nature* 189 (1961), 176–179.

93 Levine, N., S.N. Sheftel, T. Eytan, R.T. Dorr, M.E. Hadley, J.C. Weinrach, G.A. Ertl, K. Toth, D.L. McGee, V.J. Hruby: Induction of skin tanning by subcutaneous administration of a potent synthetic melanotropin. *JAMA* 266 (1991), 2730–2736.

94 Luger, T.A., A. Kock, R. Kirnbauer, T. Schwarz, J.C. Ansel: Keratinocyte-derived interleukin 3. *Ann. NY Acad. Sci.* 548 (1988), 253–261.

95 Luger, T.A., E. Schauer, F. Trautinger, J. Krutmann, J. Ansel, A. Schwarz, T. Schwarz: Production of immunosuppressing melanotropins by keratinocytes. *Ann. NY Acad. Sci.* 680 (1993), 567–570.

96 Luger, T.A., T. Schwarz: Effect of UV light on cytokine and neuroendocrine hormone production by epidermal cells. In: *Biologic Effects of Light*, (1993), p. 594. Eds.: E.G. Jung, M.F. Holick, Walter de Gruyter, Berlin.

97 Luger, T.A., B.M. Stadler, S.I. Katz, J.J. Oppenheim: Epidermal cell (keratinocyte)-derived thymocyte-activating factor (ETAF). *J. Immunol.* 127 (1981), 1493–1498.

98 Matsue, J., P.D. Cruz Jr, P.R. Bergstresser, A. Takashima: Cytokine expression by epidermal subpopulations. *J. Invest. Dermatol.* 99 (1992), 42S–45S.

99 Matsue, H., P.D. Cruz Jr, P.R. Bergstresser, A. Takashima: Langerhans cells are the major source of mRNA for IL-1 beta and MIP-1 alpha among unstimulated mouse epidermal cells. *J. Invest. Dermatol.* 99 (1992), 537–541.

100 Matsushima, K., J.J. Oppenheim: Interleukin 8 and MCAF: novel inflammatory cytokines inducible by IL 1 and TNF. *Cytokine* 1 (1989), 2–13.

101 McKenzie, R.C., C. Feliciani, T.J. Venner, D.N.

Sauder: Induction of melanoma growth factors in keratinocytes by UVB and interleukin-1 (Abstract). *J. Invest. Dermatol.* 98 (1992), 565.

102 McKenzie, R.C., R.G. Lea, D.N. Sauder: Ultraviolet radiation induces expression of interleukin 8 and transforming growth factor beta in human keratinocytes. *J. Invest. Dermatol.* 96 (1991), 610.

103 McMahan, C.J., J.L. Slack, B. Mosley, D. Cosman, J.D. Lupton, L.L. Brunton, C.E. Grubin, K. Huebner, C.M. Croce, L.A. Cannizzano, D. Benjamin, S.K. Dower, M. Spriggs, J.E. Sims: A novel IL-1 receptor cloned from B cells by mammalian expression, is expressed by many cell types. *EMBO J.* 10 (1991), 2821–2832.

104 Metcalf, D.: The colony stimulating factors. Discovery, development, and clinical applications. *Cancer* 65 (1990), 2185–2195.

105 Miyajima, A., T. Kitamura, N. Harada, T. Yokota, K. Arai: Cytokine receptors and signal transduction. *Annu. Rev. Immunol.* 10 (1992), 295–331.

106 Mizutani, H., R. Black, T. Kupper: Post-translation modification of IL-1 differ markedly in keratinocytes and monocytes: implications for epidermal biology (Abstract). *J. Invest. Dermatol.* 92 (1989), 483.

107 Moore, K.W., P. Vieira, D.F. Fiorentino, M.L. Trounstine, T.A. Khan, T.R. Mosmann: Homology of cytokine synthesis inhibitory factor (IL-10) to the Epstein-Barr virus gene BCRFI. *Science* 248 (1990), 1230–1234.

108 Moore, M.A.S.: Interleukin-3, an overview. In: *Lymphokines*, vol. 15, (1988), pp. 219–280. Ed.: J.W. Schrader, Academic Press, San Diego.

109 Morhenn, V.B.: The physiology of scratching: involvement of proopiomelanocortin gene-coded proteins in Langerhans cells. *Prog. Neuro. Endo. Immun.* 4 (1991), 265–267.

110 Mosmann, T.R., K.W. Moore: The role of IL-10 in crossregulation of TH1 and TH2 responses. *Immunol. Today* 12 (1991), A49–A53.

111 Mountjoy, K.G., L.S. Robbins, M.T. Mortrud, R.D. Cone: The cloning of a family of genes that encode the melanocortin receptors. *Science* 257 (1993), 1248–1251.

112 Murphy, G.M., P.M. Dowd, B.N. Hudspith, J. Brostoff, M.W. Greaves: Local increase in interleukin-1-like activity following UVB irradiation of human skin *in vivo*. *Photodermatol.* 6 (1989), 268–274.

113 Murphy, G.M., D.G. Quinn, R.D. Camp, J.I. Hawk, M.W. Greaves: *In vivo* studies of the action spectrum and time course for release of transforming growth-factor-α ultraviolet irradiation in man. *Br. J. Dermatol.* 125 (1991), 566–568.

114 Neote, K., D. DiGregorio, J.Y. Mak, R. Horuk, T.J. Schall: Molecular cloning, functional expression, and signaling characteristics of a C-C chemokine receptor. *Cell* 72 (1993), 415–425.

115 Neuner, P., B. Charvat, R. Knobler, R. Kirnbauer, A. Schwarz, T.A. Luger, T. Schwarz: Cytokine release by peripheral blood mononuclear cells is affected by 8-methoxypsoralen plus UV-A. *Photochem. Photobiol.* 59 (1994), 182.

116 Neuner, P., A. Möller, R. Knobler, C. Grünwald, A. Urbanski, R. Kirnbauer, T.A. Luger, T. Schwarz: Interleukin-6 (IL-6) release by PBMC and epidermal cells is affected by 8-methoxypsoralen (8-MOP) plus UVA (Abstract). *J. Invest. Dermatol.* 96 (1991), 584.

117 Neuner, P., A. Urbanski, F. Trautinger, A. Moller, R. Kirnbauer, A. Kapp, E. Schopf, T. Schwarz, T.A. Luger: Increased IL-6 production by monocytes and keratinocytes in patients with psoriasis. *J. Invest. Dermatol.* 97 (1991), 27–33.

118 Nicola, N.A., D. Metcalf: Subunit promiscuity among hemopoietic growth factor receptors. *Cell* 67 (1991), 1–4.

119 Noonan, F.P., E.C. DeFabo: Immunosuppression by ultraviolet B radiation: initiation by urocanic acid. *Immunol. Today* 13 (1992), 250–254.

120 Nozaki, S., J.S. Abrams, M.K. Pearce, D.N. Sauder: Augmentation of granulocyte/macrophage colony-stimulating factor expression by ultraviolet irradiation is mediated by interleukin 1 in Pam 212 keratinocytes. *J. Invest. Dermatol.* 97 (1991), 10–14.

121 Nozaki, S., D.N. Sauder: Ultraviolet irradiation downregulates IL-1 receptor in Pam 212 keratinocytes (Abstract). *J. Invest. Dermatol.* 96 (1991), 609.

122 Oxholm, A., P. Oxholm, B. Staberg, K. Bendtzen: Immunohistological detection of interleukin 1-like molecules and tumour necrosis factor in human epidermis before and after UVB-irradiation *in vivo*. *Br. J. Dermatol.* 118 (1988), 369–376.

123 Pawelek, J.M.: Is human melanogenesis stimulated by cyclic AMP? *J. Invest. Dermatol.* 94 (1990), 499–500.

124 Pears, R.S., R.T. Jung, W. Bartlett, M.C.K. Browning, K. Kenicer, A.J. Thody: A case of skin hyperpigmentation due to alpha-MSH hypersecretion. *Br. J. Dermatol.* 126 (1992), 286–289.

125 Pentland, A.P., M.G. Mahoney: Keratinocyte prostaglandin synthesis is enhanced by IL-1, *J. Invest. Dermatol.* 94 (1990), 43–46.

126 Petersen, M.J., C. Hansen, S. Craig: Ultraviolet A irradiation stimulates collagenase production in cultured human fibroblasts. *J. Invest. Dermatol.* 99 (1992), 440–444.

127 Piguet, P.F., G.E. Grau, B. Allet, P. Vassalli: Tumor necrosis factor/cachectin is an effector of skin and gut lesions of the acute phase of graft-vs.-host disease. *J. Exp. Med.* 166 (1987), 1280–1289.

128 Piguet, P.F., G.E. Grau, C. Hauser, P. Vassalli: Tumor necrosis factor is a critical mediator in hapten induced irritant and contact hypersensitivity reactions. *J. Exp. Med.* 173 (1991), 673–679.

129 Rheins, L.A., A.L. Cotleur, R.S. Kleier, W.B. Hoppenjans, D.N. Sauder, J.J. Nordlund: Alpha-melanocyte stimulating hormone modulates contact hypersensitivity responsiveness in C57/BL6 mice. *J. Invest. Dermatol.* 93 (1989), 511–517.

130 Richmond, A., E. Balentien, H.G. Thomas, G. Flaggs, D.E. Barton, J. Spiess, R. Bordoni, U. Francke, R. Derynck: Molecular characterization and chromosomal mapping of melanoma growth stimulatory activity, a growth factor structurally related to beta-thromboglobulin. *EMBO J.* 7 (1988), 2025–2033.

131 Rivas, J.M., S.E. Ullrich: Sytemic suppression of delayed-type hypersensitivity by supernatants from UV-irradiated keratinocytes. An essential role for keratinocyte-derived IL-10. *J. Immunol.* 149 (1992), 3865–3871.

132 Rivas, J.M., S.E. Ullrich: Keratinocyte derived IL-10 (Abstract). *J. Invest. Dermatol.* 98 (1992), 578.

133 Robertson, B., K. Dostal, R.A. Daynes: Neuropeptide regulation of inflammatory and immunologic responses. The capacity of alpha-melanocyte-stimulating hormone to inhibit tumor necrosis factor and IL-1-inducible biologic responses. *J. Immunol.* 140 (1988), 4300–4307.

134 Ross, R.: Platelet-derived growth factor. *Lancet* i (1989), 1179–1182.

135 Rothe, M., V. Falanga: Growth factors. Their biology and promise in dermatologic diseases and tissue repair. *Arch. Dermatol.* 125 (1989), 1390–1398.

136 Sauder, D.N., C.S. Carter, S.I. Katz, J.J. Oppenheim: Epidermal cell production of thymocyte activating factor (ETAF). *J. Invest. Dermatol.* 79 (1982), 34–39.

137 Scarborough, D.E.: Cytokine modulation of pituitary hormone secretion. In: *Neuropeptides and Immunopeptides: Messengers in a Neuroimmune Axis*, (1990), pp. 169–187. Eds.: M.S. O'Dorisio, A. Panerai, Ann. NY Acad. Sci., New York.

138 Scharffetter, K., M. Wlaschek, A. Hogg, K. Bolsen, A. Schothorst, G. Goerz, T. Krieg. G. Plewig: UVA irradiation induces collagenase in human dermal fibroblasts *in vitro* and *in vivo*. *Arch. Dermatol. Res.* 283 (1991), 506–511.

139 Schauer, E., F. Trautinger, A. Köck, R.S. Bhardwaj, M. Simon, A. Schwarz *et al.*: Proopiomelanocortin derived peptides are synthesized and released by human keratinocytes. *J. Clin. Invest.* 93 (1994), 2258.

140 Schrader, J.W.: Colony stimulating factors and the skin. In: *Epidermal Growth Factors and Cytokines*, (1993), p. 147. Eds.: T.A. Luger, T. Schwarz. Marcel Dekker Inc., New York.

141 Schreiber, S., O. Kilgus, E. Payer, R. Kutil, A. Elbe, C. Mueller, G. Stingl: Cytokine pattern of Langerhans cells isolated from murine epidermal cell cultures. *J. Immunol.* 149 (1992), 3525–3534.

142 Schreiber, S., O. Kilgus, E. Payer, R. Kutil, A. Elbe, C. Mueller, G. Stingl: Cytokine pattern of Langerhans cells isolated from murine epidermal cell cultures. *J. Immunol.* 149 (1992), 3525.

143 Schroder, J.M.: Chemotactic cytokines in the epidermis. *Exp. Derm.* 1 (1992), 12–19.

144 Schroder, J.M., H. Gregory, J. Young, E. Christophers: Neutrophil-activating proteins in psoriasis. *J. Invest. Dermatol.* 98 (1992), 241–247.

145 Schwarz, T., T.A. Luger: Effect of UV irradiation on epidermal cell cytokine production. *J. Photochem. Photobiol. B.* 4 (1989), 1–13.

146 Schwarz, T., T.A. Luger: Pharmacology of cytokines in the skin. In: *Pharmacology of the Skin*, (1992), pp. 283–313. Ed.: H. Muhktar, CRC Press, Boca Raton, FL.

147 Schwarz, T., A. Urbanska, F. Gschnait, T.A. Luger: UV-irradiated epidermal cells produce a specific inhibitor of interleukin 1 activity. *J. Immunol.* 138 (1987), 1457–1463.

148 Schwarz, T., A. Urbanski, F. Gschnait, T.A. Luger: Inhibition of the induction of contact hypersensitivity by a UV-mediated epidermal cytokine. *J. Invest. Dermatol.* 87 (1986), 289–291.

149 Schwarz, T., A. Urbanski, R. Kirnbauer, A. Kock, F. Gschnait, T.A. Luger: Detection of a specific inhibitor of interleukin 1 in sera of UVB-treated mice. *J. Invest. Dermatol.* 91 (1988), 536–540.

150 Schwarz, A., C. Krone, F. Trautinger, Y. Aragne, P. Neuner, T.A. Luger, T. Schwarz: Pentoxifylline suppresses irritant and contact hypersensitivity reactions. *J. Invest. Dermatol.* 101 (1993), 549.

151 Sehgal, P.B.: Interleukin-6: molecular pathophysiology. *J. Invest. Dermatol.* 94 (1990), 2S–6S.

152 Shimizu, H., K. Mitomo, T. Watanabe, S. Okamoto, K.R. Yamamoto: Involvement of a NF-kB-like transcription factor in the activation of the interleukin-6 gene by inflammatory cytokines. *Mol. Cell Biol.* 10 (1990), 561–568.

153 Simon, M.M., Y. Aragane, A. Schwarz, T.A. Luger, T. Schwarz: UVB light induces nuclear factor kappa B (NF kappa B) activity independently from chromosomal DNA damage in cell-free cytosolic extracts. *J. Invest. Dermatol.* 102 (1994), 422.

154 Sims, J.E., C.J. March, D. Cosman, M.B. Widmer, H.R. MacDonald, C.J. McMahan, C.E. Grubin, J.M. Wignall, J.L. Jackson, S.M. Call *et al.*: cDNA expression cloning of the IL-1 receptor, a member of the immunoglobulin superfamily. *Science* 241 (1988), 585–589.

155 Slominski, A., R. Paus, J. Wortsman: Can some melanotropins modulate keratinocyte proliferation? *J. Invest. Dermatol.* 97 (1991), 747.

156 Slominski, A., R. Paus, J. Wortsman: On the potential role of proopiomelanocortin in skin physiology and pathology. *Mol. Cell Endocrinol.* 93 (1993), 1–6.

157 Snyder, D.S., W.H. Eaglstein: Topical indomethacin

and sunburn. *Br. J. Dermatol.* 90 (1974), 91–93.

158 Solari, R., N. Smithers, S. Grenfell, K. Page, B. Champion. Differential expression of IL-1 receptor forms. In: *Molecular and Cellular Biology of Cytokines,* (1990), pp. 117–124. Eds.: J.J. Oppenheim, M.C. Powanda, M.J. Kluger, E.A. Dinarello, John Wiley-Liss, New York.

159 Spiro, J., S. Parker, I. Oliver, C. Fraser, J.M. Marks, A.J. Thody: Effect of PUVA on plasma and skin immunoreactive alpha-melanocyte stimulating hormone concentrations. *Br. J. Dermatol.* 117 (1987), 703–707.

160 Sterenborg, H.J., J.C. van der Leun: Tumorigenesis by a long wavelength UV-A source. *Photochem. Photobiol.* 51 (1990), 325–330.

161 Stingl, G., K.C. Gunter, E. Tschachler, H. Yamada, R.I. Lechler, W.M. Yokoyama, G. Steiner, R.N. Germain, E.M. Shevach: Thy-1+ dendritic epidermal cells belong to the T-cell lineage. *Proc. Natl. Acad. Sci. USA* 84 (1987), 2430–2434.

162 Stingl, G., K. Tamaki, S.I. Katz: Origin and function of epidermal Langerhans cells. *Immunol. Rev.* 53 (1980), 149–174.

163 Streilein, J.W., P.R. Bergstresser: Genetic basis of ultraviolet-B effects on contact hypersensitivity. *Immunogenetics* 27 (1988), 252–258.

164 Streilein, J.W., I. Kurimoto, M. Mammolenti: IL-3 production as an *in vitro* marker of the genetically determined traits of UVB susceptibility and UVB resistance. *J. Invest. Dermatol.* 99 (1992), 74S–76S.

165 Strickland, P.T.: Photocarcinogenesis by near-ultraviolet (UVA) radiation in Sencar mice. *J. Invest. Dermatol.* 87 (1986), 272–275.

166 Strieter, R.M., D.G. Remick, P.A. Ward, R.N. Spengler, J.P. Lynch, J. Larrick, S.L. Kunkel: Cellular and molecular regulation of tumor necrosis factor-alpha production by pentoxifylline. *Biochem. Biophys. Res. Commun.* 155 (1988), 1230–1236.

167 Swartz, R.P.: Role of UVB-induced serum factor(s) in suppression of contact hypersensitivity in mice. *J. Invest. Dermatol.* 83 (1984), 305–307.

168 Symons, J.A., J.A. Eastgate, G.W. Duff: Purification and characterization of a novel soluble receptor for interleukin-1. *J. Exp. Med.* 174 (1991), 1251–1254.

169 Tavernier, J., R. Devos, S. Cornelis, T. Tuypens, J. Van der Heyden, W. Fiers, G. Plaetinck: A human high affinity interleukin-5 receptor (IL5R) is composed of an IL5-specific alpha chain and a beta chain shared with the receptor for GM-CSF. *Cell* 66 (1991), 1175–1184.

170 Teschenmacher, H., G. Koch, H. Scheffler, A. Hildebrand, V. Brantl: Opioid peptides: immunological significance? In: *Neuropeptides and Immunopeptides: Messengers in a Neuroimmune Axis,* (1990), pp. 66–77. Eds.: M.S. O'Dorisio, A. Panerai. Ann. NY Acad. Sci., New York.

171 Thody, A.J., K. Ridley, R.J. Penny, R. Chalmers, C. Fisher, S. Shuster: MSH peptides are present in mammalian skin. *Peptides* 4 (1983), 813–816.

172 Thomas, K.A.: Fibroblast growth factors. *FASEB J.* 1 (1987), 434–440.

173 Toews, G.B., P.R. Bergstresser, J.W. Streilein: Epidermal Langerhans cell density determines whether contact hypersensitivity or unresponsiveness follows skin painting with DNFB. *J. Immunol.* 124 (1980), 445–450.

174 Tokura, Y., J. Yagi, R.L. Edelson, F.P. Gasparro: Inhibitory effect of 8-methoxypsoralen plus ultraviolet-A on interleukin-1 production by murine keratinocytes. *Photochem. Photobiol.* 53 (1991), 517–523.

175 Trefzer, U., M. Brockhaus, H. Lötscher, F. Parlow, A. Kapp, E. Schopf, J. Krutmann: 55-kd tumor necrosis factor receptor is expressed by human keratinocytes and plays a pivotal role in regulation of human keratinocyte ICAM-1 expression. *J. Invest. Dermatol.* 97 (1991), 911–916.

176 Trefzer, U., M. Brockhaus, H. Lötscher, F. Parlow, A. Budnik, M. Grewe, H. Christoph, E. Schöpf, T.A. Luger, J. Krutmann: The 55-kd tumor necrosis factor receptor on human keratinocytes is regulated by tumor necrosis factor alpha and by ultraviolet B radiation. *J. Clin. Invest.* 92 (1993), 462–470.

177 Tron, V.A., M.D. Coughlin, D.E. Jang, J. Stanisz, D.N. Sauder: Expression and modulation of nerve growth factor in murine keratinocytes (PAM 212). *J. Clin. Invest.* 85 (1990), 1085–1089.

178 Tsuji, K., K.M. Zsebo, M. Ogawa: Murine mast cell colony formation supported by IL-3, IL-4, and recombinant rat stem cell factor, ligand for c-kit. *J. Cell Physiol.* 148 (1991), 362–369.

179 Tsujimura, T., S. Hirota, S. Nomura, Y. Niwa, M. Yamazaki, T. Tono, E. Morii, H.M. Kim, K. Kondo, Y. Nishimune *et al.*: Characterization of Ws mutant allele of rats: a 12-base deletion in tyrosine kinase domain of c-kit gene. *Blood* 78 (1991), 1942–1946.

180 Ullrich, S.E., B.W. McIntyre, J.M. Rivas: Suppression of the immune response to alloantigen by factors released from ultraviolet-irradiated keratinocytes. *J. Immunol.* 145 (1990), 489–498.

181 Urbanski, A., T. Schwarz, P. Neuner, J. Krutmann, R. Kirnbauer, A. Kock, T.A. Luger: Ultraviolet light induces increased circulating interleukin-6 in humans. *J. Invest. Dermatol.* 94 (1990), 808–811.

182 Urbanski, A., F. Trautinger, B. Brückler, P. Neuner, T.A. Luger, T. Schwarz: Injection of TNFα antiserum reduces sunburn cell formation in murine skin (Abstract). *Arch. Dermatol. Res.* 284 (1992), 13.

183 Wahl, S.M., N. McCartney Francis, S.E. Mergenhagen: Inflammatory and immunomodulatory roles of TGF-beta. *Immunol. Today* 10 (1989), 258–261.

184 Wang, C.Q., M.L. Kripke, I.J. Fidler: Ultraviolet-B-irradiated mouse epidermis releases mediators that stimulate melanoma cells. *Photodermatol. Photoimmunol. Photomed.* 7 (1990), 128–135.
185 Williams, D.E., J. Eisenman, A. Baird, C. Rauch, K. Van Ness, C.J. March, L.S. Park, U. Martin, D.Y. Mochizuki, H.S. Boswell *et al.*: Identification of a ligand for the c-kit proto-oncogene. *Cell* 63 (1990), 167–174.
186 Wlaschek, M., K. Bolsen, G. Herrmann, A. Schwarz, F. Wilmroth, P.C. Heinrich *et al.*: UVA-induced autocrine stimulation of fibroblast derived collagenase by IL-6: a possible mechanism in dermal photodamage. *J. Invest. Dermatol.* 101 (1993), 164.
187 Wong, G.G., S.C. Clark: Multiple actions of interleukin 6 within a cytokine network. *Immunol. Today* 9 (1988), 137–139.
188 Yates, R.A., L.B. Nanney, R.E. Gates, L.E. King Jr: Epidermal growth factor and related growth factors. *Int. J. Dermatol.* 30 (1991), 687–694.
189 Yoshikawa, T., J.W. Streilein: Genetic basis of the effects of ultraviolet light B on cutaneous immunity. Evidence that polymorphism at the Tnfa and Lps loci governs susceptibility. *Immunogenetics* 32 (1990), 398–405.
190 Yoshikawa, T., J.W. Streilein: TNA-alpha released by UVB-treated keratinocytes impairs induction of contact hypersensitivity (Abstract). *J. Invest. Dermatol.* 94 (1990), 593.

6: Adhesion molecules as molecular targets in ultraviolet B radiation (290–315 nm)-induced immunomodulation: UVB radiation effects on the expression of intercellular adhesion molecule-1 in human cells

Uwe Trefzer & Jean Krutmann

Introduction

The capacity of different cell types to physically interact with each other is an indispensable prerequisite for the generation of inflammatory/immunological reactions, and thereby of central importance to the pathogenesis of a large variety of diseases. In recent years, substantial progress has been made in the attempt to identify and characterize molecules on the cell surface, which function to mediate intercellular adhesion processes. These so-called adhesion molecules include members of several superfamilies, e.g. integrins and immunoglobulins [81]. With regard to adhesion processes in murine and human skin, binding of the immunoglobulin intercellular adhesion molecule-1 (ICAM-1; CD54) to the integrin lymphocyte function-associated antigen 1 (LFA-1; CD11a/CD18) has been shown to represent the major pathway through which epidermal keratinocytes and Langerhans cells are capable of physically interacting with leukocytes [21, 42]. It should be noted that, in addition to ICAM-1, epidermal cells were reported to express several other adhesion molecules including ICAM-3 [1, 24, 25], LFA-3 [42], B7 [86] and a B7-like molecule [3, 23, 61], but the functional relevance of these cell surface receptors for intraepidermal adhesion processes remains to be elucidated.

Over the past decade, great progress has been made in understanding the mechanisms which account for the immunomodulatory effects associated with ultraviolet B (UVB) irradiation. From these studies it appears that epidermal keratinocytes and Langerhans cells represent crucial cellular targets in UVBR-induced immunomodulation. Keratinocyte and Langerhans cell immune function is affected by UVBR in several ways. Of particular importance is the capacity of UVBR to induce the intraepidermal production of soluble mediators such as cytokines, prostanoids, and neuropeptides [28, 40, 71, 88]. There is, however, increasing evidence that UVBR may also affect the capacity of epidermal cells to directly interact with leukocytes by altering the expression of cell surface receptors, in particular of adhesion molecules. The most detailed information on how adhesion molecule expression may be altered by UVBR has been provided by studies on UVBR effects on the expression of ICAM-1 in human epidermal keratinocytes. By employing the ICAM-1 molecule as a photoimmunological model system, substantial progress has been made in identifying some of the molecular mechanisms which are operative in photoimmunology. The purpose of this review is to summarize the current status of our knowledge about the mechanisms by which UVBR regulates the expression of the human ICAM-1 molecule.

Expression of ICAM-1 by human keratinocytes

Interest in keratinocyte ICAM-1 expression has mainly been stimulated by the *in vitro* observation of Dustin *et al.*, that stimulation of long-term cultured, normal human keratinocytes (HNK) with recombinant human (rh) interferon (IFN) γ enabled these cells to bind to activated T lymphocytes [21]. Interferon γ-induced keratinocyte/T-cell adhesion was associated with a marked upregulation of keratinocyte surface expression of ICAM-1, and the physical interaction between the two cell types could be effectively blocked by adding anti-ICAM-1 antibodies to keratinocytes or anti-LFA-1 antibodies to T cells. These observations indicated a major role for the ICAM-1/LFA-1 adhesion pathway in intraepidermal adhesion processes and prompted intensive studies to define the regulatory mechanisms which determine keratinocyte ICAM-1 expression.

Functional relevance of human keratinocyte ICAM-1 expression

The adhesion molecule ICAM-1 is an 85–110 kd transmembrane glycoprotein with a wide tissue distribution [32, 78, 84]. ICAM-1 has been mapped to human chromosome 19 and is encoded by an inducible 3.3 kb mRNA [37]. The molecule has a structure composed of five tandemly linked domains, each of which fulfills the criteria for membership of the immunoglobulin superfamily [84]. One function of ICAM-1 is to permit antigen-independent adhesion between lymphocytes and their targets, with the integrin LFA-1 on the surface of the leukocyte interacting with ICAM-1 on the opposing cell prior to antigen-dependent activation and cell–cell signaling [32, 42, 66, 69, 75, 81, 82]. Several *in vitro* experiments have proved that the interaction between LFA-1 and ICAM-1 is not only required for cell adhesion and migration, but also plays a role in other leukocyte functions such as Ag presentation, lymphocyte activation, killer T-cell-mediated target lysis, and immunoglobulin production through T-cell-dependent humoral immune responses [31, 81]. In addition to its capacity to bind LFA-1, ICAM-1 was found to be a ligand for another leukocyte integrin, MAC-1 (CD11b/CD18) [19] and for the leukocyte integrin sialophorin (CD43) [68]. Although the dermatological relevance of these observations is not currently known, it should be noted that ICAM-1 may also function as a receptor for certain viruses including the major group of human rhinoviruses [26, 72, 85], and has been shown to serve as a sequestration antigen for plasmodium falciparum [7].

In contrast to ICAM-1 molecules, LFA-1 molecules have been shown to possess transmembrane signaling properties as detected by phosphoinositol hydrolysis and rise in intracellular Ca^{2+}, suggesting that the LFA-1 molecule interacting with ICAM-1 may provide the T cells with costimulatory signals in Ag-specific responses [13, 14]. Recently, the importance of this adhesion system in the induction of allograft rejection was demonstrated. Treatment with anti-ICAM-1 mAb prolonged the survival of renal allografts in monkeys [15] and the combination of mAb to ICAM-1 and LFA-1 induced MHC-specific acceptance of cardiac and skin grafts [33]. In addition, the ICAM-1 molecule on endothelial cells has a major role in leukocyte transmigration from the blood to the tissues in inflammatory sites. This is emphasized by the fact that bacterial lipopolysaccharide and the inflammatory cytokines IL-1 and TNFα upregulate ICAM-1 levels on endothelium [32].

In addition to membrane-associated ICAM-1 molecules (mICAM-1), a soluble form of this molecule (sICAM-1) has been found to exist [70, 73]. Soluble ICAM-1 molecules, which were first detected in normal human serum, contain most, if not all, of the five extracellular domains of mICAM-1 and are capable of binding to LFA-1 molecules in a specific fashion [73]. As cellular sources for sICAM-1 molecules, peripheral blood mononuclear cells, melanoma cells and human keratinocytes have recently been identified [9]. Studies employing HNK have revealed that production of sICAM-1 molecules is due to proteolytic cleavage of mICAM-1 molecules, which may be inhibited by treating HNK with selected protease inhibitors or an oligopeptide containing the potential cleavage site within the ICAM-1 molecule, which was termed ICAM cleavage inhibitory peptide [8]. Although the *in vivo* function of sICAM-1 is not currently known, there is evidence from *in vitro* studies that sICAM-1 molecules are functionally active. For example, sICAM-1 molecules were found to block ICAM-1/LFA-1 mediated NK cell cytotoxicity directed against ICAM-1 expressing human melanoma cells, indicating that binding of sICAM-1 molecules to LFA-1 receptors may promote deadhesion [6, 34]. In addition, binding of sICAM-1 molecules to LFA-1 molecules on the surface of T lymphocytes may at least to some degree induce T-cell activation, and thus represent an important mechanism by which sICAM-1 producing cells are capable of modulating immunological responses [56].

Recent studies indicate that binding of keratinocyte ICAM-1 molecules to LFA-1 molecules on the surface of T cells provides a costimulatory activity required for keratinocyte accessory cell function in superantigen-induced T-cell activation [60]. In addition, keratinocyte ICAM-1 expression is thought to play a pivotal role in the generation and maintenance of an intraepidermal infiltrate by forming the molecular matrix, to which LFA-1 expressing leukocytes bind in inflammatory skin diseases. Immunohistochemical studies indicate that in skin diseases which are characterized by an intraepidermal infiltrate (e.g. psoriasis, allergic contact dermatitis, atopic dermatitis), keratinocytes strongly express ICAM-1 on their surface, and a close colocalization of intraepidermal leukocytes and ICAM-1$^+$ keratinocytes has been detected [54, 76, 77, 91]. In contrast, in normal human skin no significant keratinocyte ICAM-1 expression may be detected. Functional evidence for the *in vivo* relevance of ICAM-1 expression in skin diseases has recently been provided by animal studies employing a gene targeting approach which has resulted in the production of mice deficient in ICAM-1 [79]. In these studies, homozygous ICAM-1 deficient mice were found to develop normally, to be fertile and to have a moderate granulocytosis. On the other hand, ICAM-1 deficient mice exhibited striking abnormalities in inflammatory/immunological reactions including decreased contact hypersensitivity to 2,4-dinitrofluorobenzene.

Keratinocyte ICAM-1 expression may also be of pathogenetic relevance for skin diseases, characterized by keratinocyte cytotoxicity, such as lichen planus, subacute cutaneous lupus erythematodes, erythema multiforme and pemphigus vulgaris [63]. Immunohistochemical studies have demonstrated ICAM-1$^+$ keratinocytes in lesional skin from patients with these skin diseases, and *in vitro* studies indicate that keratinocyte cytotoxicity by LAK cells or by cytotoxic T cells is mediated at least in part via the ICAM-1/LFA-1 adhesion pathway.

Human keratinocytes have also been implicated in intraepidermal trapping of T lymphocytes in malignant cutaneous lymphomas. This concept is supported by immunohistochemical studies revealing the presence of strongly ICAM-1$^+$ keratinocytes in patch- or plaque-stage cutaneous T-cell lymphoma [91]. Moreover, patients with Sezary syndrome, which is the leukemic rather than epidermotropic form of a cutaneous T-cell lymphoma, were found to lack ICAM-1$^+$ keratinocytes [59]. Taken together, these studies strongly indicate that adhesion of LFA-1$^+$ leukocytes to ICAM-1 expressing keratinocytes is of pathophysiological relevance for a variety of skin diseases.

Regulation of keratinocyte ICAM-1 expression by cytokines

Several studies have shown that ICAM-1 can be strongly induced upon stimulation of cells with a number of inflammatory cytokines [20, 29, 43, 55, 58]. This is in contrast to its ligand, LFA-1, which is constitutively expressed in most cell types. It has therefore been proposed that ICAM-1/LFA-1-mediated adhesion is mainly controlled via regulation of the expression of ICAM-1. It should however be noted that the avidity of LFA-1 may be rapidly and transiently increased by a mechanism which involves a selected epitope on the β-subunit of LFA-1, and that this activation of LFA-1 is in part achieved by an induced fit of the ligand itself, that is ICAM-1 [10, 22].

Long-term cultured, normal human keratinocytes do not constitutively express ICAM-1 on their surface. Expression of ICAM-1, however, can be rapidly induced by stimulating HNK *in vitro* with rh IFN-γ and, to a lesser extent, with rh TNFα or rh TNFβ [12, 21, 48, 90]. The same cytokines are also effective in inducing ICAM-1 expression on transformed keratinocytes, which differ from HNK by showing low constitutive levels of ICAM-1 expression. The combination of rh IFN-γ with rh TNFα or rh TNFβ was found to induce human keratinocyte ICAM-1 expression in a synergistic fashion [5, 21, 48]. These *in vitro* results are corroborated by *in vivo* findings, in which intradermal injection of rh IFN-γ into human skin was associated with both the induction of keratinocyte ICAM-1 expression and the close spatial apposition of LFA-1$^+$ mononuclear cells to ICAM-1$^+$ keratinocytes [4]. Increased ICAM-1 surface expression can be observed in cytokine stimulated KC as early as 12 h after stimulation, reaching a peak after 16–24 h. The changes in ICAM-1 surface expression have been found to be always associated with corresponding alterations of steady state levels of ICAM-1 specific mRNA, indicating that upregulation of ICAM-1 mRNA content is a prerequisite for ICAM-1 surface expression [12, 95]. Transcriptional control mechanisms relevant for ICAM-1 expression have therefore attracted considerable interest. The 5′ region of the human ICAM-1 gene has been cloned independently by several groups, and subsequent analysis of the ICAM-1 promotor region has revealed the presence of potential *cis*-acting enhancer elements including IFN-responsive elements and binding sites for AP1, AP-2 and NFkB [17, 83, 93].

Although a large body of information has been generated describing the mechanisms by which proinflammatory stimuli such as cytokines are capable of upregulating keratinocyte ICAM-1 expression, little is currently known about regulatory principles, by which ICAM-1 expression may be negatively controlled. In this regard it is of particular interest that the ICAM-1 molecule was found to be associated with the f-actin cytoskeleton network [11], and that disruption of f-actin filaments was associated with an enhancement of cytokine-induced ICAM-1 transcription [62]. These studies indicate that the cytoskeleton is involved in the regulation of human keratinocyte ICAM-1 expression by providing a negative feedback signal, which may serve to limit the level of ICAM-1 expression inducible upon cytokine stimulation of cells.

Ultraviolet B radiation effects on human keratinocyte ICAM-1 expression

Since the early eighties, keratinocytes have been known to be capable of producing cytokines including IL-1α, and this function was soon found to be profoundly altered upon exposure of cells to UVBR. Recent studies indicate that UVBR also has the capacity to affect the regulation of ICAM-1 expression in human keratinocytes. These modulatory effects were found to be biphasic, that is immediately after UVBR exposure ICAM-1 expression is suppressed, whereas at later time points upregulation of ICAM-1 expression may be observed. Studies on the underlying immunological mechanisms have revealed that at least some of these effects are a consequence of UVBR-induced modulation of keratinocyte-derived cytokines and their respective receptors.

UVBR-induced modulation of cytokine-regulated ICAM-1 expression

Exposure of transformed (KB cells) or normal human keratinocytes (HNK) to increasing doses of UVBR (0–100 J/m^2) inhibited cytokine (IFN-γ, TNFα)-induced ICAM-1 surface expression in a dose-dependent manner [48, 50, 64]. In these studies [48], cells were exposed to UVB radiation from FS20 sun lamps, which represent the UV radiation source most commonly used in photoimmunological research and which emit primarily within the UVB range, but also UVAR [80]. Filter experiments established that modulation of cytokine-induced keratinocyte ICAM-1 expression was caused by UVBR (<315 nm) (H. Christoph, J. Krutmann, unpublished observation). The UVBR doses employed in these studies did not affect cell viability, and were previously shown to induce selected functions in human keratinocytes, e.g. the production of cytokines, neuropeptides and prostanoids [28, 71, 88]. Ultraviolet B radiation-induced inhibition of cytokine-mediated ICAM-1 expression could only be observed if cells were exposed to UVBR prior to cytokine stimulation. Similarly, Khan *et al.* observed that UVBR exposure did not affect constitutive expression of MHC class II molecules in IFN-γ-stimulated transformed human keratinocytes, but completely suppressed the induction of class II MHC molecule expression, if these cells were UBV-irradiated prior to cytokine stimulation [38].

This novel immunosuppressive effect of UVBR is thought to be of therapeutic interest, since it points to a thus far undescribed mechanism by which the therapeutic effectiveness of UVB phototherapy may at least partially be explained [48]. Further studies were therefore designed to examine the underlying molecular mechanisms more closely. It is generally accepted that cytokine-mediated ICAM-1 expression is controlled at the level of gene transcription. We therefore examined in further studies the effects of UVBR on ICAM-1 mRNA expression and found that exposure of transformed human keratinocytes to increasing doses of UVBR inhibited IFN-γ- or TNFα-induced upregulation of ICAM-1 mRNA steady state levels in a dose-dependent fashion [45]. A dose of 100 J/m^2 UVBR was sufficient to reduce the IFN-γ-induced increase in ICAM-1 mRNA expression to background levels. Since equivalent doses of UVBR inhibited cytokine-induced ICAM-1 surface expression, it was concluded that the UVBR-induced inhibition of cytokine-mediated ICAM-1 upregulation is caused by alterations which act at a pretranslational level. This observation is not specific for keratinocytes, since essentially identical results were obtained if primary human skin fibroblasts [44] or cells from a human melanoma cell line [39] were used.

In studies employing human peripheral blood monocytes as a model for human APC, we have previously demonstrated that UVBR-induced inhibition of human APC function is due to inhibition of monocyte ICAM-1 surface expression [47]. As a consequence, UVB-irradiated human APC are incapable of physically interacting with T lymphocytes and thus fail to provide the complete panel of costimulatory signals which is required for complete T-cell activation to occur. In monocytes, ICAM-1 surface expression is upregulated either upon stimulation with ICAM-1-inducing cytokines, or during adhesion, for example to fibronectin [46]. Preliminary studies indicate that UVBR inhibits adhesion-induced ICAM-1 surface expression by a posttranscriptional mechanism, indicating that UVBR-induced downregulation of monocyte ICAM-1 and keratinocyte ICAM-1 expression occur via different mechanisms [49].

Because UVBR-induced inhibition of cytokine-mediated keratinocyte ICAM-1 upregulation may be of phototherapeutic relevance, it was critical to determine whether this anti-inflammatory effect was transient in nature. This question has been addressed in a series of experiments, in which UVB-irradiated keratinocytes and/or fibroblasts were incubated for various periods of time following UVBR exposure, and were subsequently stimulated with optimal concentrations of rh IFN-γ. Restoration of cytokine responsiveness was determined by measuring ICAM-1 mRNA expression and could be observed in IFN-γ-stimulated cells as early as 12 h after UVBR exposure [44, 45]. The therapeutic relevance of this observation is emphasized by the notion that UVBR reexposure of irradiated, IFN-γ-stimulated cells after 24 h reinduced IFN-γ unresponsiveness, and that after an additional 24 h recovery phase, irradiated keratinocytes could again be stimulated to express ICAM-1 mRNA (F. Parlow, J. Krutmann, unpublished observation).

Substantial progress has been made recently in understanding the molecular mechanisms which may be responsible for the transient inhibition of cytokine-induced ICAM-1 upregulation in UVB-irradiated cells. Because UVBR is known to induce photoproducts in cellular DNA [2, 53], the potential role of UVBR-induced DNA damage in the inhibition of IFN-γ-mediated ICAM-1 mRNA expression was assessed. In these studies, a comparison of UVBR-induced inhibition of IFN-γ-mediated ICAM-1 mRNA expression, and the restoration of IFN-γ responsiveness in UVB-irradiated normal and DNA excision repair defective fibroblasts from patients with xeroderma pigmentosum (XP) was undertaken [44]. The rationale for employing XP cells was based on the following speculation: if lower UVBR doses are sufficient to retain an equivalent number of DNA lesions at a given time point post-exposure in XP cells, as compared to normal cells, then UVBR dose–response curves as well as the capacity to restore IFN-γ responsiveness between DNA repair deficient and normal cells should differ correspondingly [30]. These studies revealed that in XP fibroblasts, two- to threefold lower UVBR doses were required to achieve inhibition of IFN-γ-induced ICAM-1 mRNA expression, which was equivalent to that observed in normal fibroblasts [44]. In addition, in UVB-irradiated normal cells, restoration of IFN-γ responsiveness could be observed 12 h after UVBR exposure, whereas in XP cells restoration of IFN-γ responsiveness required a 24 h recovery phase (XPC cells) or could not be observed for up to 48 h after UVB irradiation (XPD cells) [44]. Based on these studies it has been concluded that UVBR-induced inhibition of IFN-γ-mediated ICAM-1 expression involves the generation of DNA photoproducts of a type which may not be removed by XP cells. Restoration of IFN-γ responsiveness in normal human cells may thus reflect their capacity to remove DNA photoproducts via the nucleotide excision repair pathway [53]. Ultraviolet B radiation is known to induce a variety of DNA lesions with pyrimidine dimers, 6-4 photoadducts and Dewar isomers being the major products [2]. The relevance of a specific DNA lesion for UVBR-induced modulation of cytokine-mediated ICAM-1 expression remains to be determined. Recent studies in the mouse system however, indicate that the induction of pyrimidine dimers in keratinocytes is of major importance in UVBR-induced immunomodulation [41].

Ultraviolet B radiation-induced DNA damage has previously been found to be involved in gene regulation of human cells [36, 92, 94]. Exposure of human epithelial cells to cytotoxic doses of UVC (<280 nm) radiation induced the expression of the collagenase, metallothionein and c-fos genes. The induction of these genes by UVCR involved DNA damage, which led to the activation of nuclear transcription factors and subsequent upregulation of the corresponding mRNA levels. Transcription factors do not only increase, but may also suppress, the transcriptional activation of selected genes [74]. Evidence has been provided for the existence of a DNA binding protein, which inhibits IFN-γ-mediated transcriptional activation of ICAM-1 [57]. Further studies are currently underway to determine whether UVBR-induced DNA damage may directly or indirectly affect the activation of transcription factors, which may either enhance or suppress ICAM-1 transcription in UVB-irradiated cells.

Restoration of responsiveness was not specific for IFN-γ, but could also be observed for TNFα [44]. Interestingly, responsiveness was restored differentially for these two cytokines, that is responsiveness for TNFα required a 24–48 h rather than a 12 h restoration period. In this regard it is of interest that human keratinocytes express the 55 kd TNFR, which serves to mediate TNFα-induced ICAM-1 expression in this cell type [89]. Irradiation of keratinocytes with 100 J/m^2 UVBR caused binding of UVB-induced, endogenously produced TNFα to 55 kd TNFR on the keratinocyte surface, which was followed by rapid downregulation of TNFR surface and mRNA expression at 6–12 h post-exposure [88]. Reexpression of TNFR in irradiated keratinocytes occurred 24–48 h after UVBR exposure, and this time point was identical to that at which TNFα-induced ICAM-1 mRNA expression could be detected in irradiated keratinocytes [88] Biphasic modulation of TNFα-induced keratinocyte ICAM-1 expression and of keratinocyte 55 kd TNFR expression were caused by identical doses of UVBR, and the time courses of both UVBR-induced effects are consistent with a cause–effect relationship. It is therefore most likely that UVBR-induced modulation of 55 kd TNFR expression in human keratinocytes may at least partially account for the failure of UVB-irradiated keratinocytes to respond to TNFα, since downregulation of 55 kd TNFR expression may contribute to the failure of UVB-irradiated cells to upregulate ICAM-1 expression 0–12 h after exposure, whereas TNFR reexpression at 24–48 h may explain restoration of TNF responsiveness in irradiated cells. Since human keratinocytes do not produce IFN-γ, no such autocrine, receptor-mediated mechanism would cause a time delay in restoration of IFN-γ responsiveness in UVB-irradiated cells.

Ultraviolet B radiation-induced modulation of constitutive ICAM-1 expression

In addition to suppressing cytokine-induced ICAM-1 expression, UVBR has been found to induce constitutive ICAM-1 expression in human keratinocytes [64]. Specifically, 48 h after UVBR exposure of transformed [45] or normal human keratinocytes [64], a significant upregulation of constitutive ICAM-1 surface and mRNA expression may be observed. The UVBR doses required to inhibit IFN-γ-induced ICAM-1 expression in keratinocytes were found to be identical to those which upregulate constitutive ICAM-1 expression, indicating that a single exposure to UVBR may exert both anti-inflammatory and proinflammatory effects on keratinocyte ICAM-1 expression. The potential clinical relevance of UVBR-induced upregulation of keratinocyte ICAM-1 expression is suggested by observations made in the photosensitive skin disease subacute cutaneous lupus erythematodes, which is characterized by ICAM-1^+ keratinocytes and cytotoxic damage to keratinocytes. Since ICAM-1 expression is a prerequisite for effective lysis of keratinocytes by LAK cells and/or cytotoxic T cells [35, 87], it has been proposed that UVBR-induced keratinocyte ICAM-1 expression may promote binding of cytotoxic cells to epidermal keratinocytes, which would facilitate keratinocyte cytotoxicity [18, 63]. Moreover, UVBR was previously shown to up- as well as downregulate contact hypersensitivity reactions in the skin [67]. This biphasic response may at least partially be explained by initial down- and subsequent upregulation of keratinocyte ICAM-1 expression.

Over the last 4 years, the groups of Norris *et al.* and Krutmann *et al.* have extensively studied the mechanisms by which UVBR is capable of inducing keratinocyte ICAM-1 expression. Both authors agree that UVBR-induced ICAM-1 upregulation is not due to direct effects of UVBR on ICAM-1 transcription, but rather is mediated in an autocrine fashion via UVBR-induced cytokines. Direct evidence for this mechanism has been provided by Krutmann *et al.* in studies in which UVBR-induced keratinoycte ICAM-1 expression could be mimicked by stimulating unirradiated cells with conditioned medium, that is supernatants harvested from irradiated keratinocytes [45]. Initial studies from Norris *et al.* indicated that the ICAM-1-inducing activity present in supernatants derived from irradiated HNK was identical to TNFα, since incubation of irradiated cells in the presence of neutralizing anti-TNFα antibodies was reported to prevent UVBR-induced ICAM-1 upregulation [63]. In contrast, in studies conducted by Krutmann *et al.*, it was observed that neither in transformed human keratinocytes, nor in HNK addition of anti-TNFα, do antibodies lead to significant inhibition of UVBR-induced ICAM-1 expression [45], although identical antibody concentrations completely inhibited binding of endogenously produced TNFα to 55 kd TNFR on the surface of irradiated HNK or KB cells [88]. In addition, anti-TNFα antibodies also caused partial inhibition of UVBR-induced, TNFα-mediated PGE_2 production in KB cells [28].

Recent studies by Krutmann *et al.* strongly imply a major role for UVBR-induced, keratinocyte-derived IL-1α in this system. Incubation of transformed human keratinocytes (KB cells) as well as HNK with neutralizing anti-IL-1α antibodies almost completely inhibited UVBR-induced ICAM-1 expression [51]. This finding was somewhat surprising, since until this point the dogma had been that IL-1 is not capable of significantly inducing ICAM-1 expression in keratinocytes, although this cytokine had been reported to do so in several other cell types [21]. The capacity of a given cell to respond to IL-1 is mainly determined by the degree of IL-1 receptor expression [16, 52]. In recent studies we were able to demonstrate that HNK exclusively express the type I IL-1 receptor (IL-1RI), and that IL-1RI expression is significantly enhanced in this cell type 24–48 h after exposure to 100 J/m^2 UVBR [27]. Ultraviolet B radiation-induced upregulation of IL-1RI expression in HNK could be mimicked by stimulating unirradiated cells with rh IL-1α and could be blocked by the addition of neutralizing anti-IL-1α antibodies, or functionally blocking anti-IL-1RI antibodies, indicating an autocrine IL-1α-mediated mechanism. Increased expression of

IL-1RI in IL-1α-stimulated HNK was of functional relevance for the capacity of HNK to upregulate ICAM-1 expression upon IL-1α stimulation [27]. Accordingly, resting HNK failed to express ICAM-1 molecules on the cell surface 48 h after a single stimulation with rh IL-1α. In marked contrast, if HNK were first IL-1α stimulated for 24 h to achieve upregulation of IL-1RI expression, and were then stimulated a second time for another 24 h with an identical concentration of IL-1α, induction of ICAM-1 expression in HNK could be observed, which was essentially identical to that present in UVB-irradiated, 48-h cultured cells.

Based on these findings, we propose the following two step model which is operative in UVBR-induced upregulation of ICAM-1 expression in human keratinocytes: in a first step, UVBR induces the production of IL-1α, which, in an autocrine fashion, leads to enhanced expression of IL-1RI expression on the surface of irradiated keratinocytes. These cells, which exhibit an increased sensitivity to IL-1α as compared to resting keratinocytes, will now respond in a second step to endogenously produced IL-1α by upregulating ICAM-1 mRNA and surface expression.

It should be noted however, that depending on the wavelengths used to induce keratinocyte ICAM-1 expression, different cytokines may be involved in this autocrine system. Accordingly, exposure of HNK to UVA1R (340–400 nm) was recently found to induce ICAM-1 surface and mRNA expression in a manner essentially identical to that observed after UVBR exposure [13]. Analysis of the underlying molecular mechanism, however, indicates that UVA1R and UVBR induce human keratinocyte ICAM-1 expression via different pathways. It is therefore not presently possible to exclude the suggestion that keratinocyte-derived TNFα may play a role in UVR-induced ICAM-1 expression if a significantly greater proportion of wavelengths longer than 315 nm are emitted by the irradiation device employed.

In vivo UVB radiation effects on keratinocyte ICAM-1 expression

A question of obvious importance is that of the *in vivo* relevance of UVBR-induced modulation of human keratinocyte ICAM-1 expression. Recent studies from our laboratory indicate that UVBR-induced inhibition of cytokine-mediated ICAM-1 expresion may also occur *in vivo* in a manner very similar to that observed *in vitro*. In these immunohistochemical studies, normal human skin was intradermally injected with rh IFN-γ, which led to upregulation of keratinocyte ICAM-1 expression as previously described [4]. Exposure of human skin to 3 MED of UVBR prior to IFN-γ injection almost completely inhibited IFN-γ-induced keratinocyte ICAM-1 expression (H. Stege, J. Krutmann, manuscript in preparation). Studies employing T4 endonuclease 5 containing liposomes to determine the potential role of UVBR-induced DNA damage in this *in vivo* model are currently underway.

Information on the *in vivo* relevance of UVBR-induced upregulation of constitutive keratinocyte ICAM-1 expression has mainly been provided by studies from Norris *et al.* [63–65]. Accordingly, sequential exposure of normal human skin to UVR from a solar simulator (0.5–200 J/cm^2 UVBR) on three consecutive days was found to significantly induce keratinocyte ICAM-1 expression in a dose-dependent fashion. To achieve keratinocyte ICAM-1 expression *in vivo* it may be important that several consecutive UVBR exposures are applied to overcome UVBR-induced cytokine unresponsiveness, since a single exposure of human skin to 2–8 MED of UVBR did not affect constitutive keratinocyte ICAM-1 expression [65].

Concluding remarks

Ultraviolet radiation regulates expression of the human ICAM-1 molecule in a biphasic manner: in the early phase, that is immediately after irradiation, cytokine-induced ICAM-1 expression is inhibited via a pretranslational mechanism, which involves DNA damage as an intermediate. This early phase is followed by a second, later phase, which is characterized by two events: restoration of cytokine responsiveness 12–24 h after UVR exposure, and upregulation of constitutive ICAM-1 expression via an autocrine mechanism, which is mediated by IL-1α and which requires induction of IL-1RI expression on the cell surface. The present chapter has demonstrated that substantial progress has been made in explaining at least some of the regulatory steps which are involved in this biphasic regulation. Ultraviolet radiation-induced regulation of ICAM-1 expression may thus be regarded as a reasonably well defined photoimmunological model system. It may be anticipated that this model may be very useful in future studies for analysing the molecular mechanisms which are involved in UVR-induced immunomodulation. The ICAM-1 model should allow us to study in greater detail the relevance of DNA photoproduct generation and the resulting regulatory consequences for inhibition and induction of the transcriptional control of ICAM-1 gene expression. Information obtained from these studies will be of direct importance for the pathophysiology of photosensitive skin diseases, but will also help us to better understand the mechanisms by which UVR phototherapy works.

Acknowledgment

The authors would like to thank Drs M. Grewe and S. Grether-Beck, Freiburg, for critically reading the manuscript. The continuous support of Prof. Dr E. Schöpf, Department of Dermatology, University of Freiburg, is greatly appreciated. This work was supported by a grant from the Deutsche Forschungsgemeinschaft Kr 871/3-3.

References

1 Acevedo, A., M.A. del Pozo, A.G. Arroyo, P. Snachez-Mateos, R. Gonzales-Amaro, F. Sanchez-Madrid: Distribution of ICAM-3 bearing cells in normal human tissues. Expression of a novel counter-receptor for LFA-1 in epidermal Langerhans cells. *Am. J. Pathol.* 143 (1993), 774–783.

2 Ananthaswamy, H.N., W.E. Pierceall: Molecular mechanisms of ultraviolet radiation carcinogenesis. Yearly review. *Photochem. Photobiol.* 52 (1990), 1119–1136.

3 Augustin, M., A. Dietrich, R. Niedner, A. Kapp, E. Schöpf, J.A. Ledbetter, W. Brady, P.S. Linsley, J.C. Simon: Phorbol-12-myristate-13-actetate-treated human keratinocytes express B7-like molecules that serve a costimulatory role in T-cell activation. *J. Invest. Dermatol.* 100 (1993), 275–281.

4 Barker, J.N.W.N., M.H. Allen, D.M. MacDonald: The effect of *in vivo* interferon-gamma on the distribution of LFA-1 and ICAM-1 in normal human skin. *J. Invest. Dermatol.* 93 (1989), 439–442.

5 Barker, J.N.W.N., V. Sarma, R.S. Mitra, V.M. Dixit, B.J. Nickoloff: Marked synergism between tumor necrosis factor-α and interferon-γ in regulation of keratinocyte-derived adhesion molecules and chemotactic factors. *J. Clin. Invest.* 85 (1990), 605–608.

6 Becker, J.C., R. Dummer, A.A. Hartmann, G. Burg, R.E. Schmidt: Shedding of ICAM-1 from human melanoma cells lines induced by IFN-γ and tumor necrosis factor-α. Functional consequences on cell-mediated cytotoxicity. *J. Immunol.* 147 (1991), 4398–4401.

7 Behrendt, A.R., D.L. Simmons, J. Tansey, C.I. Newbold, K. Marsh: Intercellular adhesion molecule-1 is an endothelial cell receptor for plasmodium falciparum. *Nature* 341 (1989), 57–59.

8 Budnik, A., M. Grewe, K. Gyufko, J. Krutmann: Identification of proteolytic cleavage as the mechanism responsible for the production of soluble ICAM-1 molecules by human kerationocytes (Abstract). *J. Invest. Dermatol.* 102 (1994), 566.

9 Budnik, A., U. Trefzer, F. Parlow, A. Kapp, E. Schöpf, J. Krutmann: Human epidermal keratinocytes are a source of soluble ICAM-1 molecules. *Exp. Dermatol.* 1 (1992), 27–30.

10 Cabanas, C., N. Hoog: Ligand intercellular adhesion molecule 1 has a necessary role in activation of integrin lymphocyte function-associated molecule 1. *Proc. Natl. Acad. Sci. USA* 90 (1993), 5838–5842.

11 Carpen, O., P. Pallai, D.E. Staunton, T.A. Springer: Association of intercellular adhesion molecule-1 (ICAM-1) with actin-containing cytoskelton and α-actinin. *J. Cell Biol.* 118 (1992), 1223–1234.

12 Caughman, S.W., L. Lian-Jie, K. Degitz: Characterization and functional analysis of interferon-γ-induced intercellular adhesion molecule-1 expression in human keratinocytes and A-431 cells. *J. Invest. Dermatol.* 94 (1990), 22S–26S.

13 Christoph, H., F. Parlow, A. Budnik, R. Block, E. Schöpf, J. Krutmann: Ultraviolet A1-induced immunomodulation: UVA1- and UVB-radiation both affect human keratinocyte expression of intercellular adhesion molecule-1 (ICAM-1), although by different mechanisms (Abstract). *Arch. Dermatol. Res.* 285 (1993), 56.

14 Cleaver, J.E.: Xeroderma pigmentosum: a human disease in which an initial stage of DNA repair is defective. *Proc. Natl. Acad. Sci. USA* 63 (1969), 428.

15 Cosmi, A.B., D. Conti, F.L. Delmonico, F.I. Preffer, S.-L. Wee, R. Rothlein, R. Faanes, R.B. Colvin: *In vivo* effects of monoclonal antibody to ICAM-1 (CD54) in

nonhuman primates with renal allografts. *J. Immunol.* 144 (1990), 4604–4612.

16 Curtis, B.M., B. Gallis, R.O. Overell, C.J. McMahan, P. Deross, R. Oreland, J. Eisenman, S. Dower, J.E. Sims: T cell interleukin 1 receptor expressed in Chinese hamster ovary cells regulates functional responses to interleukin 1. *Proc. Natl. Acad. Sci. USA* 86 (1989), 3045–3049.

17 Degitz, K., L. Lian-Jie, S.W. Caughman: Cloning and characterization of the 5′-transcriptional regulatory region of the human intercellular adhesion molecule-1 gene. *J. Biol. Chem.* 266 (1991), 14024–14030.

18 Deleo, V.A., H. Horlick, D. Hanson, M. Eisenger, L.C. Harber: Ultraviolet radiation induced changes in membrane metabolism of human keratinocytes in culture. *J. Invest. Dermatol.* 83 (1984), 323–326.

19 Diamond, M.S., D.E. Staunton, S.D. Marlin, T.A. Springer: Binding of the integrin MAC-1 (CD11b/CD18) to the third immunoglobulin-like domain of ICAM-1 (CD54) and its regulation by glycosylation. *Cell* 65 (1991), 961–971.

20 Dustin, M.L., R. Rothlein, A.K. Bhan, C.A. Dinarello, T.A. Springer: Induction by IL-1 and interferon-γ: tissue distribution, biochemistry, and function of a natural adherence molecule (ICAM-1). *J. Immunol.* 137 (1986), 245–254.

21 Dustin, M.L., K.H. Singer, D.T. Tuck, T.A. Springer: Adhesion of T lymphoblasts to epidermal keratinocytes is regulated by interferon-γ and is mediated by intercellular adhesion molecule 1 (ICAM-1). *J. Exp. Med.* 167 (1988), 1323–1340.

22 Dustin, M.L., T.A. Springer: T-cell receptor cross-linking transiently stimulates adhesiveness through LFA-1. *Nature* 341 (1989), 619–624.

23 Fleming, T.E. W.S. Mirando, U. Trefzer, K.A. Tubesing, C.A. Elmets: *In situ* expression of a B7-like adhesion molecule on keratinocytes from human epidermis. *J. Invest. Dermatol.* 101 (1993), 754–758.

24 de Fougerolles, A.R., L.B. Klickstein, T.A. Springer: Cloning and expression of intercellular adhesion molecule 3 reveals strong homology to other immunoglobulin family counter-receptors for lymphocyte function-associated antigen 1. *J. Exp. Med.* 177 (1993), 1187–1192.

25 de Fougerolles, A.R., S.A. Stacker, R. Schwarting, T.A. Springer: Characterization of ICAM-2 and evidence for a third counter-receptor for LFA-1. *J. Exp. Med.* 174 (1992), 253–267.

26 Greve, J.M., G. Favis, A.M. Meyer, C.P. Forte, S.C. Yost, C.W. Marlow, M.E. Kamarck, A. McClelland: The major human rhinovirus receptor is ICAM-1. *Cell* 56 (1989), 839–847.

27 Grewe, M., K. Gyufko, A. Budnik, J. Krutmann: Analysis of the expression, regulation and function of the interleukin-1 type I and type II receptors in normal and transformed human keratinocytes (Abstract). *J. Invest. Dermatol.* 102 (1994), 566.

28 Grewe, M., U. Trefzer, A. Ballhorn, K. Gyufko, H. Henninger, J. Krutmann: Analysis of the mechanism of ultraviolet B radiation-induced prostaglandin E_2 synthesis by human epidermoid carcinoma cells. *J. Invest. Dermatol.* 101 (1993), 528–531.

29 Griffith, C.E.M., J.J. Vorhees, B.J. Nickoloff: Characterization of intercellular adhesion molecule-1 and HLA-DR expression in inflamed skin: modulation by recombinant gamma interferon and tumor necrosis factor. *J. Am. Acad. Dermatol.* 20 (1989), 617–629.

30 Herrlich, P., H. Ponta, H.J. Rahmsdorf: DNA damage induced gene expression: signal transduction and relation to growth factor signaling. *Rev. Physiol Biochem. Pharmacol.* 119 (1992), 187–223.

31 Hibbs, M.L., H. Xu, S.A. Stacker, T.A. Springer: Regulation of adhesion to ICAM-1 by the cytoplasmic domain of LFA-1 integrin β subunit. *Science* 251 (1991), 1611–1613.

32 Hogg, N., P.A. Bates, J. Harvey: Structure and function of intercellular adhesion molecule-1. *Chem. Immunol.* 50 (1991), 95–115.

33 Isobe, M., H. Yagita, K. Okumara, A. Ihara: Specific acceptance of cardiac allograft after treatment with antibodies to ICAM-1 and LFA-1. *Science* 255 (1992), 1125–1127.

34 Johnson, J.P., B.G. Stade, B. Holzmann, W. Schwaeble, B.G. Riethmueller: *De novo* expression of intercellular-adhesion molecule-1 in melanoma correlates with increased risk of metastasis. *Proc. Natl. Acad. Sci. USA* 86 (1989), 641.

35 Kalish, R.A.: Non-specifically activated human peripheral blood mononuclear cells are cytotoxic for human keratinocytes. *J. Immunol.* 142 (1989), 74–80.

36 Karin, M., P. Herrlich: Cis- and trans-acting elements responsible for induction of specific genes by tumor promotors, serum factors, and stress. In: *Genes and Signal Transduction in Multistage Carcinogenesis*, (1989), pp. 415–440. Ed.: N.H. Colburn, Marcel Dekker Inc., New York.

37 Katz, F.E., M. Parker, K. Stanley, L.J. Murray, E.A. Clark, M.F. Greaves: Chromosome mapping of cell membrane antigens expressed on activated B cells. *Eur. J. Immunol.* 15 (1985), 103–106.

38 Khan, I.U., K.D. Boehm, C.A. Elmets: Modulation of IFNγ-induced HLA-DR expression on the human keratinocyte cell line SCC-13 by ultraviolet radiation. *Photochem. Photobiol.* 57 (1993), 285–290.

39 Kirnbauer, R., B. Charvat, E. Schauer, A. Koeck, A. Urbanski, E. Foerster, P. Neuner, I. Assmann, T.A. Luger, T. Schwarz: Modulation of intercellular adhesion molecule-1 expression on human melanocytes and melanoma cells: evidence for a regulatory role of IL-6, IL-7, TNF-β, and UVB light. *J. Invest. Dermatol.* 98 (1992), 320–326.

40 Koeck, A., T. Schwarz, R. Kirnbauer, A. Urbanski, P.

Perry, J.C. Ansel, T.A. Luger: Human keratinocytes are a source for tumor necrosis factor α: evidence for synthesis and release upon stimulation with endotoxin or ultraviolet light. *J. Exp. Med.* 172 (1990), 1609–1614.

41 Kripke, M.L., P.A. Cox, L.G. Alas, D.B. Yarosh: Pyrimidine dimers in DNA initiate systemic immunosuppression in UV-irradiated mice. *Proc. Natl. Acad. Sci. USA* 89 (1992), 7516–7520.

42 Krutmann, J: Pharmacological aspects of ICAM-1 regulation in the skin. In: *Pharmacology of the Skin*, (1992), pp. 316–328. Ed.: H. Mukthar, CRC Press, Boca Raton, FL.

43 Krutmann, J.: Regulatory interactions between epidermal cell adhesion molecules and cytokines. In: *Epidermal Cytokines and Growth Factors*, (1994), pp. 415–432. Eds.: T.A. Luger, T. Schwarz, Marcel Dekker Inc., New York.

44 Krutmann, J., E. Bohnert, E.G. Jung: Evidence that DNA damage is a mediate in Ultraviolet B radiation-induced inhibition of human gene expression: ultraviolet B radiation effects on intercellular adhesion molecule-1 expression. *J. Invest. Dermatol.* 102 (1994), 428–432.

45 Krutmann, J., W. Czech, F. Parlow, U. Trefzer, A. Kapp, E. Schoepf, T.A. Luger: Ultraviolet radiation effects on human keratinocyte ICAM-1 expression: UV-induced inhibition of cytokine induced ICAM-1 mRNA expression is transient, differentially restored for IFN-γ versus TNF-α, and followed by ICAM-1 induction via a TNF-α-like pathway. *J. Invest. Dermatol.* 98 (1992), 923–928.

46 Krutmann, J., U. Jonas, F. Parlow, U. Trefzer, M. Berger, E. Schöpf, J. Bauer: Analysis of the regulation of intercellular adhesion molecule 1 expression in human antigen presenting cells (Abstract). *J. Invest. Dermatol.* 98 (1992), 539.

47 Krutmann, J., I.U. Khan, R.S. Wallis, F. Zhang, K.A. Koehler, E.A. Rich, J.J. Ellner, C.A. Elmets: The cell membrane is a major locus for ultraviolet-B-induced alterations in accessory cells. *J. Clin. Invest.* 85 (1990), 1529–1536.

48 Krutmann, J., A. Koeck, E. Schauer, F. Parlow, A. Moeller, A. Kapp, E. Foerster, E. Schoepf, T.A. Luger: Tumor necrosis factor β and ultraviolet radiation are potent regulators of human keratinocyte ICAM-1 expression. *J. Invest. Dermatol.* 95 (1990), 127–131.

49 Krutmann, J., F. Parlow, E. Schöpf, C.A. Elmets: Posttranscriptional modification of ICAM-1 expression on human antigen presenting cells by ultraviolet B (UVB) radiation. *Clin. Res.* 39 (1991), 1029.

50 Krutmann, J., E. Schoepf, F. Parlow, A. Kapp, W. Czech, U. Trefzer, T.A. Luger: Modulation of intercellular adhesion molecule-1 expression and gene transcription by ultraviolet light. *Pharmacol. Skin* 4 (1992), 221–223.

51 Krutmann, J., U. Trefzer, A. Kapp, E. Schöpf, T.A. Luger: Interleukin 1 α, but not tumor necrosis factor α, mediates ultraviolet B radiation-induced ICAM-1 expression in human keratinocytes (Abstract). *J. Invest. Dermatol.* 98 (1992), 578.

52 Kupper, T.S., F. Lee, N. Birchall, S. Clark, S. Dower: Interleukin 1 binds to specific receptors on human keratinocytes and induces granulocyte macrophage colony-stimulating factor mRNA and protein. *J. Clin. Invest.* 82 (1988), 1787–1792.

53 Lambert, H.C., P.C. Hanawalt: DNA repair mechanisms and their biologic implications in mammalian cells. *J. Am. Acad. Dermatol.* 22 (1990), 299–308.

54 Lisby, S., E. Ralfkier, R. Rothlein, G.L. Veijsgard: Intercellular adhesion molecule-1 (ICAM-1) expression correlated to inflammation. *Br. J. Dermatol.* 120 (1989), 479–484.

55 Luger, T.A., T. Schwarz: Epidermal derived cytokines. In: *Skin Immune System*, (1990), pp. 257–292. Ed.: J.D. Bos, CRC Press, Boca Raton, FL.

56 McCabe, S.M., L. Ridle, G.R. Nakamura, H. Prashad, A. Mehta, P.W. Berman, P. Jardieu: sICAM-1 enhances cytokine production stimulated by alloantigen. *Cell Immunol.* 150 (1993), 364–375.

57 Naik, S., L.J. Li, T. Nguven, N. Shibagaki, S.W. Caughman: Identification of the specific region responsible for cycloheximide (CHX)-sensitive repression of ICAM-1 gene transcription (Abstract). *J. Invest. Dermatol.* 100 (1993), 554.

58 Nickoloff, B.J.: The role of IFN-γ in cutaneous trafficking of lymphocytes with emphasis on molecular and cellular events. *Arch. Dermatol.* 124 (1988), 1835–1843.

59 Nickoloff, B.J., C. Griffiths, O. Baadsgard, J.J. Vorhees, C.A. Hanson, K.D. Cooper: Markedly diminished epidermal keratinocyte expression of intercellular adhesion molecule-1 (ICAM-1) in Sezary syndrome. *JAMA* 261 (1989), 2217–2221.

60 Nickoloff, B.J., R.S. Mitra, J. Green, X.G. Zheng, Y. Shimizu, C. Thompson, L.A. Turka: Accessory cell function of keratinocytes. Dependence on lymphocyte function-associated antigen-1/intercellular adhesion molecule-1 interaction. *J. Immunol.* 150 (1993), 2148–2159.

61 Nickoloff, B.J., R.S. Mitra, K. Lee, L.A. Turka, J. Green, C. Thompson, Y. Shimizu: Discordant expression of CD28 ligands, BB-1, and B7 on keratinocytes *in vitro* and psoriatic cells *in vivo*. *Am. J. Pathol.* 142 (1993), 1029–1040.

62 Norgauer, J., F. Parlow, U.M. Göhring, E. Kownatzki, J. Krutmann: Actin cytoskeleton involvement in the regulation of intercellular adhesion molecule-1 expression by epidermoid carcinoma cells (Abstract). *J. Invest. Dermatol.* 102 (1994), 535.

63 Norris, D.A.: Cytokine modulation of adhesion

molecules in the regulation of immunologic cytotoxicity of epidermal targets. *J. Invest. Dermatol.* 95 (1990), 111S–120S.

64 Norris, D.A., B. Lyons, M.H. Middleton, J.Y. Yohn, M. Kashihara-Sawami: Ultraviolet radiation can either suppress or induce expression of intercellular adhesion molecule-1 (ICAM-1) on the surface of cultured human keratinocytes. *J. Invest. Dermatol.* 95 (1990), 132–138.

65 Norris, P., R.N. Poston, S.D. Thomas, M. Thornhill, J. Hawk, D.O. Haskard: The expression of endothelial leukocyte adhesion molecule-1 (ELAM-1), intercellular adhesion molecule-1 (ICAM-1), and vascular cell adhesion molecule-1 (VCAM-1) in experimental cutaneous inflammation: a comparison of ultraviolet B erythema and delayed hypersensitivity. *J. Invest. Dermatol.* 96 (1991), 763–770.

66 Pardi, R., J.R. Bender, C. Dettori, E. Giannazza, E.G. Engleman. Heterogenous distribution and transmembrane signaling properties of lymphocyte function antigen LFA-1 in human lymphocyte subsets. *J. Immunol.* 143 (1989), 3157–3160.

67 Polla, L., R. Margolis, C. Goulston, J.A. Parrish, R.D. Granstein: Enhancement of the elicitation phase of the murine contact hypersensitivity response by prior exposure to local ultraviolet radiation. *J. Invest. Dermatol.* 86 (1986), 13–17.

68 Rosenstein, Y., J.K. Park, W.C. Hahn, F.S. Rosen, B.E. Bierer, S.J. Burakoff: CD43, a molecule defective in Wiskott–Aldrich syndrome, binds ICAM-1. *Nature* 354 (1991), 233–235.

69 Rothlein, R., M.L. Dustin, S.D. Marlin, T.A. Springer: A human intercellular adhesion molecule (ICAM-1) distinct from LFA-1. *J. Immunol.* 137 (1986), 1270–1274.

70 Rothlein, R., E.A. Mainolfi, M. Czajkowski, S.D. Marlin: A form of circulating ICAM-1 in human serum. *J. Immunol.* 147 (1991), 3788–3793.

71 Schauer, E., F. Trautinger, A. Köck, A. Urbanski, M. Bodemer, J.C. Ansel, T. Schwarz, T.A. Luger: Production of proopiomelanocortin derived peptides by human keratinocytes. *J. Clin. Invest.* 93 (1994), 2258–2262.

72 Selinka, H.C., A. Zibert, E. Wimmer: Poliovirus can enter and infect mammalian cells by way of an intercellular adhesion molecule 1 pathway. *Proc. Natl. Acad. Sci. USA* 88 (1991), 3598–3602.

73 Seth, R., F.D. Raymond, M.W. Magkoba: Circulating ICAM-1 isoforms: diagnostic prospects for inflammatory and immune disorders. *Lancet* 338 (1991), 83–84.

74 Seto, E., A. Usheva, G.P. Zambetti, J. Momand, N. Horikoshi, R. Weinmann, A.J. Levine, T. Shenk: Wild-type p53 binds to the TATA-binding protein and represses transcription. *Proc. Natl. Acad. Sci. USA* 89 (1992), 12028–12032.

75 van Seventer, G.A., Y. Shimizu, K.J. Horhan, S. Shaw: The LFA-1 ligand ICAM-1 provides an important costimulatory signal for T cell receptor mediated activation of resting T cells. *J. Immunol.* 144 (1990), 4579–4586.

76 Shiohara, T: Lichenoid tissue reaction: an immunological perspective. *Am. J. Dermatopathol.* 10 (1988), 252–256.

77 Singer, K.H., D.T. Tuck, H.A. Sampson, R.P. Hall: Epidermal keratinocytes express the adhesion molecule intercellular adhesion molecule-1 in inflammatory dermatoses. *J. Invest. Dermatol.* 92 (1989), 746–750.

78 Simmons, D., M.W. Makgoba, B. Seed: ICAM-1, an adhesion ligand of LFA-1, is homologous to the neural cell adhesion molecule NCAM. *Nature* 331 (1988), 624–627.

79 Sligh Jr, J.E., C.M. Ballantyne, S.S. Rich, H.K. Hawkins, C.W. Smith, A. Bradley, A.L. Beaudet: Inflammatory and immune responses are impaired in mice deficient in intercellular adhesion molecule 1. *Proc. Natl. Acad. Sci. USA* 90 (1993), 8529–8533.

80 Spikes, J.D.: Comments on light, light sources, and light measurements. In: *Experimental and Clinical Photoimmunology*, vol. 1, (1983), pp. 5–26. Eds.: R.A. Daynes, J.D. Spikes, CRC Press, Boca Raton, FL.

81 Springer, T.A.: Adhesion receptors of the immune system. *Nature* 346 (1990), 425–434.

82 Springer, T.A., M.L. Dustin, T.K. Kishimoto, S.D. Marlin: The lymphocyte function-associated LFA-1, CD2, and LFA-3 molecules: cell adhesion molecules of the immune system. *Ann. Rev. Immunol.* 5 (1987), 223–252.

83 Stade, B.G., G. Messer, G. Rietmüller, J.P. Johnson: Structural characteristics of the 5′ region of the human ICAM-1 gene. *Immunobiol.* 182 (1990), 79–87.

84 Staunton, D.E., S.D. Marlin, C. Stratowa, M.L. Dustin, T.A. Springer: Primary structure of intercellular adhesion molecule-1 (ICAM-1) demonstrates interaction between members of the immunoglobulin and integrin superfamily. *Cell* 52 (1988), 925–933.

85 Staunton, D.E., V.J. Merluzzi, R. Rothlein, R. Barton, S.D. Marlin, T.A. Springer: A cell adhesion molecule, ICAM-1, is the major surface receptor for rhinoviruses. *Cell* 56 (1989), 849–853.

86 Symington, F.W., W. Brady, P.S. Linsley: Expression and function of B7 on human epidermal Langerhans cells. *J. Immunol.* 150 (1993), 1286–1295.

87 Symington, F.W., B.E. Santos: Lysis of human keratinocytes by allogeneic HLA class I-specific cytotoxic T cells. *J. Immunol.* 146 (1991), 2169–2175.

88 Trefzer, U., M. Brockhaus, H. Loetscher, F. Parlow, A. Budnik, M. Grewe, H. Christoph, A. Kapp, E. Schoepf, T.A. Luger, J. Krutmann: The 55-kd tumor necrosis factor receptor on human keratinocytes is regulated by tumor necrosis factor alpha and by

ultraviolet B radiation. *J. Clin. Invest.* 92 (1993), 462–470.

89 Trefzer, U., M. Brockhaus, F. Parlow, H. Loetscher, A. Kapp, E. Schoepf, J. Krutmann: The 55 kd tumor necrosis factor receptor is expressed by human keratinocytes and plays a pivotal role in regulation of human keratinocyte ICAM-1 expression. *J. Invest. Dermatol.* 97 (1991), 911–916.

90 Tsujisaki, M., K. Igarahsi, M. Eisinger, M. Herlyn, S. Ferrone: Immunochemical and functional analysis of HLA class II antigens by recombinant immune interferon on normal human melanocytes. *J. Immunol.* 138 (1987), 1310–1316.

91 Vejlsgaard, G.L., E. Ralfkiaer, C. Avnstrop, M. Czajkowski, S.D. Marlin, R. Rothelin: Kinetics and characterization of intercellular adhesion molecule-1 (ICAM-1) expression on keratinocytes in various inflammatory skin lesions and malignant cutaneous lymphomas. *J. Am. Acad. Dermatol.* 20 (1989), 782–790.

92 de Vos, S., M. Brach, A. Budnik, M. Grewe, F. Herrmann, J. Krutmann: Posttranscriptional regulation of interleukin-6 gene expression in human keratinocytes. *J. Invest. Dermatol.* 103 (1994), 92–96.

93 Warwryk, S.O., P.N. Cockerill, I.P. Wicks, A.W. Boyd: Isolation and characterization of the promotor region of the human intercellular adhesion molecule-1 gene. *Int. Immunol.* 3 (1991), 83–93.

94 Yarosh, D.B.: The role of DNA damage and UV-induced cytokines in skin cancer. *J. Photochem. Photobiol. B: Biol.* 16 (1992), 91–94.

95 Yohn, J.J., M. Critelli, M.B. Lyons, D.A. Norris: Modulation of melanocyte intercellular adhesion molecule-1 by immune cytokines. *J. Invest. Dermatol.* 90 (1990), 233–237.

7: UVB-induced unresponsiveness and tolerance induction: a central role for epidermal Langerhans cells

Ponciano D. Cruz Jr

Among environmental agents, ultraviolet B (UVB) radiation is unique in its ability to produce antigen-specific suppression of T-cell-mediated responses. This special property has been documented at an *in vivo* level for contact allergy in mice and humans, for delayed-type hypersensitivity (DTH) in mice, and for immunosurveillance against cutaneous photocarcinogenesis, also in mice [4, 10, 16]. More recently, both *in vivo* and *in vitro* studies have demonstrated UVB radiation to be capable of converting epidermal Langerhans cells (LC) from immunogenic to tolerogenic antigen-presenting cells (APC) for the subset of T cells (i.e., T_h1 cells) responsible for mediating DTH responses [5, 21]. Taken together, these observations indicate strongly that the suppressive influence of UVB radiation is due to altered antigen presentation for T-cell responses initiated in skin.

Antigen presentation and T-cell responses

Induction of T-cell responses to antigen introduced through the skin requires the development of sequential sets of cellular events [3]. Antigen has to be presented by cutaneous antigen-presenting cells (APC) in a form that can be recognized by the T-cell receptor (TCR) of precursor (naive) T lymphocytes specific for the particular antigen. In addition, antigen-bearing APC must be able to migrate to draining lymph nodes in order to interact with the responder T cells. Presentation of antigen by APC and its recognition by T cells must then trigger a series of molecular events within the T cells that culminates in the production of appropriate lymphokines, the expression of relevant cell surface receptors, and finally proliferative cell division. The end product is an expanded clone of activated, antigen-specific T cells that are poised to respond in a more rapid manner to subsequent antigenic challenges. Despite this expansion, however, antigen-specific T cells remain a small minority of the cellular infiltrate in either lymph nodes or affected skin sites. The vast majority of the infiltrate is made up of non-specific leukocytes that have been recruited to these tissues during the course of inflammation.

Epidermal LC are targets of UVB radiation

LC are the principal, if not the sole, APC in normal epidermis [3]. In fact the potency of LC as stimulator cells for antigen-specific proliferation of naive T cells is equalled only by lymphoid dendritic cells, with whom LC share many closely related properties including a common hematopoietic lineage derivation [8, 22]. This potent capacity of LC to act as APC for T-cell responses has been shown for a wide spectrum of antigens, including nominal proteins, virus-derived proteins, tumor antigens and reactive haptens [3, 15, 22].

The idea that epidermal LC serve as targets of UVB radiation was first suggested in experiments with C57BL/6 mice [24], in which exposure of abdominal skin to relatively low doses of UVB ($200\,J/m^2/d$ on four consecutive days) led to two parallel effects. The first effect was a reduction in the density of LC within epidermis of irradiated skin, as judged by loss of cell dendricity and of membrane ATPase activity. The second effect was the induction of ear-swelling unresponsiveness following elicitation with hapten and of sub-

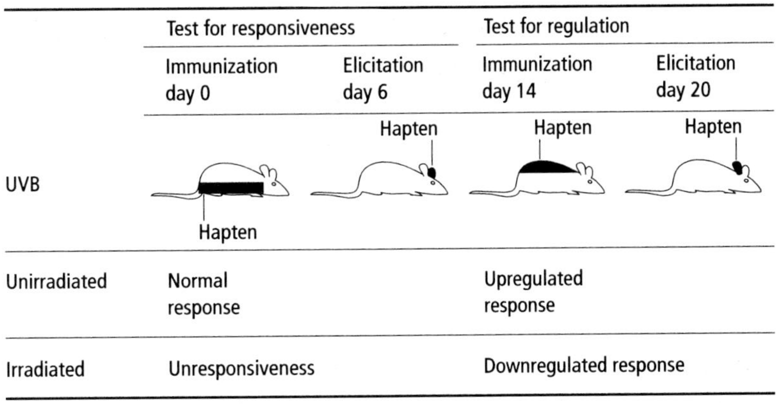

Fig. 7.1 *In vivo* UVB irradiation induces initial unresponsiveness and subsequent downregulation (tolerance) for contact allergy.

sequently downregulated responses to reimmunization with hapten through unirradiated skin (Fig. 7.1).

In other studies, *in vitro* UVB irradiation (single doses of 25–600 J/m^2) was shown to abrogate the ability of human or mouse epidermal cells (containing LC) to stimulate proliferation of antigen-specific T cells [23].

UVB converts LC from immunogenic to tolerogenic APC *in vivo*

Physical deletion or functional inactivation of LC by UVB can readily account for both the induction of ear-swelling unresponsiveness in mice painted with DNFB through UVB-irradiated skin and the loss of LC capacity to stimulate responder T cells *in vitro*. On the other hand, neither of these mechanisms can explain the failure of irradiated mice to respond normally to a second immunization with the same hapten through unirradiated skin (Fig. 7.1). We thus considered an alternative interpretation – the effect of low dose UVB on LC was not to kill these cells or merely inhibit its APC function, but rather to convert LC from immunogenic to tolerogenic APC.

To this point the evidence that UVB exerted its immunosuppressive effect on LC was largely circumstantial, awaiting the development of techniques by which epidermal cells could be sorted into relatively pure populations. Once this had been achieved, we then demonstrated a single dose of 200 J/m^2 UVB *in vitro* to abrogate the capacity of hapten-conjugated, FACS-purified LC to induce contact allergic responses, when infused intravenously into syngeneic mice (Fig. 7.2) [4]. Even more significantly, we had shown these same mice to be incapable of mounting normal ear-swelling responses, even after the same hapten was painted on unirradiated normal skin at a later date (Fig. 7.2) [4]. The conclusion derived from these studies was that UVB radiation could indeed convert the function of LC from one that induces an immunogenic signal to one that is tolerogenic.

CD4$^+$ T cells are responsible for UVB-induced suppression of contact allergy *in vivo*

Antigen-specific suppressor T cells have been shown to mediate the immunosuppression induced by UVB radiation [7]. To identify the T-cell subsets responsible for the suppression observed in our experimental system, we employed monoclonal antibodies against CD4 and/or CD8 to deplete C3H/HeN mice of T-cell subsets *in vivo* [1]. As expected, mice depleted of both CD4$^+$ and CD8$^+$ T cells were unable to mount ear-swelling responses following immunization by intravenous infusion of hapten-conjugated LC. By contrast, mice depleted of CD4$^+$ or CD8$^+$ T cells developed full responses to either form of immunization, indicating that either of the two subsets can mediate contact allergic responses. Of greatest

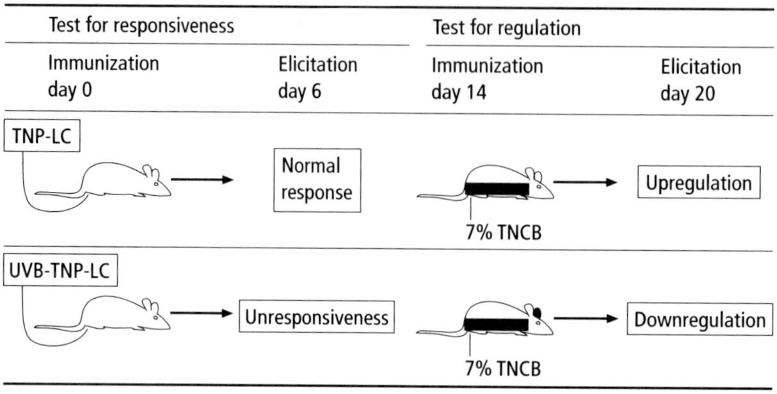

interest was the finding that the contact allergic response induced by infusion of UVB-irradiated, hapten-conjugated LC was suppressed in CD8-depleted (but not in CD4-depleted) mice. This observation indicated that $CD4^+$ (but not $CD8^+$) T cells are the mediators of the suppression of contact allergy responses induced by UVB radiation *in vivo*.

UVB radiation reverses the preferential capacity of LC to activate T_h1 over T_h2 cells

With the realization that $CD4^+$ cells are the principal controlling population for T-cell-mediated responses, our attention focused on subsets within this population as the basis for UVB-induced suppression of T-cell responses. Following the now widely accepted paradigm that was initially proposed by Mosmann and his colleagues [12], T_h1 cells secrete interleukin-2 (IL-2), interferon γ (IFN-γ), and tumor necrosis factor (TNF), whereas T_h2 cells produce IL-4, IL-5, IL-6, and IL-10. This divergence in cytokine profiles is significant because IL-2, IFN-λ, and TNF are proinflammatory cytokines that can mediate DTH responses, whereas IL-4 and IL-5 serve primarily to promote antigen-specific antibody production by B cells. The two $CD4^+$ T-cell subsets also differ with respect to growth requirements: T_h1 cells utilize IL-2, whereas T_h2 cells use IL-4 as autocrine growth factors. Finally, the two subsets display built-in counterregulatory mechanisms, since IFN-γ (from T_h1) can inhibit T_h2 proliferation, whereas IL-10 (from T_h2) can block T_h1 proliferation.

We therefore set out to compare the outcome of antigen presentation by unirradiated and UVB-irradiated LC on KLH-specific T_h1 and T_h2 clones [18]. Unirradiated LC pulsed with KLH were observed to induce vigorous proliferation of T_h1, and to a lesser extent, of T_h2. The capacity of LC to preferentially activate T_h1 cells over T_h2 cells was also noted for several other clones specific for pigeon cytochrome c, conalbumin, or DNFB (P.D. Cruz Jr, unpublished observations). By contrast, LC irradiated with a single dose of $200\,J/m^2$ UVB lost their ability to stimulate T_h1, while fully retaining their capacity to stimulate T_h2 [18]. These findings are consistent with the observation that immune responses that are *initiated* in normal epidermis are predominantly mediated by 'cellular' rather than 'humoral' immunity. They also support the conclusion that UVB radiation can reverse the normal capacity of LC to activate T_h1 preferentially over T_h2 cells.

UVB-irradiated LC induce clonal anergy in T_h1 cells

The now familiar question was again raised: did the proliferative unresponsiveness observed when UVB-irradiated LC were employed as APC for antigen-specific T_h1 clones constitute prevention of antigen presentation or active induction of

Table 7.1 Induction of tolerance (anergy) in T_h1 clones by UVB-irradiated LC

Initial incubation of T_h1 with	Subsequent incubation of T_h1 with	T_h1 proliferation
Unirradiated LC and antigen	Unirradiated LC and antigen	+++
	IL-2 alone	+++
UVB-irradiated LC and antigen	Unirradiated LC and antigen	–
	IL-2 alone	+++

tolerance? To address this issue, we subjected T_h1 clones to two sequential stimulations, initially to KLH in the presence of unirradiated or UVB-irradiated LC, and then (after 3 or 16 d) to KLH and unirradiated LC, or to exogenous IL-2 alone (Table 7.1) [21]. Our results indicated that exposure to UVB-irradiated LC rendered T_h1 cells tolerant to subsequent stimulation with KLH and LC. More importantly, the finding that such tolerant T_h1 retained the capacity to proliferate in response to exogenous IL-2 indicated that they were not physically deleted, but rather made functionally anergic.

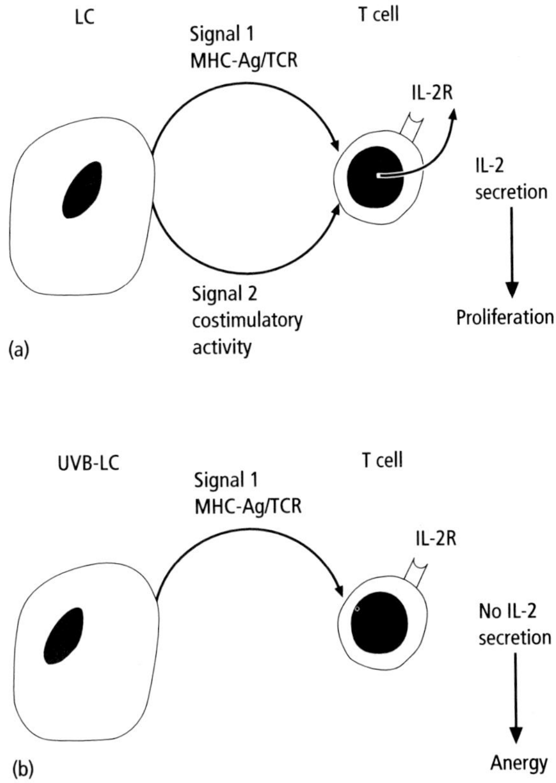

Fig. 7.3 (a) Concurrent transduction of (antigen-specific) signal 1 and (costimulatory) signal 2 results in T_h1 cell secretion of IL-2, which is an autocrine growth factor. (b) Transduction of signal 1 alone causes failure of IL-2 production, leading to a state of anergy (tolerance).

Molecular mechanisms for T_h1 cell anergy

Effective antigen presentation requires the interaction of several different APC-derived, membrane-bound or soluble ligands with appropriate receptors on responder T cells. These ligand–receptor interactions transduce signals to T cells that eventuate in their activation (Fig. 7.3). Two signals are essential. The first signal is an MHC-restricted, antigen-specific signal from APC that is transmitted to the TCR-CD3 complex on T cells. The second signal requires interaction of a costimulatory activity from APC to its corresponding receptor on T cells. Three scenarios are possible. When both signals are simultaneously transduced, T cells are activated to secrete IL-2, which acts as an autocrine growth factor that fuels their proliferation. In the absence of either signal, T cells fail to proliferate. Finally, if the TCR-CD3 complex is stimulated alone (in the absence of the costimulatory signal), T cells are rendered anergic due to a failure to secrete IL-2 [17].

Several characteristics of the tolerance induced by UVB-irradiated LC suggest that it is due to the third scenario just described. T_h1 cells rendered tolerant by antigen-pulsed, UVB-treated LC fail to secrete IL-2, but can proliferate in response to exogenous IL-2 [21]. Addition of allogeneic and unirradiated APC to the original combination of UVB-treated LC, antigen and T_h1 cells prevents the induction of tolerance [21]. Since MHC-disparate APC cannot by themselves stimulate T_h1

proliferation, it was concluded that these allogeneic cells provided the costimulatory activity that was missing from UVB-irradiated LC.

ICAM-1 and B7: candidate molecules for UVB-sensitive costimulatory factors on LC

The UVB-sensitive costimulatory activity on LC for T_h1 has yet to be identified. Preliminary studies suggest that the factor is membrane bound and not encoded by MHC genes [21]. Attractive candidates include two molecules, both of which are members of the immunoglobulin gene superfamily, namely ICAM-1 and B7 (also known as BB-1). ICAM-1 on APC interacts with its ligand, LFA-1, on T cells. ICAM-1 expression has been shown to be upregulated on LC as they become more potent APC, either in tissue culture or *in vivo* [22]. In addition, UVB radiation has been reported to diminish ICAM-1 expression on LC, concurrent with UVB-induced abrogation of the capacity of these APC to induce T-cell proliferation [11]. To date, however, no studies have shown that deletion of the signal transduced by ICAM-1 leads to tolerance in T_h1 cells.

The B7 molecule on APC interacts with its ligand, CD28, on T cells. The level of B7 expression on different mouse APC appears to correlate with their ability to function as costimulatory accessory cells. Thus, spleen dendritic cells (which are very potent costimulators) express several times more B7 than activated spleen cells (which are less potent costimulators) and in turn, activated B cells express more B7 than resting spleen B cells (which are the least potent costimulators) [9]. On the other hand, there has been some difficulty and controversy in documenting B7 expression on LC; some laboratories have reported absence of expression, whereas others have claimed considerable expression after short-term culture.

Are T_h2 cells responsible for the active suppression induced by UVB?

The finding that UVB-irradiated LC exert disparate effects on activation of T_h1 and T_h2 cells *in vitro* may explain several *in vivo* observations. For example, the tolerogenic effects of irradiated LC on T_h1 cells are consistent with the long-lasting unresponsiveness induced by UVB in contact allergy and DTH assays in mice. Furthermore, we speculate that T_h2 cells, whose proliferative capacity remains intact despite interaction with irradiated LC, may exert active suppression of DTH responses via secretion of cytokines such as IL-10.

Modulation of LC function by soluble mediators

Previously we have discussed direct effects of UVB radiation on the APC function of LC. An alternative or concurrent possibility is that UVB may exert its influence on LC indirectly, via UVB-induced secretion by keratinocytes of immunomodulatory cytokines. In fact, it is now clear that keratinocytes are a rich source of soluble factors, many of which have been reported to induce immunosuppression. Among such factors are prostaglandin E_2, IL-1α, contra-IL-1α (a substance that appears to inhibit the effects of IL-1α), TNFα, TGFβ, and IL-10 [2]. What is not certain is how any of these molecules produce antigen-specific suppression of T-cell responses. The most reasonable supposition is that one or a combination of these factors exert suppression by altering the function of APC or of relevant T cells, almost certainly during the time of antigen presentation.

Preliminary work from our laboratory has shown UVB-irradiation (single dose of 200 J/m^2) of a mouse keratinocyte line (Pam 212) to produce soluble mediators that are capable of abrogating the APC function of LC for T_h1, while preserving analogous function for T_h2 [6]. Although studies performed in other laboratories have promoted TNFα and IL-10 as the keratinocyte-derived factors likely to perturb APC function relevant to UVB-induced immunosuppression [14, 25], we have found neither of these cytokines to be the relevant mediators in our experimental system [19].

Urocanic acid (UCA) is an abundant, UV-absorbing component of the stratum corneum, formed by the enzymatic deamination of histidine during keratinization. Two effects of UV light on

UCA are: (i) induction of *trans*-to-*cis* isomerization, both *in vitro* and in mammalian skin; and (ii) formation of photoadducts with DNA *in vitro* [13]. Support for the idea that *trans*-UCA is the chromophore for, and that *cis*-UCA is the mediator of, UVB-induced immunosuppression derives from two sets of investigations. First, the action spectrum for UVB-induced suppression of contact allergy in BALB/c mice was shown to be congruent with the *in vivo* absorption spectrum of *trans*-UCA. Secondly, several laboratories have demonstrated topically applied or parenterally administered *cis*-UCA to inhibit DTH responses in mice, including those against *Herpes simplex* virus type I [13].

Recent studies from our laboratory have suggested that *cis*-UCA can also suppress the capacity of mouse LC to stimulate Ag-specific proliferation of T_h1 cells (S. de Kossodo & P.D. Cruz Jr, unpublished observations). We have yet to determine, however, whether this T_h1 unresponsiveness induced by *cis*-UCA-treated LC constitutes tolerance induction. Nonetheless, these preliminary results hold promise for integrating the different, apparently disparate, mechanisms proposed to explain induction of antigen-specific suppression of T-cell responses by UVB radiation.

Summary

UVB radiation induces a very selective form of immunosuppression, one that is exerted specifically at T-cell responses for antigens introduced during the time of irradiation. Our studies have focused on the central role played by epidermal LC in this form of immunosuppression. We propose that UVB radiation can perturb the APC function of LC, both directly and indirectly. The direct UVB effect causes a reversal of the preferential capacity of LC to activate $CD4^+$ T_h1 over $CD4^+$ T_h2 cells. In this scenario, the APC function as unirradiated LC, which are normally poised to stimulate DTH reactions mediated by T_h1 cells, is profoundly altered. UVB irradiation causes LC to lose the capacity to trigger T_h1-mediated reactions, but remain fully capable of stimulating T_h2-mediated responses (e.g. antigen-specific antibody production by B cells). The molecular basis for this direct effect of UVB on APC function for T_h1 cells appears to be the inactivation of a membrane-bound costimulatory activity on LC for T_h1 cells. On the one hand, presentation of antigen in the absence of this UVB-sensitive costimulatory factor leads to clonal anergy in T_h1 cells. On the other hand, the preserved proliferative capacity of T_h2 cells renders these cells as potential mediators of active suppression *in vivo*. Finally, indirect UVB effects on LC may be exerted via the influence of *cis*-UCA or one or more keratinocyte-derived cytokines on APC function. Preliminary studies suggest that *cis*-UCA or supernatant from UVB-irradiated keratinocytes can indeed abrogate the APC function of LC for T_h1 cells, without changing analogous function for T_h2 cells.

References

1 Cruz, P.D.: Langerhans cells are initiators of the immunosuppressive effect of ultraviolet B radiation. *Springer Sem. Immunopathol.* 13 (1992), 281–288.

2 Cruz, P.D.: UVB-induced immunosuppression: biologic, cellular, and molecular effects. *Adv. Dermatol.* 8 (1993), 79–95.

3 Cruz, P.D., P.R. Bergstresser: Antigen processing and presentation by epidermal Langerhans cells. Induction of immunity or unresponsiveness. *Dermatol. Clin.* 8 (1990), 633–647.

4 Cruz, P.D., P.R. Bergstresser: Immunosuppressive effects of UVB radiation on cutaneous photocarcinogenesis and contact hypersensitivity. In: *Photomedicine for Clinical Dermatologists*, (1993), pp. 137–151. Eds.: H.W. Lim, N. Soter, Marcel Dekker Inc., New York.

5 Cruz, P.D., J. Nixon-Fulton, R.E. Tigelaar, P.R. Bergstresser: Disparate effects of *in vitro* low-dose UVB irradiation on intravenous immunization with purified epidermal cell subpopulations or the induction of contact hypersensitivity. *J. Invest. Dermatol.* 92 (1989), 160–165.

6 Eerkes, K.J., A. Takashima, P.R. Bergstresser, P.D. Cruz: UVB-irradiated keratinocytes secrete a factor that suppresses the antigen presenting function of Langerhans cells (Abstract). *J. Invest. Dermatol.* 98 (1992), 557.

7 Elmets, C.A., P.R. Bergstresser, R.E. Tigelaar, P.J. Wood, J.W. Streilein: Analysis of the mechanism of unresponsiveness produced by haptens painted on skin exposed to low dose ultraviolet radiation. *J. Exp. Med.* 158 (1983), 781–794.

8 Inaba, K., R.M. Steinman, M. Witmer Pack, H. Aya, M. Inaba, T. Sudo, S. Wolpe, G. Schuler: Identification of proliferating dendritic cell precursors in mouse blood. *J. Exp. Med.* 175 (1992), 1157–1167.

9 Johnson, J.G., M.K. Jenkins: Costimulatory functions of antigen presenting cells. *J. Invest. Dermatol.* 99 (1992), 62–65.

10 Krutmann, J., C.A. Elmets: Recent studies on mechanisms in photoimmunology. *Photochem. Photobiol.* 6 (1988), 787–798.

11 Krutmann, J., I.U. Kahn, R.S. Wallis, F. Zhang, E.A. Rich, J.J. Ellner, C.A. Elmets: Cell membrane is a major locus for ultraviolet B-induced alterations in accessory cells. *J. Clin. Invest.* 85 (1990), 1529–1536.

12 Mosmann, T.R., J.H. Schumacher, N.F. Street, R. Budd, A. O'Garra, T.A.T. Fong, M.W. Bond, K.W.M. Moore, A. Sher, D.F. Fiorentino: Diversity of cytokine synthesis and function of mouse $CD4^+$ T cells. *Immunol. Rev.* 123 (1991), 209–229.

13 Noonan, F.P., E.C. DeFabo: Immunosuppression by ultraviolet B radiation. Initiation by urocanic acid. *Immunol. Today* 13 (1992), 250–254.

14 Rivas, J.M., S.E. Ullrich: Systemic suppression of delayed-type hypersensitivity by supernatants from UV-irradiated keratinocytes. *J. Immunol.* 149 (1992), 3865–3871.

15 Romani, N., S. Koide, M. Crowley, M. Witmer-Pack, A.M. Livingstone, C.G. Fathman, K. Inaba, R.M. Steinman: Presentation of exogenous protein antigens by dendritic cells to T cell clones. Intact protein is presented best by immature, epidermal Langerhans cells. *J. Exp. Med.* 169 (1989), 1169–1180.

16 Romerdahl, C.A., H. Okamoto, M. Kripke: Immune surveillance against cutaneous malignancies in experimental animals. In: *Immune Mechanisms in Cutaneous Disease*, (1989), pp. 749–769. Ed.: D.A. Norris, Marcel Dekker Inc., New York.

17 Schwartz, R.H.: A cell culture model for T lymphocyte clonal anergy. *Science* 15 (1990), 1349–1356.

18 Simon, J.C., P.D. Cruz Jr, P.R. Bergstresser, R.E. Tigelaar: Low-dose UVB-irradiated Langerhans cells preferentially activate CD4 cells of the T_h1 subset. *J. Immunol.* 145 (1990), 2087–2091.

19 Simon, J.C., D. Edelbaum, P.R. Bergstresser, P.D. Cruz Jr: Distorted antigen presenting function of Langerhans cells induced by tumor necrosis factor α via a mechanism that appears different from that induced by ultraviolet B radiation. *Photodermatol. Photoimmunol. Photomed.* 8 (1991), 190–194.

20 Simon, J.C., J. Krutmann, C.A. Elmets, P.R. Bergstresser, P.D. Cruz Jr: Ultraviolet B-irradiated antigen presenting cells display altered accessory signaling for T cell activation. Relevance to immune responses initiated in skin. *J. Invest. Dermatol.* 98 (1992), 66–69.

21 Simon, J.C., R.E. Tigelaar, P.R. Bergstresser, D. Edelbaum, P.D. Cruz: UVB radiation converts Langerhans cells from immunogenic to tolerogenic antigen presenting cells. Induction of specific clonal anergy in $CD4^+$ T helper 1 cells. *J. Immunol.* 146 (1991), 485–491.

22 Steinman, R.M.: The dendritic cell system and its role in immunogenicity. *Annu. Rev. Immunol.* 9 (1991), 271–296.

23 Stingl, G., L.A. Gazze-Stingl, W. Aberer, K. Wolff: Antigen presentation by murine epidermal Langerhans cells and its alteration by UVB light. *J. Immunol.* 127 (1981), 1707–1713.

24 Toews, G.B., P.R. Bergstresser, J.W. Strelein: Epidermal Langerhans cell density determines whether contact hypersensitivity or unresponsiveness follows skin painting with DNFB. *J. Immunol.* 134 (1980), 445–453.

25 Yoshikawa, T., J.W. Streilein: Tumor necrosis factor-alpha and ultraviolet light have similar effects on contact hypersensitivity in mice. *Reg. Immunol.* 3 (1990), 139–146.

PART 2
PHOTOIMMUNOLOGY AND THE PATHOGENESIS OF DISEASE

8: Photoimmunology of UV-induced skin cancer

Stephan Grabbe & Richard D. Granstein

Introduction

Since the 1920s it has been known that ultraviolet radiation (UVR), especially UVB (280–320 nm) radiation, can cause skin cancer in humans and experimental aminals [12, 29]. The mechanisms involved in UV carcinogenesis and especially the modulation of the cutaneous immune system by UVR, however, have only relatively recently become understood and are still the subject of active investigation. Classical experiments performed by Kripke, Daynes and others less than 20 years ago established the concept that UV irradiation not only acts as a potent carcinogen, but also induces specific alterations within the immune system which lead to increased growth of cutaneous neoplasms in UV irradiated animals that would be immunologically rejected by normal subjects [19, 68, 69, 122, 125]. These studies greatly influenced the field of tumor immunology, provided insight into the regulation of immune responses in general, and led to the establishment of photoimmunology as a discipline. The first part of this chapter will provide some background information on the mechanisms involved in tumor immune responses and summarize briefly some of the classical findings about UV-induced alterations of the cutaneous immune system in ultraviolet carcinogenesis. Since several excellent reviews of this subject are available [19, 41, 71, 72], only the basic concepts will be summarized in this chapter. The second part then reviews more recent data, mostly concerning the presentation of tumor antigens and its modulation by UVR or UV-inducible cytokines.

Do tumor immune surveillance mechanisms exist?

The finding that an alteration of the immune status of the host results in increased tumor growth and incidence implies that there are at least some connections between the immune system and the outgrowth of neoplasms. However, more than 20 years after its generation, the concept that mechanisms exist that are able to recognize malignantly transformed cells and to induce an immunological defense against them [14] is still controversial. Table 8.1 summarizes some clinical and experimental data that support or challenge the immune surveillance theory. It is the authors' view that convincing evidence for the existence of tumor immune surveillance mechanisms *in general* is lacking, but it is quite clear that some tumors frequently elicit tumor immune responses. In a substantial proportion of cases, these immune phenomena are actually down-regulatory and promote rather than inhibit tumor growth (Fig. 8.1). Furthermore, there is significant data suggesting that immune surveillance mechanisms against emerging tumors exist *within the skin* (perhaps as opposed to other organs). This view is mainly supported by three facts:

1 Immunosuppressed patients and experimental animals have a vastly increased incidence of cutaneous neoplasms, but – with the exception of lymphomas and a few other, mostly virally induced tumors – not of other malignancies [42, 51, 53, 58, 64, 132].

2 UV-irradiated animals accept tumors that would be immunologically rejected in normal animals, due to a specific alteration of the immune response

Table 8.1 Immune surveillance – clinical and experimental data

Pro	Contra
• Tumor-specific and tumor-associated antigens can be demonstrated on the surface of a wide variety of spontaneously arisen and carcinogen-induced tumors • Immune reactivity against tumor cells can be demonstrated in a number of experimental systems *in vivo* and *in vitro* • Both type and extent of tumor immune reactions vary between different tumor systems, indicating differential recognition of tumor antigens by different parts of the immune system • In a number of experimental systems, specific or non-specific activation of tumor immune responses can lead to tumor destruction *in vivo* and *in vitro* • Immunosuppressed patients have an increased incidence of lymphomas and skin tumors • Spontaneous regression of tumors has been observed in some cases • Tumors frequently evoke a mononuclear infiltrate; tumor-infiltrating lymphocytes can be isolated, expanded and used as a therapeutic agent • The presence or absence of certain tumor antigens can serve as a prognostic factor in some tumors • Many cutaneous carcinogens also alter the function of cutaneous APC • Tumor infiltration with dendritic cells indicates better prognosis in some tumor systems	• Athymic nude mice (which have almost no functional T cells) do not have an increased susceptibility to the carcinogenic effects of UVR compared to normal controls (with regard to tumor incidence, the degree of malignancy and metastatic potential, as well as antigenicity of emerging tumors). • There is no significant increase in incidence of extracutaneous or non-lymphatic malignancies in immunosuppressed patients • The degree of lympho-histiocytic infiltrate is not a prognostic factor (in melanoma) • Partial regression of a tumor is often a negative prognostic factor (in melanoma) • The presence of certain tumor antigens on the tumor cell surface is a negative prognostic factor in some tumors • So far, specific or non-specific immunotherapy of tumors has not generally been of significant therapeutic value

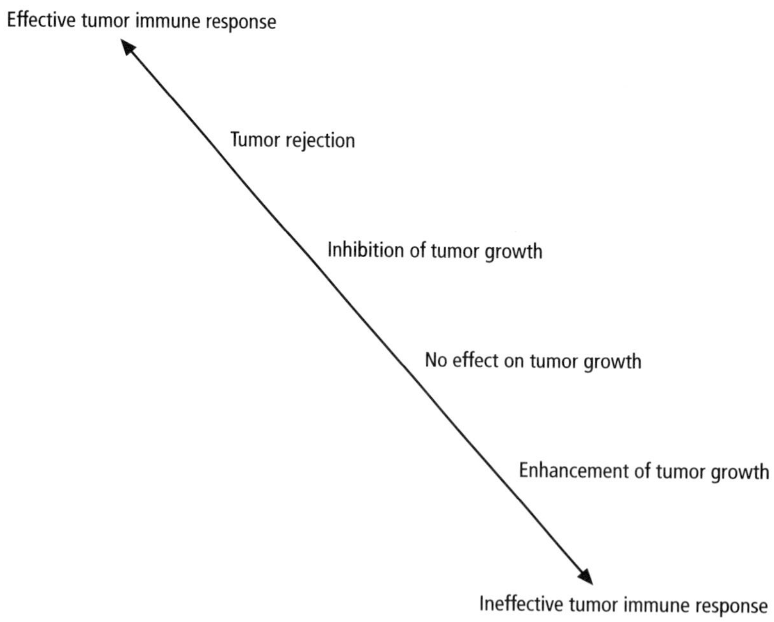

Fig. 8.1 Effects of immune responses on tumor growth.

in UV-irradiated animals (discussed in detail below).

3 Mechanisms exist within the skin that are capable of recognizing tumor-associated antigens (TAA) and inducing protective tumor immunity in naive mice as well as eliciting specific antitumor delayed-type hypersensitivity responses (DTH). This will also be discussed in detail below.

Furthermore, there is preliminary evidence in humans that individuals who display decreased cellular immunocompetence after UVR exposure may have an increased incidence of skin tumors [127, 143]. In experimental animals most cutaneous carcinogen-induced neoplasms are antigenic, while spontaneously arising tumors are often non-antigenic. In humans most neoplasms are non-antigenic, but antigenic structures on many squamous cell carcinomas, and especially on melanomas, are readily detectable [73, 101]. Interestingly, the resistance of normal mice to tumor development decreases with age in a fashion similar to the decline of general immune reactivity with age, i.e. with age mice lose the capacity to mount both idiotypic and anti-idiotypic responses which are responsible for controlling tumor growth [34]. A possible hypothesis to explain these phenomena is that immune surveillance mechanisms exist within the skin which detect and eliminate immunogenic neoplasms before they become clinically apparent, and that only non-immunogenic tumors are able to grow progressively under normal conditions *in situ* [106]. It could then be speculated that carcinogens, which often also inhibit tumor immune recognition [19, 27, 47] overwhelm immunological defense mechanisms and thereby lead to the emergence of antigenic neoplasms. Furthermore, a dysfunction of cutaneous tumor immune surveillance may lead to the active suppression of tumor immune responses. The following sections of this chapter intend to provide some evidence in support of this concept.

Table 8.2 Elements of tumor immunity

Afferent side of tumor immune responses
Non-specific tumor immunity
NK cells/LAK cells/NC cells
Macrophages (?)
Local tissue cells capable of producing cytokines, growth factors and inflammatory mediators
Antigen-specific tumor immunity
Antigen-presenting cells:
Dendritic cells
Macrophages, monocytes
MHC class II positive bystander cells
B cells
Antigen-specific B lymphocytes:
Antibodies
Antigen-specific T lymphocytes:
T helper cells (T_h1, T_h2)
CTL
T suppressor cells
Immune-effector mechanisms
T lymphocytes
Macrophages
Granulocytes, mononuclear phagocytes
Mast cells/eosinophils
Platelets
Vascular endothelial cells
Cells forming the tumor stroma
Cytokines, mediators of inflammation
Cytotoxic enzymes, other cytotoxic cytokines and factors
Result
Direct cellular cytotoxicity
Antibody-dependent cytotoxicity
Release of cytotoxic or proinflammatory mediators
Destruction of tumor vasculature

Mechanisms of antitumor immune reactivity

In general, antigen-specific and non-specific tumor immune responses can be distinguished. Elements of non-specific and specific defense mechanisms against tumors are listed in Table 8.2. Non-specific tumor immunity will not be discussed in detail, although there is some evidence that NK cells are involved in the immune response to UV-induced murine tumors, since UVR induces surface expression of an NK target structure on some tumor cell lines, and UVR exposure has been demonstrated to inhibit NK cell activity in mice and humans [8, 9, 28, 50, 126, 129, 130].

Of foremost importance for the immune regulation of UVR-induced carcinogenesis are T-cell-mediated immune responses [19, 73, 113, 114]. Therefore, some of the principles of T-cell-mediated antitumor immune responses are summarized in this section.

As in all specific immune responses, several conditions must be fulfilled for the establishment of tumor immune reactivity:

1 Tumor-specific antigen(s) must be present on the surface of malignant cells.

2 Antigen-presenting cells (APC) must be present in the vicinity of the tumors and able to acquire, potentially process and present the relevant antigen(s) in the context of MHC class I/II molecules.

3 Antigen-carrying APC must establish contact with naive T cells and be able to sensitize them to tumor antigens.

4 Antigen-carrying APC must also be able to elicit effector T-cell responses that lead to the destruction of the tumor cells.

5 Effector T cells must potentially be able to recruit other proinflammatory or immunocompetent cells.

6 Regulatory mechanisms must exist that control type, extent and duration of the immune response.

The various elements of tumor immune responses appear to be activated in totally different ways, depending on the individual tumor, the organ in which it arises, the immune status of the host, the mechanism of tumor induction and especially the experimental conditions that are employed to investigate tumor immunity. Therefore, it is often difficult to combine data from different experimental systems or to make general statements, which underlines the complexity and variability of tumor–host immune interactions.

Carcinogenic potential of UVR and alteration of the immune status by UVR

Characteristics of UV carcinogenesis

UV-induced tumors begin to appear in UV-irradiated skin several weeks after irradiation, usually applied in doses of 20 kJ/m^2 or more of UVB radiation, three times weekly for 10–20 weeks onto the shaved dorsum of mice. Studies by DeFabo and Kripke demonstrated that the cumulative UVR dose is critical for the generation of malignancies in mice and that the number of UVR exposures, the duration of treatment and the relative intensity of the UV irradiation play only a minor role in this regard [20, 21]. Clinical observations suggest that this applies to humans as well [135].

Histologically, UVR-induced skin tumors are usually squamous cell carcinomas, basal cell carcinomas (in humans) or epitheliomas (in mice), but occasional melanomas are also observed in mice and other experimental animals [23, 70, 84, 107]. With increasing doses of UVR, fewer differentiated and more undifferentiated tumors are generated [124], many of which lose their histotypic markers (keratins), become spindle-shaped and express vimentin (alone or in addition to keratins) [7, 15]. Thus, dedifferentiated UV-induced cutaneous tumors were thought to be of dermal origin and were termed 'fibrosarcomas'. Since most of these undifferentiated UV- or carcinogen-induced tumors appear to be, in fact, of epithelial origin [92], they should be referred to as spindle cell tumors.

A higher incidence of metastases of UV-induced tumors has been reported after i.v. injection into UV-treated mice [74]. UVB-irradiated mice have an increased incidence of primary leukemias [24], but not of mammary tumors or sarcomas [82]. Spikes *et al.* [124] found that the tumor incidence in mice after UVR is slightly higher in males than in females. Young mice were more susceptible to the induction of skin tumors from FS40 sun lamps than were old animals, and more tumors developed in skin grafted to young recipients as compared to old recipients, regardless of the age of the skin donor [25]. Interestingly, these findings are contrary to the general increased risk of tumor development in older as opposed to younger animals. The reasons for this are unclear.

Effector mechanisms of specific tumor immunity

In general, almost all possible effector pathways of immune reactivity which the immune system is able to exert can be employed during tumor immune responses [49]. Therefore these are not principally different from immune responses to other cellular elements. However, the variability and complexity of the immune response and the *regulation of tumor immunity* are possibly unique at least for some aspects of tumor immunity. As mentioned above, the regulation of tumor immunity is of foremost interest for UV-induced tumors. It is important to note that the different effector cell mechanisms generated against tumors may vary with regard to their susceptibility to immunomodulation by UVR and that different types of suppressor T cells with differential effects on various immune effector mechanisms exist [36, 81, 133, 134].

Regulation of T-cell-mediated tumor immunity – role of suppressor T cells and cytokines

UVR-induced cutaneous tumors in some strains of mice are usually highly antigenic and subject to immunological rejection upon transplantation to a normal mouse. The poor transplantability of UV-induced skin cancer was first reported by Rusch and Bauman in 1939, and was thought to be due to bacterial contamination [110]. Studies performed by Kripke and Daynes and their collaborators, however, clearly showed that this poor transplantability of UV-induced skin cancers was due to immunological recognition and rejection of the tumor cells by a normal host [32, 105, 133]. Moreover, further studies showed that these tumors grow progressively in UV-irradiated or immunosuppressed mice [19, 122, 125]. Subcarcinogenic doses of UVB were sufficient to induce susceptibility for syngeneic UV-induced tumors, and to prevent the expression of tumor cytotoxicity by splenocytes *in vitro* [35, 69]. However, UVR does not interfere with secondary rejection of syngeneic UV-induced tumors in animals that have been specifically immunized before UV exposure, implying that a specific step (or steps) in acquiring immunity is affected by UVR [31, 69, 72]. This finding led to a series of investigations to determine how antigenic tumors are able to escape immunological destruction in the primary UV-irradiated host as well as in UV-irradiated or immunosuppressed secondary hosts. It was found that during the course of UV irradiation, mice lost their ability to reject transplants of highly antigenic UV-induced skin cancers and that this phenomena was not restricted to the site of UV irradiation [72, 74]. Immunological alterations occurred before the appearance of skin cancers, and it was concluded that UV irradiation produced a systemic alteration of the host's immune response [30, 75]. Studies performed by Kripke and co-workers demonstrated that various cutaneous tumors grow better in the skin of UV-irradiated, but not of T-cell-depleted mice, indicating that immunologic mechanisms play a role in the outgrowth of immunogenic murine tumors [71]. Later it was found that suppressor T lymphocytes specific for UV-induced tumors (recognizing UV-induced tumors as a group) could be found in the lymphoid organs of UV-irradiated mice [30–32]. Thus, the occurrence of suppressor cells with specificity for a set of different UVR-induced tumors suggest that common, UV-associated regulatory antigens or determinants may be present on many or most UV-induced skin cancers. Specific characteristics of suppressor T cells in UV-irradiated hosts are discussed by P.D. Cruz in Chapter 7 of this book. However, this suppressor cell activity is only directed against a portion of the TAA on UV-induced tumors and leaves the host's response towards other tumor antigens unaltered. Urban *et al.* [136] demonstrated that UV-irradiated mire are still able to mount a partial immune response against certain tumor antigens of UV-induced tumors, so that small tumor inoculates, which are not rejected by nude mice, are rejected. Furthermore, the immune alteration induced by the use of long-wave UVR (UVA, 320–400 nm) in combination with the administration of the photosensitizer 8-methoxypsoralen (PUVA) appears to differ from that of UVB irradiation [133]. PUVA treatment does not appear to alter the immunologic rejection of UVB-induced tumors [80].

In addition to its carcinogenic activity, UVR is capable of modifying various immunological reactions. Immunologic alterations induced in mice by UV irradiation include both local and distant effects [10, 17, 19, 37, 43, 69, 76–78, 82, 104, 122, 125]. Local alterations result from a direct effect of UV irradiation on an immune reaction that takes place at the site of the irradiation and affects mostly cutaneous antigen-presenting cells [17, 131]. Distant effects are those in which exposure of skin to UV irradiation at one site modifies an immune reaction occurring at the distant, unexposed site. Based on a number of studies, these two immune alterations appear to be different [17–19, 104]. Local effects of UV irradiation include cutaneous inflammation, suppression of delayed-type hypersensitivity and suppression of graft versus host reactions [10, 11, 26]. Systemic UV-induced immunomodulation probably involves soluble mediators and is manifested by the systemic suppression of contact hypersensitivity and the formation of antigen-specific suppressor T lymphocytes [19, 31, 35, 71, 82, 94]. Yee and Kripke found that soluble suppressor factors play a role in the regulation of immune responses by UV-induced suppressor T cells [141, 142]. Both in the high dose as well as in the low dose UV-irradiation model, antigen-specific suppressor cells appear in the spleens of UV-irradiated animals [30, 35]. Elements inducing tolerance to UV-induced skin tumors are, however, also present within UV-irradiated skin itself, since immunologic tolerance to UV-induced tumors could be transferred by transplantation of UV-irradiated skin onto normal syngeneic mice [99]. Furthermore, a deficiency in splenic and regional antigen-presenting cell

Table 8.3 Effects of UVB on immunity in mice

Induction of UV-specific antigenic determinants on malignantly transformed as well as on normal cells
Inhibition of tumor antigen presentation by epidermal APC for both induction and elicitation of tumor immunity *in vitro*
Inhibition of antigen-presenting cell function *in vivo:*
Low dose UVB ($<1000\,J/m^2$) *in vivo*: number of LC decreases, local suppression of cutaneous LC function (defective presentation of hapten in unprimed, but not in most primed systems), number of LC recovers with time even during chronic irradiation
High dose UVB ($20–40\,kJ/m^2$) *in vivo*: systemic suppression of cutaneous LC function (defective presentation of hapten in unprimed systems)
Generation of transferrable T suppressor cell activity (high-dose UVB):
Heterogeneous phenotype of UVB-induced UV-T_s (Most are thy-1^+, Lyt1^-, L3T4^-, I-A^-, and I-J^+, others are Thy-1^+, Lyt1^+, Lyt2^-, L3T4^+, I-A^-, and also functionally I-J restricted)
UV-T_s are present in the spleen, lymph nodes and thymus of UV-irradiated ice
UV-T_s appear before clinically visible tumors develop
UV-induced immunosuppression is non-sepecific with regard to the appropriate conditions of dose and regimen of radiation and timing of antigen administration, and the immune response to many different antigens, (contact sensitizers, UV tumor antigens, bacterial antigens). Once induced, however, the suppressor T cells are antigen specific
Cloned UV-T_s lines have been established *in vitro*
UV-T_s activity can be transferred to unirradiated or x-irradiated mice, rendering them incapable of rejecting otherwise immunogenic UV-induced tumors, but leaving immunity towards other neoplasms (e.g. chemically induced) unaltered
UV-T_s recognize UV-induced tumors as a group. They do not affect the incidence of extracutaneous malignancies (e.g. mammary carcinomas)
UV-T_s inhibit antitumor CTL and/or DTH responses (depending on the tumor cell line)
UV-T_s specifically block the generation of tumor-specific T_h cells
UV-T_s do not affect already established T_h cell activity
Generation of T_s is not prevented by thymectomy or splenectomy prior to UV irradiation
Suppression is MHC-restricted
Inhibition of the sensitization phase of CHS and DTH responses to haptens and protein antigens
No inhibition of macrophage function in UV-irradiated mice
No inhibition of inflammatory responses to irritants, antibody responses or allograft rejection

capacity was demonstrated in UV-irradiated mice [5, 76, 94]. The systemic suppression of contact hypersensitivity by UVB irradiation is, however, unrelated to the UV-induced alterations of the morphology and number of Langerhans cells, whereas local effects correlate with altered morphology and function of Langerhans cells at the site of irradiation, as well as a disappearance of thy-1^+ dendritic epidermal T cells at the site of UV irradiation [91, 131] (for a detailed review on UV effects on APC, see Chapter 4). Some of the immunomodulatory effects of UVB radiation in mice are listed in Table 8.3.

Table 8.4 Tumor-associated antigens (TAA)

Structural classification
Embryonic antigens
Viral antigens
Aberrant cell surface glycoproteins or carbohydrates
Mutated or (qualitatively or quantitatively) abnormal MHC antigens
Others
Functional classification
Tumor-specific transplantation antigens (TSTA) – specific for each individual tumor
Tumor antigens that are specific for the mechanism of tumor induction, e.g. tumor antigens shared by all UV-induced tumors as a group
Tumor antigens that induce predominantly CTL-mediated immune responses (e.g. MHC antigens, some viral antigens)
Tumor antigens that induce predominantly T_h-mediated immune responses (e.g. some viral antigens)
Tumor antigens that induce T_s-mediated immune responses, leading to enhanced tumor growth
Tumor-specific epitopes recognized by NK cells or – as currently discussed – the lack of 'self' antigens, leading to NK-mediated lysis

Tumor antigens in ultraviolet carcinogenesis

A variety of antigenic structures on the surface of tumor cells has been described so far, most of which are characterized biochemically or as antigenic determinants by specific monoclonal antibodies [44, 45, 59, 86, 117, 118, 128]. Those tumor-associated antigens (TAA) which are identified on the molecular level generally belong to either viral antigens, embryonic antigens, aberrant cell surface glycoproteins or MHC-like molecules (Table 8.4). In UVR carcinogenesis, two types of TAA can be discriminated functionally: tumor-specific transplantation antigens (TSTA) which are characteristic for each individual tumor, and antigens that appear to be expressed by most or all UV-induced tumors as a group [55, 57, 103, 116, 121]. Most importantly, some UV-specific TAA are already induced on non-transformed normal keratinocytes after UV irradiation [52, 54, 122, 123, 125]. PUVA-induced tumors do not appear to have the same antigenic properties as UVB-induced tumors, since they do not grow better in UVB-irradiated hosts [79–81, 90]. Quality and quantity of MHC molecules on tumor cells appear to be of significance for recognition by various elements of the immune system [13, 16, 62, 65, 83, 102]. Non-specific recognition of tumor cells by NK cells may be induced by the lack of self-MHC structures on malignant cells [63, 85, 138]. Transfection studies performed by Kaba *et al.* [59] showed that at least one TSTA is a novel MHC class I-like molecule, which is selectively expressed in the UVR-induced tumor UV-1591 and is responsible for its antigenicity. On the other hand, Roberts and Daynes demonstrated that genes or gene products encoded within the MHC complex itself did not influence either the latency period, histologic type or immunogenicity of UV-induced skin tumors [102], but at least some cutaneous tumors can be induced by cytokines to upregulate their MHC molecules in number [89, 100] making them more susceptible to immune recognition. Functional studies indicated that MHC-related TAA induce predominantly cytotoxic T lymphocyte (CTL)-mediated immune responses [98]. Indeed, it has been suggested that the deficient ability of UV-irradiated mice to upregulate MHC molecules on tumor cells *in vivo* may contribute to their increased tumor incidence [13].

Further studies have revealed that the regulation of TAA expression in UV-induced tumors appears to be independent of neoplastic transformation [52, 54, 122, 123, 125]. The expression of UV-specific TAA varies even between cells within a single tumor [56], and was also found to be independent of host factors since it takes place both *in vivo* and *in vitro* in a similar fashion [7, 33]. Urban *et al.* demonstrated that UV-generated tumors seem to undergo a sequence of changes in immunogenicity leading to the ultimate outgrowth of poorly immunogenic tumor cell variants [137]. Resistance to immunological attack begins to develop with insensitivity to cytocidal effects of activated macrophages, followed by loss of one or more tumor-specific antigens that are recognized by cytolytic T cells. UV tumors passed through UV-irradiated mice lost certain antigenic determinants but retained others [136]. In melanoma, clonal immunological heterogeneity seems to be relevant for the outgrowth of poorly immunogenic and clinically aggressive metastases [23, 62, 66, 93, 108].

Role of antigen-presenting cells in cutaneous tumor immunity

Effect of UVR on cutaneous antigen-presenting cells

As described above, a significant body of information exists about UVR-induced alterations of immune phenomena on the T-cell level with regard to effector T cells, cytotoxic T cells, helper T cells and suppressor T cells. All of these phenomena, however, are in general antigen-specific, and therefore require the presentation of antigen in association with MHC molecules on the surface of antigen-presenting cells. Other chapters in this book review the effects of UVR on cutaneous APC function in detail, especially with regard to differences between local and systemic UVR effects. Most of these data have been generated using haptens or protein antigens; studies using tumor antigens will be discussed below.

Initial studies by Toews *et al.* found alterations in the number and morphology of cutaneous LC after UVR *in vivo* [131]. Woodward and Daynes reported that I-A$^+$ macrophages were necessary to induce cytotoxic T lymphocyte (CTL) responses against UVR-induced tumors and that the generation of tumor-specific CTL may be suppressed by suppressor T cells that are induced by UVR [140]. Norbury and Kripke demonstrated that treatment of mice with macrophage cytotoxic agents (silica) shortened the latency period of skin cancer formation during UV irradiation [95, 96]. Furthermore, Alcalay and Kripke demonstrated that cutaneous Langerhans cells have impaired antigen-presenting capacity during the latency period of UV carcinogenesis, and that the decrease in the number of identifiable epidermal Langerhans cells early in the course of chronic UV-irradiation carcinogenesis correlated with a decrease in antigen-presenting activity after sensitization through the UV-irradiated skin [2–5]. These authors also investigated the alterations in Langerhans cells during the evolution of UVR-induced skin cancer and showed that Langerhans cells disappear during carcinogenic UVB radiation treatment, but reappear in the late latency period before the actual emergence of UV-induced tumors [5]. On the other hand, Thy-1$^+$ dendritic epidermal cells were totally depleted and appeared only very late after PUVA or UVB radiation treatment [2, 5]. Halliday *et al.* reported recently that Langerhans cells migrate into cutaneous neoplasms, and that the migration is probably due to chemotactic effects of a tumor-derived cytokine. These authors also presented data indicating that the concentration of Langerhans cells within cutaneous tumors does not correlate with the type and degree of tumor immunity [46, 48]. Furthermore, no significant difference in the average reduction of Langerhans cells by UVB irradiation was found between patients with histologically confirmed basal cell carcinoma and normal healthy volunteers, and patients with basal cell carcinoma were not found to have a lower MED than cancer-free patients [3]. Similarly, no correlation was found between the percentage decrease in ATPase$^+$ epidermal Langerhans cells and the dose of UV irradiation required to produce erythema [3]. These data suggest that erythematogenic, carcinogenic and immunosuppressive effects of UVR may be mediated by different mechanisms and have different dose–response characteristics.

APC may also be involved in tumor immunity to non-UV-induced tumors. Kosugi *et al.* demonstrated that I-A$^+$ APC are required for the development of both CTL as well as T_h-mediated tumor immunity [67], and other groups have shown that APC also control the regulation of various tumor immune responses in non UV-induced neoplasms [60, 111, 119, 120, 144].

Tumor antigen presentation by I-A$^+$ epidermal APC

The above data indicated that APC may play a significant role in the induction and modulation of tumor-specific immune responses, but direct evidence of tumor antigen presentation by cutaneous APC was still lacking. Therefore our own investigations focused on the ability of epidermal APC to present tumor-associated antigens for the induction of specific tumor immunity, and on its modulation by ultraviolet light and various cytokines [38, 40]. Principal questions addressed in this regard were:

1 Are epidermal APC capable of inducing specific tumor immunity in naive mice (antigen presentation to unprimed T cells)?

2 Are epidermal APC capable of presenting tumor-associated antigens in tumor-immune mice (antigen presentation to primed T cells)?

3 How do cytokines and ultraviolet radiation modulate the ability of epidermal APC to induce and elicit tumor-specific immune responses?

To address these questions, two different assay systems were employed. To assess the ability of epidermal antigen-presenting cells to induce *de novo* tumor immunity in syngeneic hosts, epidermal cells were generated, pulsed with tumor fragments (TF) derived from the cutaneous spindle cell tumor S1509a (syngeneic to A/J, H-2^a) as a source of tumor-associated antigens (TAA), and injected into CAF_1 (H-2$^{a/d}$) recipient mice. This immunization protocol was repeated twice at weekly intervals. One week after the last immunization, the mice were challenged with live S1509a cells, injected subcutaneously at a dose that would lead to progressive tumor growth in naive mice. Mice immunized with tumor antigen-pulsed epidermal APC were capable of rejecting this tumor challenge, indicating that they developed complete protective tumor immunity against S1509a. Control mice immunized with untreated epidermal cells showed progressive tumor growth leading to the eventual death of the animals. A second assay system was used to assess tumor antigen presentation in a primed system. Mice were immunized against S1509a by subcutaneous injection of dead S1509a tumor cells, and the elicitation of tumor-specific delayed-type hypersensitivity (DTH) response was measured after injection of tumor antigen-pulsed epidermal antigen-presenting cells into hind footpads of tumor-immune mice.

Using these experimental protocols we found that fresh tumor antigen-pulsed epidermal Langerhans cells were only weak inducers of tumor immunity when injected into native animals. However, preincubation of epidermal cells in 50 U/ml GM-CSF for 18 h before exposure to tumor antigens induced the capacity of epidermal cells to mediate potent protective tumor immunity in naive animals. This capacity of epidermal cells to initiate tumor immunity was restricted to I-A$^+$ epidermal APC, since deletion of I-A$^+$ epidermal cells with anti-I-A and complement-mediated lysis abrogated the capacity of epidermal cells to induce protective tumor immunity (Fig. 8.2). Further studies investigated the genetic restriction of tumor antigen presentation. When CAF_1 (BALB/c × A/J) F_1 mice were immunized against S1509a using epidermal antigen-presenting cells from A/J mice, these mice were only capable of recognizing S1509a tumor antigen in association with the I-A^k MHC class II molecules on the surface of epidermal Langerhans cells from A/J mice, but not in association with I-A^d MHC class II molecules on the surface of epidermal cells from BALB/c mice when tested for delayed-type hypersensitivity by footpad injection of tumor antigen-pulsed epidermal APC (A/J or BALB/c) into a hind footpad of immunized mice. Taken together, these studies indicate that there are cells within the epidermis that have the capacity to recognize tumor-associated

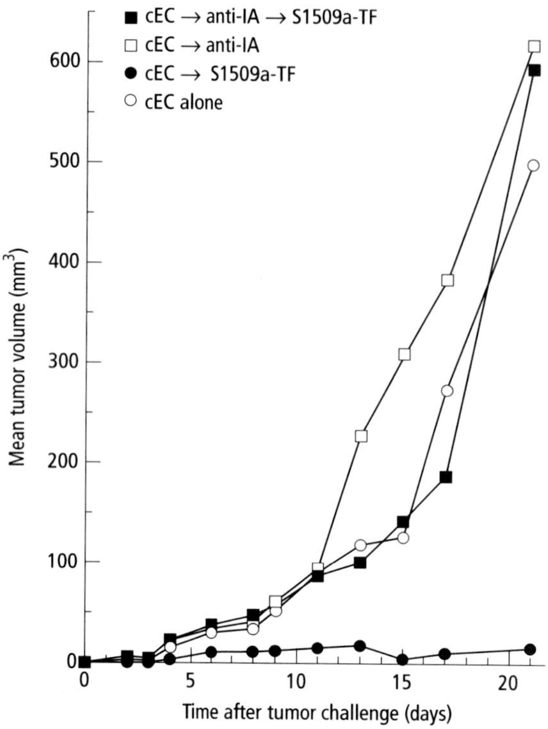

Fig. 8.2 I-A$^+$ epidermal APC present TAA for the induction of protective tumor immunity. EC were generated, enriched for LC by treatment with anti-Thy 1.2 + C′, and incubated in 50 U/ml GM-CSF for 18 h. Some groups of EC (■ □) were treated with anti-I-A^k + C′ to delete I-A$^+$ LC. The cells were then exposed to either S1509a-tumor fragments (TF, as a source of TAA) (● ■) or medium alone (□ ○) and injected s.c. into naive mice three times at weekly intervals. One week after the last immunization, all mice were challenged with live S1509a tumor cells and tumor growth was measured. n = 6/group.

antigens and present them to immunocompetent cells, leading to the induction of protective tumor immunity.

The inability of UV-irradiated mice to reject an immunogenic UVR-induced tumor prompted us to investigate the effect of UVB irradiation on the ability of Langerhans cells to present tumor-associated antigens in the S1509a system. We found that UV irradiation *in vitro*, before or after exposure of GM-CSF-incubated epidermal APC to tumor-associated antigens, reduced the ability to generate protective immunity in naive recipient mice. Doses larger than 200 J/m^2 abrogated their ability to present tumor-associated antigens. In the past similar observations have been made using haptens or protein antigens. These data indicate that UV irradiation *in vitro*, even at relatively small doses, profoundly disturbs the ability of epidermal APC to induce *de novo* protective tumor immunity.

Since epidermal Langerhans cells were found to be able to recognize tumor-associated antigens and to induce effective tumor immunity, the question arises why this apparently does not happen in a tumor-bearing host *in situ*. Several investigators have shown an alteration of Langerhans cell morphology, and in some instances also of Langerhans cell number, in the vicinity of growing tumors *in situ*, but quantitative differences are quite unlikely to account for the fact that progressive tumor growth is seen within the skin despite the direct proximity of APC that are capable of recognizing these tumor cells as antigens. These data suggested to us that mechanisms exist in tumor-bearing mice, or in the vicinity of growing tumors, that are capable of downregulating the ability of epidermal antigen-presenting cells to induce potent protective tumor immunity. Possible levels at which these regulatory processes operate include the recognition of tumor antigen, its processing and reassociation in conjunction with class II molecules, the antigen-presenting function of Langerhans cells, including costimulatory signals, and immunoregulatory mechanisms on T-cell and effector-cell level. Therefore we investigated whether cytokines which are known to be present within the skin or in the vicinity of growing tumors, and which are produced by epidermal cells or infiltrating lymphocytes, are able to modulate the ability of Langerhans cells to present tumor-associated antigens for the induction of tumor immunity. As mentioned earlier, incubation of epidermal cells in GM-CSF before exposure to tumor antigen was found to be necessary to induce their ability to present tumor-associated antigens effectively. Incubation of epidermal cells in TNFα alone did not induce significant protective tumor immunity, and coincubation of epidermal cells in GM-CSF and 100 U/ml TNFα had a moderate downregulatory effect when compared to epidermal cells incubated with GM-CSF alone. However, epidermal cells incubated in TNFα for 2 h after overnight exposure in GM-CSF completely abrogated the upregulatory effect of GM-CSF on Langerhans cell tumor antigen presentation in this system. Conversely, when TNFα was given for 2 h before incubation in GM-CSF overnight, no inhibition of tumor antigen presentation was seen. In fact, some experiments have suggested that this treatment even upregulates the ability of epidermal antigen-presenting cells to present tumor-associated antigens in unprimed systems. These studies indicate that incubation of epidermal cells in TNFα after culture in GM-CSF reversed the GM-CSF-induced upregulation of their ability to induce tumor immunity in unprimed systems, and identified TNFα as a cytokine that was capable of downregulating the induction of tumor immunity by epidermal APC. Incubation of EC in 1–10 ng/ml TGFβ had no inhibitory effect in this system, despite its capacity to downregulate antigen presentation by other APC [6, 22, 115]. The effects of these cytokines on LC function were confirmed by using cytokine-incubated epidermal cells as stimulators in the allogeneic primary mixed epidermal lymphocyte reaction, and thus seem to apply for LC function in unprimed systems in general.

Further studies revealed that other cytokines were also capable of modulating tumor antigen presentation by epidermal APC. To our surprise, IL-1α inhibited tumor antigen presentation in this system, despite the fact that IL-1 is generally a pro-inflammatory and immunostimulatory cytokine, and has been shown to augment antigen presentation by monocytes, macrophages and dendritic cells [1, 61, 87, 98, 109, 112, 139]. Coincubation of epidermal cells in GM-CSF plus 100 U/ml IL-1α led to a dose-dependent inhibition of the immunostimulatory effect on tumor antigen presentation by Langerhans cells induced by GM-CSF. Incubation of epidermal cells in IL-1 alone did not augment their ability to present tumor-associated antigens in an unprimed system, and sequential exposure of epidermal cells to IL-1 first and then GM-CSF (or vice versa) only led to a minor inhibition of the effects induced by GM-CSF alone. Coincubation of epidermal cells in IL-1α, GM-CSF, and neutralizing anti-TNF antibody, however, led to a complete abrogation of the effect of IL-1 on this system, i.e. no inhibition of tumor antigen presentation was seen after addition of anti-TNFα antibody to the culture system. These results indicate that GM-CSF upregulates the ability of epidermal antigen-presenting cells to present tumor-associated antigens in this system, and that TNFα as well as IL-1α downregulate it, with IL-1α acting at least partially through induction of TNFα-secretion. Further studies confirmed that exposure of epidermal cells of IL-1α significantly upregulates the secretion of TNFα into the culture system. These results may also be of relevance for the understanding of UV-induced alterations of cutaneous APC function, since both IL-1 and TNFα are released in significant quantities after UVR. However, additional experiments have demonstrated that UV-induced secretion of IL-1 and/or TNF cannot fully account for the inhibition of cutaneous LC function after UVR, since addition of antisera to IL-1α and/or TNFα after UV irradiation of EC did not fully restore their ability to present TAA in this system (Fig. 8.3). Therefore we conclude that UVR has both direct as well as indirect, i.e. cytokine-mediated, effects on the ability of epidermal APC to present TAA in unprimed systems.

A second set of experiments was aimed at determining the capacity of epidermal APC to present tumor-associated antigens in primed systems. For this purpose epidermal cells were pulsed with TAA and injected into the hind footpads of tumor-immune mice which had already rejected challenge with live S1509a cells. Injection of TAA-pulsed epidermal cells, but not of epidermal cells that were not exposed to TAA, resulted in significant delayed-type hypersensitivity responses in tumor-immune mice, as measured by specific footpad swelling. Again, this response was due to I-A$^+$ APC within the epidermal cell population and genetically restricted. These experiments indicate that freshly prepared epidermal antigen-presenting cells are fully capable of presenting TAA to generate significant immune reactivity in the form of a DTH response. This experimental system was

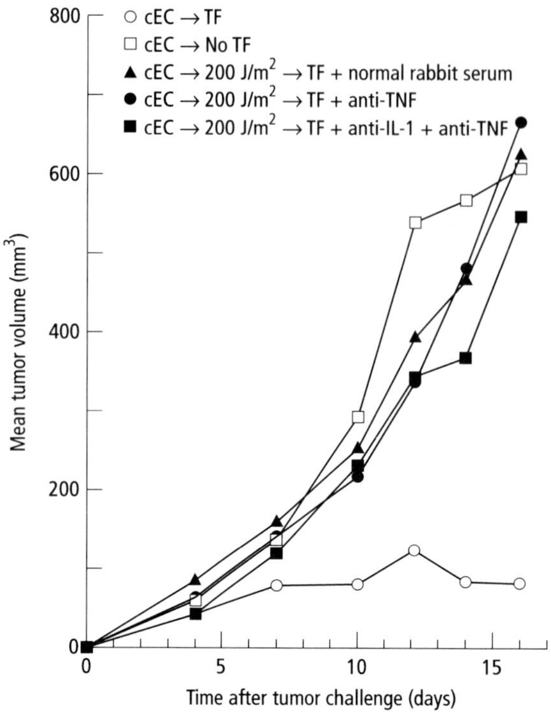

Fig. 8.3 Effects of UVR on the induction of tumor immunity by TAA-pulsed epidermal APC. EC were generated, enriched for LC by treatment with anti-Thy 1.2 + C′, and incubated in 50 U/ml GM-CSF for 18 h. Some groups of EC (▲ ● ■) were UV irradiated with 200 mJ/m^2 before exposure to either S1509a-TF (○ ▲ ● ■) or medium alone (□). Immediately following irradiation and throughout all following incubations, cells were in the presence of anti-TNF antibody (●), anti-TNF + anti-IL-1 antibodies (■), or control serum (▲). Cells were then injected s.c. into naive mice three times at weekly intervals. One week after the last immunization, all mice were challenged with live S1509a tumor cells and tumor growth was measured.

then used to determine the effects of UVR and cytokines on tumor antigen presentation in a primed system. In a manner similar to the effect of UVR on the presentation of TAA in unprimed systems, UVR also profoundly inhibited the ability of epidermal APC to elicit a tumor-specific DTH response in tumor immune mice (Fig. 8.4). TNFα, which was found to profoundly inhibit the ability of GM-CSF-pulsed EC to induce a tumor-specific immune response, augmented the ability of fresh EC to elicit tumor-specific DTH. These results further emphasize that the regulation of primary and secondary immune responses by cytokines and accessory signals differs significantly. No significant difference was observed in the effects of TNFα when either fresh or GM-CSF-cultured EC were used, although in general, culture of EC in GM-CSF for either 3 or 18 h seemed to decrease their ability to induce tumor-specific DTH. IL-1α was also capable of suppressing the DTH response elicited by TAA-pulsed EC significantly, both when fresh or GM-CSF-cultured EC were used to elicit tumor-specific DTH. Therefore the effects of IL-1α and TNFα differed significantly with regard to induction and elicitation of immunity. TGFβ did not inhibit the presentation of TAA, and appeared to upregulate the ability of GM-CSF-cultured epidermal APC to elicit tumor-specific DTH in this system. These data are summarized in Table 8.5.

Taken together, these data indicate that UVR and UV-induced cytokines have the ability to modulate the presentation of TAA in unprimed as well as in primed systems. Whether the observed effects of these agents on cutaneous antigen presentation merely explain the lack of generation of tumor immunity, or whether they also account for the altered quality of antigen presentation, leading to Ag-specific suppression of immunity, still needs to be established.

Table 8.5 Effects of cytokines on LC function in a primed system: *elicitation* of tumor immunity

	fLC*	cLC†
TNFα	Augmentation	Augmentation
IL-1α	Inhibition	Inhibition
TGFβ	No effect or slight augmentation	Augmentation

*fLC, freshly prepared Langerhans cells. †cLC, Langerhans cells after culture in 50 U/ml GM-CSF for 18 h. Reprinted by permission of Elsevier Science Publishing Co., Inc. from: Effects of immunomodulatory cytokines on the presentation of tumor-associated antigens by epidermal Langerhans cells., by S. Grabbe, S. Bruvers and R.D. Granstein. *The Journal of Investigative Dermatology*, vol. 99, pp. 665–685. Copyright 1992 by the Society for Investigative Dermatology, Inc.

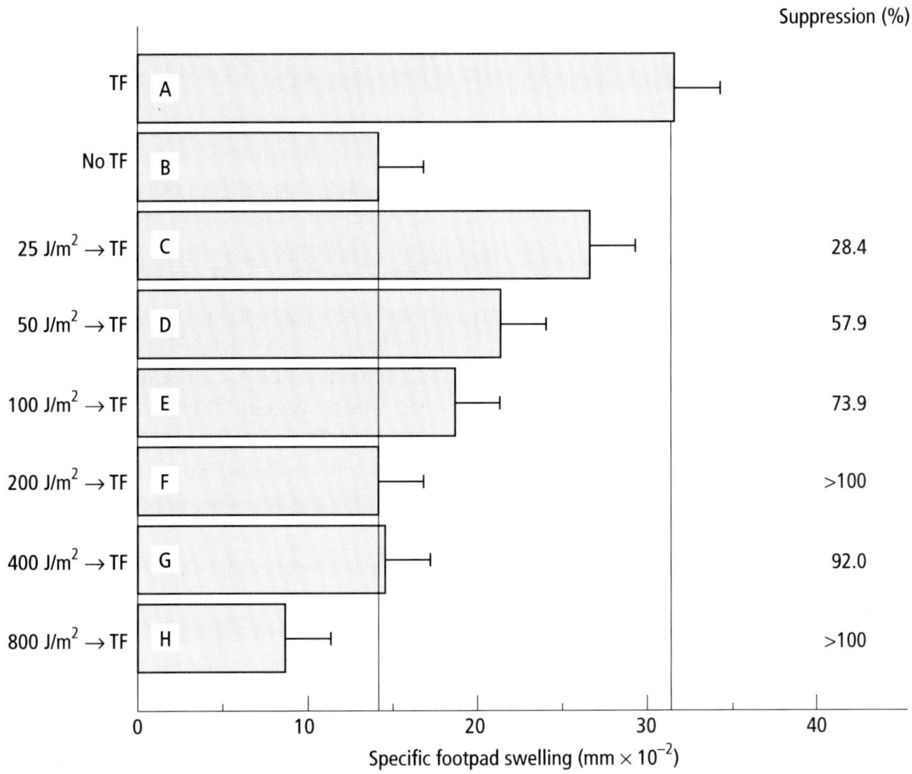

Fig. 8.4 Effects of UVR on the ability of TF-pulsed EC to elicit a tumor-specific DTH response. Tumor-immune mice were prepared by repeated injections of dead S1509a cells. LC-enriched EC were generated, irradiated with varying doses of UVR (C-H) and incubated in complete medium for 1 h before exposure to S1509a-TF (A, C-H) or medium alone (B). Cells were then injected s.c. into hind footpads of tumor-immune mice, and the specific footpad swelling response was measured 36 h after injection. Reprinted by permission of the Society for Leukocyte Biology from [39]. Tumor antigen presentation by epidermal antigen-presenting cells in the mouse: modulation by granulocyte-macrophage colony-stimulating factor, tumor necrosis factor α, and ultraviolet radiation, by S. Grabbe, S. Bruvers, A.M. Lindgren, J. Hosoi, K.C. Tan, and R.D. Granstein. *Journal of Leukocyte Biology*, vol. 52, pp. 209–217, 1992. Copyright 1992 by The Society for Leukocyte Biology. n = S/group.

Summary

The ability of UVR to modify immunity *in vivo* has been well established by a number of investigators in recent years. The possible relevance to cutaneous oncogenesis was established quite early with the observation that UVR-induced tumors were highly antigenic and that their growth and development depended on the appearance of a set of T suppressor cells within chronically UV-exposed mice. Subsequent studies using contact hypersensitivity responses have helped to define the immunological deviations induced by UVR. These data indicate that UVR induces a number of immunologic alterations within the irradiated host that eventually aid the survival and progressive growth of antigenic UV-induced malignancies, and may also lead to the selection of poorly antigenic, highly malignant tumor forms. More recent work, described in detail above, has shown that murine Langerhans cells can indeed present TAA both for the induction and elicitation of tumor immunity in the S1509a system. Importantly, the ability of epidermal APC to subserve these functions appears to be regulated by a number of polypeptide cytokines, which can be produced by keratinocytes or by recruited inflammatory cells. Therefore the local cytokine microenvironment may affect both the tumor cells themselves (tumor promotion from benign to malignant tumor growth [88]) and the ability of the host to mount a protective immune response against the tumor. Production and secretion of a number of cytokines by UVR-exposed keratinocytes has been well established, and the finding that exposure of epidermal cell populations to UVR abrogates their ability to present tumor antigen for induction or elicitation of immunity provides further evidence for the apparent deleterious effect of this modality on cutaneous immunity *in vivo*.

Although much progress has been made in understanding these events, a number of interesting questions remain. The subcellular and molecular events by which exposure to UVR alters antigen presentation remain obscure, as do the reasons why evolution has allowed these responses to UVR to develop. One possibility, of course, is that UVR-induced immunologic changes did not evolve but rather are an unavoidable consequence of other evolutionary choices, and that animals have simply evolved only partial or incomplete mechanisms to prevent these UVR effects. A second possibility is that these down-regulatory effects have evolved to inhibit the development of autoimmunity after UVR-induced damage to the skin, with release of various potential autoantigens, and/or to limit inflammation induced in the skin by UVR.

The most important unanswered question may be the relation of the various findings from animal experiments, discussed above, to the human situation. Exposure of humans to modest doses of UVB radiation has indeed been shown to alter contact hypersensitivity responses and the proportions of different phenotypes of circulating lymphocytes. In addition, the role of the human immune system in regulating UVR-induced skin cancer is highlighted by the epidemiologic observations that persons immunosuppressed therapeutically have much greater rates of nonmelanoma and probably melanoma skin cancer. Very recent observations have suggested that skin cancer patients may have an increased sensitivity to UVR-induced suppression of contact hypersensitivity compared to age-matched controls without skin cancer [143]. These studies provide some circumstantial evidence that many of the mechanisms discerned in animal models may apply to humans as well.

Many of the mechanisms involved in the regulation of tumor growth by the immune system remain very poorly understood. The investigation of the regulation of immune effector mechanisms during the development and growth of cutaneous malignancies remains a subject of intense study which may lead to new and more effective means of preventing and treating tumors.

References

1 Aiello, F.B., D.L. Longo, R. Overton, L. Takacs, S.K. Durum: A role for cytokines in antigen presentation: IL-1 and IL-4 induce accessory functions of antigen-presenting cells. *J. Immunol.* 144 (1990), 2572–2581.

2 Alcalay, J., J. Craig, M. Kripke: Alterations in Langerhans cells and Thy-1^+ dendritic epidermal cells in murine epidermis during the evolution of ultraviolet radiation-induced skin cancers. *Cancer Res.* 49 (1989), 4591–4596.

3 Alcalay, J., L.H. Goldberg, M.L. Kripke, J.E.J. Wolf: The sensitivity of Langerhans cells to simulated solar radiation in basal cell carcinoma patients. *J. Invest. Dermatol.* 93 (1989), 746–750.

4 Alcalay, J., L. Goldberg, J. Wolf, M. Kripke: Variations in the number and morphology of Langerhans cells in the epidermal component of squamous cell carcinomas. *Arch. Dermatol.* 125 (1989), 917–920.

5 Alcalay, J., M.L. Kripke: Antigen-presenting activity of draining lymph node cells from mice painted with a contact allergen during ultraviolet carcinogenesis. *J. Immunol.* 146 (1991), 1717–1721.

6 Altman, D.J., S.L. Schneider, D.A. Thompson, H.-L. Cheng, T.B. Tomasi: A transforming growth factor β2 (TGF-β2) like immunosuppressive factor in amniotic fluid and localization of TGF-β2 mRNA in the pregnant uterus. *J. Exp. Med.* 172 (1990), 1391–1401.

7 Ananthaswamy, H.N., M.L. Kripke: *In vitro* transformation of primary cultures of neonatal BALB/c mouse epidermal cells with ultraviolet B radiation. *Cancer Res.* 41 (1981), 2882–2890.

8 Begovic, M., R.B. Herberman, E. Gorelik: Increase in immunogenicity and sensitivity to natural cell-mediated cytotoxicity following *in-vitro* exposure of MCA105 tumor cells to ultraviolet radiation. *Cancer Res.* 51 (1991), 5133–5139.

9 Begovic, M., R.B. Herberman, E. Gorelik: Ultraviolet light-induced increase in tumor cell susceptibility to TNF-dependent and TNF-independent natural cell-mediated cytotoxicity. *Cell. Immunol.* 138 (1991), 349–359.

10 Bergstresser, P.R., J.W. Streilein: Ultraviolet radiation produces selective immune incompetence. *J. Invest. Dermatol.* 81 (1983), 85–86.

11 Bergstresser, P.R., G.B. Toews, J.W. Streilein: Natural and perturbed distributions of Langerhans cells. Responses to ultraviolet light, heterotopic skin grafting, and dinitrofluorobenzene sensitization. *J. Invest. Dermatol.* 75 (1980), 73–77.

12 Blum, H.F.: *Carcinogenesis by Ultraviolet Light* (1959). Princeton University Press, Princeton, NJ.

13 Burnham, D.K., C.K. Mak, R.J. Webster, R.A. Daynes: Relationship between inducible H-2 expression and the immunogenicity of murine skin neoplasms. I. Evidence that the immunogenicity of ultraviolet radiation- and chemically induced tumors is associated with their susceptibility to gamma-interferon-mediated enhancement of H-2^{Kk} expression. *Transplantation* 47 (1989), 533–542.

14 Burnet, F.M.: *Immunologic Surveillance.* (1970) Pergamon Press, Oxford.

15 Canfield, P.J., G.E. Greenoak, E.N. Macasaet, V.E. Reeve, C.H. Gallagher: The characterization of squamous cell carcinoma induced by ultraviolet irradiation in hairless mice. *Pathology* 20 (1988), 109–117.

16 Choi, Y., P. Marrack, J.W. Kappler: Structural analysis of a mouse mammary tumor virus superantigen. *J. Exp. Med.* 175 (1992), 847–852.

17 Cruz, P.D., P.R. Bergstresser: The low dose model of UVB induced immunosuppression. *Photodermatology* 5 (1988), 151–161.

18 Cruz, P.D., J.L. Nixon-Fulton, R.E. Tigelaar, P.R. Bergstresser: Disparate effects of *in vitro* low-dose UVB irradiation on intravenous immunization with purified epidermal cell subpopulations for the induction of contact hypersensitivity. *J. Invest. Dermatol.* 92 (1989), 160–165.

19 Daynes, R.A., D.K. Burnham, C.W. Dewitt, L.K. Roberts, G.G. Drueger: The immunobiology of ultraviolet-radiation carcinogenesis. *Cancer Surv.* 4 (1985), 51–99.

20 DeFabo, E.C., M.L. Kripke: Dose–response characteristics of immunologic unresponsiveness to UV-induced tumors produced by UV irradiation of mice. *Photochem. Photobiol.* 30 (1979), 385–390.

21 DeFabo, E.C., M.L. Kripke: Wavelength dependence and dose-rate independence of UV radiation induced immunologic unresponsiveness of mice to a UV induced fibrosarcoma. *Photochem. Photobiol.* 32 (1980), 183–188.

22 Demidem, A., J.R. Taylor, S.F. Grammer, J.W. Streilein.: Comparison of effects of transforming growth factor-beta and cyclosporin A on antigen-presenting cells of blood and epidermis. *J. Invest. Dermatol.* 96 (1991), 401–407.

23 Donawho, C., M.L. Kripke: Immunogenicity and cross-reactivity of syngeneic murine melanomas. *Cancer Commun.* 2 (1990), 101–107.

24 Ebbesen, P.: Enhanced lymphoma incidence in BALB/c mice after ultraviolet light treatment. *J. Natl. Cancer Inst.* 67 (1981), 1077–1078.

25 Ebbesen, P., M.L. Kripke: Influences of age and anatomical site on ultraviolet carcinogenesis in BALB/c mice. *J. Natl. Cancer Inst.* 68 (1982), 691–694.

26 Elmets, C.A., P.R. Bergstresser: Ultraviolet radiation effects on immune processes. *Photochem. Photobiol.* 36 (1982), 715–722.

27 Elmets, C.A., W.A. Khan, J.C. Klemme, H. Mukthar: Impaired immunological surveillance by 7,12-dimethylbenz(a)anthracene augments its skin tumorigenicity in C3H mice. *Biochem. Biophys. Res. Commun.* 151 (1988), 148–152.

28 Elmets, C.A., K. Larson, G.A. Urda, B. Schacter: Inhibition of postbinding target cell lysis and of lymphokine-induced enhancement of human natural killer cell activity by *in vitro* exposure to ultraviolet B radiation. *Cell. Immunol.* 104 (1987), 47–58.

29 Findlay, G.M.: Ultraviolet light and skin cancer. *Lancet* ii (1928), 1070–1073.

30 Fisher, M.S., M.L. Kripke: Systemic alteration induced in mice by ultraviolet light irradiation and its relationship to ultraviolet carcinogenesis. *Proc. Natl. Acad. Sci. USA* 74 (1977), 1688–1692.

31 Fisher, M.S., M.L. Kripke: Further studies on the tumor-specific suppressor cells induced by ultraviolet radiation. *J. Immunol.* 121 (1978), 1139–1144.

32 Fisher, M.S., M.L. Kripke: Suppressor T lymphocytes control the development of primary skin cancer in ultraviolet-irradiated mice. *Science* 216 (1982), 1133–1134.

33 Fisher, M.S., M.L. Kripke, G.L. Chan: Antigenic similarity between cells transformed by ultraviolet radiation *in vitro* and *in vivo*. *Science* 223 (1984), 593–594.

34 Flood, P.M., J.L. Urban, M.L. Kripke, H. Schreiber: Loss of tumor-specific and idiotype-specific immunity with age. *J. Exp. Med.* 154 (1981), 275–290.

35 Fortner, G.W., M.L. Kripke: *In vitro* reactivity of splenic lymphocytes from normal and UV-irradiated mice against syngeneic UV-induced tumors. *J. Immunol.* 118 (1977), 1483–1487.

36 Fujiwara, H., M. Fukuzama, T. Yoshioka, H.

Nakajima, T. Hamaoka: The role of tumor-specific LYT-1^+2^- T cells in eradicating tumor cells *in vivo*. I. Lyt-1^+2^- T cells do not necessarily require recruitment of host's cytotoxic T cell precursors for implementation of *in vivo* immunity. *J. Immunol.* 133 (1984), 1671–1678.

37 Glass, M.J., P.R. Bergstresser, R.E. Tigelaar, J.W. Streilein: UVB radiation and DNFB skin painting induce suppressor cells universally in mice. *J. Invest. Dermatol.* 94 (1990), 273–278.

38 Grabbe, S., S. Bruvers, R.L. Gallo, T.L. Knisely, R. Nazareno, R.D. Granstein: Tumor antigen presentation by murine epidermal cells. *J. Immunol.* 146 (1991), 3656–3661.

39 Grabbe, S., S. Bruvers, R.D. Granstein: Effects of immunomodulatory cytokines on the presentation of tumor-associated antigens by epidermal Langerhans cells. *J. Invest. Dermatol.* 99 (1992), 66S–68S.

40 Grabbe, S., S. Bruvers, A.M. Lindgren, J. Hosoi, K.C. Tan, R.D. Granstein: Tumor antigen presentation by epidermal antigen-presenting cells in the mouse: modulation by granulocyte-macrophage colony stimulating factor, tumor necrosis factor α, and ultraviolet radiation. *J. Leukoc. Biol.* 52 (1992), 209–217.

41 Granstein, R.D.: Photoimmunology. *Semin. Dermatol.* 9 (1990), 16–24.

42 Greene, M.H., T.I. Young, J.W.H. Clark: Malignant melanoma in renal transplant patients. *Lancet* i (1981), 1196–1198.

43 Gurish, M.F., D.H. Lynch, R.A. Daynes: Changes in antigen-presenting cell function in the spleen and lymph nodes of ultraviolet-irradiated mice. *Transplantation* 33 (1982), 280–284.

44 Hakomori, S.: Biochemical basis and clinical application of tumor-associated carbohydrate antigens: current trends and future perspectives. *Gan To Kagaku Ryoho* 19 (1989), 715–731.

45 Hakomori, S.: New directions in cancer therapy based on aberrant expression of glycosphingolipids: anti-adhesion and ortho-signaling therapy. *Cancer Cells* 3 (1991), 461–470.

46 Halliday, G.M., A.D. Lucas, R.S.C. Barnetson: Control of Langerhans cell density by a skin tumour-derived cytokine. *Immunology* 77 (1992), 13–18.

47 Halliday, G.M., K.A. Odling, J.C. Ruby, H.K. Muller: Suppressor cell activation and enhanced skin allograft survival after tumor promotor but not initiator induced depletion of cutaneous Langerhans cells. *J. Invest. Dermatol.* 90 (1988), 293–297.

48 Halliday, G.M., V.E. Reeve, R.S.C. Barnetson: Langerhans cell migration into ultraviolet light-induced squamous skin tumors is unrelated to anti-tumor immunity. *J. Invest. Dermatol.* 97 (1991), 830–834.

49 Herberman, R.B.: *Basic and Clinical Tumor Immunology* (1983). Martinus Nijhoff Publishers, Boston, MA.

50 Hersey, P., M. MacDonald, C. Henderson, S. Schibeci, G. D'Alessandro, M. Pryor, F.J. Wilkinson: Suppression of natural killer cell activity in humans by radiation from solarium lamps depleted of UVB. *J. Invest. Dermatol.* 90 (1988), 305–310.

51 Hill, B.H.: Immunosuppressive drug therapy as a potentiator of skin tumors in five patients with lymphoma. *Australas. J. Dermatol.* 17 (1976), 46–48.

52 Hong, S.R., L.K. Roberts: Cross-reactive tumor antigens in the skin of mice exposed to subcarcinogenic doses of ultraviolet radiation. *J. Invest. Dermatol.* 88 (1987), 154–160.

53 Hoover, R., J.F. Fraumeni: Risk of cancer in renal-transplant recipients. *Lancet* ii (1973), 55–57.

54 Hostetler, L.W., H.N. Ananthaswamy, M.L. Kripke: Generation of tumor-specific transplantation antigens by UV radiation can occur independently of neoplastic transformation. *J. Immunol.* 137 (1986), 2721–2725.

55 Hostetler, L.W., M.L. Kripke: Origin and significance of transplantation antigens induced on cells transformed by UV radiation. In: *Development and Recognition of the Transformed Cell*, (1987), pp. 307–329. Eds.: M.I. Greene, T. Hamaoka, Plenum Publishing Corp., New York.

56 Hostetler, L.W., M.L. Kripke: Proportion of antigenic variants induced by *in vitro* UV irradiation differs in clones derived from a single tumor. *J. Immunol.* 140 (1988), 666–670.

57 Hostetler, L.W., C.A. Romerdahl, M.L. Kripke: Specificity of antigens on UV radiation-induced antigenic tumor cell variants measured *in vitro* and *invivo*. *Cancer Res.* 49 (1989), 1207–1213.

58 Hoxtell, E.O., J.S. Mandel, S.S. Murray, L.M. Schuman, R.W. Goltz: Incidence of skin cancer after renal transplantation. *Arch. Dermatol.* 113 (1977), 436–438.

59 Kaba, D.S., B.W. McIntyre, M.L. Kripke, R.S. Goodenow, H. Schreiber, H.N. Ananthaswamy: Generation of antigenic variants from a nonantigenic murine tumor cell line by transfection with a gene encoding a novel tumor-specific transplantation antigen. *Cancer Res.* 49 (1989), 38–43.

60 Kato, T., K. Inaba, Y. Ogawa, M. Inaba, K.-I. Kakihara, S. Shimizu, S. Ikehara, T. Sudo, S. Muramatsu: Granulocyte-macrophage colony stimulating factor enhances macrophage accessory function in ConA stimulated T cell proliferation. *Cell Immunol.* 130 (1990), 490–500.

61 Kawakami, K., Y. Yamamoto, K. Karimoto, K. Onoue: Requirement for delivery of signals by physical interaction and soluble factors from accessory cells in the induction of receptor mediated T cell proliferation. *J. Immunol.* 142 (1989), 1818–1825.

62 Kawakami, Y., R. Zakut, S.L. Topalian, H. Stotter, S.A. Rosenberg: Shared human melanoma antigens. Recognition by tumor infiltrating lymphocytes in HLA-A2. 1 transfected melanomas. *J. Immunol.* 148 (1992), 638–643.

63 Kern, D., J. Klarnet, M. Jensen, P. Greenberg: Requirement for recognition of class II molecules and processed tumor antigen for optimal generation of syngeneic tumor-specific class I-restricted CTL. *J. Immunol.* 136 (1986), 4303–4310.

64 Kinlen, L.J., A.G. Sheil, J. Peto, R. Doll: Collaborative United Kingdom – Australia study of cancer in patients treated with immunosuppressive drugs. *Br. J. Med.* 2 (1979), 1461–1466.

65 Kinochi, T., H. Fujiwara, T. Hamaoka: Expression of differentiation antigens cross-reactive with Ly 6.2 alloantigens on MM102 mouse mammary tumor cells. *GANN* 73 (1982), 758–765.

66 Klein, C.E., T. Steinmayer, D. Kaufmann, L. Weber, E.B. Brocker: Identification of a melanoma progression antigen as integrin VLA-2. *J. Invest. Dermatol.* 96 (1991), 281–284.

67 Kosugi, A., T. Yoshioka, T. Suda, H. Sano, Y. Takahama, H. Fujiwara, T. Hamaoka: The activation of L3T4$^+$ helper T cells assisting the generation of anti-tumor Lyt-2$^+$ cytotoxic T lymphocytes. *J. Leukoc. Biol.* 42 (1987), 632–641.

68 Kripke, M.L.: Antigenicity of murine skin tumors induced by ultraviolet light. *J. Natl. Cancer Inst.* 53 (1974), 1333–1336.

69 Kripke, M.L.: Immunologic parameters of ultraviolet carcinogenesis. *J. Natl. Cancer Inst.* 57 (1976), 211–215.

70 Kripke, M.L.: Latency, histology and antigenicity of tumors induced by ultraviolet light in three inbred mouse strains. *Cancer Res.* 37 (1977), 1395–1400.

71 Kripke, M.L.: Immunologic mechanisms in UV-radiation carcinogenesis. *Adv. Cancer Res.* 34 (1981), 69–106.

72 Kripke, M.L.: Immunological unresponsiveness induced by ultraviolet radiation. *Immunol. Rev.* 80 (1984), 87–102.

73 Kripke, M.L.: Effects of UV radiation on tumor immunity. *J. Natl. Cancer Inst.* 82 (1990), 1392–1395.

74 Kripke, M.L., I.J. Fidler: Enhanced experimental metastasis of ultraviolet light-induced fibrosarcomas in ultraviolet light-irradiated syngeneic mice. *Cancer Res.* 40 (1980), 625–629.

75 Kripke, M.L., J.S. Lofgreen, J. Beard, J.M. Jessup, M.S. Fisher: *In vivo* immune responses of mice during carcinogensis by ultraviolet irradaiation. *J Natl. Cancer Inst.* 59 (1977), 1227–1230.

76 Kripke, M.L., E. McClendon: Studies on the role of antigen-presenting cells in the systemic supression of contact hypersensitivity by UVB radiation. *J Immunol.* 137 (1986), 443–447.

77 Kripke, M.L., W.L. Morison: Studies on the mechanism of systemic suppression of contact hypersensitivity by UVB radiation. *Photodermatol.* 3 (1986), 4–14.

78 Kripke, M.L., W.L. Morison: Studies on the mechanism of systemic suppression of contact hypersensitivity by UVB radiation. II. Differences in the suppression of delayed and contact hypersensitivity in mice. *J. Invest. Dermatol.* 86 (1986), 543–549.

79 Kripke, M.L., W.L. Morison, J.A. Parrish: Differences in the immunologic reactivity of mice treated with UVB or methoxsalen plus UVA radiation. *J. Invest. Dermatol.* 76 (1981), 445–448.

80 Kripke, M.L., W.L. Morison, J.A. Parrish: Induction and transplantation of murine skin cancers induced by methoxsalen plus ultraviolet (320–400 nm) radiation. *J. Invest. Dermatol.* 68 (1982), 685–690.

81 Kripke, M.L., W.L. Morison, J.A. Parrish: Systemic suppression of contact hypersensitivity in mice by psoralen plus UVA radiation (PUVA). *J. Invest. Dermatol.* 81 (1983), 87–92.

82 Kripke, M.L., R.M. Thorn, P.H. Lill, P.I. Civin, N.H. Pazmino, M.S. Fisher: Further characterization of immunological unresponsiveness induced in mice by ultraviolet radiation. Growth and induction of non-ultraviolet-induced tumors in ultraviolet-irradiated mice. *Transplantation* 28 (1979), 212–217.

83 Leshem, B., E. Kedar: Cytotoxic T lymphocytes reactive against a syngeneic murine tumor and their specific suppressor T cells are both elicited by *in vitro* allosensitization. *J. Exp. Med.* 171 (1990), 1057–1071.

84 Ley, R.D., L.A. Applegate, R.S. Padilla, T.D. Stuart: Ultraviolet radiation-induced malignant melanoma in *Monodelphis domestica*. *Photochem. Photobiol.* 50 (1989), 1–5.

85 Liao, N.-S., M. Bix, M. Zijlstra, R. Jaenisch, D. Raulet: MHC class I deficiency: susceptibility to natural killer (NK) cells and impaired NK activity. *Science* 253 (1991), 199–201.

86 Lloyd, K.O.: Humoral immune responses to tumor-associated carbohydrate antigens. *Semin. Cancer Biol.* 2 (1991), 421–431.

87 Mantovani, A., B. Bottazzi, F. Colotta, S. Sozzani, L. Ruco: The origin and function of tumor-associated macrophages. *Immunol. Today* 13 (1992), 265–269.

88 Marks, F., G. Fürstenberger: Mechanism of skin tumor promotion (Abstract). In: *Environmental Skin Cancer*, international meeting, Oct. 1992, Cleveland, OH.

89 Messadi, D.V., J.S. Pober, G.F. Murphy: Effects of recombinant γ-interferon on HLA-DR and DQ expression by skin cells in short-term organ culture. *Lab. Invest.* 58 (1988), 61–67.

90 Morison, W.L.: *In vivo* effects of psoralens plus longwave ultraviolet radiation immunity. *Natl Cancer Inst. Monogr.* 66 (1984), 243–246.

91 Morison, W.L., C. Bucana, M.L. Kripke: Systemic suppression of contact hypersensitivity by UVB radiation is unrelated to the UVB-induced alterations in the morphology and number of Langerhans cells. *Immunology* 52 (1984), 299–306.

92 Morison, W.L., M.S. Jerdan, T.L. Hoover, E.R. Farmer: UV radiation-induced tumors in haired mice: identification as squamous cell carcinomas. *J. Natl Cancer Inst.* 77 (1986), 1155–1162.

93 Van Muijen, G.M., L.M. Cornelissen, C.F. Jansen, C.G. Figdor, J.P. Johnson, E.B. Brocker, D.J. Ruiter: Antigen expression of metastasizing and non-metastasizing human melanoma cells xenografted into nude mice. *Clin. Exp. Metastasis* 9 (1991), 259–272.

94 Noonan, F.P., E.C. DeFabo, M.L. Kripke: Suppression of contact hypersensitivity by UV radiation and its relationship to UV-induced suppression of tumor immunity. *Photochem. Photobiol.* 34 (1981), 683–689.

95 Norbury, K.C., M.L. Kripke: Ultraviolet induced carcinogenesis in mice treated with silica, trypan blue or pyran copolymer. *J. Reticuloendothelial Soc.* 26 (1979), 827–837.

96 Norbury, K.C., M.L. Kripke, M.B. Budmen: *In vitro* reactivity of macrophages and lymphocytes from ultraviolet-irradiated mice. *J. Natl. Cancer Inst.* 59 (1977), 1231–1235.

97 North, R.J., R.H. Neubauber, J.J.H. Huang, R.C. Newton, S.E. Loveless: Interleukin 1-induced, T cell mediated regression of immunogenic murine tumors. *J. Exp. Med.* 168 (1988), 2031–2043.

98 Ogata, M., J. Shimizu, H. Kosaka, R. Maekawa, K. Shimizu, H. Fujiwara, T. Hamaoka: Expression of H-2 antigens and inducibility of antitumor immune responses in various tumor cell clones established from methylcholanthrene-induced fibrosarcomas. *Jpn. J. Cancer Res.* 77 (1986), 1134–1143.

99 Palaszynski, E.W., M.L. Kripke: Transfer of immunological tolerance to ultraviolet-radiation-induced skin tumors with grafts of ultraviolet-irradiated skin. *Transplantation* 36 (1983), 465–466.

100 Pfizenmaier, K., P. Scheurich, C. Schlüter, M. Krönke: Tumor necrosis factor enhances HLA-A,B,C and HLA-DR gene expression in human tumor cells. *J. Immunol.* 138 (1987), 975–980.

101 Reisfeld, R.: Pre-clinical models for immunotherapy of melanoma. *Immun. Cancer* 11 (1989), 183–193.

102 Roberts, L.K., E.J. Bernhard, R.A. Daynes: Experimental ultraviolet radiation carcinogenesis: I. Relationship of the major histocompatibility complex to tumor latency and immunogenicity. *Photodermatol.* 1 (1984), 57–64.

103 Roberts, L.K., D.H. Lynch, R.A. Daynes: Evidence for two functionally distinct cross-reactive tumor antigens associated with ultraviolet light and chemically induced tumors. *Transplantation* 33 (1982), 352–360.

104 Roberts, L.K., W.E. Samlowski, R.A. Daynes: Immunological consequences of ultraviolet radiation exposure. *Photodermatology* 3 (1986), 284–298.

105 Roberts, L.K., C.W. Spellman, N.L. Warner: Establishment of a continuous T cell line capable of suppressing anti-tumor immune responses *in vivo*. *J. Immunol.* 131 (1983), 514–519.

106 Romerdahl, C.A., H. Okamoto, M.L. Kripke: Immune surveillance against cutaneous malignancies in experimental animals. *Immunol. Ser.* 46 (1989), 749–767.

107 Romerdahl, C.A., L.C. Stephens, C. Bucana, M.L. Kripke: The role of ultraviolet radiation in the induction of melanocytic skin tumors in inbred mice. *Cancer Commun.* 1 (1989), 209–216.

108 Ruiter, D.J., V. Mattijssen, E.B. Broecker, S. Ferrone: MHC antigens in human melanomas. *Semin. Cancer Biol.* 2 (1991), 35–45.

109 Ruppert, J., J.H. Peters: IL-6 and IL-1 enhance the accessory activity of human blood monocytes during differentiation to macrophages. *J. Immunol.* 146 (1991), 144–149.

110 Rusch, H.P., C.A. Baumann: Tumor production in mice with ultraviolet irradiation. *Am J. Cancer* 35 (1939), 55–62.

111 Sakamoto, K., H. Nakajima, J. Shimizu, T. Katagiri, C. Kiyotaki, H. Fujiwara, T. Hamaoka: The mode of recognition of tumor antigens by noncytolytic-type antitumor T cells: role of antigen-presenting cells and their surface class I and class II H-2 molecules. *Cancer Immunol. Immunother.* 27 (1988), 261–266.

112 Sauder, D.N., F.P. Noonan, E.C. DeFabo, S.I. Katz: Ultraviolet radiation inhibits alloantigen presentation by epidermal cells: partial reversal by the soluble epidermal cell products, epidermal cell-derived thymocyte-activation factor (ETAF). *J. Invest. Dermatol.* 80 (1983), 485–489.

113 Schatten, S., J. Drebin, L. Perry, W. Chung, M. Greene: Regulation of the immune response to tumor antigens. X. Activation third-order suppressor T cells that abrogate anti-tumor immune responses. *J. Immunol.* 133 (1984), 1064–1069.

114 Schatten, S., R.D. Granstein, J.A. Drebin, M.I. Greene: Suppressor T cells and the immune response to tumors. *CRC Crit. Rev. Immunol.* 4 (1984), 335–379.

115 Schluessner, H.J.: Transforming growth factors type β_1 and β_2 suppress rat astrocyte autoantigen presentation and antagonize hyperinduction of class II major histocompatibility complex expression by interferon-γ and tumor necrosis factor-α. *J. Neuroimmunol.* 27 (1990), 41–47.

116 Schmitt, M., R.A. Daynes: Heterogeneity of tumorigenicity phenotype in murine tumors. Characterization of tumor clones isolated from primary 3-methylcholanthrene-induced fibrosarcomas. *Transplantation* 33 (1982), 387–392.

117 Sell, S.: Cancer-associated carbohydrates identified by monoclonal antibodies. *Hum. Pathol.* 21 (1990), 1003–1019.

118 Seregni, E., E. Bombardieri, L. Maffioli, G.L. Buraggi: Mucinic tumor markers recognized by monoclonal antibodies. *J. Nucl. Med. Allied Sci.* 34 (1990), 314–320.

119 Shimizu, J., T. Suda, T. Yoshioka, A. Kosugi, H. Fujiwara, T. Hamaoka: Induction of tumor-specific *in vivo* protective immunity by immunization with tumor antigen-pulsed antigen-presenting cells. *J. Immunol.* 142 (1989), 1053–1059.

120 Shimizu, J., J.-P. Zou, K. Ikegame, T. Katagiri, H. Fujiwara, T. Hamaoka: Evidence for the functional binding *in vivo* of tumor rejection antigens to antigen-presenting cells in tumor-bearing hosts. *J. Immunol.* 146 (1991), 1708–1714.

121 Simcik, W.J., M.L. Kripke, T.L. Sheu, S.J. LeGrue Differential extraction of tumor-specific antigens from two ultraviolet light-induced murine fibrosarcomas with the use of 1-butanol. *Cancer Res.* 48 (1988), 7004–7008.

122 Spellman, C.W., R.A. Daynes: Modification of immunological potential by ultraviolet radiation. II. Generation of suppressor cells in short-term UV-irradiated mice. *Transplantation* 24 (1977), 120–126.

123 Spellman, C.W., R.A. Daynes: Cross-reactive transplantation antigens between UV-irradiated skin and UV-induced tumors. *Photodermatol.* 1 (1984), 164–169.

124 Spikes, J.D., M.L. Kripke, R.J. Connor, E.J. Eichwald: Time of appearance and histology of tumors induced in the dorsal skin of C3Hf mice by ultraviolet radiation from a mercury arc lamp. *J. Natl. Cancer Inst.* 59 (1977), 1637–1643.

125 Spellman, C.W., J.G. Woodward, R.A. Daynes: Modification of immunological potential by ultraviolet radiation. I. Immune status of short-term UV-irradiated mice. *Transplantation* 24 (1977), 112–119.

126 Streeter, P.R., G.W. Fortner: Immune response to ultraviolet-induced tumors. II. Effector cells in tumor immunity. *Transplantation* 46 (1988), 250–255.

127 Streilein, J.W.: Immunogenetic factors in skin cancer. *N. Engl. J. Med.* 325 (1992), 884–886.

128 Suda, T., J. Shimizu, M. Muramatsu, K. Kimura, T. O. Yoshida, Y. Fukami, H. Fujiwara, T. Hamaoka: The tumor rejection antigen separated from Rous sarcoma virus-murine fibrosarcoma exhibits a molecular weight of approximately 60 kD but differs from functional pp60src. *Jpn. J. Cancer Res.* 80 (1989), 879–886.

129 Toda, K., Y. Miyashi, K. Kuribayashi, S. Imamura: Decreased NK activity of nude mice receiving long-term ultraviolet irradiation: in relation to tumor induction. *J. Clin, Lab. Immunol.* 20 (1986), 129–131.

130 Toda, K., Y. Miyashi, N. Nesumi, J. Konishi, S. Imamura: UVB/PUVA-induced suppression of human natural killer cell activity is reduced by superoxide dismutase and/or interleukin 2 *in vitro*. *J. Invest. Dermatol.* 86 (1986), 519–522.

131 Toews, G.B., P.R. Bergstresser, J.W. Streilein: Epidermal Langerhans cell density determines whether contact hypersensitivity or unresponsiveness follows skin painting with DNFB. *J. Immunol.* 124 (1980), 445–453.

132 Tucker, M.A., D. Misfeldt, C.N. Coleman, J.W.H. Clark: Cutaneous malignant melanoma after Hodgkins disease. *Ann. Intern. Med.* 102 (1985), 37–41.

133 Ullrich, S.E., M.L. Kripke: Mechanisms in the suppression of tumor rejection produced in mice by repeated UV irradiation. *J. Immunol.* 133 (1984), 2786–2790.

134 Ullrich, S.E., G.K. Yee, M.L. Kripke: Suppressor lymphocytes induced by epicutaneous sensitization of UV-irradiated mice control multiple immunological pathways. *Immunology* 58 (1986), 185–190.

135 Urbach, F., P.D. Forbes: Photocarcinogenesis. In. *Dermatology in General Medicine*, 3rd edn, (1987), pp. 1475–1480. Eds.: T.B. Fitzpatrick, A. Eizen, K. Wolff, I.M. Freedman, K.F. Austen, McGraw-Hill, New York.

136 Urban, J.L., R.C. Burton, J.M. Holland, M.L. Kripke, H. Schreiber: Mechanisms of syngeneic tumor rejection. Susceptibility of host-selected progressor variants. *J. Exp. Med.* 155 (1982), 557–573.

137 Urban, J.L., J.M. Holland, M.L. Kripke, H. Schreiber: Immunoselection of tumor cell variants by mice suppressed with ultraviolet radiation. *J. Exp. Med.* 156 (1982), 1025–1041.

138 Versteeg, R.: NK cells and T cells: mirror images? *Immunol. Today* 13 (1992), 244–247.

139 Weaver, C.T., L.M. Duncan, E.R. Unanue: T cell induction of macrophage IL-1 during antigen presentation. Characterization of a lymphokine mediator and comparison of TH1 and TH2 subsets. *J. Immunol.* 142 (1989), 3469–3476.

140 Woodward, J.G., R.A. Daynes: Cell-mediated immune response to syngeneic UV-induced tumors. I. The presence of tumor-associated macrophages and their possible role in the *in vitro* generation of cytotoxic lymphocytes. *Cell. Immunol.* 41 (1978), 304–319.

141 Yee, G.K., J.G. Levy, M.L. Kripke, S.E. Ullrich: The role of suppressor factors in the regulation of immune responses by ultraviolet radiation-induced supressor T lymphocytes. III. Isolation of a suppressor factor with the B16G monoclonal antibody. *Cell. Immunol.* 126 (1990), 255–267.

142 Yee, G.K., S.E. Ullrich, M.L. Kripke: The role of suppressor factors in the regulation of immune responses by ultraviolet radiation-induced suppressor T lymphocytes. II. Activation of suppressor

cell culture sonicates. *Cell Immunol.* 121 (1989), 88–98.

143 Yoshikawa, T., V. Rae, W. Bruins-Slot, J.W. v.d. Berg, R. Taylor, J.W. Streilein: Susceptibility to effects of UVB radiation on induction of contact hypersensitivity as a risk factor for skin cancer in humans. *J. Invest. Dermatol.* 95 (1990), 530–536.

144 Zou, J., J. Shimizu, K. Ikegame, N. Yamamoto, S. Ono, H. Fujiwara, T. Hamaoka: Tumor-bearing mice exhibit a progressive increase in tumor antigen-presenting cell function and reciprocal decrease in tumor antigen-responsive $CD4^+$ T cell activity. *J. Immunol.* 148 (1992), 648–655.

9: Photoimmunology of cutaneous melanoma

Cherrie Donawho & Margaret L. Kripke

Introduction

Considerable circumstantial evidence implicates UV radiation as a contributing factor to the incidence of and mortality from cutaneous melanoma in Caucasian populations [1, 12]. Nonetheless, the roles played by UV radiation in the induction and pathogenesis of melanoma remain poorly defined. In recent years, interest in the relationship between UV radiation and melanoma has intensified because of the alarming, worldwide increases in the incidence and mortality rates of this potentially lethal form of skin cancer. Furthermore, the threat of increased ambient UVB (280–320 nm) radiation resulting from depletion of stratospheric ozone raises additional concern that the upward trend in melanoma incidence will continue and possibly accelerate over the next century [14].

Circumstantial evidence supporting a role for UV radiation in melanoma development derives from several diverse types of observations. First, there is a latitude gradient for melanoma in North America and Australia, in which the incidence of cutaneous melanoma increases with increasing proximity to the equator and increasing availability of solar UVB radiation [15]. Second, the body distribution of primary melanomas is non-random, with areas of the body that are normally doubly clothed or covered by a bathing suit having a lesser incidence of primary melanomas [25]. Third, there is an excess incidence of melanomas in persons with xeroderma pigmentosum, a genetic disease characterized by sun sensitivity, skin cancer, and defects in the ability of somatic cells to repair UV-induced lesions in DNA [19]. Fourth, case–control epidemiologic studies indicate that sensitivity to sunlight and exposure to sunlight are risk factors for the development of melanoma [12]. Fifth, experiments performed with opossums [23], tropical fish [30] and cultured cells from transgenic mice containing the SV40 transforming gene linked to the tyrosinase promoter [22] have demonstrated that UV radiation can induce primary melanomas in these model systems. In other murine models, UV radiation has been shown to contribute to the development of primary melanomas induced or initiated by chemical carcinogens [13, 18, 27].

One important conclusion drawn from recent epidemiologic studies is that the incidence of most melanomas is associated with the intermittent exposure of untanned skin to moderate doses of UV radiation; this is in direct contrast to the induction of basal and squamous cell carcinomas, which are associated with a cumulative, high dose of UV radiation [12]. Such differences in the epidemiology of melanoma and the other skin cancers suggest that UV radiation may be playing different roles in their induction. In fact, various authors have suggested that melanomas are etiologically heterogeneous and multifactorial, and that UV radiation may play different roles (including no role) in their induction in different individuals [17, 20].

Studies on skin cancers induced in mice by chronic exposure to UV radiation have demonstrated that UVB has immunosuppressive effects that contribute to the induction and pathogenesis of non-melanoma skin cancers [21] (see also Chapter 8). This finding has raised the possibility that one way in which UV radiation might contribute to melanoma induction is by means of its immunosuppressive activity. This is an attractive hypothesis because both human [16] and murine

[5] melanomas are highly immunogenic, relative to most other types of cancers. Thus, in addition to any direct carcinogenic effect UV radiation may have in the formation of melanomas, theoretically it could also contribute to the development and progression of melanomas by means of its photoimmunologic activity. Although there is no evidence in humans to support this hypothesis, recent studies in a murine system, as detailed below, suggest that the immunomodulatory action of UV radiation can contribute to melanoma growth and progression.

Photoimmunology of murine melanomas

Studies with primary melanomas

In our laboratory we have investigated the role of UV radiation in the induction of melanomas in C3H mice using a system developed by Berkelhammer *et al.* [4]. In this system melanomas are induced by applying a single dose of 7,12-dimethylbenzanthracene (DMBA) to the dorsal skin of four-day-old mice, and beginning two weeks later, applying croton oil (CO) twice per week until tumors develop. This initiation–promotion procedure induced skin tumors in 100% of C57BL/6 mice, approximately 15% of which were cutaneous melanomas [3, 4].

Using this approach we investigated whether UV radiation could serve as the initiator, the promoter, or a cocarcinogen for melanoma development in C3H mice [27]. When UV radiation was used as the initiator with CO as the promoter, very few mice developed tumors (6/42), but half of them were melanomas, giving an incidence of 7% melanomas. In mice given DMBA as the initiator and UV radiation as the promoter, 92% developed skin cancer, 6% of which (3/53) were melanomas. Although the incidence of melanomas in these groups was much lower than that observed in mice treated with DMBA and CO (25%), these results demonstrate that UV radiation can serve as both the initiator and the promoter for the induction of primary murine melanomas. A similar study employing Skh-hr2 pigmented hairless mice reported a 25–33% incidence of melanomas in mice given an initiating dose of DMBA and then treated with UVB, UVA, or UVB + UVA radiation [18]. These studies confirm and extend the original report by Epstein *et al.* [13] that melanomas could arise following UV irradiation of 'blue nevi' initiated in the skin of pigmented hairless mice with DMBA.

We also investigated the effect on the incidence of melanomas of UV irradiating the site of DMBA and CO treatment [27]. Combining chronic UV irradiation with the chemical treatments on dorsal skin dramatically accelerated the appearance of the melanomas; for example, at 1 year, the incidence of melanoma in the UV-irradiated mice was 38%, whereas that in the mice treated with DMBA and CO only was 14% ($P < 0.005$). This effect of UV irradiation on the development of melanomas did not result from a systemic effect of the UV treatment; mice exposed to UV radiation on ventral skin developed melanomas on the dorsal skin (the site of application of the chemicals) at the same rate as unirradiated mice treated with DMBA and CO.

It cannot be discerned from the forgoing studies whether the immunosuppressive action of UV irradiation contributes to its effects on melanoma carcinogenesis. However, subsequent studies using transplantable melanomas suggest that at least the latter, cocarcinogenic, action of UV radiation may be due to a photoimmunologic effect.

Studies with transplantable melanomas

The finding that UV irradiation accelerated the appearance of melanomas induced by DMBA and CO raised the question of whether UV radiation was affecting the growth of the melanoma cells directly or was influencing melanoma development by interacting with the surrounding host tissues. To distinguish between these possibilities, we performed a series of experiments with transplantable melanoma cell lines [26]. Injection of syngeneic melanoma cells subcutaneously into the external ears of C3H mice exposed previously or concurrently to UV radiation resulted in an

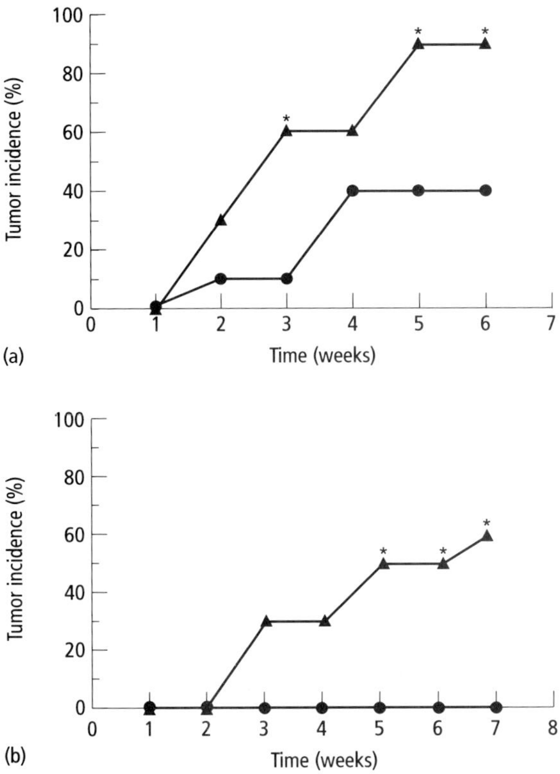

Fig. 9.1 Effect of UVR exposure on the outgrowth of melanoma. A group of ten mice (triangles) was UV irradiated with 4.8 kJ/m^2 twice a week for 3 weeks. Control mice (circles) received no UV treatment. One day after the final UV exposure, 2.5×10^4 syngeneic melanoma cells were injected into the external ears. (a) K1735 is a C3H/HeN melanoma induced by initiation with UVB radiation and promotion with croton oil. (b) CM3205 is a C3H/HeN melanoma induced by initiation with DMBA and promotion with phorbol ester. Tumor incidence is the cumulative number of palpable tumors (≥2 mm in diameter) divided by the total number of tumor cell injections in each group. $*P \leq 0.05$, Chi square test.

earlier appearance and higher final incidence of melanomas (Fig. 9.1). The optimal irradiation regimen for producing this effect was to expose unshaved mice to 4.8 kJ/m^2 UVB radiation from FS40 sun lamps twice per week for 3 weeks. Studies of this phenomenon demonstrated that it was an effect of UV radiation on the host environment, not on the tumor cells; it was localized mainly to the site of UV irradiation; it was transient and disappeared by 5 days after cessation of the irradiation; and it was most clearly apparent using a small tumor cell inoculum. Injection of radiolabeled tumor cells demonstrated, in addition, that the effect was not simply due to greater retention of the tumor cells at the site of inoculation in the UV-irradiated ears [26].

Studies to determine the UV waveband mediating enhanced melanoma development suggested that wavelengths in the UVB region of the spectrum were primarily responsible for the effect of the FS40 sun lamps. Experiments were performed using a mylar filter to remove wavelengths below 320 nm. Whereas melanoma incidence was increased in the ears of mice exposed to unfiltered FS40 sun lamps, an equivalent dose of UV delivered through the mylar filter had no effect on melanoma development. This result indicates that wavelengths below 320 nm are necessary for enhanced melanoma development. Because the sun lamps emit little energy in the UVC (200–280 nm) region (<5%), and because UVC radiation has little capacity to penetrate tissues, the activity of the sun lamps is most likely due to the presence of the UVB radiation [11].

There was no correlation between the effects of UV radiation on melanoma growth and its ability to induce acute inflammation in the skin. Acute inflammation, as detected both by microscopic examination of histologic sections of ear skin and by measurement of ear thickness did not correlate with the effect of UV on melanoma development either temporally or in terms of UV dose–response. In fact, there was a better temporal correlation with epidermal hyperplasia and increased melanization [11]. These findings suggest that inflammation and enhanced melanoma development are two unrelated effects of UV irradiation.

In addition to influencing the outgrowth of the local tumors, exposure to UV radiation during the growth of small tumor fragments implanted into the ears enhanced the metastatic dissemination of melanomas [9]. In these experiments, cells of an immunogenic, highly metastatic variant (M2) of the K1735 melanoma cell line were implanted into the ears of syngeneic mice, and the resulting local tumors were excised 3 weeks later. Some of these

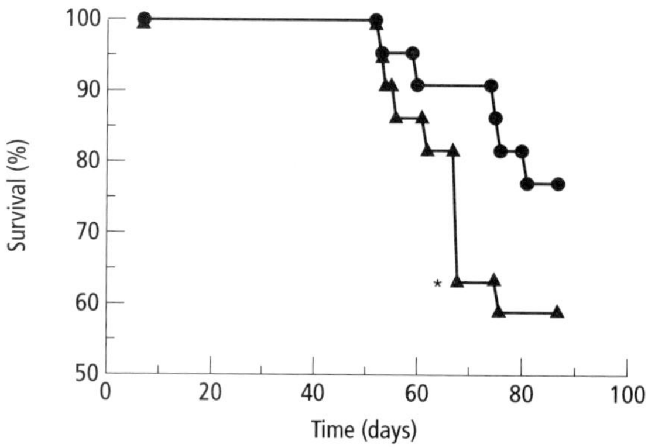

Fig. 9.2 Effect of UV irradiation on death from melanoma metastases. Tumor fragments ($1\,mm^3$) from the spontaneously metastatic cell line K1735 var. M2 were implanted into one ear of non-irradiated mice (circles) or mice whose ears were UV irradiated twice weekly for 3 weeks ($4.8\,kJ/m^2$ each exposure) beginning 1 day after implantation (triangles). Tumors were removed 1 day after the final UV irradiation by excision of the ears. Mice were observed daily, and killed and examined for metastases when moribund. $^*P \leqslant 0.05$, Chi square test.

mice subsequently died from pulmonary metastases. When the ears of the mice were exposed to UV irradiation during the 3-week period of local tumor growth the mice died more rapidly (Fig. 9.2), and they had an increase in the average number of pulmonary metastases. These results indicate that UV irradiation can affect the progression, as well as the local growth, of melanomas.

Immunologic basis for the UV effect

To address the possibility that the effect of UV on the outgrowth of melanomas was immunologically mediated, we tested the ability of UV irradiation to influence the outgrowth of a variety of syngeneic tumors [6]. After first establishing the minimal tumorigenic dose for each cell line, we tested the effect of UVR on the incidence of C3H melanomas, sarcomas, and carcinomas. Of 13 tumors tested, we observed an increased incidence of all three melanomas and four of six fibrosarcomas, demonstrating that the effect was not specific for melanomas, and therefore not likely to be due to a melanocyte-specific growth factor. We then tested the immunogenicity of the 13 tumors by an *in vivo* immunization and challenge assay, and found that the seven tumors affected by UV irradiation were immunogenic, whereas the six tumors that were unaffected were not detectably immunogenic [5, 6]. This finding supported the hypothesis that the effect of UV irradiation on tumor growth was immunologically mediated.

We also investigated the ability of UV irradiation to influence the growth of melanoma cells in immunocompromised mice. We reasoned that if UV radiation was affecting immunological resistance to tumor growth, then the effect would only be observed in immunologically normal mice. We found that UV irradiation of nude mice, 450R X-irradiated mice, and mice depleted of T lymphocytes by *in vivo* injection of anti-Thy-1 antibody, did not cause enhancement of tumor development, even when submaximal doses of tumor cells were injected [6]. These results also supported the hypothesis that the enhanced development of the tumors resulted from an effect of UV irradiation on local cutaneous immunity.

One possible mechanism by which UV radiation might impair local immunological function is by means of its effects on cutaneous antigen presentation. UV radiation reduces the number and alters the morphology and function of epidermal Langerhans cells and interferes with the afferent arm of the immune response to contact sensitizers. Applying a contact sensitizer to UV-irradiated murine skin induces suppressor T lymphocytes instead of a contact hypersensitivity response (for details, see Chapters 7, 8 and 10). We therefore postulated that UV irradiation might interfere with the ability of the skin to initiate an immune response to tumor antigens, thereby decreasing host resistance to the development of antigenic tumors. We tested this hypothesis using two experimental approaches. First, mice were treated with doses of UV known to reduce the number of Langerhans cells and impair the induction of contact hypersensitivity, and then the effect of this regimen on the growth of melanoma cells was determined. Second, mice were given a course of

UV irradiation that increased the incidence of melanomas in the UV-irradiated ears; the ear skin was then examined for the presence of Langerhans cells and tested for its ability to support the generation of a contact hypersensitivity response. No correlation between these effects of UV irradiation could be detected. Furthermore, implantation of melanoma cells into UV-irradiated ear skin resulted in the generation of systemic immunity against subsequent tumor challenge [8]. We therefore concluded that the ability of UV irradiation to modify melanoma development was probably unrelated to its effects on the afferent arm of the contact hypersensitivity response, and that enhanced melanoma development was not due to an impairment in the induction of tumor immunity.

Since UV irradiation seemed to have little effect on the *induction* of a primary immune response to melanoma cells, we investigated its effect on the *expression* of immunity against melanomas implanted into UV-irradiated ears of immunized mice. Groups of mice were immunized by implantation and excision of the K1735 melanoma. When the tumor-immunized mice were challenged with the same tumor in UV-irradiated ear skin, the tumors grew progressively, even though these mice exhibited resistance to tumor growth when challenged at a non-irradiated site [7]. These studies imply that UV irradiation interferes with the elicitation of the antitumor immune response rather than its induction. This conclusion was supported futher in experiments demonstrating that local UV irradiation inhibited the efferent arm of the delayed-type hypersensitivity (DTH) response to alloantigen (Donawho and Kripke, unpublished data).

Mechanisms of the UV effect

The reduced expression of cell-mediated immunity within chronically UV-irradiated skin could result from an absence of immune effector cells within the challenge site, from inhibition of immune cell activity within the site, or both. To determine whether lymphoid cells were able to gain access to tumors during their growth in UV-irradiated ears, melanomas were removed at weekly intervals from groups of UV-irradiated and control mice. Histologic sections of the tumors were stained with immunogold reagents to detect $CD3^+$, $CD4^+$, and $CD8^+$ T lymphocytes, and $F4/80^+$ and $MOMA^+$ (activated) macrophages. During the first 3 weeks of tumor development, there was a marked increase in the number of T lymphocytes in tumors from the unirradiated control mice. In contrast, these increases were considerably less in tumors recovered from the UV-irradiated mice (Figs 9.3 & 9.4). There was a gradual increase in the number of macrophages during the 5 weeks of tumor growth, and no difference was detected between the UV-irradiated and control groups [10]. These findings support the hypothesis that UV irradiation may lead to an increased incidence of melanomas, at least in part, by inhibiting the influx of antitumor T effector cells into the tumor.

Whether this phenomenon is related to the ability of UV irradiation to alter the expression of adhesion molecules on cells is not known (see Chapter 6). This is an attractive idea because the expression of adhesion molecules determines the homing properties of lymphocytes and controls their ability to move out of blood and lymphatic vessels.

Another possible mechanism by which UV irradiation could decrease the expression of immunity locally is by causing the local production or release of cytokines (see Chapter 5). UV irradiation has been shown to increase the production of immunomodulatory cytokines by epidermal cells. Furthermore we have demonstrated that murine keratinocytes release factors that stimulate melanoma growth *in vivo*. Serum-free supernatants from cells of a murine keratinocyte line exposed to UV radiation *in vitro* stimulated tumor growth *in vivo* when mixed with melanoma cells prior to their inoculation. Similar results were obtained with supernatants derived from keratinocytes treated with psoralen plus UVA (320–400 nm) radiation (PUVA), and PUVA treatment of mice also caused enhanced growth of melanoma cells *in vivo* [2].

Although many steps in the process require substantiation, we can speculate, based on

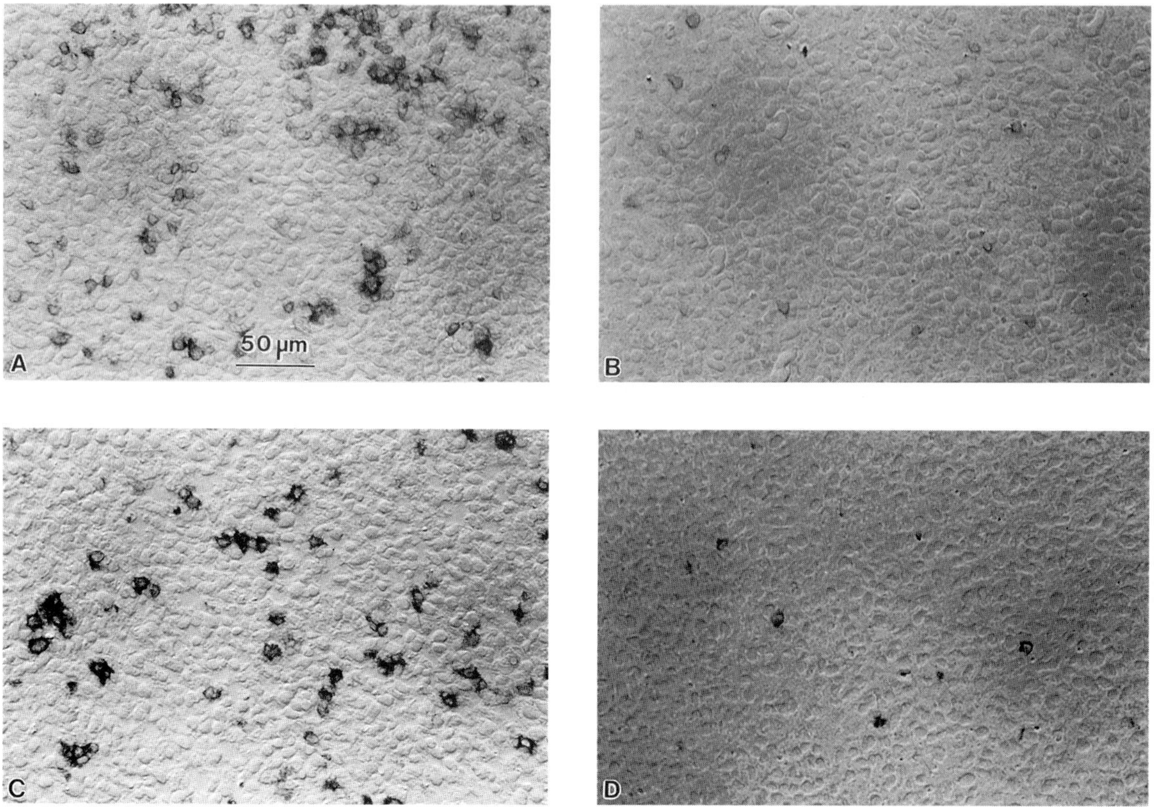

Fig. 9.3 T lymphocytic infiltrates in melanomas. T lymphocytes were assessed in melanomas 2 weeks after injection of 5×10^5 K1735 cells into the external ears. Control mice received no UV treatment (A, C). UV-irradiated mice received 4.8 kJ/m^2 twice a week for 3 weeks (B, D). Tumor cells were injected 1 day after the last UV exposure, and tumors were excised from each group and frozen sections were treated with immunogold reagents to stain T lymphocytes within the tumors. Sections stained for CD3 antigen are shown in (A) and (B); sections stained for CD8 are shown in (C) and (D).

the available information, that UV irradiation enhances the outgrowth of melanomas in naive mice in the following manner. Chronic UV irradiation of the skin causes both acute effects, such as inflammation and release of certain cytokines, and late effects, such as epidermal hyperplasia and increased melanization. Cells arising from the late effects of UV irradiation produce factors that inhibit the activity of immune cells within the local, UV-irradiated site. UV irradiation may also alter the expression of adhesion molecules, either by direct action on the target cells, or indirectly through the local production of cytokines. Altered expression of adhesion molecules on endothelial cells within the UV-irradiated site would serve to limit the ability of circulating T lymphocytes to enter the affected area; altered expression on lymphoid cells within the area might inhibit their ability to interact with other cells, thereby reducing their activity. Thus we speculate that UV irradiation enhances melanoma development by inhibiting immunological effector mechanisms within the irradiated site. This local, immunosuppressive milieu could also be responsible for the impaired ability of immune mice to reject tumors upon challenge in the UV-irradiated site.

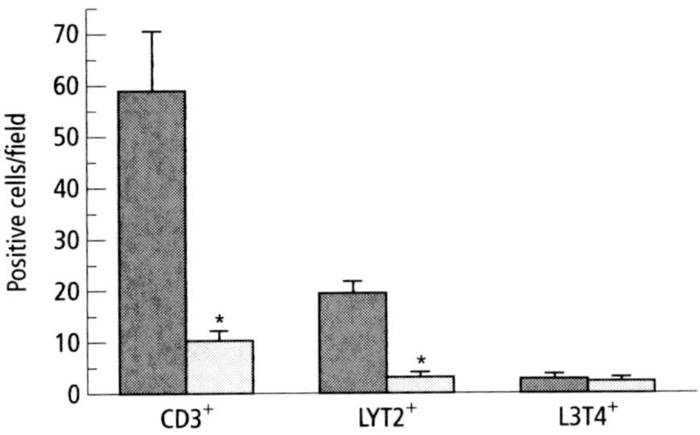

Fig. 9.4 Effect of UVR on the number of T lymphocytes within the K1735 melanomas 2 weeks after injection of tumor cells. Mice were UV irradiated with 4.8 kJ/m^2 twice a week for 3 weeks (light grey); 1 day after the final UV exposure, 5×10^5 K1735 cells were injected into the pinna. Control mice received no UV treatment (dark grey). Two weeks after injection, tumors were excised from each group, and frozen sections were treated with immunogold reagents to stain CD3$^+$, CD8$^+$ (Lyt2), and CD4$^+$ (L3T4) cells within the tumors. Sections from the tumors shown in Fig. 3 were examined microscopically, and the number of positive cells per field was counted in at least five random fields. The mean number of positive cells/field ± SE is shown for the two treatment groups. $*P \leq 0.01$, Student's *t*-test.

Significance and future directions

Studies on the photoimmunology of transplantable murine melanomas have provided a new concept as to how UV radiation may participate in the development of cutaneous melanoma. They present evidence for a previously unsuspected role of UV radiation in the induction and pathogenesis of this disease. Aside from any role it might play in the transformation of normal melanocytes into malignant melanoma cells, UV irradiation appears to aid the establishment of foci of melanoma cells in the skin by means of a local effect on the host environment. Furthermore, this effect of chronic UV irradiation also facilitates the metastatic spread of melanomas from the UV-irradiated site to internal organs, and accelerates death from disseminated disease. Studies of this phenomenon may help to clarify the role of local host factors, particularly cutaneous immunity, in melanoma growth and progression, and identification of the mechanisms involved may suggest approaches to circumvent it.

These studies have also helped to define a new, immunomodulatory effect of chronic UV irradiation that is distinct from its well-characterized acute effects on antigen presentation and the afferent arm of the contact hypersensitivity reponses. This late effect of UV irradiation inhibits the efferent arm of the immune response, as evidenced by decreased tumor rejection and delayed hypersensitivity responses. It seems to result from the creation of a local, immunosuppressive milieu within the UV-irradiated site and may involve the inhibition of lymphocyte migration and the production of locally acting cytokines. Although their relationship is not yet clear, this phenomenon is reminiscent of that described by O'Dell *et al.* [24] in studies of immune reactivity in chronically sun-exposed human skin. They demonstrated that elicitation of delayed hypersensitivity responses to recall antigens was impaired in skin exhibiting signs of chronic sun damage.

These studies on the photoimmunology of murine melanomas may also help to explain the curious observation that in the United States a preponderance of cutaneous melanomas in humans, particularly those arising on the extremities, are diagnosed during the summer months [28, 29]. Although this finding is generally attributed to the fact that more of the skin is available for inspection during warm weather, studies with the murine model raise another possibility. If UV irradiation inhibits local immunity in human skin, perhaps exposing the skin to environmental UV radiation stimulates the outgrowth of existing, quiescent melanoma cells by modifying the local immunological environment. This effect of UV radiation could also account for the more frequent summer diagnosis of cutaneous melanoma.

Further studies are obviously required to define the mechanisms involved in the UV-induced enhancement of melanoma development in the murine model. In addition, studies are needed to determine whether these mechanisms are relevant to the induction and pathogenesis of cutaneous melanoma in humans. Studies on immune reactivity of sun-exposed skin in melanoma patients, analyses of immune cells in melanomas occurring on sun-exposed versus non-sun-exposed sites, and comparison of the prognoses of such melanomas may provide insight into this issue.

References

1 Armstrong, B.K., D.R. English: Epidemiologic studies. In: *Cutaneous Melanoma*, (1992), pp. 12–26. Eds.: C.M. Balck, A.N. Houghton, G.W. Milton, A.J. Sober, S.-j. Soong, J.B. Lippincott Company, Philadelphia.

2 Aubin, F., C.K. Donawho, M.L. Kripke: Effect of psoralen plus ultraviolet A radiation on *in vivo* growth of melanoma cells. *Cancer Res.* 51 (1991), 5893–5897.

3 Berkelhammer, J., R.W. Oxenhandler: Evaluation of premalignant and malignant lesions during the induction of mouse melanomas. *Cancer Res.* 47 (1987), 1251.

4 Berkelhammer, J., R.W. Oxenhandler, R.R. Hook, J.M. Hennessy: Development of a new melanoma model in C57BL/6 mice. *Cancer Res.* 42 (1982), 3157–3163.

5 Donawho, C.K., M.L. Kripke: Immunogenicity and cross-reactivity of syngeneic murine melanomas. *Cancer Commun.* 2 (1990), 101–107.

6 Donawho, C.K., M.L. Kripke: Evidence that the local effect of UV radiation on the growth of murine melanomas is immunologically mediated. *Cancer Res.* 41 (1991), 4176–4181.

7 Donawho, C.K., M.L. Kripke: Photoimmunology of experimental melanoma. *Cancer and Metastasis Reviews* 10 (1991), 177–188.

8 Donawho, C.K., M.L. Kripke: Lack of correlation between UV-induced enhancement of melanoma development and local suppression of contact hypersensitivity. *Exp. Dermatol.* 1 (1992), 20–26.

9 Donawho, C.K., M.L. Kripke: Effect of UV-B radiation on the metastasis of murine melanomas. *J. Cell Biochem.* Supplement 16b (1992), 316.

10 Donawho, C.K., M.L. Kripke, C. Bucana, P. Cox, H.K. Muller: Reduction in tumor infiltrating lymphoid cells in murine melanomas growing in UV-irradiated skin (Abstract 2913). *Proc. Am. Assoc. Cancer Res.* 34 (1993), 488.

11 Donawho, C.K., P. Wolf, M.L. Kripke: Enhanced development of murine melanoma in UV-irradiated skin: UV dose response, waveband dependence, and relation to inflammation. *Med. Res.* 4 (1994), 93–100.

12 Elwood, J.M.: Melanoma and ultraviolet radiation. *Clinics in Dermatology* 10 (1992), 41.

13 Epstein, J.H., W.L. Epstein, T. Nakai: Production of melanomas from DMBA-induced 'blue nevi' in hairless mice with ultraviolet light. *J. Natl. Cancer Inst.* 38 (1967), 19–30.

14 Grin-Jorgensen, C.M., D.S. Rigel, R.J. Friedman: The worldwide incidence of malignant melanoma. In: *Cutaneous Melanoma*, (1992), pp. 27–39. Eds.: C.M. Balch, A.N. Houghton, G.W. Milton, A.J. Sober, S.-j. Soong, J.B. Lippincott Company, Philadelphia.

15 Hoffman, J.: *Assessing the Risks of Trace Gases that Can Modify the Stratosphere* (1987). Office of Air and Radiation, US Environmental Protection Agency, Washington, DC.

16 Houghton, A.N., M. Herlyn, S. Ferrone: Melanoma antigens. In: *Cutaneous Melanoma*, (1992), pp. 130–143. Eds.: C.M. Balch, A.N. Houghton, G.W. Milton, A.J. Sober, S.-j. Soong, J.B. Lippincott Company, Philadelphia.

17 Houghton, A.N., M.V. Viola: Solar radiation and malignant melanoma of the skin. *J. Am. Acad. Dermatol.* 5 (1981), 477.

18 Husain, Z., M.A. Pathak, T. Flotte, M.M. Wick: Role of ultraviolet radiation in the induction of melanocytic tumors in hairless mice following 7,12-dimethylbenz (a)anthracene application and ultraviolet irradiation. *Cancer Res.* 51 (1991), 4964–4970.

19 Kraemer, K.H., M.M. Lee, J. Scotto: DNA repair protects against cutaneous and internal neoplasia: evidence from xeroderma pigmentosum. Carcinogenesis 5 (1984), 511–514.

20 Kripke, M.L.: Speculations on the role of ultraviolet radiation in the development of malignant melanoma. *J. Natl. Cancer Inst.* 63 (1979), 541–545.

21 Kripke, M.L.: Immunologic mechanisms in UV radiation carcinogenesis. *Adv. Cancer Res.* 34 (1981), 69–106.

22 Larue, L., N. Dougherty, B. Mintz: Genetic predisposition of transgenic mouse melanocytes to melanoma results in malignant melanoma after exposure to a low ultraviolet B intensity nontumorigenic for normal melanocytes. *Proc. Natl. Acad. Sci. USA* 89 (1992), 9534–9538.

23 Ley, R.D., L.A. Applegate, R.S. Padilla, T.D. Stuart: Ultraviolet radiation-induced malignant melanoma in *Monodelphis domestica*. *Photochem. Photobiol.* 50 (1989), 1–5.

24 O'Dell, B.L., R.T. Jessen, L.E. Becker: Diminished immune response in sun damaged skin. *Arch. Dermatol.* 116 (1980), 559–561.

25 Rhodes, A.R., M.A. Weinstock, T.T. Fitzpatrick, M.C. Mihm, A.J. Sober: Risk factors for cutaneous melanoma. *JAMA* 248 (1987), 3146–3154.

26 Romerdahl, C.A., C. Donawho, I.J. Fidler, M.L. Kripke: Effect of UV-B radiation on the *in vivo* growth of murine melanoma cells. *Cancer Res.* 48 (1988), 4007–4010.

27 Romerdahl, C.A., L.C. Stephens, C. Bucana, M.L. Kripke: The role of ultraviolet radiation in the in-

duction of melanocytic skin tumors in inbred mice. *Cancer Comm.* 1 (1989), 209–216.

28 Schwartz, S.M., B.K. Armstrong, N.S. Weiss: Seasonal variation in the incidence of cutaneous malignant melanoma: an analysis by body site and histologic type. *Am. J. Epidemiol.* 126 (1987), 104.

29 Scotto, J., J.M. Nam: Skin melanoma and seasonal patterns. *Am. J. Epidemiol.* 111 (1980), 309–314.

30 Setlow, R.B., A.D. Woodhead, E. Grist: Animal model for ultraviolet radiation-induced melanoma: platyfish-swordtail hybrid. *Proc. Natl. Acad. Sci. USA* 86 (1989), 8922–8926.

10: Immunogenetics of sunlight-induced skin cancer

J. Wayne Streilein

Introduction

It has been firmly established that prolonged sun exposure is the single most important environmental factor in the pathogenesis of skin cancer in humans, particularly tumors of the basal and squamous cell types. The evidence is based almost exclusively on numerous epidemiologic studies, and includes the oft-repeated observation that the incidence of skin cancer is highest among Caucasian individuals living in geographic regions of the world where sunlight is intense and oriented perpendicularly to the Earth's surface during the midday [46, 83, 112, 117, 135, 140]. It is generally believed that the critical component of sunlight that is central to cancer pathogenesis is electromagnetic energy in the so-called ultraviolet B range (290–320 nm). Both biophysical considerations, which quantify the ability of the various wavelengths of sunlight to penetrate through the layers of the surface of the skin [12, 30, 63], and physiologic considerations, which examine the acute biologic effects of sun exposure on skin, support this belief [42]. The usual conclusion reached from these studies is that UVB is virtually the only segment of the ultraviolet range that penetrates to the appropriate level of the epidermis (where cells with mitotic potential reside). Moreover, UVB radiation creates within DNA a set of alterations that are stereotypic and easy to identify at the molecular genetic level [35, 77]. Quite recently, support for the importance of UVB radiation in squamous cell cancer came from molecular analysis of the tumor suppressor gene, p53, in cells derived from human skin cancers [16]. It was found that p53 was defective in these cells – and therefore could not suppress autonomous tumor cell proliferation. More importantly, analysis of the nucleotide sequences of the p53 genes in cancer cells from several skin tumors revealed the same type of stereotypic mutations known to be caused by exposure of DNA to UVB.

Claims have also been made that sun exposure, and presumably UVB radiation, is also an important etiologic factor in malignant melanoma [33, 55, 75, 85, 110]. However, in this disease, epidemiologic evidence suggests that one, or perhaps only a few extreme, episode(s) of sun exposure causing severe sunburning of the skin is (are) critical to the malignant degeneration of melanocytes. There is not universal agreement about this proposed etiologic relationship. One reason for the lack of consensus on this matter may be that malignant melanoma is not a single disease phenotypically, and therefore UVB may be important in the pathogenesis of only certain forms of the disease.

Surveys of humans living in areas of the world where sun exposure is of the type, duration and intensity known to promote the development of skin cancer reveal that not every individual receiving comparable levels of sun exposure develops the disease. The most obvious example is the common knowledge that the incidence of basal and squamous cell cancers among black-skinned individuals is extremely small, unlike that found in Caucasians. As an extension of this observation, it is generally believed that the intensity of genetically determined cutaneous melanization is inversely proportional to skin cancer susceptibility, although the epidemiologic evidence in favor of this notion is weak [117]. Nonetheless, the degree to which an individual's epidermis is melanized genetically is a polymorphic trait, governed by allelic differences at the several loci that determine type and extent of pigment formation. In principle these facts indicate that human beings vary in their susceptibility to the carcinogenic effects of UVB radiation and sunlight, and that one important cause of this variation is an individual's genetic make-up. Stated in another way, it appears that polymorphic gene loci in man can act to render some individuals more susceptible to sunlight-induced skin cancer than others. The loci that dictate skin color are important, but certainly not the only genes relevant to this issue. The evidence summarized in this chapter points to polymorphisms at genetic loci governing various aspects of the immune response as important contributors to skin cancer susceptibility.

Effects of ultraviolet B light on the immune system

Evidence in humans

More than 30 years ago, Thomas [130] and Burnet [17] separately proposed that the immune system plays an essential role in the body's war against malignancy – termed by Burnet the 'immune surveillance' hypothesis. By then it was known that immunocompetent lymphocytes express recognition structures for antigens on their surfaces, and that in the aggregate the potential range of antigens that could be recognized by lymphocytes is enormous. It was also known, or strongly suspected, that transformed and neoplastic cells express neoantigens on their surfaces and lymphocytes might possess recognition structures which could therefore be recognized as alien. These authors proposed that the heterogenous mass of antigen-detecting lymphocytes which constitutively recirculate through the interstices of virtually all tissues of the body was ideally suited to detect newly emerging malignant cells. Upon recognition of alien antigen-bearing cells it was proposed that the attacking lymphocytes proceeded to eliminate the malignant cellular degenerates. This seminal idea has been rethought, reformed and rejuvenated numerous times during the past few decades, but its power remains intact and still drives much of the research in the field of tumor imunology.

Perhaps some of the most convincing evidence that has been marshalled to support the validity of the immune surveillance hypothesis comes from the study of patients who are maintained on long-term chronic immunosuppression, for example patients whose solid organ transplants are sustained by immunosuppressive drugs, or whose presumed autoimmune diseases are treated with similar agents [54]. Kinlen *et al.* [59] first reported from Australia that long-term immunosuppressed individuals generally experienced a significantly greater risk of developing cancers of many different tissue origins. More particularly, chronically immunosuppressed patients living in Australia, Micronesia, and other regions of Southeast Asia displayed an especially large increase in the incidence of skin cancers of the basal and squamous cell types, when compared with non-immunosuppressed controls. Similar data has been reported from other areas of the world which lends credence to the belief that immunity is an important host defense against malignancy. However, the data from Australia, where sun exposure is particularly intense, make the additional point that sunlight-induced skin cancers are even more prevalent than other types of cancer when the immune system is ineffective or damaged.

Thus, there seems to be strong circumstantial evidence to support the theoretical contentions: (i) that the immune system is an important defense against malignancy; and (ii) that when the immune system is impaired, sunlight-induced skin cancers are especially common. It is relevant to point out at this juncture that, even when the immune system is suppressed, not all individuals of comparable skin melanization and sun exposure develop skin cancer. This observation suggests that host genetic factors – beyond the genes governing skin pigmentation – are responsible for individual differences in susceptibility and, in this context, polymorphic genes dictating immune function may be involved.

Evidence in laboratory mice

The field of experimental photoimmunology, at least as it applies to cutaneous tumors, was launched more than 20 years ago by the pioneering studies of Kripke [66] and Daynes [24]. These investigators exposed inbred strains of mice to UVB radiation of high intensity for prolonged intervals of time (months). They observed that after several months of UVB exposure an increasing number of animals developed cutaneous neoplasms. Eventually the incidence of malignant tumors proved to be very high. These studies formally demonstrated that UVB is an oncogenic agent, at least for the skin. Kripke and her various associates then examined the exposed animals for evidence of immune responses to the tumors. It was learned that: (i) UVB-induced tumors are highly immunogenic and are promptly rejected by normal (unirradiated) recipients syngeneic with the donor of the tumor [64, 65]; (ii) tumor-bearing mice can accept tumors derived from syngeneic mice if the tumors are UVB-induced, but not if the tumors are induced by other carcinogens [69]; (iii) UVB-treated mice can accept UVB-induced tumors even *before* tumors arise in their own skin. These findings strongly pointed to a profound, but non-global, defect in the immune responses of UVB-irradiated mice. Kripke and her collaborators demonstrated that mice exposed to the chronic high-dose UVB regimen lacked antigen-presenting cell function in spleen and lymph nodes [46, 92] and that their spleens contained suppressor T cells that can adoptively transfer the ability to accept a UVB-induced tumor implant to naive, syngeneic recipients [68, 69]. In addition, UVB-irradiated mice were unable to develop and display contact hypersensitivity (CH) when a reactive hapten was painted on either irradiated or unirradiated skin [68]. Despite these cellular and systemic immune abnormalities, UVB-exposed mice were fully able to reject orthotopic skin allografts, and to make serum antibody responses to conventional antigens. Moreover, their lymphocytes responded briskly *in vitro* in mixed lymphocyte reactions. These dramatic findings go far beyond making the point that UVB radiation can cause skin cancer. They indicate that UVB radiation, at least in high dosage and extended regimen, has profound effects on the immune system, and not just at the level of the skin. The systemic deficiency of APC and the presence of splenic suppressor cells, combined with the inability of UVB-exposed mice to become sensitized to epicutaneously applied hapten, even if it is painted on *unirradiated* skin, point to a widespread systemic impact of UVB radiation.

To address the issue of genetic susceptibility to UVB effects implied by epidemiologic studies in humans, the chronic, high dose UVB protocol has been tested in several different genetically inbred strains of mice. The outcome has been virtually identical in all, i.e. all mice appear to have the same level of susceptibility to skin cancer induction by this regimen of UVB radiation, irrespective of their genetic origins and disparities [65]. For whatever reasons, this finding indicates that host genes, including important polymorphic systems that influence the immune system differentially, are relatively unimportant in the pathogenesis of cutaneous tumors – at least when exposure to UVB is chronic and high intensity. Since the epidermis of normal mouse skin is not pigmented, the potential role of melanization in protecting against UVB-induced skin cancer can not be evaluated in mice. Thus, the classic experimental studies, which have proved that UVB can induce skin cancer in laboratory animals, and that it does so in a setting where UVB compromises the immune system independently of its oncogenic effects, fail to give evidence that host genetic factors participate in the tumorigenic process. It should be admitted that the chronic, high dose protocol bears little relationship to common human experiences with sun exposure, and so it is important not to over interpret the evidence derived from the use of this model.

Effects of UVB radiation on contact hypersensitivity in mice

The phenomenon of contact hypersensitivity (CH) is a cell-mediated immune response elicited by highly reactive haptens painted on the cutaneous surface [19, 27]. When applied epicutaneously, haptens diffuse into the epidermis and dermis, rapidly forming covalent conjugates with indigenous proteins present in the interstitium and on the surfaces of resident cells. Each conjugate has a certain probability of becoming a 'neoantigen' that can be recognized by specific T lymphocytes. Following sensitization, effector T cells can initiate a vigorous intracutaneous inflammation if they encounter the hapten-dependent 'neoantigen' again at a cutaneous site. In a real sense, the 'neoantigens' created by haptens resemble the neoantigens produced by tumor cells, and for this reason the study of the cellular and molecular bases of CH is believed to be representative of the manner in which the immune system recognizes and responds to neoantigens on tumor cells.

Following Silberberg–Siniakin's critical observation that lymphocytes cluster around epidermal Langerhans cells (LC) in contact hypersensitivity reactions [115, 116] the idea was born that LC might function as antigen-presenting cells in the skin. Stingl and his various collaborators then demonstrated that epidermal cells highly enriched for Langerhans cells functioned as antigen-presenting cells *in vitro* [120, 121]. It had previously been shown by Macher and Chase [84] that the induction of CH through the skin required an obligatory intracutaneous event which occurred shortly after hapten was applied. It was proposed that cells within the skin might be picking up the hapten, and then acting as hapten-presenting cells. More recent evidence from Knight and her colleagues [82] and from Munn *et al.* [88] have proved that cells derived from hapten-painted skin actually carry an immunogenic, hapten-specific signal to the draining lymph node. Highly immunogenic class II-positive cells bearing the hapten are found at this site within 24 h of epicutaneous hapten application. It has been assumed that the skin-derived cells carrying this signal to the node are epidermal Langerhans cells, fulfilling the prediction of Macher and Chase, and providing an *in vivo* meaning to the *in vitro* evidence by Stingl *et al.* that epidermal LC can function as APC.

Streilein and his numerous collaborators sought via *in vivo* experiments to determine whether the obligatory intracutaneous event in CH induction was related to epidermal Langerhans cells [125]. Bergstresser *et al.* [8, 9] identified cutaneous sites with abnormally low densities of Langerhans cells: murine tail skin, hamster cheek pouch, central cornea. Tail skin and hamster cheek pouch were then found to be unsuitable as sites through which to induce CH by direct hapten application [123, 133]. Previous workers had demonstrated that UVB radiation had a damaging effect on LC morphology [51], and it was suggested that UVB might interfere with CH expression in guinea pigs. *In vitro* studies by Stingl *et al.* revealed that UVB radiation impaired the ability of epidermal Langerhans cells to activate T cells [119]. Capitalizing on this observation, Toews *et al.* [133] devised an acute, low dose regimen of UVB radiation that was designed to eliminate LC from the epidermis of mice prior to application of a sensitizing dose of hapten. In this regimen, shaved, abdominal skin of C57BL/6 mice received 40 J/m^2 UVB at the same skin site for four consecutive days. At the completion of the fourth dose virtually all normal-appearing Langerhans cells were found to be depleted from the irradiated epidermis, but there were no changes among LC at unexposed skin sites. When a sensitizing dose of the hapten, dinitrofluorobenzene (DNFB), was applied to the irradiated skin surface, the mice failed to become sensitized. Ear challenge with dilute DNFB 5–6 days later elicited only feeble ear swelling responses. These results were similar to those mentioned above when DNFB was painted on murine tail skin and hamster cheek pouch, further emphasizing the importance of Langerhans cells in the induction of CH. As anticipated, the effects of the acute, low dose UVB regimen were found to be strictly local; hapten applied to unirradiated skin in naive mice induced vigorous CH. An important, unsuspected dimension to the effects of local UVB radiation on skin immunity was revealed when these mice were reexposed to the hapten. Mice that first experienced DNFB via UVB-exposed skin were painted epicutaneously with a second sensitizing dose of DNFB on unirradiated skin. When ear challenged subsequently, they failed to display CH. Since the failure to become sensitized was found to be hapten-specific, it was apparent that animals that encounter hapten first via UVB-exposed skin are rendered specifically tolerant. (See below for more detailed discussion of tolerance induced by acute, low dose UVB.) Several other laboratories have used a similar UVB radiation protocol in mice, and have reported essentially similar results [81].

Evidence that tumor necrosis factor-alpha mediates the effects of UVB on contact hypersensitivity induction

Daynes was the first to suggest that the ability of acute, low dose UVB radiation to inhibit CH induction to haptens might reside in cytokines produced by UVB-perturbed epidermal cells [5, 40]. Interleukin-1 (IL-1) is constitutively produced by keratinocytes in unperturbed skin, and Daynes and others [3, 104] discovered that *in vitro* UVB radiation of keratinocytes markedly enhanced IL-1 production. However, several different experimental approaches to determining whether IL-1 might be the mediator of UVB effects on CH induction came to nothing. If anything, IL-1 tended to enhance, rather than inhibit, CH induction by hapten.

Streilein and his colleagues then explored the possible participation of tumor necrosis factor-alpha (TNFα). It had been known that when macrophages are stimulated with bacterial lipopolysaccharide, there is a concomitant upregulation of three cytokines: IL-1, IL-6, and TNFα [109]. This observation raised the possibility that TNFα might be a mediator of UVB-induced immunotoxicity in the skin. In support of this idea was the demonstration by Köck *et al.* that UVB radiation of keratinocytes *in vitro* induces TNFα secretion, although the *Tnfα* gene is silent in unstimulated keratinocytes [61]. Moreover, Köck *et al.* reported that exposure of humans to whole body UVB radiation resulted in detectable levels of TNFα in serum within a few hours of the exposure [61, 62]. Thus there was direct evidence that UVB could induce TNFα production by skin cells. Of equal importance to this issue was the discovery by Yoshikawa and Streilein that acute, low dose exposure of C3H/HeJ mice to UVB failed to impair CH induction by DNFB, although the same regimen produced hapten-specific unresponsiveness in C3H/HeN mice [145]. These two strains are almost genetically identical, differing solely at the *Lps* locus. LPS treatment of macrophages of C3H/HeJ mice fails to enhance the production of IL-1, IL-6 and TNFα, whereas LPS routinely induces C3H/HeN macrophages to produce all three cytokines [141]. Thus the stage was set for studies designed to test whether TNFα might be the mediator of the UVB effect on CH induction.

To test this proposal, Yoshikawa and Streilein injected small amounts (50 ng) of recombinant murine TNFα intradermally into mouse skin [145]. Shortly thereafter the surface immediately above the injection site was painted with a sensitizing dose of DNFB. When the ears of these mice were challenged subsequently, they failed to express CH. Of equal importance was the finding that systemic administration of neutralizing anti-TNFα antibodies prior to the fourth UVB exposure and prior to DNFB painting completely prevented the deleterious effects of UVB; mice treated in this manner displayed vigorous CH upon subsequent ear challenge. These results make a convincing case for the hypothesis that TNFα is the prime, if not the only, mediator of the deleterious effects of UVB on CH induction in mice.

It should be noted at this point that if TNFα is the primary mediator of the UVB effects on cutaneous immunity, the relevant cellular source of TNFα within UVB-treated skin remains unclear. A prime candidate is the keratinocyte. However, in addition to keratinocytes, Langerhans cells can secrete TNFα – at least under pathologic conditions [91]. Moreover, cells in the dermis should also be considered. Dermal macrophages can produce TNFα, and, perhaps more importantly, TNFα is constitutively present in dermal mast cells [43]. In fact, it has been demonstrated in both *in vitro* and *in vivo* experiments that UVB can trigger the release of TNFα from mast cells [74]. In human skin, the erythema induced by UVB radiation has been ascribed to factors released from the mast cells that are located high up within the dermal papillae [42]. It is therefore probable that UVB radiation of mouse skin also reaches the dermis and could therefore directly trigger the release of TNFα from mast cells.

But another possibility exists. Noonan and DeFabo have proposed that the cutaneous photoreceptor of UVB radiation that is responsible for immunosuppression is the deaminated form of histidine, urocanic acid, which is found in high concentration in the stratum corneum of the epidermis [6, 25, 26, 53, 96, 107]. When the native form of deaminated histidine (*trans*-UCA) is bombarded with electromagnetic energy in the UV range, it is converted to the *cis* isoform. DeFabo and Noonan [25, 26], as well as other investigators [108], have presented circumstantial evidence that *cis*-UCA is itself immunosuppressive. Kurimoto and Streilein [72] have extended this inquiry recently by demonstrating that: (i) epicutaneous application of DNFB to skin into which an intradermal injection of *cis*-UCA was placed 5 h previously fails to induce CH; and (ii) the capacity of intradermally injected *cis*-UCA to impair CH induction in mice is abrogated if the murine hosts are treated systemically with anti-TNFα antibodies. Thus it appears that *trans*-UCA in murine epidermis is a photoreceptor for UVB, and that when this compound is racemized by UV radiation, the isoform, *cis*-UCA, impairs CH induction through a TNFα-dependent process. The molecular basis of the *cis*-UCA/TNFα connection is unknown. It has been shown that UCA has a measurable affinity for histamine receptors [97], and through this mechanism it is conceivable that *cis*-UCA produced in the epidermis by UVB radiation may diffuse into the upper dermis and interact with receptors on mast cells, triggering the release of TNFα. Of course, it is also possible that *cis*-UCA acts directly on keratinocytes, generating increased levels of intracutaneous TNFα by rendering the *Tnfa* gene transcriptionally active. *cis*-UCA may possess an LPS-like effect, which could conceivably promote TNFα production by cutaneous cells with LPS receptors – dermal macrophages, or even Langerhans cells.

The mechanism by which TNFα impairs CH induction in mice is of considerable interest, but still unclear. Several groups [70, 94, 95] have demonstrated that treatment of keratinocytes and Langerhans cells with TNFα induces the expression of ICAM-1, an outcome that one would expect to promote, rather than interfere with, induction of CH. However, the induction of ICAM-1 expression is time-dependent and does not appear until many hours after TNFα treatment. Macher and Chase [84], as well as Streilein [122], have produced evidence that the most important intracutaneous steps in the induction of CH take place within only a few hours of hapten application. These temporal considerations led us to anticipate that the deleterious effects of UVB, and TNFα on CH induction might be optimally examined in the immediate interval post-hapten application. Based on a series of studies using *in vivo* immune experiments and microscopic evaluations, Yoshikawa *et al.* [146] and Vermeer and Streilein [138] have suggested that TNFα impairs CH induction to hapten by a direct action on epidermal Langerhans cells. It was found that low dose UVB radiation (a single exposure) produces morphologic changes in epidermal Langerhans cells that were apparent within 1 h: the cells withdrew their dendrites, leaving only a few short, stubby cellular processes, and their cell bodies appeared swollen and rounded. However, the cells were not killed by this limited exposure to UVB; instead they recovered and assumed their conventional morphology within 12–24 h. Similar changes were observed in Langerhans cells of skin that received intradermal injections of TNFα or *cis*-UCA [72]; however, the morphologic changes appeared within minutes of injection in these experiments. Based on circumstantial evidence, UVB appears to immobilize Langerhans cells, at least transiently, within the epidermis. After a single exposure to UVB radiation, hapten application induces significantly fewer LC to leave the epidermis. Support for this notion comes from the studies of Bigby *et al.* [11] and from Okamoto and Kripke [98]. Thus it seems reasonable to propose that one reason that hapten fails to induce CH when painted on UVB-exposed skin is that UVB creates an intracutaneous microenvironment that is rich in

TNFα, and this cytokine prevents local antigen-presenting cells from carrying hapten to the draining lymph node in time for conventional sensitization to occur. There is reason to expect that other UVB-dependent mechanisms will be found, and that these may well contribute to failed cutaneous immunity, but at least a start has been made in understanding this important process at the mediator/cytokine level.

Studies on the mechanism of tolerance induced by hapten painted on UVB-irradiated skin

There seems to be general agreement that experimental analysis of the effects of UVB radiation on CH induction should yield information of relevance to the UVB-dependent pathogenesis of sunlight-induced skin cancer. However, UVB impairs not only the induction of CH to epicutaneously applied haptens in mice, but also leads to the induction of hapten-specific tolerance [133]. One could easily argue that tolerance induction to the neoantigens on malignant cells might be a critically important contributor to the success of cutaneous neoplasms. Moreover, one might expect that host genetic factors would contribute to the induction of tolerance, and it is therefore pertinent to this chapter to consider briefly the evidence concerning hapten-specific tolerance induced by acute, low dose UVB radiation in mice. It was mentioned previously that hapten application to a cutaneous site following UVB radiation produces a state of hapten-specific unresponsiveness [133]. Elmets *et al.* [28] demonstrated that this form of tolerance is mediated in part by hapten-specific suppressor T cells that can be harvested from lymph nodes and spleens of tolerant mice. The downregulatory capacities of these cells were assayed by adoptively transferring the cells to syngeneic naive recipients. The infectious tolerance revealed by these studies resembles other forms of peripheral tolerance induced in adult animals because the responsible T cells are chiefly $CD8^+$. Very recently, Shimizu, Kurimoto, Yoshikawa and Streilein have discovered that hapten-specific tolerance can be induced in mice by painting DNFB on skin into which TNFα or *cis*-UCA had been injected intradermally (unpublished observations). Moreover, TNFα- and *cis*-UCA-induced unresponsiveness can be transferred with spleen cells from tolerant donors, implying that a cellular mechanism similar to that induced by UVB radiation is operative. However, in preliminary experiments, systemic administration of anti-TNFα antibodies prior to painting hapten on TNFα- or *cis*-UCA-treated skin has *failed* to prevent tolerance induction. If confirmed, this observation would suggest that factors in addition to TNFα are important in tolerance induction by UVB radiation. Thus we anticipate that the cellular and molecular mechanisms that lead to the creation of hapten-specific tolerance may be partially distinct from those that impair CH induction, implying that different sets of host genetic elements may also be involved.

This outcome should not be too surprising. Whereas induction of CH appears to depend heavily on epidermal and dermal antigen-presenting cells of the Langerhans cell type, tolerance induction through murine skin has been achieved in a number of different ways, involving quite different cell types and approaches: UVB-irradiated Langerhans cells [127], unirradiated thy-1^+ dendritic T cells [127] and free (unconjugated, excess) hapten. Moreover, it is likely that this list is incomplete. Thus it appears that at this point in time we are much further from an acceptable understanding of the mechanisms responsible for tolerance after UVB radiation than we are from knowing the mechanisms that impair CH induction.

Identification of genetic loci determining the deleterious effects of UVB on CH induction in mice

Although the evidence presented in preceding sections of this chapter implies that the capacity of UVB to impair CH induction in mice is a universal effect, quite the opposite is actually the case. Approximately 10 years ago, Streilein and his collaborators in Dallas discovered that an acute low dose regimen of UVB that impaired CH induction in C57BL/6 mice, utterly failed to impair

CH induction in BALB/c mice [126]. Since by microscopic criteria this dose of UVB radiation was equally damaging to epidermal Langerhans cells from both strains of mice [10], these investigators concluded: (i) that cutaneous antigen-presenting cells *other than Langerhans cells* (and presumably beyond the reach of UVB's damage) can substitute for LC in CH induction in some strains of mice (such as BALB/c) [124]; and (ii) that polymorphic host genetic factors must interact with UVB to produce the deleterious effects of UVB that are observed in some, but not other, strains of mice. Strong support for this idea was also deduced from the experiments already described in which UVB impaired CH induction in C3H/HeN mice, but not in C3H/HeJ; after all, the only known genetic differences between these strains resides on the fourth murine chromosome and encodes sensitivity to LPS [141].

Identifying the genetic loci responsible for the deleterious effects of UVB radiation on CH induction in certain genetically defined mice became a high priority, because of the likelihood that this knowledge would advance our understanding of the phenomenon. Accordingly, numerous inbred strains of mice were subjected to the acute low dose UVB protocol, followed by epicutaneous DNFB in order to determine which strains are susceptible to the deleterious effects of UVB on CH induction (and are therefore termed UVB susceptible; UVB-S), and which are resistant, i.e. develop vigorous CH even when DNFB is painted on UVB-exposed skin (termed UVB resistant; UVB-R). Different strains of genetically defined mice were selected so as to sample: (i) a wide range of background genes, i.e. representing different genetic stocks and geographic origins; and (ii) a wide range of disparate major histocompatibility complex (MHC) haplotypes, on the presumption that class I and/or class II genes within the MHC might be particularly important in governing immunity induced through the skin. In addition, F_1 hybrids were produced by mating: (i) UVB-S and UVB-R strains in order to determine which phenotype was dominant; and (ii) two different UVB-S, as well as two different UVB-R, strains to unmask a recessive, complementing gene loci. A summary of results [126, 145] (Kurimoto and Streilein, manuscript submitted for review) of these experiments is presented in Tables 10.1 & 10.2. It is important to point out that all of the mice used in these studies contained the LPS-sensitive allele, *Lps*^{n}. Table 10.2 displays data derived from strains of mice that are polymorphic at the *Lps* locus, as well as at *H-2*, the murine major histocompatibility locus. Clearly, mice that are homozygous for the mutant *Lps*^{d} allele are UVB resistant; similarly, mice that are homozygous for the *d* haplotype of the MHC are UVB resistant. All other *H-2* haplotypes dictate UVB-S, irrespective of background genes. Background genes from B10, C3H, BALB/c, SJL, SWR and NZW were included in this survey. Therefore, with the exception of the *Lps* locus (on murine chromo-

Table 10.1 Genetics of susceptibility and resistance to deleterious effects of ultraviolet B radiation on contact hypersensitivity induction

Strain	Allele	UVB phenotype
Polymorphism at Lps		
C3H/HeJ, B10.ScN	*d*	Resistant
C3H/HeN, C57BL/10	*n*	Susceptible
Polymorphism at H-2		
BALB/c, DBA/2, B10.D2, NZB	*d*	Resistant
C57BL/6	*b*	Susceptible
C3H/HeN, B10.BR	*k*	Susceptible
SJL, A.SW, B10.S	*s*	Susceptible
SWR, B10.Q	*q*	Susceptible
NZW	*z*	Susceptible

Table 10.2 Dominance-recessive relationships among UVB susceptible and UVB resistant strains of mice

Strains	Alleles: *Lps*	Alleles: *H-2*	UVB phenotype
(BALB/c × C57BL/6)F_1	*n/n*	*d/b*	Susceptible
(BALB/c × C3H/HeJ)F_1	*n/d*	*d/k*	Resistant
(C3H/HeJ × C3H/HeN)F_1	*d/n*	*k/k*	Susceptible
(C3H/HeN × C57BL/6)F_1	*n/n*	*k/b*	Susceptible

some 4), the only other locus governing UVB susceptibility appears to reside within *H-2*.

Table 10.2 makes the further point that UVB susceptibility is a dominant trait. When either parent of an F_1 hybrid is UVB-S, then the hybrid is also UVB-S. Moreover, F_1 hybrids derived from phenotypically identical parents also displayed the parental phenotypes, implying that a single genetic locus was determining the UVB trait, and no recessive alleles that function as complementing loci were present. The findings that C3H/HeN is UVB-S, while BALB/c and A/J are UVB-R clearly indicate that the locus dictating UVB-S cannot reside at the class I or II loci that encode *H-2 K*, *A*, or *E*, but must exist in the *H-2* segment that is telomeric to *H-2E*. This was a surprising observation because it was known that hapten-conjugated peptides cleaved by proteases of cutaneous antigen-presenting cells must be presented on class I or class II molecules in order to activate CH-producing T cells, and therefore it was regarded as likely that UVB would interfere with this antigen-processing and presentation pathway.

Consequently it was necessary to conduct additional studies designed to explore the so-called class III region of *H-2* in which reside genes encoding several serum complement components, a cytochrome P450 enzyme, heat shock proteins, and the tumor necrosis factor cytokines. Because our studies had already implicated TNFα in the effects of UVB on CH induction [145], our selection of recombinant inbred strains within the *H-2* complex was biased in favor of combinations that discriminate among the different *Tnfα* alleles that have been described [73]. Data from these experiments (Vincek, Kurimoto, Medema and Streilein, manuscript submitted for review) are summarized in Table 10.3. The findings pinpoint the locus dictating the UVB-dependent phenotypes to the chromosomal region that includes *Tnfα*, but excludes the region of *H-2* toward the centromere, i.e. including and beyond *Bat-5*. However, we cannot map the UVB effect exclusively to *Tnfα* because no *H-2* recombinant strains exist that permit us to separate *Tnfα* from *H-2D*. But given the overwhelming functional evidence that TNFα mediates the effects of UVB on CH induction, we feel comfortable in interpreting the genetic evidence as compatible with the hypothesis that polymorphisms at *Tnfα* are critical to determining whether mice will be UVB-S or UVB-R.

The existence of mutant mouse strains that are relatively unresponsive to LPS has made it possible for us to identify a second, non-MHC, locus contributing to UVB susceptibility. In addition to strains C3H/HeN and C3H/HeJ, we have studied strain C57BL/10, which is LPS sensitive, and its congenic partner B10ScN, which possesses an abnormal allele at the *Lps* locus that renders these mice LPS resistant (Kurimoto and Streilein, manuscript submitted for review). B10ScN mice proved to be UVB-R, whereas C57BL/10 mice are very susceptible to the effects of UVB on CH induction (see Table 10.1). Therefore it can be concluded that

Table 10.3 Localization within *H-2* of UVB-S trait to *Tnfα*

Strain	*Eα*	*C4*	*Bf*	*Hsp70*	*Bat-5*	*Tnfα*	*D*	UVB*
B10.D2	*d*	*d*	*d*	*d*	*d*	*d*	*d*	R
A/J	*k* –>†	*d*	*d*	*d*	*d*	*d*	*d*	R
B10.D1(R106)	*b*	*b*	*b* –>	–	*d*	*d*	*d*	R
A.TH	*s*	*s*	*s*	*s* –>	*d*	*d*	*d*	R
B10.T(6R)	*q*	*q*	*q*	*q*	*q* –>	*d*	*d*	R
B10.BR	*k*	*k*	*k*	*k*	*k*	*k*	*k*	S
C57BL/10	*b*	*b*	*b*	*b*	*b*	*b*	*b*	S
B10.Q	*q*	*q*	*q*	*q*	*q*	*q*	*q*	S
A.SW	*s*	*s*	*s*	*s*	*s*	*s*	*s*	S

* Indicates whether phenotype is UVB-R or UVB-S. † Indicates critical crossover point within *H-2* chromosome.

mice that are resistant to LPS are also resistant to UVB effects on CH induction, and therefore a polymorphism at the *Lps* locus can also dictate the UVB-dependent phenotypes. To our knowledge only these two genetic loci – *Tnfα* and *Lps* – contain alleles that influence the effects of UVB radiation on CH induction in mice. By knowing the genotype of mice at these loci, one can accurately predict whether UVB radiation will impair CH induction or not.

Before leaving this consideration of the link between polymorphisms at the *Lps* locus and the deleterious UVB effects on CH induction, the recent work of Elmets *et al.* [29] should be mentioned. These investigators appear to have generated data that is at odds with our evidence concerning the biologic relevance of LPS-responsiveness and skin cancer. Dimethylbenzanthracene (DMBA) is a chemical carcinogen that induces local tumors when painted on mouse skin. The chemical is known to form adducts with DNA and in this regard may induce mutations that are important to malignant generation, similar to UVB radiation. In fact, Halliday and his colleagues have reported that skin treated with DMBA is unable to support the induction of CH when DNFB is applied to the same surface, an outcome very similar to the consequences of exposure to UVB radiation [47, 48]. Elmets *et al.* have demonstrated that the incidence of skin tumors is markedly greater when DMBA is painted on the skin of C3H/HeJ mice compared to C3H/HeN mice, and they have produced evidence that the mumber of DNA adducts induced by DMBA was also greater in HeJ than in HeN mice [29]. Other chemicals, some of which are tumorigenic, have also been associated with LPS sensitivity [20, 80, 129]. At present we are at a loss to explain why one tumorigenic agent, UVB radiation, is deleterious to CH induction in LPS-sensitive mice, whereas another tumorigenic agent, DMBA, promotes a higher incidence of tumors in LPS-resistant mice. On the surface, these results are contradictory. However, we are mindful that tumorigenesis is a multistep process, the sequence of which is only partially understood. It is reasonable to expect that a factor that appears to promote tumors at one step along this pathway may interfere with tumor development at another step. Perhaps these disparate results will be reconciled when more information has been gathered.

Molecular genetic basis of UVB susceptibility and resistance

Numerous studies have contributed to our understanding of the genetic basis of LPS sensitivity [51, 106, 141]. As mentioned previously, *in vitro* treatment of murine macrophages with LPS leads to markedly increased production of IL-1, IL-6 and TNFα. There is evidence that increased production results both from increased transcriptional activity at each of these cytokine loci as well as from stabilization of mRNA for each cytokine [50]. At present the best interpretation of the experimental evidence is that the wild-type allele at the *Lps* locus (Lps^n) encodes a protein that binds to a UAAUUAAU sequence that is distinctive and present in the 3′ region of these mRNAs; by binding to this sequence, the *Lps*-directed protein protects the message from rapid degradation by cytoplasmic RNAases which use the same sequence for recognition. The mutant Lps^d appears to encode an inactive (or no) protein and therefore cells homozygous for this gene are unable to stabilize mRNA for IL-1, IL-6, and TNFα. There is also evidence to suggest that LPS can stimulate transcriptional activity at the *Tnfα* locus in macrophages, but this effect is of a lesser magnitude compared to the 3′ UT effect of LPS. This is pertinent because the *Tnfα* locus is constitutively active in macrophages but is believed to be transcriptionally silent in keratinocytes and other cells in normal 'resting' skin [61, 71]. These same authors suggest that LPS can initiate transcriptional activation of *Tnfα* in epidermal cells.

The putative polymorphisms at the *Tnfα* locus in mice are of considerable interest, and may reside within non-coding sequences at the 5′ or 3′ untranslated regions of the gene, or even within introns. It is believed that all mice produce identical TNFα molecules and therefore it is unlikely that polymorphisms of interest will be present among coding sequences in the four exons of

Table 10.4 Polymorphisms within *Tnfα* define UVB-S and UVB-R mice

Strains	BamHI*	$(CA)_n$†	UVB‡
BALB/c	12.0	14	R
DBA/2	12.0	14	R
A/J	12.0	14	R
B10.T(6R)	12.0	14	R
B10.D2(R106)	12.0	14	R
C57BL/6	10.5	20	S
C3H/HeN	10.5	17	S
B10.Q	10.5	19	S
B10.S	10.5	20	S

* Size of fragment (kb) revealed by digestion with BamHI and probing with *Tnfα*. † Number of CA repeats found at microsatellite within 5′ untranslated region of *Tnfα*. ‡ Indicates whether phenotype is UVB-R or UVB-S.

this gene. In order to gain information on this matter our laboratory has used restriction fragment length analysis, and we have cloned and sequenced eight different murine *Tnfα* alleles (Vincek *et al.*, manuscript submitted for review). Several laboratories [56, 103], including ours, have described a restriction fragment length polymorphism at the *Tnfα* locus, when BamHI is used to digest murine DNA. As shown in Table 10.4, BamHI digestion reveals a 10.5 kb *Tnfα* fragment in the DNAs of mice that have the UVB-R phenotype, whereas DNAs from all UVB-S strains of mice contain a 12.0 kb *Tnfα* fragment. Thus, this polymorphism accurately discriminates the two UVB phenotypes. Analysis of the nucleotide data derived from sequencing the introns of *Tnfα* alleles from eight different mouse strains has revealed the BamHI site to reside at position 329 in the first *Tnfα* intron, where a C is found in UVB-S mice, and a T has been found in UVB-R mice. However, very little is known about the possibility that regulation of TNFα production can be exerted within the first intron of the gene. It is relevant that Jacob and his colleagues [56] and Richter *et al.* [103] have used evidence of BamHI polymorphism at *Tnfα* to correlate with lupus-like syndromes in mice [32, 56, 103], and with toxoplasma encephalitis [37]. In fact, these investigators have claimed that mice that are susceptible to toxoplasma encephalitis possess the BamHI allele that we find to be associated with UVB-S, and they have demonstrated that the brains of infected mice with encephalitis have measurable TNFα, whereas the brains of resistant mice do not.

Transcriptional regulation of genes is usually focused within the 5′ untranslated region upstream from the start site. Jongeneel *et al.* [114] first provided the complete sequence of a murine *Tnfα* gene, and these investigators detected several segments likely to be responsive to cytokines within the 5′ UT region. In addition, a microsatellite was found to exist immediately 5′ to the Y box that is commonly found in genes within the MHC [58]. We have cloned and sequenced the 5′ UT region of numerous UVB-S and UVB-R *Tnfα* alleles. As summarized in Table 4 (Vincek *et al.*, manuscript submitted for review) the only differences that distinguish the different alleles reside at the microsatellite. *Tnfα* alleles from UVB-R mice have precisely 14 (CA) repeats at this region. By contrast the alleles of UVB-S mice never contain $(CA)_{14}$; instead, there may be >14 or <14 (CA)s at the microsatellite site in the UVB-S alleles. We should also note in passing that we have cloned and sequenced the 3′ UT regions of *Tnfα* alleles of these same UVB-S and UVB-R mice, and have found no evidence of nucleotide differences (unpublished data).

Thus if polymorphism at *Tnfα* is responsible for distinguishing UVB-S from UVB-R responses to UVB, then either the genetic differences at the BamHI site in the first intron, or at the $(CA)_n$ site in the 5′ UT region must be responsible. Since no functional evidence examining these regions has been reported, there is little cause to speculate on the biologic meaning of these nucleotide differences. However, it may be worth pointing out that microsatellites have been claimed to influence transcriptional activity in other systems [7, 49, 105], so it is possible that the $(CA)_n$ differences we and others have found may actually be relevant to the UVB-S and UVB-R phenotypes.

We are hopeful that studies currently underway in various laboratories, including ours, will unravel the transcriptional and translational control mechanisms that govern TNFα production by both macrophages and epidermal- and dermal-derived cells. We anticipate that this knowledge will reveal one of the mechanisms by which UVB alters cutaneous immunity. Our evidence that *trans*-UCA is a photoreceptor for the effects of UVB, and that the link is via TNFα, places *cis*-UCA in this molecular loop. Perhaps *cis*-UCA mimics LPS and promotes stabilization of TNFα mRNA – in mast cells, where the *Tnfα* gene is usually transcriptionally active [43]. Alternatively, or in addition, *cis*-UCA, isomerized by UVB radiation, may interact with cytoplasmic or nuclear proteins that have the ability to initiate *Tnfα* transcription in keratinocytes (or other cutaneous cells where the gene is usually silent). In either event, our expectations are that it will turn out that the *Tnfα* and *Lps* alleles found in UVB-R mice are relatively unable to respond to these UVB-dependent stimuli, and as a consequence, the intracutaneous production/availability of TNFα after UVB radiation is insufficient to interfere with CH induction.

Effects of UVB radiation on CH induction and tolerance in humans

Experience in normal Caucasian subjects

The realization that the deleterious consequences of UVB radiation on CH induction in mice are genetically determined made it imperative that similar studies be conducted in humans. If a similar situation were found in humans, then it should be possible to define a novel set of genes important in the pathogenesis of skin cancer. Accordingly our laboratory initiated an experimental plan 6 years ago to determine whether the traits of UVB-R and UVB-S exist in humans, and whether these traits were related to skin cancer. The UVB treatment protocol employed was fashioned after that used in mice, except that the total dose of UVB was increased [101]. This modification was necessary in order to achieve virtual depletion of Langerhans cells from human skin. In the regimen currently in use, a circular site (2 cm in diameter) of buttock skin of adult volunteers is treated on four consecutive days with 144 mJ/cm^2 UVB radiation. Within minutes after the fourth exposure, 2000 μg dinitrochlorobenzene (DNCB), a universally sensitizing dose in humans [18], is applied exclusively to the irradiated surface [144]. Subsequently, usually 4 weeks later, the forearm skin of these subjects is challenged with 50 μg DNFB to determine whether CH had been induced. In the initial study group approximately 35% of the Caucasian volunteers displayed no evidence of CH when challenged at 30 days; the remaining subjects displayed vigorous CH [134, 144]. Now, after experiments have been conducted on 50 normal volunteers, the incidence of the non-reactivity to DNCB painted on UVB-exposed skin remains at about 40% (see Table 10.5). We have designated the non-responder individuals as UVB-S, and the responders as UVB-R. Based on this cumulative experience we conclude that humans have an affinity with mice in that some individuals are able to develop CH when a hapten is painted on skin acutely exposed to UVB radiation, whereas other individuals fail to respond. Results recently reported by Cooper *et al.* [22] using a rather different hapten sensitization protocol, confirm our findings that UVB impairs CH induction in humans. Perhaps because of differences in the sensitization procedures these authors believe that the effect of UVB is dependent upon hapten dose, although their published data indicate that among normal individuals subjects exist that we would define as UVB-S.

Our studies have also revealed that humans have an affinity with mice with respect to the induction of hapten-specific tolerance [134]. Many of our UVB-S volunteers have received a second sensitizing dose of DNCB applied to unirradiated normal skin after the completion of the first experiment. Approximately 6% of these individuals never developed CH to the second attempt at DNCB sensitization (see Table 5). Since the dose of DNCB employed, 2000 μg, is known to universally sensitize all normal humans (>98%, our own data) [18], the unresponsiveness observed in this subset of UVB-S individuals implies that a form of tolerance was induced. In support of this implication, these subjects have subsequently been exposed to a universally sensitizing dose (500 μg)

Table 10.5 Effect of ultraviolet B radiation on induction of contact hypersensitivity and tolerance with dinitrochlorobenzene

Subjects	UVB treatment	Contact hypersensitive	Tolerant
Caucasian – healthy	No	12/12 (100%)	NA
Caucasian – healthy	Yes	28/50 (56%)	1/16 (6%)
Pigmented – healthy	No	10/10 (100%)	NA
Pigmented – healthy	Yes	10/19 (53%)	1/8 (12%)
Skin cancer	Yes	3/41 (7%)	11/26 (50%)

Volunteers were painted with DNCB (2000 μg) on untreated (No) or UVB-treated (yes) skin, then assessed for contact hypersensitivity 30 days later with 50 μg DNCB applied to forearm skin. Individuals with negative responses were reexposed to 2000 μg DNCB and reassessed for contact hypersensitivity. Those who failed to display contact hypersensitivity after second exposure to 2000 μg DNCB were designated as tolerant. Data presented as number positive or tolerant/number tested. NA, not applicable.

of the non-crossreactive hapten, diphencypone (DPCP), and in every instance, intense CH was detected (data not shown). Thus, the unresponsiveness to DNCB displayed upon reimmunization of a minority of UVB-S subjects appears to be hapten-specific, and therefore is very likely to represent immunologic tolerance. Cooper *et al.* [22] have also generated evidence to indicate that some individuals exposed to hapten through UVB-exposed skin develop tolerance.

Experience in patients with skin cancer

Perhaps the most important findings from our studies to date concern the effects of UVB on CH induction in patients with skin cancer. Using the typical acute low dose UVB radiation protocol, patients with biopsy-proven basal/squamous cell skin cancer were assessed for their ability to become sensitized with 2000 μg DNCB [134, 144]. More than 40 patients have now been studied. To date only three of these individuals have proved to be UVB-R, i.e. have developed DNCB-specific CH (approximately 7%); this should be compared with the frequency of the UVB-R trait among normal individuals (56%; see Table 10.5). Further experiments have revealed that this outcome does not result from an age effect since the UVB-R trait was detected in >60% of normal volunteers that were age-matched with our skin cancer patients. Thus patients with skin cancer, who at the time of testing had no evidence of systemic malignancy or any other serious disease, were found almost invariably to be UVB-S. This finding strongly suggests that the UVB-S phenotype is a risk factor for skin cancer in humans. It is equally important to note that hapten-specific tolerance was also detected in our skin cancer patients, using the same approach described in UVB-S normal subjects (see Table 10.5). However, the incidence of tolerance was found to be considerably higher among the cancer subjects, approaching 50%. Extrapolating to the pathophysiology of skin cancer, this result implies that tolerance induction by neoantigens expressed by malignant cells in UVB-exposed skin may also be important to the ultimate success of the tumor, and may be achieved by separate regulatory processes.

Experience in humans with deeply pigmented skin

Unlike murine skin, human skin contains melanocytes within the epidermis, and humans vary considerably in the amount and type of melanin that is deposited in epidermal keratinocytes, a variation that is under polymorphic, polygenic control. We have studied the impact that epidermal melanization might have on the expression of the UVB-S and UVB-R traits in humans and the answer we have obtained is rather surprising [137]. In a group of normal volunteers with genetically pigmented skin ranging from dark tan to black, UVB-S was detected with a frequency *comparable* to that found in Caucasian skin (see Table 10.5). In addition, UVB-S was also detected as a trait in Caucasian volunteers with deeply tanned skin secondary to prolonged sun exposure. These findings indicate that melanin deposition in the epidermis does not protect against the deleterious effects of UVB radiation on CH induction in UVB-S humans. Although this finding might be viewed with surprise, it is important to recall that the heaviest melanin deposits in human epidermis are localized within the basilar epidermal cells. By contrast, the heaviest concentration of *trans*-UCA is in the most superficial layers of the epidermis, and Langerhans cells typically reside in the suprabasilar levels of the epidermis, a region that is virtually unprotected by melanin. Thus, Langerhans cells can be directly exposed to UVB radiation; moreover, if *cis*-isomerization of *trans*-UCA is important in initiating the deleterious effects of UVB on CH induction in humans (as it is in mice), then because of its basilar location melanin would be irrelevant to this immune outcome. However, the location of melanin primarily within the basal epidermis serves to protect the DNA of these generative keratinocytes from UVB damage. It is probably the dense deposition of melanin in basilar keratinocytes in black-skinned individuals that accounts for the observation that skin cancer is rare in these persons.

Are UVB-S and UVB-R traits genetically determined in humans?

Studies of the type we used to demonstrate a genetic basis for UVB susceptibility in mice are not possible in humans. Although a single human *Tnfa* gene has been cloned, sequenced and published, it is unknown whether the locus is polymorphic in humans [86, 90]. We have attempted to answer this important question by indirect means. The HLA phenotypes of all of our subjects have been assessed by serologic typing, but attempts to find class I or class II alleles that are particularly associated with either the UVB-S or UVB-R traits have failed (unpublished observations). This was a disappointing result because a group of Dutch investigators, led by Bouwes Bavinck, have reported that HLA-A11 is negatively associated with skin cancer in patients who are at increased risk of the disease because of prolonged immunosuppressive therapy associated with renal allografts [14, 15]. If HLA-A11 is protective against skin cancer, and if the protection is relevant to the UVB-R trait as we define it, we would have hoped to find an enrichment of HLA-A11 among our normal UVB-R subjects. Unfortunately, no such enrichment was found. Moreover, A11 was detected at a normal frequency in our skin cancer patients, the vast majority of whom were UVB-R. We should keep in mind that the skin cancer patients we have studied have *not* been under immunosuppressive therapy, and therefore the finding of A11 among our patients does not necessarily contradict the Dutch data. A11 may only be protective when the immune system is compromised. In addition, the skin cancer in patients in South Florida is manifestly a consequence of chronic, intense sun exposure. In the Netherlands the climate and level of sun exposure is not usually thought to be of the type needed to promote a high incidence of skin cancer.

It has been reported that humans are polymorphic with regard to the sensitivity of their peripheral blood monocytes to LPS stimulation, that a locus exists within the human HLA region that dictates different levels of responsiveness, and that the LPS-non-reponsive allele is strongly associated with HLA-DR2 [87]. We have examined our data with this in mind. While the incidence of the HLA-DR2 allele is highly represented in our UVB-R subjects, it is also present in some UVB-S subjects, and the differences in these incidences are not statistically significant. Most humans are heterozygous at HLA, and it is relevant to this line of reasoning to recall that LPS resistance in mice is a *recessive* trait. If a similar situation applies to humans then the presence of a single HLA-DR2 allele putatively associated with an LPS-resistant allele in a UVB-S human would not be expected to mitigate against expression of the susceptibility phenotype. Therefore, to date, HLA typing has not been of help in linking the UVB-S and UVB-R traits to the human MHC.

Several years ago an RFLP was claimed to be revealed in the human *Tnfα* locus by digestion with NcoI [39]. To examine this possibility, we analysed DNA from our subjects to determine whether the NcoI RFLP correlates with the UVB-S and UVB-R traits. We found no evidence of a correlation. Moreover, it has recently been demonstrated that the location of the polymorphic nucleotide sequences that dictate this RFLP is actually within the *Tnfβ* gene, not within *Tnfα* [136, 143]. In summary, at present we have no direct evidence that UVB-S and UVB-R are genetically determined, nor that these traits are linked to HLA or a human LPS-responsive locus.

We have, however, begun to collect functional evidence that bears indirectly on the genetic hypothesis. In mice, unpublished evidence suggests that the UVB-S and UVB-R phenotypes remain the same whether the eliciting hapten is DNFB or TNCB. Such an outcome would be the prediction of a genetically based phenomenon. That is, if UVB-S results from a genetic polymorphism at the *Tnfα* locus, then the UVB-S trait should be observed in mice with susceptibility *Tnfα* alleles, irrespective of which hapten is used for testing. Therefore, we have begun to determine whether the UVB-dependent phenotypes of our human subjects, as delineated with DNCB, remain the same (concordant) or not (discordant) when their skin is exposed to UVB and painted with the non-crossreacting hapten DPCP. The results of these experiments, which have been conducted primarily by Dr C. Tie and Ms C. Colomb, and which are preliminary because the study is incomplete at the time of writing, are presented in Table 10.6. The level of concordance between the UVB-S/UVB-R phenotypes revealed by DNCB and the phenotypes revealed by DPCP approaches 80%. Moreover, with one exception, all subjects that are UVB-R with DNCB are also UVB-R with DPCP. The discordances, which are actually rather modest in number, are found in subjects that are UVB-S when tested with DNCB, but UVB-R when tested with DPCP. In our hands DPCP is a more potent immunogen than is DNCB, and we suspect that this accounts for the higher frequency of the UVB-R trait when the hapten employed is DPCP. Whatever the explanation, a high level of concordance exists between the UVB-dependent phenotypes found in human subjects exposed to DNCB and DPCP, and this finding supports the hypothesis that these traits are genetically, rather than environmentally, determined.

Table 10.6 Comparison of UVB-S and UVB-R traits in volunteers tested with DNCB and DPCP

Group	(n)	DNCB	DPCP
1	(10)	+	+
2	(1)	+	−
3	(9)	−	−
4	(6)	−	+

Volunteers were treated twice with four doses of UVB radiation; after the first, DNCB (2000 μg), and after the second, DPCP (500 μg) was painted on irradiated sites. Challenge with dilute haptens on forearm skin is listed as + (positive) or − (negative).

Immunogenetics and skin cancer in humans

Immunogenetics defines that segment of genetics that is concerned with genes and molecules closely associated with the immune system. Therefore, while certain genetic loci are obviously pertinent to the immune system, the margins of this field are imprecise, just as the interface between the immune system and other organ systems is imprecise. The field of immunogenetics has been dominated by studies of the major histocompatibility complex which began more than 60 years ago. As a consequence, it is expected that an important dimension to any immunogenetic consideration is the issue of polymorphisms, especially with regard to susceptibility and resistance to environmental insults, ranging from pathogens to allografts. By definition, polymorphism refers to the existence of two or more alleles at a single genetic locus, and each allele is present in the gene pool with a frequency of 1% or greater. Thus, a discourse on immunogenetics of sunlight-induced skin cancer must attempt to list the genetic loci most important in generating immune responses to tumors, and must emphasize the participation of loci where polymorphisms are known or suspected to exist and which might therefore explain the most critical clinical observation: some, but not all, individuals equally exposed to sunlight develop skin cancer. The report of Czarnecki *et al.* [23] gives hope that a genetic explanation will eventually prevail; these investigators have determined that multiple non-melanoma skin cancers occur in family constellations consistent with the hypothesis that the risk for developing these types of cancers may be inherited.

Sunlight, of the duration and intensity that is typical of common human experience, penetrates into the epidermis and upper dermis [12, 30, 63]. Therefore UVB radiation of normal human skin has the potential to alter directly genetic material in epidermal cells (keratinocytes, Langerhans cells, melanocytes) and cells within the dermal papillae (mast cells, dermal dendritic cells, fibroblasts, endothelial cells). When skin is inflamed, as occurs following an initial UVB exposure [42], then genes within invading blood-borne cells, such as lymphocytes and monocytes, can also be impacted by subsequent exposure to UVB radiation. Yet evidence suggests that only tumors of keratinocytes [46, 83, 112, 117, 135, 140] and melanocytes [33, 55, 75, 85, 110] are causally linked to UVB radiation. This should not be construed to mean that only genes within keratinocytes and melanocytes are involved in the process by which UVB damages cutaneous immune responses. It is altogether possible that polymorphisms, which could interact differentially with UVB in creating the UVB-S and UVB-R phenotypes, as well as in conferring differential susceptibilities to sunlight-induced skin cancer, might be found among genetic loci uniquely expressed in cells of the immune system, in keratinocytes, in pigment-producing cells, or in fixed dermal cells, especially mast cells (Table 10.7).

Genes of the immune system

Only two sets of genes are known to be uniquely activated and expressed in cells of the immune system: those that encode the heavy and light chains of immunoglobulins, and those that encode the alpha, beta, gamma, and delta chains comprising T-cell receptors for antigen. No evidence exists to implicate these genes in the immune pathogenesis of skin cancer in particular, or in any form of cancer for that matter. However, few studies have actually addressed this issue, and therefore this should not be considered a definitive statement. For many years, polymorphisms have been known to exist at loci encoding immunoglobulin genes, and some of these polymorphisms have been linked to certain autoimmune diseases [111]. Recently, polymorphisms have been documented to exist in the loci encoding beta chains of the human Tcr, and a convincing case has been advanced that there is an association between these polymorphisms and multiple sclerosis, a disease strongly suspected of an autoimmune pathogenesis [113]. Since it is easy to imagine how polymorphic differences at human immunoglobulin and Tcr loci could have profound effects quantitatively and qualitatively on immune responses to antigens, it would not be surprising if polymorphisms at these loci are eventually found to contribute to cancer development by modifying the vigor and/or specificity of immunity directed at tumor-specific antigens. Whether UVB radiation could influence this process is completely speculative.

The class I and class II genes of the MHC are not uniquely expressed in immune system cells but their expression is closely regulated by cytokines secreted by cells of the immune system. These cell surface molecules are enormously polymorphic, play definitive roles in antigen processing and presentation to T cells, and the peptides that become associated with the groove of their most external domains, as well as the molecular configuration of the domains themselves, shape the T-cell repertoire in the thymus, and are requisite signals for T-cell activation in immunity. Thus speculation is rife that class I and II MHC deter-

Table 10.7 Immunogenetic loci that may contribute to sunlight-induced skin cancer

Locus	Cellular source	Polymorphism exists	Effects on immunity	Link with UVB-induced skin cancer
Ig	B cells			
Kappa		No	–	–
Lambda		Yes	No	No
Heavy		Yes	Yes	No
Tcr	T cells			
Alpha/beta		Yes	Yes	No
Gamma/delta		No	–	–
Class I MHC	Ubiquitous	Yes	Yes	Yes
Class II B cells	APC	Yes	Yes	Yes (?LPS)
Tnfα	Many	Yes	Yes	Yes
HSP-68	Many	?	–	–
Complement	Many	Yes	Yes	No
21-OHase	Adrenal	Yes*	No	No
Mn-SOD	Many	Yes	No	No
alpha-MSH	KC, lymphs	No		
Lps	Macrophages	Yes*	Yes	Yes
Enzymes of excis/repair	Many	Yes*	No	Yes

*Mutant alleles exist, but whether they exist at polymorphic frequencies is unknown.

minants will turn out to be critical in determining whether, and which, tumor specific neoantigens are detected by the immune system. Our evidence suggests that class I and II MHC molecules play little or no role in the process by which UVB impairs CH induction in mice [126]. However, it remains to be determined whether polymorphisms at class I and/or II MHC loci participate in the hapten-specific tolerance induced by UVB radiation. Although we have no evidence relevant to this suggestion, we anticipate that the MHC molecules that restrict T-cell activation are likely to determine whether tolerance of tumor antigens emerges after UVB radiation.

It came as a surprise to us that the MHC genes that were found to dictate the UVB-S and UVB-R traits actually reside within the class III region of this chromosomal region in mice [60]. To date these traits appear to be dictated by polymorphisms within the regulatory segments of alleles at the *Tnfα* locus. But it should be kept in mind that there are other loci within this region that could be suspected of participating in UVB-initiated changes in the skin. Examples include the locus that encodes heat shock protein-70, one or more of the complement components, and even the cytochrome P450 enzyme involved in the pathogenesis of the adrenogenital syndrome. Polymorphisms of unknown functional significance are already described at many of these loci. In mice, the *manganese-SOD* locus is located immediately centromeric to the *H-2* complex and several alleles exist, as defined by RFLPs [31]. This locus is of interest because the enzyme it encodes is located within mitochondria where it acts to eliminate oxygen free radicals. It is known that UVB induces free radicals [100, 129], as does TNFα [139], and that these agents can create mutations in DNA. It is further known that the nuclear enzymes that normally repair these mutations are not present in mitochondria. Thus DNA within mitochondria is particularly vulnerable to UVB- and TNFα-related free radical damage. It is possible therefore that defective *Mn-SOD* alleles could render mitochondria especially vulnerable to UVB radiation, and could theoretically lead indirectly to cell death by respiratory failure. Whether this might increase the likelihood that CH induction would fail, or that tumors would develop, is unknown.

Other genes that are uniquely associated with the immune system and that are undoubtedly important in immune surveillance and in rejection of tumors are interleukin-2, interleukin-4, and gamma-interferon. However, at present there is no evidence that polymorphisms exist at these crucial loci. In addition, the *Lps* locus which strongly influences the production of various cytokines associated with immunogenic inflammation and effector responses must also be mentioned.

Therefore the genes that are associated primarily with the immune system, and which may be considered to be immunogenetically involved in the pathogenesis of skin cancer, include loci encoding immunoglobulins, T-cell receptors for antigen, class I and II MHC molecules, certain class III MHC loci – such as *Tnfa*, *Hsp68*, and nearby loci, such as *Mn-SOD*, and (perhaps) *Lps*. But at present only *Tnfa* and *Lps* are unequivocally linked to the UVB-S and UVB-R traits in mice. In addition, circumstantial evidence suggests that one or more class I HLA antigens can influence susceptibility to sunlight-induced skin cancer.

Genes of non-immune cells of the skin

As the major cellular constituents of the epidermis, keratinocytes contain genes that are particularly well situated to be altered by UVB radiation directly – via mutation – and indirectly – via factors generated by UVB radiation that alter gene transcription and translation within keratinocytes. It has already been amply described that UVB induces the production of cytokines by keratinocytes, and considerable evidence supports the view that one of these cytokines, TNFα, actually mediates the deleterious effects of UVB on CH induction – at least in mice. Another cytokine whose production is induced in the epidermis by UVB radiation is alpha-melanocyte stimulating hormone (α-MSH) [13]. It is of interest that α-MSH has been found to inhibit contact hypersensitivity in mice [102], raising the possibility that UVB may influence resistance to skin cancer via α-MSH. However, no experimental evidence has been reported that would suggest that production of this cytokine is regulated in a polymorphic fashion.

There are other types of genes whose expression in keratinocytes might be altered by UVB radiation in a manner that could influence the cutaneous immune response. A series of genes exists that encode enzymes that act in concert within the nucleus to excise and repair abnormalities in DNA, including the damage caused by UVB radiation. Mutants of these genes have been demonstrated to cause the various forms of xeroderma pigmentosum [21, 38, 57, 93, 142]. Visible light has been shown to activate and enhance these genes in lower vertebrates [2, 76], and may well do so in humans. Freeman [34, 36] has generated evidence that there may be individual differences in the ability of UVB-irradiated cells to repair UVB-induced DNA aberrations. Thus there may be polymorphisms at the gene loci that encode excision and repair enzymes, and these polymorphisms may influence skin cancer risk. Kripke has speculated that an unexpected feature of UVB-dependent mutations might be to promote the development of immunological tolerance to antigens arising in skin after UVB radiation [4, 67]. Her laboratory has produced circumstantial evidence to support this view. One could also argue that excision/repair enzymes may themselves have a role to play in tolerance induction. It will be interesting to watch this speculation unfold under experimental scrutiny, and it will be particularly interesting to learn whether these genes are polymorphic or whether the control of their transcription and translation is polymorphically regulated.

Concluding remarks

We have much to learn about the pathogenesis of sunlight-induced skin cancer in humans. Even the individual steps in the multistage process that begins with UVB-induced damage to DNA and leads eventually (and not inevitably) to the emergence of a malignant tumor are unclear. Since skin cancer is not a universal outcome when humans experience similar sun exposure histories, host genetic factors are undoubtedly important in the pathogenesis of the disease. The genetic polymorphisms which are known to exist at the multiple loci that govern skin pigmentation qualitatively and quantitatively offer an excellent example of the validity of this idea: black-skinned people suffer much less sunlight-induced skin cancer than do their lighter-skinned counterparts, even when both occupy the same geographic region.

In this chapter evidence has been marshalled and an argument has been advanced to the effect that polymorphisms at genetic loci governing immune responses to cutaneous antigens are important in the pathogenesis of human skin cancer. The two major phenotypes of UVB-S and UVB-R, which have been detected in mice and in humans whose skin has been painted with a hapten after exposure to acute low dose UVB radiation, reveal that the capacity of UVB radiation to impair cutaneous immunity is genetically determined. In mice, the relevant polymorphic genetic loci appear to be concerned with regulation of intracutaneous cytokine production – TNFα. Since polymorphisms affecting TNFα production cannot alone account for UVB-induced tolerance, and since tolerance induction also occurs in only a subset of UVB-exposed mice and humans, it is very likely that other polymorphic genetic loci will be found to participate in the induction of tolerance of tumor-specific antigens. Although the identity of these genes is conjectural, it is anticipated that some of these loci will encode molecules uniquely used or regulated by the immune system.

The existence of genetic polymorphisms, especially the extensive polymorphisms at genetic loci associated with the immune system, has led to the speculation that polymorphism is 'good' and serves an important evolutionary end. It is presumed that the 'end' in question is preservation of individual species. In other words, polymorphisms at immunogenetically defined loci are thought to confer upon mice, humans and other vertebrates with advanced immune systems an advantage that insures that each species will prevail in its own environmental niche, despite the niche being filled with hostile pathogens. Despite the obvious importance of human skin cancer, at the individual level and at the societal level, it should be kept in mind that this disease usually affects individuals who have passed the reproductive years and are no longer contributing to the human gene pool. Therefore, if and when polymorphisms are defined that influence whether sunlight induces skin cancer or not, it is very likely that the 'true' evolutionary significance of these polymorphisms will not be revealed by describing their relationship to skin cancer. It is beyond the scope of this chapter to consider what the true meaning might be of polymorphisms at immunogenetic loci that are found to influence skin cancer development inadvertently. But I cannot avoid mentioning that TNFα has been suspected of playing an important role in the susceptibility to, as well as the immunopathogenesis of, a number of pathogens, such as the organisms that cause leishmaniasis [41, 78, 79, 132], schistosomiasis [1], salmonellosis [89, 131], malaria [44] and leprosy [99]. It may turn out that the evolutionary reason that our species retains polymorphisms at the *Tnfα* locus is to maintain a balanced human response to those pathogens that threaten life when humans are immature and reproductively active. Viewed from this perspective, susceptibility to UVB in relation to skin cancer may be an accident of nature, revealed because the lifespan of most humans now exceeds reproductive importance by many years.

Acknowledgment

The author is deeply grateful to the large group of individuals with whom he has had the opportunity to collaborate over the past 20 years. An incomplete list of these gifted individuals includes Rubert Billingham, Paul Bergstresser, Robert Tigelaar, Galen Toews, Craig Elmets, Joy Rico, Y.-H. Chin, Churchill McKinney, J.Richard Taylor, Vladimir Vincek and a set of exceptional trainees including Sabra Sullivan, Warner Bruins-Slot, Takeshi Yoshikawa, Jan-Willem van den Berg, Aicha Demidem, Martin Vermeer, G.J. Schmieder, Carroll Tseng, Iwao Kurimoto, Tadamichi Shimizu, Ruoping Dai, Benjamin Hoffman, Cindy Golomb and Cynthia Tie. This work was supported in part by USPHS grant AI-22072, and by American Cancer Society Grant PT-383.

References

1 Amiri, P., R.M. Locksley, T.G. Parslow, M. Sadick, E. Rector, D. Ritter, J.H. McKerrow: Tumour necrosis factor α restores granulomas and induces parasite egg-laying in schistosome-infected SCID mice. *Nature* 356 (1992), 604–607.

2 Ananthaswamy, H.N., M.S. Fisher: Photoreactivation of ultraviolet radiation-induced pyrimidine dimers in neonatal BALB/c mouse skin. *Cancer Res.* 41 (1981), 1829.

3 Ansel, J.C., T.A. Luger, I. Green: The effect of *in vitro* and *in vivo* UV irradiation on the production of ETAF activity by human and murine keratinocytes. *J. Invest. Dermatol.* 81 (1983), 519.

4 Applegate, L.A., R.D. Ley, J. Alcalay, J.L. Kripke: Identification of the molecular target for the suppression of contact hypersensitivity by ultraviolet radiation. *J. Exp. Med.* 170 (1989), 117.

5 Araneo, B.A., T. Dowell, H.B. Moon, R.A. Daynes: Regulation of murine lymphokine production *in vivo*. Ultraviolet radiation exposure depresses IL-2 and enhances IL-4 production by T cells through an IL-1-dependent mechanism. *J. Immunol.* 143 (1989), 1737.

6 Baden, H.P., M.A. Pathak: The metabolism and function of urocanic acid in skin. *J. Invest. Dermatol.* 48 (1967), 11.

7 Berg, D., K. Walls, A. Reifel-Miller, B. Grinnell: E1A-induced enhancer activity of the Poly (dT-dG), Poly (dC-dA) element (GT element) and interactions with the GT-specific nuclear factor. *Mol. Cell Biol.* 9 (1989), 5248.

8 Bergstresser, P.R., C.A. Elmets, J.W. Streilein: Local effects of ultraviolet radiation on immune function in mice. In: *The Effect of Ultraviolet Radiation on the Immune System*, (1983), p. 73. Ed.: J.A. Parrish, Johnson & Johnson and the New Mexico Health Coalition.

9 Bergstresser, P.R., C. Fletcher, J.W. Streilein: Surface densities of Langerhans cells in relation to rodent epidermal sites with special immunologic properties. *J. Invest. Dermatol.* 74 (1979), 77.

10 Bergstresser, P.R., G.B. Toews, J.W. Streilein: Natural and perturbed distributions of Langerhans cells: responses to ultraviolet light, heterotopic skin grafting, and dinitrofluorobenzene sensitization. *J. Invest. Dermatol.* 75 (1980), 73.

11 Bigby, M., R. Vargas, M.-S. Sy: Production of hapten-specific T cell hybridomas and their use to study the effect of ultraviolet B irradiation on the development of contact hypersensitivity. *J. Immunol.* 143 (1989), 3867.

12 Blum, H.F.: Wavelength dependence of tumor induction by ultraviolet radiation. *J. Natl. Cancer Inst.* 3 (1943), 533.

13 Bolognia, J., M. Murray, J. Pawelek: UVB-induced melanogenesis may be mediated through the MSH-receptor system. *J. Invest. Dermatol.* 92 (1989), 651.

14 Bouwes Bavinck, J.N., B.J. Bermeer, F.J. van der Woude, J.P. Vandenbroucke, G.M.T. Schreuder, J. Thorogood, G.G. Persijn, F.H.J. Class: Relation between skin cancer and HLA antigens in renal-transplant recipients. *N. Eng. J. Med.* 325 (1991), 843.

15 Bouwes Bavinck, J.N., A.M.M. Kootte, F.J. van der Woude, J.P. Vendenbroucke, B.J. Bermeer, F.H.J. Class: On a possible protective effect of HLA-A11 against skin cancer and keratotic skin lesions in renal transplant recipients. *J. Invest. Dermatol.* 97 (1991), 269.

16 Brash, D.E., J. Rudolph, J. Simon, A. Lin, G. McKenna, H. Baden, A. Halperin, J. Ponten: A role for sunlight in skin cancer: UV-induced p53 mutations in squamous cell carcinoma. *Proc. Natl. Acad. Sci. USA* 88 (1991), 10124.

17 Burnet, F.M.: In: *Immunological Surveillance* (1970). Pergamon Press, Oxford.

18 Catalona, W.J., P.T. Taylor, P.B. Chretien: Quantitative dinitrochlorobenzene contact sensitization in a normal population. *Clin. Exp. Immunol.* 12 (1972), 325.

19 Claman, H.N., S.D. Miller, M.-S. Sy, J.W. Moorhead: Suppressive mechanisms involving sensitization and tolerance in contact allergy. *Immunol. Rev.* 50 (1980), 105.

20 Clark, G.C., M.J. Taylor, A.M. Tritscher, G.Q. Lucier: Tumor necrosis factor involvement in 2,3,7,8-tetrachlorodibenzo-*p*-dioxin-mediated endotoxin hypersensitivity in C57BL/6 mice congenic at the Ah locus. *Toxicol. Appl. Pharmacol.* 111 (1991), 422.

21 Cleaver, J.E., K.H. Kraemer: Xeroderma pigmentosum. In: *The Metabolic Basis of Inherited Disease*, vol. II, (1989), p. 2949. Eds.: C.R. Scriver, A.L. Beaudet, D. Valle, McGraw-Hill and Co., New York.

22 Cooper, K.D., L. Oberhelman, T.A. Hamilton, O. Baadsgaard, M. Terhune, G. LeVee, T. Anderson, H. Koren: UV exposure reduces immunization rate and promotes tolerance to epicutaneous antigens in humans: relationship to dose, CD1a+DR$^+$ epidermal macrophage induction, and Langerhans cell depletion. *Proc. Natl Acad. Sci. USA* 89 (1992), 8497.

23 Czarnecki, D., J. Zalcberg, C. Meehan, T. O'Brien, S. Leahy, A. Bankier, C.G. Nash: Familial occurrence of multiple nonmelanoma skin cancer. *Cancer Genet. Cytogenet.* 61 (1992), 1–5.

24 Daynes, R.A., E.J. Bernhard, M.F. Gurish, D.H. Lynch: Experimental photoimmunology: immunologic ramifications of UV-induced carcinogenesis. *J. Invest. Dermatol.* 77 (1981), 77.

25 DeFabo, E.C., F.P. Noonan: Mechanism of immune suppression by ultraviolet irradiation *in vivo*. I. Evidence for the existence of a unique photoreceptor in skin and its role in photoimmunology. *J. Exp. Med.* 157 (1983), 84098.

26 DeFabo, E., F.P. Noonan, M. Fisher, J. Burns, H. Kaiser: Further evidence that the photoreceptor mediating UV-induced systemic immune suppression is urocanic acid. *J. Invest. Dermatol.* 80 (1983), 319.

27 DeWeck, A., N. Hunziker: Allergic contact dermatitis. In: *Clinical Immunology*, (1980), p. 1190. Ed.: C.W. Parker, W.B. Saunders CO, Philadelphia.

28 Elmets, C.A., P.R. Bergstresser, R.E. Tigelaar, P.J. Wood, J.W. Streilein: Analysis of the mechanism of unresponsiveness produced by haptens painted on skin exposed to low dose ultraviolet radiation. *J. Exp. Med.* 158 (1983), 781.

29 Elmets, C.A., S.I.A. Zaidi, D.R. Bickers, H. Mukhtar: Immunogenetic influences on the initiation stage of the cutaneous chemical carcinogenesis pathway. *Cancer Res.* 52 (1992), 6106.

30 Everett M, Yeargers E, Sayre R, Olson R: Penetration of epidermis by ultraviolet rays. *Photochem. Photobiol.* 5 (1966), 533.

31 Figueroa, F., V. Vincek, M. Kasahara, G.I. Bell, J. Klein: Mapping of the *Sod*-2 locus into the *t* complex on mouse chromosome 17. *Immunogen.* 28 (1988), 260.

32 Fitzpatrick, T.B.: Ultraviolet induced pigmentary changes: benefits and hazards. *Curr. Probl. Dermatol.* 15 (1985), 25.

33 Fitzpatrick, T.B., A.J. Sober: Sunlight and skin cancer (Editorial). *New. Eng. J. Med.* 313 (1985), 818.

34 Freeman, S.E.: Variations in excision repair of UVB-induced pyrimidine dimers in DNA of human skin *in situ*. *J. Invest. Dermatol.* 90 (1988), 814.

35 Freeman, S.E., H. Hacham, R.W. Gange, D.J. Maytum, J.C. Sutherland: Wavelength dependence of pyrimidine dimer formation in DNA of human skin irradiated *in situ* with ultraviolet light. *Proc. Natl Acad. Sci. USA* 86 (1989), 5605.

36 Freeman, S.E., R.D. Ley: Action spectrum for the induction of pyrimidine dimers in DNA of UV-irradiated human skin: variations detected in individuals of different skin types. *Trends Photochem. Photobiol.* 1 (1988), 31.

37 Freund, Y., G. Sgarlato, C. Jacob, Y. Suzuki, J. Remington: Polymorphisms in the tumor necrosis factor α (TNFα) gene correlate with murine resistance to development of toxoplasmic encephalitis and with levels of TNFα mRNA in infected brain tissue. *J. Exp. Med.* 175 (1992), 683.

38 Friedberg, E.C.: *DNA Repair* (1985). W.H. Freeman and Co, San Francisco.

39 Fugger, L., N. Morling, L.P. Ryder *et al.*: NcoI restriction fragment length polymorphism (RFLP) of the tumor necrosis factor region in primary biliary cirrhosis and in healthy Danes. *Scand. J. Immunol.* 30 (1989), 185.

40 Gahring, L., M. Baltz, M.B. Pepys, R. Daynes: Effect of ultraviolet radiation on production of epidermal cell thymocyte-activating factor/interleukin-1 *in vivo* and *in vitro*. *Proc. Natl Acad. Sci. USA* 81 (1984), 1198.

41 Giannini, M.: Suppression of pathogenesis in cutaneous leishmaniasis by UV irradiation *Infect. Immun.* 51 (1986), 838.

42 Gilchrest, B.A., N.A. Soter, J.S. Stoff, M.C. Mihm Jr.: The human sunburn reaction: histologic and biochemical studies. *J. Am. Acad. Dermatol.* 5 (1981) 411.

43 Gordon, J.R., S.J. Galli: Mast cells as a source of both preformed and immunologically inducible TNFα/cachectin. *Nature* 346 (1990), 274.

44 Grau, G.E., L.F. Fajardo, P.-F. Piguet, B. Allet, P.-H. Lambert, P. Vassalli: Tumor necrosis factor (cachectin) as an essential mediator in murine cerebral malaria. *Science* 237 (1987), 1210–1212.

45 Green, A.E.S., G.B. Findley, K. Klenk, W. Wilson, T. Mo: The ultraviolet dose dependence of non-melanoma skin cancer incidence. *Photochem. Photobiol.* 24 (1976), 353.

46 Greene, M.I., M.-S. Sy, M. Kripke, B. Benacerraf: Impairment of antigen-presenting cell function by ultraviolet radiation. *Proc. Natl. Acad. Sci. USA* 76 (1979), 6591.

47 Halliday, G.M., H.K. Muller: Induction of tolerance via skin depleted of Langerhans cells by a chemical carcinogen. *Cell Immunol*, 99 (1986), 220.

48 Halliday, G.M., H.K. Muller: Sensitization through carcinogen-induced Langerhans cell-deficient skin activates specific long-lived suppressor cells for both cellular and humoral immunity. *Cell Immunol.* 109 (1987), 206.

49 Hamada, H., M. Seidman, B. Howard, C. Gorman:

Enhanced gene expression by the Poly (TdG). Poly (dC-dA) sequence. *Molecular Cellular Biology* 4 (1984), 2622

50 Han, J., T. Brown, B. Beutler: Endotoxin-responsive sequences control cachectin/tumor necrosis factor biosynthesis at the translational level. *J. Exp. Med.* 171 (1990), 465.

51 Haniszko, J., R.R. Suskind: The effect of ultraviolet radiation on experimental cutaneous sensitization in guinea pigs. *J. Invest. Dermatol.* 40 (1963), 183.

52 Harber, J.C., D.R. Bickers: Ultraviolet carcinogenesis. In: *Photosensitivity Diseases*, (1981), pp. 246. WB. Saunders, Philadelphia.

53 Harriott-Smith, T.G., W.J. Halliday: Suppression of contact hypersensitivity by short-term ultraviolet irradiation: II. The role of urocanic acid. *Clin. Exp. Immunol.* 72 (1988), 174.

54 Hartevelt, M.M., J.N. Bouwer Bacinck, A.M. Kootte, B.J. Vermeer, J.P. Vandenbroucke: Incidence of skin cancer after renal transplantation in the Netherlands. *Transplantation* 49 (1990), 506–509.

55 Holman, C.D.J., B.K. Armstrong: Cutaneous malignant melanoma and indicators of total accumulated exposure to the sun: an analysis separating histogenetic types. *J. Natl. Cancer. Inst.* 73 (1984), 75.

56 Jacob, C.O., H.O. McDevitt: Tumor necrosis factor-α in murine autoimmune 'lupus' nephritis. *Nature* 331 (1988), 356.

57 Johnson, J.A.: Xeroderma pigmentosum, photosensitivity and risk of skin cancer (Letter). *J. Invest. Dermatol.* 94 (1990), 366.

58 Jongeneel, C.V., H. Acha-Orbea, T. Blankenstein: A polymorphic microsatellite in the tumor necrosis factor α promoter identifies an allele unique to the NZW mouse strain. *J. Exp. Med.* 171 (1990), 2141.

59 Kinlen, L., A. Sheil, J. Peta, R. Doll: Collaborative United Kingdom – Australasian study of cancer in patients treated with immunosuppressive drugs. *Br. J. Med.* ii (1979), 1461.

60 Klein, J.: In: *Natural History of the Major Histocompatibility Complex* (1986). John Wiley & Sons, New York.

61 Kock, A., T. Schwartz, R. Kirnbauer, A. Urbanski, P. Perry, J.C. Ansel, T.A. Luger: Human keratinocytes are a source for tumor necrosis factor α: evidence for synthesis and release upon stimulation with endotoxin or ultraviolet light. *J. Exp. Med.* 172 (1990), 1609.

62 Kock, A., A. Urbanski, T.A. Luger: mRNA expression and release of tumor necrosis factor by human epidermal cells. *J. Invest. Dermatol.* 92 (1989), 462A.

63 Kollias, N., A.H. Baqer: Quantitative assessment of UV-induced pigmentation and erythema. *Photodermatology* 5 (1988), 53.

64 Kripke, M.L.: Antigenicity of murine skin tumors induced by ultraviolet light. *J. Natl. Cancer Inst.* 53 (1974) 1333.

65 Kripke, M.L.: Latency, histology, and antigenicity of tumors induced by ultraviolet light in three inbred mouse strains. *Cancer Res.* 37 (1977), 1395.

66 Kripke, M.L.: Immunobiology of photocarcinogenesis. In: *The Effect of Ultraviolet Radiation on the Immune System*, (1983), p. 87. Ed.: J.A. Parish, Johnson and Johnson Products, New Brunswick, NJ.

67 Kripke, M.L., P.A. Cox, L.G. Alas, D.B. Yarosh: Pyrimidine dimers in DNA initiate systemic immunosuppression in UV-irradiated mice. *Proc. Natl. Acad. Sci. USA* 89 (1992), 7516.

68 Kripke, M.L., W.L. Morison: Studies on the mechanism of systemic suppression of contact hypersensitivity by ultraviolet B radiation. *Photodermatology* 3 (1986), 4.

69 Kripke, M.L., R.M. Thorn, P.H. Lill, C.I. Civin, N.H. Pazmino, M.S. Fisher: Further characterization of immunological unresponsiveness induced in mice by ultraviolet radiation. Growth and induction of nonultraviolet-induced tumors in ultraviolet-irradiated mice. *Transplantation* 28 (1979), 212.

70 Krutmann, J., W. Czech, F. Parlow, U. Trefzer, A. Kapp, E. Schopf, T.A. Luger: Ultraviolet radiation effects on human keratinocyte ICAM-1 expression: UV-induced inhibition of cytokine-induced ICAM-1 mRNA expression is transient, differentially restored for IFNγ versus TNFα, and followed by ICAM-1 induction via a TNFα-like pathway. *J. Invest. Dermatol.* 98 (1992), 923.

71 Kupper, T.S.: The role of epidermal cytokines. In: *The Immunophysiology of Cells and Cytokines*, 1989. Ed.: E. Shevach, J. Oppenhaim, Oxford University Press, New York.

72 Kurimoto, I., J.W. Streilein: *cis*-Urocanic acid suppression of contact hypersensitivity induction is mediated via tumor necrosis factor-α. *J. Immunol.* 148 (1992), 3072.

73 Lafuse, W.P., D. Lanning, T. Spies, C.S. David: PFGE mapping and RFLP analysis of the S/D region of the mouse H-2 complex. *Immunogenetics* 36 (1992), 110.

74 Lavker, R.M., M.S. Kaminer, G.F. Murphy: Mast cell degranulation results in endothelial activation after acute exposure of human skin to ultraviolet irradiation. In: *Environmental Threat to the Skin*, (1992). Ed.: R. Marks, G. Plewig, Martin Dunitz Ltd, London.

75 Lew, R.A., A.J. Sober, N. Cook, R. Marvell, T.B. Fitzpatrick: Sun exposure habits in patients with cutaneous melanoma: a case control study. *J. Dermatol. Surg. Oncol.* 9 (1983), 981.

76 Ley, R.D.: Photoreactivation of UV-induced pyrimidine dimers and erythema in the marsupial

Monodelphis domestica. *Proc. Natl. Acad. Sci. USA* 82 (1985), 2409.

77 Ley, R.D., B.A. Sedita, D.D. Grube, R.J.M. Fry: Induction and persistence of pyrimidine dimers in the epidermal DNA of two strains of hairless mice. *Cancer Res.* 37 (1977), 3243.

78 Liew, F.Y., C. Parkinson, S. Millott, A. Severn, M. Carrier: Tumor necrosis factor (TNFα) in leishmaniasis. I. TNFα mediates host protection against cutaneous leishmaniasis. *Immunology* 69 (1990), 570.

79 Liew, F.Y., Y. Li, D. Yang, A. Severn, F. Cox: TNFα reverses the disease-exacerbating effect of subcutaneous immunization against murine cutaneous leishmaniasis. *Immunology* 74 (1991), 304.

80 Luster, M.I., G.A. Boorman, M.W. Harris, J.A. Moore: Laboratory studies on polybrominated biphenyl-induced immune alterations following low-level chronic or prepostnatal exposure. *Int. J. Immunopharmacol.* 2 (1980), 69.

81 Lynch, D.H., M.F. Gurish, R.A. Daynes: Relationship between epidermal Langerhans cell density, ATPase activity, and the induction of contact hypersensitivity. *J. Immunol.* 126 (1981), 1891.

82 Macatonia, S., S.C. Knight, A.J. Edwards, S. Griffiths, P. Fryer: Localization of antigen on lymph node dendritic cells after exposure to the contact sensitizer fluorescein isothiocyanate. *J. Exp. Med.* 166 (1987), 1654.

83 MacDonald, J.: The epidemiology of skin cancer. *J. Invest. Dermatol*, 32 (1959), 379.

84 Macher, E., M.W. Chase: Studies on the sensitization of animals with simple chemical compounds. XII. The influence of excision of allergenic depots on onset of delayed hypersensitivity and tolerance. *J. Exp. Med.* 129 (1969), 103.

85 Magnus, K.: Habits of sun exposure and risk of malignant melanoma: an analysis of incidence rates in Norway 1955–1977 by cohort, sex, age, and primary tumor site. *Cancer* 48 (1981), 2329.

86 Marmenout, K., L. Fransen, J. Tavernier, J. van der Heyden, R. Tizard, E. Kawashima, S. Shaw, M.-J. Johnson, D. Semon, R. Muller, M.-R. Ruysschaert, A. van Vliet, W. Fiers: Molecular cloning and expression of human tumor necrosis factor in comparison with mouse tumor necrosis factor. *Eur. J. Biochem.* 152 (1985), 515.

87 Molvig, J., L. Baek, P. Christensen, K.R. Monague: Endotoxin-stimulated human monocyte secretion of interleukin-1, tumor necrosis factor-alpha, and prostaglandin E_2 shows stable interindividual differences. *Scand. J. Immunol.* 27 (1988), 705.

88 Munn, C.G., Bucana, C., Kripke, M.L.: Antigen-binding cells from lymph nodes of mice sensitized epicutaneously with hapten have features of epidermal Langerhans cells. *J. Invest. Dermatol.* 92 (1989), 487.

89 Nakano, Y., K. Onozuka, Y. Terada, H. Shinomiua, M. Nakano: Protective effect of recombinant tumor necrosis factor-α in murine salmonellosis. *J. Immunol.* 144 (1990), 1935–1941.

90 Nedospasov, S.A., A.N. Shakhov, R.L. Turetskaya, V.A. Mett, M.M. Azizov, G.P. Georgiev, V.G. Korobko, V.N. Dobrynin, S.A. Filippov, N.S. Bystrov, E.F. Boldyreva, S.A. Chuvpilo, A.M. Chumakov, L.N. Shingarova, Y.A. Ovchinnikov: Tandem arrangement of genes coding for tumor necrosis factor (TNF-α) and lymphotoxin (TNF-β) in the human genome. *Cold Spring Harbor Symp. Quant. Biol.* 511 (1986), 611.

91 Nickoloff, B., G. Karabin, J. Barker, C. Griffiths, V. Sarma, R. Mitra, J. Elder, S. Kunkel, V. Dixit: Cellular localization of interleukin-8 and its inducer, tumor necrosis factor-alpha in psoriasis. *Am. J. Path.* 138 (1991), 129.

92 Noonan, F.P., M.L. Kripke, G.M. Pedersen, M.I. Greene: Suppression of contact hypersensitivity in mice by ultraviolet irradiation is associated with defective antigen presentation. *Immunology* 43 (1981), 527.

93 Norris, P.G., A. Limb, A.S. Hamblin, A.R. Lehmann, C.F. Arlett, J. Cole, A.P.W. Wauch, J.L.M. Hawk: Immune function, mutant frequency, and cancer risk in the DNA repair defective genodermatoses xeroderma pigmentosum, Cockayne's syndrome, and trichothiodystrophy. *J. Invest. Dermatol.* 94 (1990), 94.

94 Norris, D.A., M.B. Lyons, M.H. Middleton, J.J. Yohn, M. Kishihara-Sawami: Ultraviolet radiation can either suppress or induce expression of intercellular adhesion molecule 1 (ICAM-1) on the surface of cultured human keratinocytes. *J. Invest. Dermatol.* 95 (1990), 132.

95 Norris, P., R.N. Poston, D.S. Thomas, M. Thornhill, J. Hawk, D.O. Haskard: The expression of endothelial leukocyte adhesion molecule-1 (ELAM-1), intercellular adhesion molecule-1 (ICAM-1), and vascular cell adhesion molecule-1 (VCAM-1) in experimental cutaneous inflammation: a comparison of ultraviolet B erythema and delayed hypersensitivity. *J. Invest. Dermatol.* 96 (1991), 763.

96 Norval, M., C.R. McIntyre, T.J. Simpson, S.E.M. Howie, E. Bardshiri: Quantification of urocanic acid isomers in murine skin during development and after irradiation with UVB light. *Photoderm.* 5 (1988), 179.

97 Norval, M., T.J. Simpson, E. Bardshiri, S.E.M. Howie: Urocanic acid analogues and the suppression of the delayed type hypersensitivity response to herpes simplex virus. *Photochem. Photobiol.* 49 (1989), 633.

98 Okamoto, H., M.L. Kripke: Effector and suppressor circuits of the immune response are activated *in vivo* by different mechanisms. *Proc. Natl. Acad. Sci. USA*

84 (1987), 3841.

99 Parida, S.K., G.E. Grau, S.A. Zaheer, R. Mukherjee: Serum tumor necrosis factor and interleukin 1 in leprosy and during Lepra reactions. *Clin. Immun. Immunopath.* 63 (1992), 23–27.

100 Parish, J.A.: Responses of skin to visible and ultraviolet radiation. In: *Biochemistry and physiology of Skin*, (1983), p. 713. Ed.: L.A. Goldsmith, Oxford University Press, Oxford.

101 Rae, V., T. Yoshikawa, W. Bruins-Slot, J.W. Streilein, A. Taylor: An ultraviolet B radiation protocol for complete depletion of human epidermal Langerhans cells. *J. Dermatol. Surg. Oncol.* 15 (1989), 1199.

102 Rheins, L.A., A.L. Cotleur, R.S. Kleier, W.B. Hoppenjans, D.N. Saunder, J.J. Nordlund: Alpha-melanocyte stimulating hormone modulates contact hypersensitivity responsiveness in C57BL/6 mice. *J. Invest. Dermatol.* 93 (1989), 511.

103 Richter, G., Z. Qin, T. Diamanstein, T. Blankenstein: Analysis of restriction fragment length polymorphism in lymphokine genes of normal and autoimmune mice. *J. Exp. Med.* 170 (1989), 1439.

104 Robertson, B., L. Gehring, R. Newton, R. Daynes: *In vivo* administration of interleukin-1 to normal mice depressed their capacity to elicit contact hypersensitivity responses: prostaglandins are involved in this modification of immune function. *J. Invest. Dermatol.* 88 (1987), 380.

105 Rodriguez-Campos, A., M. Ellison, L. Perez-Graw, F. Azorin: DNA conformation and chromating organization of a $d(CA/GT)_{30}$ sequence cloned in SV 40 minichromosomes. *EMBO J.* 5 (1986), 1727.

106 Rosenstreich, D., A. Nowotny, T. Chused, S. Bergenhagen: *In vitro* transformation of mouse bone marrow-derived lymphocytes induced by the lipid component of endotoxin. *Infect. Immun.* 8 (1973), 406.

107 Ross, J.A., S.E.M. Howie, M. Norval, J.P. Maingay: Induction of suppression of delayed type hypersensitivity to herpes simplex virus by epidermal cells exposed to UV-irradiated urocanic acid *in vivo*. *Viral Immunol.* 1 (1987/1988), 191.

108 Ross, J.A., S.E.M. Howie, M. Norval, J. Maingay, T.J. Simpson: Ultraviolet-irradiated urocanic acid suppresses delayed-type hypersensitivity to herpes simplex virus in mice. *J. Invest. Dermatol.* 87 (1986), 630.

109 Ruff, M., R. Griffin: Tumor necrosis factor. In: *Lymphokines*, (1981), p. 235. Academic Press, New York.

110 Schreiber, M.M., P.D. Bozzo, T.E. Moon: Malignant melanoma in southerm Arizona. *Arch. Dermatol.* 117 (1981), 6.

111 Schwartz, R.S., N.R. Rose (Eds.): Autoimmunity: experimental and clinical aspects. *Ann. N.Y. Acad. Sci.* (1986), 475.

112 Scotto, J., A.W. Kopf, F. Urbach: Nonmelanoma skin cancer in four areas of the US. *Cancer* 34 (1974), 1331.

113 Seboun, E., M.A. Robinson, T.H. Doolittle, T.A. Ciulla, T.J. Kindt, S.L. Hauser: A susceptibility locus for multiple sclerosis is linked to the T cell receptor β chain complex. *Cell* 57 (1989), 1095.

114 Shakhov, A.N., M.A. Collart, P. Vassalli, S.A. Nedospasov, C.V. Jongeneel: kB-type enhancers are involved in lipopolysaccharide-mediated transcriptional activation of the tumor necrosis factor α gene in primary macrophages. *J. Exp. Med.* 171 (1990), 35.

115 Silberberg, I.: Apposition of mononuclear cells to Langerhans cells in contact allergic reactions. *Acta. Derm. Venerol.* 53 (1973), 1.

116 Siberberg, I., R.L. Baer, S.A. Rosenthal: The role of Langerhans cells in allergic contact hypersensitivity. A review of findings in man and guinea pig. *J. Invest. Dermatol.* 66 (1976), 210.

117 Solomon, S., D. Albritton: Time-dependent ozone depletion potentials for short- and long-term forecasts. *Nature* 357 (1992), 33.

118 Stern, R.S., K. Momtaz: Skin typing for assessment of skin cancer and acute response to UVB and oral methoxsalen photochemotherapy. *Arch. Dermatol.* 120 (1984), 869.

119 Stingl, G., L.A. Gazze-Stingl, W. Aberer, K. Wolff: Antigen presentation by murine epidermal Langerhans cells and its alteration by ultraviolet B light. *J. Immunol.* 127 (1981), 1707.

120 Stingl, G., S.I. Katz, L. Clement, I. Green, E.M. Shevach: Immunologic functions of Ia-bearing epidermal Langerhans cells. *J. Immunol.* 121 (1978), 2005.

121 Stingl, G., S.I. Katz, E.M. Shevach, E. Wolff-Schreiner, L. Green: Detection of Ia antigens of Langerhans cells in guinea pig skin. *J. Immunol.* 120 (1978), 570.

122 Streilein, J.W.: Antigen-presenting cells in the induction of contact hypersensitivity in mice: evidence that Langerhans cells are sufficient but not required. *J. Invest. Dermatol.* 93 (1989), 443.

123 Streilein, J.W., P.R. Bergstresser: Langerhans cell function dictates induction of contact hypersensitivity or unresponsiveness to DNFB in Syrian hamsters. *J. Invest. Dermatol.* 77 (1981), 272.

124 Streilein, J.W., P.R. Bergstresser: Two antigen presentation pathways, only one of which requires Langerhans cells, lead to the induction of contact hypersensitivity (Abstract). *J. Invest. Dermatol.* 80 (1983), 302.

125 Streilein, J.W., P.R. Bergstresser: Langerhans cells: antigen presenting cells of the epidermis. *Immunobiology* 168 (1984), 285.

126 Streilein, J.W., P. Bergstresser: Genetic basis of ultraviolet-B effects on contact hypersensitivity. *Immunogenetics* 27 (1988), 252.

127 Sullivan, S., P.R. Bergstresser, R.E. Tigelaar, J.W.

Streilein: Induction and regulation of contact hypersensitivity by resident bone marrow derived, dendritic epidermal cells: Langerhans cells and Thy-1^+ epidermal cells. *J. Immunol.* 137 (1986), 2460.
128 Sutherland, B.M.: Ultraviolet light-induced damage and its repair. In: *Experimental and Clinical Photoimmunology*, (1983), p. 27. Eds.: R.A. Daynes, J.D. Spikes, CRC Press, Boca Raton, FL.
129 Taylor, M.J., C. Lafagre-Frayssinet, M.I. Luster, C. Frayssinet: Increased endotoxin sensitivity following T-2 toxin treatment is associated with increased absorption of endotoxin. *Toxicol. Appl. Pharmacol.* 109 (1991), 51.
130 Thomas, L.: Reactions to homologous tissue antigens in relation to hypersensitivity. In: *Cellular and Humoral Aspects of the Hypersensitive States*, (1959), p. 529. Ed.: H.S. Lawrence, Hoeber-Harper, New York.
131 Tite, J.P., G. Dougan, S.N. Chatfield: The involvement of tumor necrosis factor in immunity to *Salmonella* infection. *J. Immunol.* 147 (1991), 3151–3164.
132 Titus, R.G., B. Sherry, A. Cerami: Tumor necrosis factor plays a protective role in experimental murine cutaneous leishmaniasis. *J. Exp. Med.* 170 (1989), 2097–2104.
133 Toews, G., P. Bergstresser, J.W. Streilein: Epidermal Langerhans cell density determines whether contact sensitivity or unresponsiveness follows skin painting with DNFB. *J. Immunol.* 124 (1980), 445.
134 Tseng, C., B. Hoffman, I. Kurimoto, T. Shimizu, G.F. Schmieder, J.R. Taylor, J.W. Streilein: Analysis of effects of ultraviolet-B radiation on induction of primary allergic reactions. *J. Invest. Dermatol.* 98 (1992), 871.
135 Urbach, F., D.B. Rose, M. Bonnem: Genetic and environmental interaction in skin carcinogenesis. In: *Environment and Cancer*, (1971), pp. 355–371. Williams and Wilkins Co, Baltimore.
136 Valtonen, E.J., J. Janne, M. Siimes: The effect of the erythemal reaction caused by ultraviolet irradiation on mast cell degranulation in the skin. *Acta. Derm. Venerol. (Stockh.)* 44 (1964), 269.
137 Vermeer, M., S. Schmieder, J.-W. van den Berg, T. Yoshikawa, M.S. Metzman, J.R. Taylor, J.W. Streilein: Effects of ultraviolet B light on cutaneous immune responses in humans with deeply pigmented skin. *J. Invest. Dermatol.* 97 (1991), 729.
138 Vermeer, M., J.W. Streilein: Ultraviolet-B-light induced alterations in epidermal Langerhans cells are mediated by tumor necrosis factor-alpha. *Photodermatol. Photoimmunol. Photomed.* 7 (1990), 258.
139 Visner, G.A., W.C. Dougall, J.M. Wilson, I.A. Burr, H.S. Nick: Regulation of manganese superoxide dismutase by lipopolysaccharide, interleukin-1, and tumor necrosis factor. *J. Biol. Chem.* 265 (1990), 2856.
140 Vitasa, B.C., H.Y. Taylor, P.T. Strickland, F.S. Rosenthal, S. West, H. Abbey, S.K. Ng, B. Bunoz, E.A. Emmett: Association of nonmelanoma skin cancer and actinic keratosis with cumulative solar ultraviolet exposure in Maryland watermen. *Cancer* 65 (1990), 2811.
141 Vogel, S., D. Rosenstreich: LPS-unresponsive mice as a model for analyzing lymphokine-induced macrophage differentiation *in vitro*. *Lymphokines* 3 (1981), 149.
142 Vusch, D., C. Greiner, K. Lewis, R. Ford, G. Adair, L.H. Thompson: Summary of complementation groups of UV-sensitive CHO cell mutants isolated by large-scale screening. *Mutagenesis* 4 (1989), 349.
143 Wedd, G.C., D.D. Chaplin: Genetic variability at the human tumor necrosis factor loci. *J. Immunol.* 145 (1990), 1278.
144 Yoshilkawa, T., V. Rae, W. Bruins-Slot, J.-W. van den Berg, J.R. Taylor, J.W. Streilein: Susceptibility to effects of UVB radiation on induction of contact hypersensitivity as a risk factor for skin cancer in man. *J. Invest. Dermatol.* 95 (1990), 530.
145 Yoshikawa, T., J.W. Streilein: Genetic basis of the effects of ultraviolet light B on cutaneous immunity. Evidence that polymorphism at the *Tnfα* and *Lps* loci governs susceptibility. *Immunogenetics* 32 (1990), 398.
146 Yoshikawa, T., J.W. Streilein: Tumor necrosis factor-alpha and ultraviolet B light have similar effects on contact hapersensitivity in mice. *Regional Immunol.* 3 (1991), 139.

11: UV and infectious diseases

Amminikutty Jeevan, Eric Brown & Margaret L. Kripke

Introduction

Studies in experimental animals have demonstrated that ultraviolet radiation (UVR) interferes with the induction and elicitation of contact and delayed-type hypersensitivity (DTH) responses to a variety of antigens. Exposure of humans to natural and artificial sources of UVR can alter immune function in similar ways. The mechanisms responsible for these effects of UVR most likely involve alterations in cutaneous immune cells and the release of soluble mediators from UV-irradiated epidermis [41, 42, 62, 64, 72] (see also Chapter 7). These findings have raised the question of whether exposure to UVR might also modify the course of some infectious diseases, particularly those in which the DTH response or the cutaneous immune system are thought to play a major protective role.

This is an important question because of the expected increase in environmental UVR resulting from reductions in the concentration of stratospheric ozone. Most of the immunological effects of UVR, including those on immunity to infectious agents, are caused by wavelengths in the UVB (280–320 nm) region of the solar spectrum. The ability of these wavelengths to reach the Earth's surface is strongly dependent on the concentration of ozone in the atmosphere, and their intensity in sunlight is expected to increase dramatically over the next several decades because of ozone depletion [23]. If UVR can adversely affect immunity to infectious agents in humans then an increase in the amount of UVR in sunlight has the potential to increase the incidence and severity of infectious diseases in human populations worldwide, in addition to increasing the incidence of skin cancer and cataracts. Therefore it is important to determine whether exposure to UVB radiation can alter immunity or resistance to infectious agents, which disease processes are likely to be affected by UVR, and the mechanisms by which UVR influences disease pathogenesis.

A second important consideration is that many diseases, such as tuberculosis, measles, and pneumonic plague, that were once considered to be under control in industrialized countries, are again rising in incidence [59, 69]. Diseases such as leprosy and tuberculosis, both caused by mycobacteria, are not only major public health problems in developing nations, but are gaining importance in industrialized countries because of increased prevalence of mycobacterial infections in AIDS patients, immigration from underdeveloped nations, and the emergence of drug-resistant strains of bacteria [52, 74]. The DTH response is thought to be of critical importance in controlling certain infectious diseases [7, 21, 38, 39], particularly those caused by mycobacteria and other intracellular pathogens. Because UVR can impair DTH responses, at least in animal model systems, increased exposure to solar UVR could further exacerbate this problem.

Effects of UVR on human diseases

Historically, UVR has been used therapeutically to treat a variety of skin diseases. Niels Finsen received the Nobel Prize for Medicine in 1903 for his studies on the uses of light for treatment of infectious diseases. He demonstrated that exposure to UVR was beneficial in healing the cutaneous lesions of skin tuberculosis and lupus vulgaris [15]. In recent years, psoriasis, acne, and other cutaneous diseases have been treated successfully with UVB radiation [67].

On the other hand, sunlight exposure was thought to exacerbate scarring from smallpox [14] and Finsen found that pulmonary tuberculosis, unlike skin tuberculosis, was actually aggravated by heliotherapy [15, 53]. There is considerable evidence indicating that exposure to sunlight and to UVR can trigger the appearance of cutaneous lesions caused by herpes simplex virus (HSV) types 1 and 2 in persons harboring latent infection [36, 56, 70]. More recently, it was reported that low doses of UVR increased the severity of skin lesions associated with herpes zoster [61]. Exposure to sunlight has also been implicated as a contributing factor in papillomavirus infections in immunosuppressed patients; such individuals have a high incidence of viral warts on sun-exposed body sites [2, 12].

Thus, in cutaneous diseases, UVR can be either beneficial or harmful, depending on the type of disease, and perhaps on the conditions of irradiation such as dose, regimen and spectral quality. However, the effects of UVR on the development of systemic immunity to infectious agents and on the pathogenesis of infections in humans have not been investigated.

Effects of UVR on experimental models of infection

Viral infections

HSV

Because of the ability of sunlight exposure to activate HSV lesions in humans, studies on the effects of UVR on HSV infections in murine models have been carried out for many years. Exposure to UVR can trigger active disease in both humans and mice with latent HSV infection [22]. In addition, exposure of mice to UVB radiation decreases resistance to the initial infection with HSV type 1; this effect of UVR is thought to result from impairment of Langerhans cells in UV-irradiated epidermis [55]. UV irradiation of the site of HSV infection decreases the DTH response to HSV types 1 and 2 and increases the severity of both the cutaneous lesions and the systemic manifestations of disease [24, 43, 71].

Extensive studies have been carried out on the suppression of the DTH response to HSV [24, 71]. This suppression was associated with the induction of two phenotypically distinct populations of suppressor T cells in the spleens of mice, $Ly1^+2^-$ cells and $Ly1^-2^+$ cells [26]. Similarly, UV radiation-induced, antigen-specific suppressor factors could be detected in spleen cells of HSV-2-infected mice; these factors suppressed HSV-2-induced proliferative responses of lymph node cells obtained from immunized mice [1]. In addition, the ability of epidermal antigen-presenting cells to present HSV antigens *in vitro* was decreased after exposure of mice to a suberythemal dose of UV radiation [25]. Epidermal application or systemic administration of UV-irradiated (*trans-*) urocanic acid (UCA) suppressed the DTH response to HSV-1 [49, 50] suggesting that UCA may be involved in this form of UV-induced immune suppression.

HIV

Interest in the possible role of UV irradiation in the pathogenesis of HIV infection was stimulated by reports that the virus could be detected in the skin of HIV^+ individuals and the suggestion that epidermal Langerhans cells might constitute a reservoir of HIV during the early, latent stage of infection [58, 63]. Previous studies had demonstrated that *in vitro* exposure to UVR of HeLa cells containing the HIV promoter region activated the viral promoter, an event that leads to virus replication in HIV-infected cells [66, 75]. More recently, UVB irradiation of a monocyte cell line was shown to activate latent HIV infection [57]. These findings have raised the possibility that exposure of HIV-infected persons to UVR early in the course of infection, when virus is present in the skin, might accelerate the course of AIDS. Preliminary reports from studies on AIDS patients treated with UVB or psoralen plus UVA radiation for psoriasis or other skin conditions have not yet provided any evidence to support this possibility [44, 45]. However, generally in these studies the subjects were already symptomatic; therefore the question of whether UV irradiation can shorten the latent period between HIV infection and active disease remains unresolved.

There have been several recent attempts to study the effect of UV irradiation on the pathogenesis of AIDS using murine models of HIV infection. The murine acquired immunodeficiency syndrome (MAIDS) model involves infection of mice with a murine retrovirus (LP-BM5 MuLV) that causes a progressive immunodeficiency disease. Exposure of C57BL/6 mice to low doses of UVR from FS40 sun lamps before inoculation with the virus accelerated both the antibody response to the virus and the onset of the reduced proliferative response of splenic lymphocytes to mitogen stimulation [3]. Although preliminary, this study suggests that UV irradiation may accelerate the course of this immunodeficiency disease.

Transgenic mice carrying the HIV promoter have also been developed. Exposure of such animals to UVR activated the expression of reporter genes in the skin that were under the control of the HIV promoter [40]. Similarly, UVB radiation induced HIV gene expression in the skin of transgenic mice whose cells contained the *cis-* and *trans-*acting elements required for virus activation and replication in infected cells [68]. Thus UV irradiation may augment virus expression and replication by direct interaction with the infected cells. Whether it can also affect disease pathogenesis by means of photoimmunological effects remains to be determined.

Parasitic infections

Leishmania major

Leishmaniasis is a tropical disease caused by a protozoan parasite and transmitted by sand flies. The protozoa are deposited intradermally into human skin, where they give rise to ulcerating, cutaneous lesions. The infection may be limited to the skin or can progress to a systemic, often fatal, form of the disease.

The effects of UV irradiation on murine models of cutaneous leishmaniasis have been explored [16–18]. In these studies, either B10.129(10 M) ScSn or C3H/HeJ mice were inoculated with *L. major* through tail skin, exposed repeatedly, before and after infection, to low doses of UV radiation from FS40 sun lamps. The size and severity of the cutaneous lesions were reduced in the UV-irradiated mice; however, the number of parasites in the skin was not decreased. Furthermore, in UV-irradiated mice the DTH response to leishmania antigen was reduced, more animals died from infection, and those surviving the infection exhibited decreased resistance to reinfection with the parasite. Thus, in this model of infection, an improved appearance of the cutaneous lesions in UV-irradiated skin did not correlate with progression of systemic disease; in fact, systemic disease was exacerbated in these animals, in spite of the seemingly beneficial effect of UV irradiation on the cutaneous lesions.

Schistosoma mansoni

Schistosoma mansoni is a parasitic trematode with a complex life cycle involving a snail vector and a mammalian host. It normally infects humans by penetrating the skin during contact with water inhabited by infected snails. After entering the skin, the cercariae migrate to the liver and develop into mature worms [5]. Experiments in a murine model of infection were carried out to determine whether modification of epidermal Langerhans cells by UV irradiation would decrease resistance to percutaneous infection by the parasites [27, 28]. Doses of UVR sufficient to reduce the number of Langerhans cells and inhibit the induction of contact hypersensitivity to hapten applied to the site of irradiation were used. UV irradiation of C3H/HeN mice before percutaneous infection with *S. mansoni* cercariae did not influence the number of worms recoverable from the internal organs 6 weeks later. Thus, in contrast to the findings with HSV and *Leishmania* infection of mice, there is no evidence that a UV-induced local alteration of cutaneous immunity influences the infectivity of *Schistosoma* or the pathogenesis of disease in this murine model.

Fungal infections

Candida albicans

To date, the only fungal pathogen studied in relation to UV irradiation is *Candida albicans*. This is an opportunistic fungus normally present on the skin, which infects mucous membranes and causes systemic disease in immunosuppressed hosts [48]. The induction of DTH to *Candida* was inhibited by exposing C3H/HeN mice to a single dose of UV radiation from FS40 sun lamps before subcutaneous immunization with live or fixed *Candida* at a non-irradiated site. In these studies the DTH response was assessed by injecting *Candida* antigen into the footpad of the mice and measuring the degree of footpad swelling 24 h later. Inhibition of the DTH response was proportional to the dose of UVB radiation administered, and approximately one minimal erythemal dose of UVR ($2.2\,kJ/m^2$) was needed to reduce the response by 50%. The reduced DTH response was associated with the appearance of splenic suppressor T lymphocytes [10].

To determine the role of DNA damage in initiating this effect of UV irradiation on DTH, liposomes containing T4 endonuclease V, a DNA repair enzyme, were applied to the skin immediately after UV irradiation [37]. Application of these liposomes to UV-irradiated skin decreased the number of pyrimidine dimers in the epidermis relative to controls given UV irradiation alone or UV irradiation followed by liposomes containing heat-inactivated endonuclease. They also completely prevented inhibition of the DTH response to *Candida* and induction of suppressor lymphocytes by UV irradiation, whereas liposomes containing inactive enzyme were ineffective. These studies demonstrate that the formation of lesions in epidermal DNA, particularly pyrimidine dimers, constitutes the primary molecular mechanism by which UV irradiation initiates this immunological effect.

Exposing *Candida*-immunized mice to UVR immediately before footpad challenge also reduced the DTH response [10]. In contrast to its effect on induction of DTH, the effect of UVR on elicitation of DTH was not associated with the formation of suppressor T lymphocytes. Also, higher doses of UVR were required, suggesting that a different mechanism was involved. Injection of radiolabeled lymphocytes from *Candida*-immune mice into UV-irradiated recipients demonstrated that UV irradiation altered the homing pattern of the immune cells (Y.M. Denkins, doctoral thesis), confirming the observation of others [54]. This result suggests that UVR may reduce the efferent arm of the DTH response by altering the body distribution of sensitized T cells.

Subcutaneous injection of viable *Candida* into mice causes a self-limiting, inapparent infection. Thus the significance of the UV-induced alterations in the DTH response to *C. albicans* for the pathogenesis of the local infection is difficult to assess. The effect of UV irradiation on an acute, lethal infection with *Candida* has been examined however [11, 27]. In this model, *C. albicans* is injected intravenously into C3H mice at a dose that causes death in 2–4 weeks. Exposing mice to a single high dose of UVR 1 day before intravenous inoculation of *Candida* significantly reduced their mean survival time. Examination of the internal organs of the mice at various times during infection revealed a greater number of organisms in the kidneys of the UV-irradiated animals.

Immunization of mice subcutaneously with fixed *Candida* before intravenous challenge extended their mean survival time and increased the percentage of survivors. Exposure of the mice to UVR either before or after this immunization suppressed the DTH response to *Candida*; however, the mean survival time was decreased only in mice exposed to UVR after immunization (i.e., 1 day before intravenous challenge) [11]. These studies indicate that UV irradiation given before intravenous inoculation of a lethal dose of *Candida* can accelerate the course of the disease, both in naive and immunized animals. They also demonstrate that immunization increases resistance to lethal challenge; however, protective immunity does not correlate with the ability of the mice to mount a DTH response, as indicated by the footpad swelling reaction. Thus the mechanism of the effect of UVR on lethal infection with *C.*

albicans is not clear, but it does not appear to be related to the ability of UVR to inhibit the DTH response.

Bacterial infections

Borrelia burgdorferi

This spirochete is the causative agent of Lyme disease and is transmitted to humans and other mammals by ticks of the genus *Ixodes*. The tick bite results in a local swelling, generally followed by flu-like symptoms. Late manifestations can include arthritis, meningitis, facial nerve palsy and cardiac disease. At present we are attempting to characterize the humoral and cell-mediated immune responses to this organism in mice in order to assess the effects of UV irradiation on immunity and pathogenesis. Immunization of C3H mice with whole, inactivated *B. burgdorferi* in complete Freund's adjuvant induced a DTH response as measured by footpad challenge. Exposing the mice to a single dose of UVR (5 kJ/m^2, approximately two minimal erythemal doses) from FS40 sun lamps 5 days before immunization inhibited the DTH response. Antibodies of isotypes IgG1, IgG1 2a, IgG1 2b, and IgG3 could be detected in the serum of immunized mice, with the IgG2a and IgG2b isotypes predominating. Preliminary results suggest that although UV irradiation of the mice does not inhibit total antibody production, the proportions of the different isotypes produced may be altered. Specifically, UV irradiation appears to decrease the amount of IgG1 and increase the amount of IgG during a secondary response. Immunization without adjuvant induced no DTH response and low, but detectable levels of IgG2a and IgG2b (E. Brown, M. Kripke, unpublished data). Whether UVR affects the pathogenesis of disease in this model has not yet been explored.

Mycobacterium bovis BCG

As mentioned earlier, infection of human populations with various species of mycobacteria is increasing. It is estimated that there are now 10 million new cases and 3 million deaths from tuberculosis every year worldwide [33, 60], and one-third of the world's population is infected with some form of mycobacteria [34]. In many countries, vaccination with avirulent strains of mycobacteria is used in an attempt to prevent the development of tuberculosis.

Mycobacterium bovis BCG is an avirulent form of the bovine tubercle bacillus that resembles *M. tuberculosis*, the main causative agent of human tuberculosis. This organism causes an acute, self-limiting infection in certain mouse strains [9] characterized by the development of an inflammatory response at the site of inoculation, multiplication of the bacilli in regional lymph nodes and spleen, and the development of a strong DTH response. The DTH response often correlates with control of the disease and resistance to systemic infection [7, 9, 38]. The effects of UV irradiation on DTH to this organism and on the pathogenesis of infection in mice have been investigated in our laboratory over the past several years [27–32].

Exposure of shaved dorsal skin of C3H/HeN and BALB/c mice to a single dose of UVR from FS40 sun lamps 3 days before injection of live BCG into the footpad delayed the development of the DTH response. The DTH response was assessed by injecting the purified protein derivative of BCG (PPD) into the uninfected footpad at monthly intervals and measuring the amount of footpad swelling 24 h later. These animals were also impaired in their ability to clear bacilli from the infected footpad, draining lymph node and spleen [30]. Lymphocytes from the draining lymph node or spleen of the UV-irradiated, BCG-infected mice exhibited decreased proliferative responses to PPD compared to BCG-infected, non-irradiated mice (A. Jeevan, unpublished data). Thus in this model a single dose of UVR given several days before inoculation of viable bacilli impaired the development of the DTH response, permitted greater multiplication of bacteria, and prolonged the duration of the infection.

UV dose–response studies have indicated that the DTH response to BCG decreased in a dose-dependent manner and that doses of 1.4 kJ/m^2 or higher significantly reduced the DTH response when tested at 3 weeks after infection [31]. The dose of UVR required to decrease the DTH response by 50% was 2.7 kJ/m^2, which is close to the minimal erythemal dose of UVR for BALB/c mice (2.3 kJ/m^2). Mice exposed chronically to low doses of UVR during BCG infection also showed a decreased DTH response to PPD and an increased number of bacilli in lymphoid organs early in the course of irradiation. However, the mice recovered from these effects later in the course of UV irradiation [31]. This finding suggests that some compensatory mechanism, such as epidermal thickening, is activated in response to UV irradiation and protects against the effects of subsequent irradiations; alternatively, an essential mediator of the effect may be depleted by repeated exposure of the skin to UVR.

Experiments have also been carried out in which mice were infected with BCG by intradermal injection into UV-irradiated skin [28]. In this model, the DTH response was not affected by UV irradiation, in contrast to the effects of UVR on infections with HSV and *Leishmania* [16, 24]. However, the rate of clearance of BCG from the regional lymph node and spleen was reduced in the UV-irradiated mice. This study suggests that epidermal Langerhans cells, which are altered by UV irradiation and play an important role in resistance to HSV infection, are probably not involved in the development of DTH to BCG injected intradermally. Antigen-presenting cells in other sites, such as the dermis or regional lymph nodes, may play a more important role in the induction of DTH to this organism. In addition the results indicate that the ability of mice to inhibit the multiplication of BCG in lymphoid organs does not correlate with their ability to make a DTH response.

Mycobacterium lepraemurium

Experiments similar to those described for BCG have been carried out with *M. lepraemurium* (MLM). This organism causes a slowly progressive, lethal infection in susceptible mouse strains and is used as a model of chronic mycobacterial infection [6]. Again, UV irradiation of dorsal skin several days before inoculation of live MLM into the footpad reduced the DTH response and impaired the clearance of bacilli from lymphoid tissues [29]. The UV dose–response for impairment of the DTH response to MLM was nearly identical to those for BCG and *Candida*. In addition the median survival time of mice infected subcutaneously via the footpad, or intravenously with MLM, was significantly shorter in mice exposed to a single high dose of UVR several days before infection [29]. This finding is particularly noteworthy because in this model death occurs between 6 months and 1 year after infection. It suggests that the initial interaction between MLM and the host has a major influence on the rate of disease progression and that exposure to UVR can alter this initial interaction, thereby accelerating the course of the disease.

Mechanisms of UV effects on BCG infection

The host response to mycobacterial infection represents a complex set of inflammatory and immunological reactions [7, 13, 21, 39]. Monocytes and macrophages play a role in clearing mycobacteria early in the course of infection and activate the proliferation and differentiation of T lymphocytes specific for mycobacterial antigens. As illustrated in Fig. 11.1, upon antigen stimulation sensitized T lymphocytes release lymphokines that cause the DTH response, including ones that activate macrophages, thereby increasing their ability to phagocytize and digest the bacilli. Granuloma formation is also immunologically mediated and represents a defensive reaction on the part of the host. The fibroblasts and reticuloendothelial cells of the granuloma, which include macrophages, histiocytes, epithelioid cells, and multinucleate giant cells serve to isolate the bacilli and inhibit their dissemination [7, 13].

How UV irradiation interferes with this process is not completely understood, but because of the systemic nature of its effects it almost certainly involves the production or release of soluble mediators from UV-irradiated epidermis. Keratinocytes and other cells in the epidermis have been shown to produce a variety of factors in

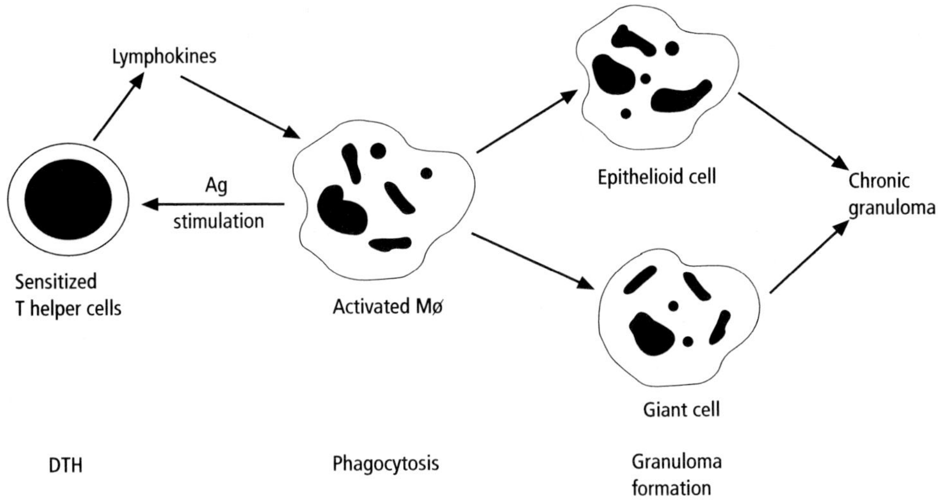

Fig. 11.1 Cell-mediated immune response against mycobacteria. This reaction involves production of lymphokines by sensitized T cells, the uptake of organisms by macrophages, and the formation of granulomatous lesions. The granuloma reaction, a major protective response to mycobacterial infection, is mediated by cells of the reticuloendothelial system, including macrophages, epithelioid cells and giant cells.

response to UV irradiation that cause systemic immune suppression [4, 8, 35, 46, 47, 51, 65, 73], including prostaglandins, interleukin-1 (IL-1), contra-IL-1, IL-10, TNFα, and *cis*-UCA (see also Chapter 5). Recently we demonstrated that supernatants from cultures of a UV-irradiated murine keratinocyte cell line mimicked the effects of *in vivo* UV irradiation on BCG infection [32]. Injection of such supernatants from UV-irradiated, but not non-irradiated, keratinocytes into mice inhibited the induction of DTH to BCG and impaired the clearance of bacilli from their lymphoid organs. Furthermore, splenic and peritoneal macrophages taken from mice injected with the supernatants or exposed to UVR *in vivo* had a reduced ability to phagocytize BCG [32] (A. Jeevan, M.L. Kripke, unpublished data). Addition of supernatants from UV-irradiated keratinocytes to cultures of normal macrophages *in vitro* also reduced their phagocytic ability. These studies strongly suggest that soluble mediators released by UV-irradiated keratinocytes are involved in the suppression of systemic immunity to BCG by UVR and that the mediators may act, at least in part, by altering macrophage function.

Preliminary studies on granuloma formation suggest that UV irradiation does not prevent the formation of granulomas in the lymphoid organs of BCG-infected mice (A. Jeevan, unpublished data). However, it is not yet clear whether the number, size and cellular composition of the granulomas are the same in UV-irradiated and non-irradiated hosts; analyses are currently in progress.

Because of the complexity of the host response to mycobacteria, there are many possible ways by which UV irradiation might influence the outcome of the infection. It may be that cytokines released from UV-irradiated epidermis impair the phagocytic function of macrophages directly. This effect may account for the reduced clearance of mycobacteria from the lymphoid organs of UV-irradiated mice. These or other immunosuppressive cytokines decrease the DTH response and may cause the induction of suppressor T lymphocytes, perhaps by altering the antigen-presenting activity of macrophages [19, 20, 32]. Clearly, much more information is needed before we fully understand how UV irradiation alters resistance to BCG infection.

Conclusions

The literature available to date, although limited, indicates that UV irradiation may have both beneficial and harmful effects on disease processes in humans and experimental animals. In conditions in which the immune response contributes to the pathogenesis of disease, UV-induced immunosuppression might help ameliorate the disease process, as in the case of psoriasis or the formation of cutaneous lesions in the murine model of leishmaniasis. On the other hand, reducing a protective immune response by UV irradiation is likely to exacerbate the disease process, leading to prolonged, more severe infections and decreased resistance to reinfection. Studies in experimental animal systems clearly indicate that under the appropriate conditions, environmentally relevant doses of UV irradiation can adversely affect the course of some infectious diseases and even result in decreased survival. Based on these findings it seems likely that UV irradiation will also adversely affect the course of some infectious diseases in human populations. However, these effects of UV irradiation may be quite subtle and difficult to detect, since a prolonged or more severe infection would be difficult to quantify in human populations.

The studies summarized here also illustrate another important point, namely that the effects of UV irradiation on the pathogenesis of disease can differ dramatically, depending upon the infectious agent, its route of entry into the host, and the immune mechanisms involved in resistance to infection. Thus, it is difficult to generalize about the effects of UVR on infectious diseases and to extrapolate from the animal models to disease processes in humans. Clearly, much more information is needed, both in humans and in animal models, before we will be able to assess, even in the most general terms, the impact of increased solar UVB radiation on human health.

References

1 Aurelian, L., S. Yasumoto, C.C. Smith: Antigen-specific immune-suppressor factor in herpes simplex virus type 2 infections of UVB-irradiated mice. *J. Virol.* 62 (1988), 2520–2524.

2 Boyle, J., R.M. MacKie, J.D. Briggs, B.J.R. Junor, T.C. Aitchison: Cancer, warts and sunshine in renal transplant patients: a case–control study. *Lancet* 1 (1984). 702–705.

3 Brozek, C.M., G.M. Shopp, D.L. Ryan, P.M. Gillespie, D.F. Kusewitt, M.S. Rajagopalan, K.D. Ley, R.D. Ley: *In vivo* exposure to ultraviolet radiation enhances pathogenic effects of murine leukemia virus LP-BM5, in murine acquired immunodeficiency syndrome. *Photochem. Photobiol.* 56 (1992), 287–295.

4 Chung, H-T., D.K. Burnham, B. Robertson, L.K. Roberts, R.A. Daynes: Involvement of prostaglandins in the immune alterations caused by the exposure of mice to ultraviolet radiation. *J. Immunol.* 137 (1986), 2478–2484.

5 Clegg, J.A., S.R. Smithers: Death of schistosome cercariae during penetration of the skin. II. Penetration of mammalian skin by *S. mansoni*. *Parasitology* 8 (1968), 111–128.

6 Closs, O.: Experimental murine leprosy: growth of *Mycobacterium lepraemurium* in C3H and C57BL/mice after footpad inoculation. *Infect. Immun.* (1975), 480–489.

7 Dannenberg, A.M. Jr: Delayed-type hypersensitivity and cell-mediated immunity in the pathogenesis of tuberculosis. *Immunol. Today* 12 (1991), 228–231.

8 De Fabo, E.C., F.P. Noonan: Mechanism of immune suppression by ultraviolet radiation *in vivo*. I. Evidence for the existence of a unique photoreceptor in skin and its role in photoimmunology. *J. Exp. Med.* 157 (1983), 84–98.

9 Denis, M., A. Forget, M. Pelletier, R. Turcotte, E. Skamene: Control of the Bcg gene resistance to infections with atypical mycobacteria and BCG substrains. *Clin. Exp. Immunol.* 63 (1986), 517–525.

10 Denkins, Y., I.J. Fidler, M.L. Kripke: Exposure of mice to UVB radiation suppresses delayed hypersensitivity to *Candida albicans*. *Photochem. Photobiol.* 49 (1989), 615–619.

11 Denkins, Y.M., M.L. Kripke: Effect of UV radiation on lethal infection of mice with *Candida albicans*. *Photochem. Photobiol.* 57 (1993), 266–271.

12 Dyall-Smith, D.G. Varigos: The malignant potential of papillomavirus. *Austral. J. Dermatol.* 26 (1985), 102–107.

13 Edwards D., C.H. Kirkpatrick: The immunology of mycobacterial disease. *Am. Rev. Respir. Dis.* 135 (1986), 1062–1071.

14 Finsen, N.R.: The chemical rays of light and smallpox. In: *Phototherapy*, (1901), pp. 1–36. Ed.: E. Arnold, London.

15 Finsen, N.R.: The treatment of lupus vulgaris by concentrated chemical rays. In: Phototherapy, (1901), pp. 73–75. Ed.: E. Arnold, London.

16 Giannini, M.S.H.: Suppression of pathogenesis in cutaneous leishmaniasis by UV radiation. *Infect. Immun.* 51 (1986), 838–843.

17 Giannini, S.H.: Effects of UVB on infectious diseases. In: *Global Atmospheric Change and Public Health,* (1990), p. 33. Ed.: J.C. White, Elsevier Science Publishing Co., Inc., New York.

18 Giannini, S.H., E.C. DeFabo: *Abrogation of Skin Lesions in Cutaneous Leismaniasis by Ultraviolet B Radiation* (1987). Plenum Publishing Company Ltd., London.

19 Greene, M.I., M.S. Sy, M.L. Kripke, B. Benacerraf: Impairment of antigen-presenting cell function by ultraviolet radiation. *Proc. Natl. Acad. Sci. USA* 76 (1979), 6591–6595.

20 Gurish, M.F., D.H. Lynch, R.A. Daynes: Changes in antigen-presenting cell function in the spleen and lymph nodes of ultraviolet-irradiated mice. *Transplantation* 33 (1982), 280–284.

21 Hahn, H., S.H.E. Kaufmann: Role of cell mediated immunity in bacterial infections. *Rev. Infect. Dis.* 3 (1981), 1221–1250.

22 Harbour, D.A., T.J. Hill, W.A. Blyth: Recurrent herpes simplex in the mouse: inflammation in the skin and activation of virus in the ganglia following peripheral stimulation. *J. Gen. Virol.* 64 (1983), 1491–1498.

23 Hoffman, J.S.: *Assessing the Risks of Trace Gases that can Modify the Stratosphere,* (1987). EPA 400/1-87/001, US Environmental Protection Agency, Washington, DC.

24 Howie, S.E.M., M. Norval, J. Maingay: Exposure to low-dose ultraviolet radiation suppresses delayed-type hypersensitivity to herpes simplex virus in mice. *J. Invest. Dermatol.* 86 (1986), 125–128.

25 Howie, S.E.M., M. Norval, J. Maingay: Alterations in epidermal handing of HSV-1 antigens *in vitro* induced by *in vivo* exposure to UV-B light. *Immunology* 57 (1986), 225–230.

26 Howie, S.E.M., M. Norval, J. Maingay, J.A. Ross: Two phenotypically distinct T cells ($Ly1^+2^-$ and $Ly1^-2^+$) are involved in ultraviolet-B light-induced suppression of the efferent DTH response to HSV-1 *in vivo*. *Immunology* 58 (1986), 653–658.

27 Jeevan, A., Y. Denkins, E. Brown, M.L. Kripke: Effects of UV radiation on infectious diseases. In: *Biologic Effects of Light,* (1992), pp. 83–96. Walter de Gruyter & Co., Berlin.

28 Jeevan, A., R. Evans, E.L. Brown, M.L. Kripke: Effect of local ultraviolet irradiation on infections of mice with *Candida albicans, Mycobacterium bovis* BCG, and *Schistosoma mansoni*. *J. Invest. Dermatol.* 99 (1992). 59–64.

29 Jeevan, A., K. Gilliam, H. Heard, M.L. Kripke: Effect of ultraviolet radiation on the pathogenesis of *Mycobacterium lepraemurium* in mice. *Exp. Dermatol.* 1 (1992), 152–160.

30 Jeevan, A., M.L. Kripke: Effect of a single exposure to ultraviolet radiation on *Mycobacterium bovis* bacillus Calmette-Guérin infection in mice. *J. Immunol.* 143 (1989), 2837–2843.

31 Jeevan, A., M.L. Kripke: Alteration of the immune response to *Mycobacterium bovis* BCG in mice exposed chronically to low doses of UV radiation. *Cell Immunol.* 130 (1990), 32–41.

32 Jeevan A., S.E. Ullrich, V.V. Dizon, M.L. Kripke: Supernatants from ultraviolet-irradiated keratinocytes decrease the resistance and delayed-type hypersensitivity response to *Mycobacterium bovis* bacillus Calmette-Guérin in mice and impair the phagocytic ability of macrophages. *Photodermatol. Photoimmunol. Photomed.* 9 (1992), 255–263.

33 Joint International Union against Tuberculosis and World Health Organization Study Group: Tuberculosis control. *Tubercle* 63 (1982), 157–169.

34 Kaufmann, S.H.E.: Towards new leprosy and tuberculosis vaccines. *Microbiol. Sci.* 4 (1987), 324–328.

35 Kim, T-Y., M.L. Kripke, S.E. Ullrich: Immunosuppression by factors released from UV-irradiated epidermal cells: selective effects on the generation of contact and delayed hypersensitivity after exposure to UVA or UVB radiation. *J. Invest. Dermatol.* 94 (1990), 26–32.

36 Klein, K.L., C.C. Linnemann Jr: Induction of recurrent genital herpes simplex virus type 2 infection by ultraviolet light. *Lancet* i (1986), 796–797.

37 Kripke, M.L., P.A. Cox, L.G. Alas, D.B. Yarosh: Pyrimidine dimers in DNA initiate systemic immunosuppression in UV-irradiated mice. *Proc. Natl. Acad. Sci. USA* 89 (1992), 7516–7520.

38 Mackaness, G.B.: The relationship of delayed hypersensitivity to acquired cellular resistance. *Br. Med. Bull.* 23 (1967), 52–54.

39 Mackaness, G.B.: The immunology of antituberculous immunity. *Am. Rev. Respir. Dis.* 97 (1968), 337–344.

40 Morrey, J.D., S.M. Bourn, T.D. Bunch, M.K. Jackson, R.W. Sidwell, L.R. Barrows, R.A. Daynes, C.A. Rosen: *In vivo* activation of human immunodeficiency virus type 1 long terminal repeat by UV type A (UV-A) light plus psoralen and UV-B light in the skin of transgenic mice. *J. Virol.* 65 (1991), 5045–5051.

41 Noonan, F.P., E.C. DeFabo, M.L. Kripke: Suppression of contact hypersensitivity by UV radiation and its relationship to UV-induced suppression of tumor immunity. *Photochem. Photobiol.* 34 (1981), 683–689.

42 Okamoto, H., M.L. Kripke: Effector and suppressor circuits of the immune response are activated *in vivo* by different mechanisms. *Proc. Natl. Acad. Sci. USA* 84 (1987), 3841–3845.

43 Otani, T., R. Mori: The effects of ultraviolet radiation of the skin on herpes simplex virus infection: altera-

tion in immune function mediated by epidermal cells and in the course of infection. *Arch. Virol.* 96 (1987), 1–15.

44 Pardo, R.J., M.A. Bogaert, N.S. Penneys, G.E. Byrne Jr, P. Ruiz: UVB phototherapy of the pruritic papular eruption of the acquired immunodeficiency syndrome. *J. Am. Acad. Dermatol.* 26 (1992), 423–428.

45 Ranki, A., P. Puska, S. Mattinen, A. Lagerstedt, K. Krohn: Effect of PUVA on immunologic and virologic findings in HIV-infected patients. *J. Am. Acad. Dermatol.* 24 (1991), 404–410.

46 Rivas, J.M., S.E. Ullrich: Systemic suppression of delayed-type hypersensitivity by supernatants from UV-irradiated keratinocytes. An essential role for keratinocyte-derived IL-10. *J. Immunol.* 149 (1992), 3865–3871.

47 Robertson, B., L. Gahring, R. Newton, R.A. Daynes: *In vivo* administration of IL-1 to normal mice decreases their capacity to elicit contact hypersensitivity responses: prostaglandins are involved in this modification of the immune response. *J. Invest. Dermatol.* 88 (1987), 380–387.

48 Rogers, T.J., E. Balish: Immunity to *Candida*. *Microbiol. Rev.* 44 (1980), 660–682.

49 Ross, J.A., S.E.M. Howie, M. Norval, J. Maingay: Systemic administration of urocanic acid generates suppression of the delayed type hypersensitivity response to herpes simplex virus in a murine model of infection. *Photodermatology* 5 (1988), 9–14.

50 Ross, J.A., S.E.M. Howie, M. Norval, J. Maingay, T.J. Simpson: Ultraviolet-irradiated urocanic acid suppresses delayed-type hypersensitivity to herpes simplex virus in mice. *J. Invest. Dermatol.* 87 (1986), 630–633.

51 Schwarz, T., A. Urbanska, F. Gschnait, T.A. Luger: UV-irradiated epidermal cells produce a specific inhibitor of IL-1 activity. *J. Immunol.* 138 (1987), 1457–1463.

52 Snider, D.E., W.L. Roper: The new tuberculosis. *New. Engl. J. Med.* 326 (1992), 703–705.

53 Sorgo, J.: Die Lichtbehandlung der Lungentuberkulose. In: *Haudbuch der Lichttherapie*, (1927), pp. 284–301. Eds.: W. Hausmann, R. Volk, Julius Springer, Vienna.

54 Spangrude, G.J., E.J. Bernhard, R.S. Ajioka, R.A. Daynes: Alterations in lymphocyte homing patterns within mice exposed to ultraviolet radiation. *J. Immunol.* 103 (1983), 2974–2981.

55 Sprecher, E., Y. Becker: Herpes simplex virus type 1 pathogenicity in footpad and ear skin of mice depends on Langerhans cell density, mouse genetics, and virus strain. *J. Virol.* 61 (1987), 2515–2522.

56 Spruance, S.L.: Pathogenesis of herpes simplex labialis: experimental induction of lesions with UV light. *J. Clin. Microbiol.* 22 (1985), 366–368.

57 Stanley, S.K., T.M. Folks, A.S. Fauci: Induction of expression of human immunodeficiency virus in a chronically infected promonocytic cell line by ultraviolet radiation. *AIDS Res. Human Retroviruses* 5 (1989), 375–384.

58 Stingl, G., K. Rappersberger, E. Tschachler, S. Gartner, V. Groh, D.L. Mann, K. Wolff, M. Popovic: Langerhans cells in HIV-1 infection. *J. Am. Acad. Dermatol.* 22 (1990), 1210–1217.

59 Stone, R.: Infectious diseases: IOM weighs in on microbial threat. *Science* 258 (1992), 540.

60 Styblo, K.: Overview and epidemiologic assessment of the current global tuberculosis situation with an emphasis on control in developing countries. *Rev. Infect. Dis.* 11 (1989), S339–S346.

61 Szigeti, B., J. Chapiro, J.M. Saint-Girons: Traitement du zona par actinotherapie ultraviolette. *J. Radiol. Electrol.* 57 (1976), 549–551.

62 Toews, G.B., P.R. Bergstresser, J.A. Streilein, S. Sullivan: Epidermal Langerhans cell density determines whether contact hypersensitivity or unresponsiveness follows skin painting with DNFB. *J. Immunol.* 124 (1980), 445–453.

63 Tschachler, E., V. Groh, M. Popovic, D.L. Mann, K. Konrad, B. Safai, L. Eron, F.D. Veronese, K. Wolff, G. Stingl: Epidermal Langerhans cells – a target for HTLV-III/LAV infection. *J. Invest. Dermatol.* 88 (1987), 233–237.

64 Ullrich, S.E., E. Azizi, M.L. Kripke: Suppression of the induction of delayed type hypersensitivity reactions in mice by a single exposure to ultraviolet radiation. *Photochem. Photobiol.* 43 (1986), 633–638.

65 Ullrich, S.E., B.W. Mclntyre, J.M. Rivas: Suppression of the immune response to alloantigen by factors released from ultraviolet-irradiated keratinocytes. *J. Immunol.* 145 (1990), 489–498.

66 Valerie, K., A. Delers, C. Bruck, C. Thiriart, H. Rosenberg, C. Debouck, M. Rosenberg: Activation of human immunodeficiency virus type 1 by DNA damage in human cells. *Nature* 333 (1988), 78–81.

67 Van Weelden, H., H. Baart de la Faille, E. Young, J.C. van der Leun: A new development in UVB phototherapy of psoriasis. *Brit. J. Dermatol.* 119 (1988), 11–19.

68 Vogel, J., M. Cepeda, E. Tschachler, L.A. Napalitano, G. Jay: UV activation of human immunodeficiency virus gene expression in transgenic mice. *J. Virol.* 66 (1992), 1–5.

69 Weiss, R.: Measles battle loses potent weapon. *Science* 258 (1992), 546.

70 Wheeler, C.E. Jr: Pathogenesis of recurrent herpes simplex infections. *J. Invest. Dermatol.* 65 (1975), 341–346.

71 Yasumoto, S., Y. Hayashi, L. Aurelian: Immunity to herpes simplex virus type 2: suppression of virus-induced immune responses in ultraviolet B-irradiated mice. *J. Immunol.* 139 (1987), 2788–2793.

72 Yoshikawa, T., V. Rae, W. Bruins-Slot, W. Van den Berg, J.R. Taylor, J.W. Streilein: Susceptibility to effects of UVB radiation on induction of contact hypersensitivity as a risk factor for skin cancer in humans. *J. Invest. Dermatol.* 95 (1990), 530–536.

73 Yoshikawa, T., J.W. Streilein: Tumor necrosis factor alpha and ultraviolet light have similar effects on contact hypersensitivity in mice. *Regional Immunol.* 3 (1990), 139–144.

74 Young, L.S., C.B. Inderlied, O.G. Berlin, M.S. Gottlieb: Mycobacterial infections in AIDS patients, with an emphasis on the *Mycobacterium avium* complex. *Rev. Infect. Dis.* 8 (1986), 1024–1033.

75 Zmudzka, B.Z., J.Z. Beer: Activation of human immunodeficiency virus by ultraviolet radiation. *Photochem. Photobiol.* 52 (1990), 1153–1162.

PART 3
CLINICAL PHOTOIMMUNOLOGY

12: Polymorphous light eruption

Erhard Hölzle

Summary

Polymorphous light eruption is the most frequent idiopathic photodermatosis with a prevalence estimated to be between 10% and 20%. The eruption is monomorphous in the same patient and varies only interindividually. Papular, plaque-type, and papulovesicular variants represent the most frequent morphological subsets. Histopathological features are characteristic and comprise perivascular lymphocytic infiltrate throughout the dermis, subepidermal edema, and some vacuolar alteration of basal cells associated with focal spongiosis and exocytosis. Skin lesions follow intensive sun exposure and resolve spontaneously without residues. Pruritus is prominent. Light hardening is a frequent phenomenon. The pathogenesis of polymorphous light eruption is unclear. A delayed type of immunoreaction induced by UV irradiation and maintained by defect immunoregulatory mechanisms seems likely. The initiating antigenic photoproduct is, however, elusive. The action spectrum is mainly in the UVA, and in some patients also in the UVB, region. Photoprotective methods in combination with hardening by graduated sun exposures provide the basis for prophylaxis. Additionally, induction of tolerance by phototherapy and, in extreme cases, photochemotherapy are useful prophylactic methods. Systemic treatment by chloroquine, beta-carotene, calcium or nicotinamide is of uncertain efficacy.

Introduction

Polymorphous light eruption is the most frequent photodermatosis: its etiology is still elusive. Clinical features, time course, and histopathologic picture suggest a delayed type immunoreaction. Firm proof of this notion has not been put forward so far.

At the end of the 19th century it was realized that solar radiation can induce skin rashes. Willan [4] was the first to designate such eruptions as 'eczeme solare'. In 1879 Sir Jonathan Hutchinson [27] described 14 patients presenting with a rash very similar to what we would perceive today as polymorphous light eruption. He used the term 'summer prurigo'. The eruption occurred in spring or early summer: predominantly young adults were afflicted. The lesions comprised inflammatory papules without ulceration or formation of pustules. The time course was chronically intermittent for many years. Synonymously the terms 'prurigo estivalis', 'prurigo adolescentium', and 'acne prurigo' were used. In 1900 Rasch [72] coined the term 'polymorphous light eruption'. At that time this historical designation included several photoinduced dermatoses. Since clear definitions and methods for differential diagnosis were lacking, patients with solar urticaria, hydroa vacciniforme, porphyria, photoallergic dermatitis, photosensitive erythema multiforme, and cutaneous lupus erythematosus were in part included in this diagnosis. During the following decades definitions of these diseases were refined. It is noteworthy that as late as 1979 Sontheimer *et al.* [75] defined subacute cutaneous lupus erythematosus as a photosensitive variant of cutaneous lupus erythematosus.

Established features of polymorphous light eruption are the clinical description, basic histopathologic criteria, and main characteristics of the action spectrum. Controversies have arisen about morphological variants and the relationship to differential diagnoses, such as actinic prurigo, inherited polymorphous light eruption in American Indians, subtypes of lupus erythematosus, and lymphocytic infiltration (Jessner–Kanof). Open questions comprise etiology, pathomechanism, and prophylaxis or treatment by systemic agents. Although a delayed-type hypersensitivity reaction is postulated as the pathomechanism, the putative photoallergen has not been identified. It also seems puzzling how a singular chromophore and a uniform mechanism should be activated by different parts of the UV spectrum and result in such a morphological diversity of lesions. Hence different mechanisms and various chromophores may play a role.

Epidemiology

Polymorphous light eruption is the most frequent photodermatosis and includes more than 90% of all patients with light-induced eruptions. The true prevalence is uncertain, it seems likely to be between 10% and 20% [5, 50]. The prevalence could be even higher, since minor transient eruptions are tolerated by many patients and never come to the attention of a physician.

Polymorphous light eruption is not restricted to a certain skin type or race. It was found in blacks, Orientals, and American Indians [15, 17]. Patients with fair skin may be more frequently afflicted [11, 43]. A familial predisposition seems to be present; reports range from 3% to 56% [11, 29, 43]. Underestimation of the familial aspect is, however, assumed.

The hereditary polymorphous light eruption in American Indians is considered a variety different from polymorphous light eruption observed in other parts of the world [39]. An autosomal–dominant inheritance with a reduced penetrance has been reported in 75% of Canadian Inuit patients [64]. This familial type of polymorphous light eruption is a specific variant and probably represents familial actinic prurigo in Indians of North and South America. In patients with polymorphous light eruption HLA-phenotyping could not establish differences to a control population [40]. In contrast, actinic prurigo in American Indians was shown to be associated with HLA-types A24 and Cw4, and negatively correlated with A3 [74]. This finding again underscores the difference between these two diseases. Hereditary polymorphous light eruption in American Indians will not be discussed in the present overview.

The age of onset is mainly in young adulthood [11, 30, 43]. Increasingly frequently children are afflicted, probably owing to changes in sun exposure habits during leisure time and vacation. Nowadays 'juvenile spring eruption' [3] is regarded as a variant of polymorphous light eruption in children and youths. Most studies report a preponderance of women [5, 11, 21, 30, 43]. In our experience the ratio is about 7:1. This observation gave rise to speculations about a hormonal etiology of polymorphous light eruption which has been refuted [55].

Clinical features and differential diagnosis

In polymorphous light eruption only the sun-exposed skin areas are involved. Skin lesions in sun-protected skin have, however, been reported [6, 30, 33] especially in patients with long-standing polymorphous light eruption [43]. A main feature of polymorphous light eruption is the occurrence in individual sites of predilection. Most frequently the upper arms, V-neck, dorsal aspects of forearms, and face are involved [5, 11, 30]. In children lesions appear most frequently on the face. There are marked interindividual differences in the distribution pattern, and only in extremely photosensitive patients are larger areas of the trunk involved. Rarely, skin lesions on lips and eyelids as well as association with systemic symptoms like chills, headache, fever and nausea have been reported [30]. The actual distribution pattern of skin lesions results from superimposition of intrinsic sites of predilection and the actual state of skin hardening. Most patients experience this hardening phenomenon following repeated eruptions. It results in complete tolerance even of intense sunshine [11, 21, 43] and is the rationale for phototherapy as a prophylactic measure.

Extensive sun exposure after longer periods of seasonal or weather-related lack of sunshine provokes pronounced and widespread polymorphous light eruption [28].

The time course of polymorphous light eruption is distinct and aids in distinguishing the disease from other differential diagnoses. It is a delayed type of reaction following intensive sun exposure. Skin lesions occur after a few hours or a few days and subside without leaving residues within several days if further sun exposure is avoided [11, 21, 28, 43]. Patients frequently report the onset of lesions following several successive days of sun exposure, e.g. during a beach vacation. In such instances the occurrence of lesions during sun exposure can falsely be interpreted as an immediate type reaction. The characteristic sequence of events is itching followed by patchy erythema and finally the emergence of distinct lesions.

The historical designation polymorphous light eruption is actually a misnomer. In the same patient the eruption is always monomorphous. It is only between different patients that the various morphological variants are observed. There is much debate about how many varieties of polymorphous light eruption are to be distinguished. These range from polymorphous light eruption 'sine eruptione' [13] as the minor variant characterized by the occurrence of pruritus only, to small and large papular, urticarial, plaque-like, erythema-multiforme-like, vesiculous, papulovesiculous, vesiculobullous, hemorrhagic, ictus-like, eczematous, and prurigo-like type [7, 14, 15, 21, 30, 38, 42]. Not all of these variants are, in our view, part of the entity polymorphous light eruption. Widespread diffuse erythema, eczematous dermatitis and prurigo should be excluded and represent differential diagnoses like sunburn, phototoxic reaction, photoallergic dermatitis, lupus erythematosus, photoaggravated atopic dermatitis, chronic actinic dermatitis, and actinic prurigo. We distinguish three major types of polymorphous light eruption: papular type, plaque type, and papulovesicular type [20]. Further variants may be grouped as subtypes under these headings (Table 12.1). The hemorrhagic type may be considered as a variant of the papular type. The erythema-multiforme-like type is regarded as a maximal variant of the plaque type. Ictus (insect bite)-like type and vesiculobullous type are subgroups of papulovesicular polymorphous light eruption. The vesiculobullous variant was first observed in American tourists to Hawaii [14]; this type seems to be rare in European latitudes [42].

Table 12.1 Morphological variants of polymorphous light eruption

Papular type
Hemmorhagic type
Plaque type
Erythema-multiforme-like type
Papulovesicular type
Ictus (insect bite)-like type
Vesiculobullous type

Several investigators agree that patients with polymorphous light eruption do not exhibit a higher susceptibility to sunburn reaction. This is documented by the normal values for minimal erythema doses of UVB [12, 31, 44]. It is also confirmed in our series in which we found normal erythemal and pigmentary responses to UVA as well as to UVB. Threshold doses for immediate pigment darkening (UVA), minimal tanning (UVA), UVA-erythema and UVB-erythema were normal.

The most important differential diagnosis, clinically and histopathologically, is cutaneous lupus erythematosus. Subacute cutaneous lupus erythematosus as well as tumidus-type of discoid lupus erythematosus can be very similar to the plaque-type of polymorphous light eruption. The time course is, however, different. In lupus erythematosus, skin lesions emerge 2–3 weeks after intensive sun exposure and usually persist for many weeks before they subside gradually. This prolonged course of lupus erythematosus was also demonstrated in experimentally induced skin lesions of lupus erythematosus patients [41]. In some patients the separation between polymorphous light eruption and cutaneous lupus erythematosus is still difficult. It was found in patients with persistent lesions of polymorphous light eruption that antinuclear antibodies were present in the serum. The authors postulated a 'forme fruste' of lupus erythematosus [67]. Consequently these patients should be categorized as lupus erythematosus. The presence of anti-SSA/Ro antibodies in patients with polymorphous light eruption was also reported [37]. On the other hand, long-term follow-up of patients with polymorphous light eruption failed to reveal a transition from polymorphous light eruption to lupus erythematosus [33].

Action spectrum

The majority of patients are also sensitive to sunlight radiating directly through window glass while travelling in planes, trains or cars [11, 24, 43]. This raises questions about a role of UVB in these patients. The lack of protection obtained from the use of UVB-absorbing sunscreens as observed by the majority of patients also implicates UVA as the eliciting wave band. In fact the exact action spectrum of polymorphous light eruption is still unknown despite various investigations. Experimental data of earlier studies are conflicting and sometimes contradictory. Factors that may relate to the reported variability include differences in individual susceptibility, seasonal variations, differences in test sites, light sources, and interpretation of test results, as well as the clinical varieties of polymorphous light eruption, including geographic differences. Many investigators report the induction of lesions with sunburn radiation; in some instances responses to UVC were also noted. On the other hand, several other researchers were able to reproduce skin lesions of polymorphous light eruption with long-wave UV. In a multitude of experimental studies, various authors concluded that the action spectrum might range from UVB to UVA into visible light. Also an additive effect of infrared radiation was described (see overview in [22]).

With the availability of high-intensity UVA sources, which allow the irradiation of large UVA doses without any UVB component, it was possible to reproduce clinically and histopathologically characteristic lesions of polymorphous light eruption in the laboratory [21]. In several series our group was able to establish that about 75% of patients were sensitive to UVA, 10% to UVB, and 15% to UVA as well as UVB [22]. Selecting the appropriate procedure for provocative phototesting (Table 12.2) in 60–90% of patients tested, typical lesions of polymorphous light eruption were induced. Similar results were obtained from other groups [11, 43, 65]. Whole body exposures in light cabins are also suitable to provoke skin lesions [56]. This procedure may determine the intrinsic sites of predilection. In few patients a relationship between activating and inhibiting wavelengths was examined [69]. The action spectrum was found in the long-wave UVA; shorter wavelengths seemed to be inhibitory.

Table 12.2 Experimental reproduction of lesions in polymorphous light eruption

Location of test area	Site of predilection
Size of test area	5 × 8 cm
UV source	UVA: UVASUN 3000 (340–460 nm) UVB: Philips TL 12 (285–350 nm)
UV dose	3 × 60–100 J/cm² UVA 3 × 1.0–1.5 MED-UVB (Irradiations are performed on successive days)
Time of testing	Before the sunny season, preferably in winter or early spring
Readings	24, 48, 72 h following irradiations

Histopathology and immunohistochemistry

The histopathologic picture of fully developed lesions in polymorphous light eruption is fairly characteristic, but not absolutely distinct [1, 21, 25]. The main features comprise a dense perivascular lymphocytic infiltrate throughout the dermis with a preponderance of the upper plexus, subepidermal edema with some vacuolar degeneration of basal cells, and sometimes focal spongiosis with some exocytosis. These changes represent the basic pattern of polymorphous light eruption and are typical for the papular type. In morphological variants certain features are pronounced or additional ones are present. In the hemorrhagic type some focal extravasation of erythrocytes in the stratum papillare is typical. The plaque type shows a band-like organization of the infiltrate in the upper dermis with conspicuous subepidermal edema. Very similar in appearance is the erythema-multiforme-like type: edema is even more pronounced. Differentiation from the plaque-type is not made with certainty. The vesiculobullous type also exhibits the basic pattern of papular polymorphous light eruption in addition to spongiotic vesicles and severe subepidermal edema resulting in blisters. Focal necrosis of keratinocytes is the additional characteristic feature of ictus (insect bite)-like type. An admixture of neutrophils and eosinophils to the dermal inflammatory infiltrate has been described [16, 25] and was found in 15% of lesions in our material. Leukocytes were especially numerous in the papulovesicular type of polymorphous light eruption.

Immunohistochemical phenotyping of the dermal infiltrate was first performed by Muhlbauer and coworkers [51]. The infiltrate was found to be mainly composed of T cells with a preponderance of T suppressor cells (CD8) versus T helper cells (CD4). Macrophages comprised less than 12% of the infiltrate. CD1-positive cells (Langerhans cells, indeterminated dendritic cells) were found to be increased in numbers. Moncada and his group [49], however, described a preponderance of T helper cells in the dermal infiltrate. In one study immunofluorescence revealed intervascular and focal perivascular deposition of fibrin, and deposition of C3 and IgM in vessel walls. The lupus band test was negative [52]. In recent investigations taking into account the evolution of lesions [62], it was found that the ratio between $CD4^+$ and $CD8^+$ cells was dependent on the stage of development of polymorphous light eruption. In the early phase, T helper cells and T suppressor cells were present in equal numbers; in later stages T suppressor cells were dominant. These findings were also confirmed by our group [34]. In contrast to polymorphous light eruption, the dermal infiltrate in lupus erythematosus is characterized by a majority of $CD4^+$ cells in combination with the presence of some B cells. The activation markers (HLA-DR, interleukin-2 receptor) are present both in polymorphous light eruption and lupus erythematosus [34]. The value of direct immunofluorescence in the differential diagnosis between lupus erythematosus and polymorphous light eruption is limited, since it is negative in polymorphous light eruption, and positive results in lupus erythematosus are only obtained in lesions persisting for more than 6 weeks [9].

Etiology and pathogenesis

The etiology and pathogenesis of polymorphous light eruption is unknown and subject to speculation. Many hypotheses are founded on the existence of an exogenous or endogenous photosensitizer as the initiator of the reaction. One of these theories focused on a disturbance of tryptophane metabolism with accumulation of cynurenic acid [6]. This represented the theoretical basis for the therapeutic use of nicotinamide [45, 57]. This concept has not been confirmed. The preponderance of female patients gave rise to the speculation that hormonal influences would be an etiologic factor in polymorphous light eruption. Oral contraceptives were discussed as possible photosensitizers; a notion which has been disproved [55]. Also reported on was the release of leukotrien B4 from leucocytes of patients with polymorphous light eruption [73] following UVA irradiation. Confirmation and conclusive interpretation of this finding regarding the pathogenesis of polymorphous light eruption is still lacking.

Currently, a delayed-type immunoreaction induced by electromagnatic irradiation and maintained by defect immunoregulatory mechanisms is proposed as the pathogenesis of polymorphous light eruption. The delayed course after initiating irradiation, gross morphology and histopathological features suggests a cell-mediated immune response [16, 61]. Additional evidence came from the investigation of expression of intercellular adhesion molecules (ICAM-1). ICAM-1 is expressed on the surface of keratinocytes in allergic contact dermatitis. Expression was also found in polymorphous light eruption [59] but not in sunburn reaction [63]. So far a relevant photoallergen has not been identified. Hence it seemed appropriate to search for functional disturbances of lymphocytes or defect immunoregulatory mechanisms. There have, however, been reports about a transient decrease [26] as well as an increase in lymphocyte numbers [49]. In another study [35] the number of T cells in the peripheral blood, as well as their response upon stimulation with phytohemagglutinin, concanavalin A, and purified protein derivative of tuberculin, were found to be normal when studied in patients with polymorphous light eruption before and after photochemotherapy. In addition, UV irradiation failed to activate lymphocytes from patients with polymorphous light eruption [71]. A recent investigation [18], however, demonstrated stimulation of autologous mononuclear cells of peripheral blood by UV-irradiated epidermal cells in patients with polymorphous light eruption. This indicates sensitization of mononuclear cells against UV-modified epidermal antigens. Currently heat shock proteins are being discussed as the possible antigens expressed in epidermis after brisk UV irradiation. In experimentally induced polymorphous light eruption, expression of heat shock proteins was found to be raised in both endothelial cells of dermal blood vessels and keratinocytes from 1 h post-irradiation up to 6 days [47]. This is intriguing, since heat shock proteins are believed to play an essential role in autoimmune diseases [36] and peculiarly in lupus erythematosus [10, 58].

Therapy and prophylaxis

Treatment of acute stage polymorphous light eruption is simple. Sun protection and the symptomatic use of topical steroids or zinc lotion ameliorate inflammation and itching and shorten the eruption. Prophylactic treatment to prevent further bouts is much more difficult.

Teaching the patient about the nature of polymorphous light eruption and giving proper advice on photoprotection form the basis of patient management. Many patients learn to avoid pronounced eruptions of polymorphous light eruption by changing their sun exposure habits with concomitant use of potent broad band sunscreens which also absorb UVA [21, 23, 46]. If this fails, prophylactic phototherapy is induced. UVA as well as UVB or a combination of both can be employed. In our experience UVA or UVA plus UVB is superior to UVB phototherapy. Other authors prefer UVB phototherapy since it is felt that UVA tends to provoke lesions during prophylactic irradiations [25, 54]. In cases of extreme photosensitivity and if the aforementioned methods fail altogether, photochemotherapy combining oral psoralens and UVA irradiation may be conducted. Beyond doubt, this treatment represents the most effective prophylaxis [2, 24, 54, 68]. The uncertain long-term risks have, however, to be weighed against the benign and self-limited disease.

From early on attempts have been made to prevent polymorphous light eruption by systemic medication. Anti-malarials, especially chloroquine, were used. In controlled studies, however, a reproducible and convincing therapeutic effect could not be demonstrated [8, 53]. Reported treatment successes may be due to faulty diagnoses. Photosensitive variants of lupus erythematosus might have been erroneously treated as polymorphous light eruption. It was as late as 1979 when Sontheimer and Gilliam [75] separated subacute cutaneous lupus erythematosus as a very photosensitive variant from other types of cutaneous lupus erythematosus.

Beta-carotene acts as a quencher of radicals and is effective in the treatment of erythropoietic protoporphyria. Its use in polymorphous light eruption has also been advocated [32]. Response rates were reported to reach from 30% to 80%. Controlled trials, however, questioned the efficacy of beta-carotene [8]. Since side-effects are virtually absent, a therapeutic attempt can always be made. The notion of a disturbance of tryptophane metabolism as the pathomechanism in polymorphous light eruption provided the basis of the therapeutic use of nicotinamide. Uncontrolled trials [45, 57] and anecdotal reports support the efficacy of that therapy. In experimental investigations [66] and controlled studies [19] nicotinamide was revealed as being largely ineffective. This also holds true for calcium. In an experimental setting, extracts of *Escherichia coli* given orally or parenterally suppressed polymorphous light eruption when provoked by artificial irradiation in test areas [70]. Reports on the clinical use of these extracts are lacking. Extremely incapacitated patients can benefit from systemic corticosteroids [48] and azathioprine [60] if phototherapy and photochemotherapy fail.

References

1 Ackerman, A.B.: *Histologic Diagnosis of Inflammatory Skin Diseases* (1978). Lea & Febiger, Philadelphia.

2 Addo, H.A., S.C. Sharma: UVB phototherapy and photochemotherapy (PUVA) in the treatment of polymorphic light eruption and solar urticaria. *Br. J. Dermatol.* 116 (1987), 539–547.

3 Anderson, D., H.J. Wallace, E.T.B. Howes: Juvenile spring eruption. *Lancet* i (1954), 755–756.

4 Bateman, T.: *A Practical Synopsis of Cutaneous Diseases According to the Arrangement of Dr. Willan* (1817). Collins & Croft, Philadelphia.

5 Berg, M: Epidemiological studies of the influence of sunlight on the skin. *Photodermatol.* 6 (1989), 80–84.

6 Bonafe, J.I., H. Chap: Lucite polymorphe et perturbation du metabolisme du tryptophanne (voie de la cynurénine). *Ann. Dermatol. Venereol. (Paris)* 107 (1980), 89–90.

7 Christiansen, J.V., H. Brodthagen: The treatment of polymorphous light eruption with chloroquine. *Br. J. Dermatol.* 68 (1956), 204–208.

8 Corbett, M.F., J.L. Hawk, A. Herxheimer, I.A. Magnus: Controlled therapeutic trials in polymorphous light eruption. *Br. J. Dermatol.* 107 (1982), 571–581.

9 Cripps, D.J., J. Rankin: Action spectra of lupus erythematosus and experimental immunofluorescence. *Arch. Dermatol.* 107 (1973), 563–567.

10 Deguchi, Y: Inhanced expression of heat shock protein gene in kidney lymphoid cells of lupus prone mice during growing process. *Autoimmunity* 10 (1991), 1–5.

11 Diepgen, T.I., M. Häberle, M. Fartasch, O. Stüben, O.P. Hornstein: Charakteristika der polymorphen Lichtdermatose – Ergebnisse einer prospektiven Befragung und Untersuchung 302 Betroffener. *Z. Hautkr.* 64 (1988), 279–285.

12 Diffey, B.I., P.M. Farr: The erythemal response to ultraviolet radiation in subjects with polymorphic light eruption. *Br. J. Dermatol.* 114 (1986), 103–108.

13 Dover, J.S., J.L.M. Hawk: Polymorphous light eruption sine eruptione. *Br. J. Dermatol.* 118 (1988), 73–76.

14 Elpern, D.J., W.L. Morison, A.F. Hood: Papulovesicular light eruption. A defined subset of polymorphous light eruption. *Arch. Dermatol.* 121 (1985), 1286–1288.

15 Epstein, J.H.: Polymorphous light eruption. *Ann. Allergy* 24 (1966), 397–405.

16 Epstein, J.H.: Polymorphous light eruption. *J. Am. Acad. Dermatol.* 3 (1980), 329–343.

17 Everett, M.A., W. Crockett, J.H. Lamp, D. Minor: Light sensitive eruptions in American Indians. *Arch. Dermatol.* 83 (1961), 243–248.

18 Gonzalez-Amaro, R., L. Baranda, J.F. Salazar-Gonzalez, D. Abud-Mendoza, B. Moncada: Immune sensitization against epidermal antigens in polymorphous light eruption. *J. Am. Acad. Dermatol.* 24 (1991), 70–73.

19 Häberle, M., M. Arnold, H.U. Koch, E. Grühn, O.P. Hornstein: Nikotinsäureamid zur Prophylaxe der polymorphen Lichtdermatose (PLD) – eine plazebokontrollierte Doppelblindstudie. *Vita. Min. Spur.* 6 (1991), 29–37.

20 Hölzle, E: Die polymorphe Lichtdermatose. In: *Jahrbuch der Dermatologie. Licht und Haut,* (1993), pp. 143–154. Eds.: E. Macher, G. Kolde, E.B. Bröcker, Biermann, Zülpich.

21 Hölzle, E., G. Plewig, C. Hofmann, E. Roser-Maass: Polymorphous light eruption. Experimental reproduction of skin lesions. *J. Am. Acad. Dermatol.* 7 (1982), 111–125.

22 Hölzle, E., G. Plewig, R.v. Kries, P. Lehmann: Polymorphous light eruption. *J. Invest. Dermatol.* 88 (1987), 325–385.

23 Hönigsmann, H.: Wirkung einer therapeutischen Lichtschutzcreme auf die Phototestreaktion bei polymorpher Lichtdermatose. *Z. Hautkr.* 64 (1989), 1065–1068.

24 Hönigsmann, H., B. Ortel: Die polymorphe Lichtdermatose – Photobiologische Diagnostik und Therapie. *Z. Hautkr.* 63 (1988), 676–678.

25 Hood, A.F., D.J. Elpern, E.L. Morison: Histopathologic findings in papulovesicular light eruption. *J. Cut. Pathol.* 13 (1986), 13–21.

26 Horkay, I., J. Krajczar, E. Bodolay, M. Debreczeni, A. Begany: A study on cell-mediated immunity in polymorphic light eruption. *Dermatologica* 166 (1983), 75–80.

27 Hutchinson, J.: *Lectures on Clinical Surgery: on Certain Rare Diseases of the Skin* (1879), Churchill, London.

28 Ippen, H.: Photodermatitis multiformis acuta. *Dermatologische Monatsschrift* 166 (1980), 145–150.

29 Jansen, C.T.: Heredity of chronic polymorphous light eruptions. *Arch. Dermatol.* 114 (1978), 188–190.

30 Jansen, C.T.: The natural history of polymorphous

light eruptions. *Arch. Dermatol.* 115 (1979), 165–169.
31 Jansen, C.J.: Erythemal and pigmentary phototest reactions in polymorphous light eruption. *Acta Derm. Venereol. (Stockh.)* 59 (1979), 499–503.
32 Jansen, C.T.: Oral carotenoid treatment in polymorphous light eruption. A cross-over comparison with oxychloroquine and placebo. *Photodermatol.* 2 (1985), 166–169.
33 Jansén, C., J. Karvonen: Polymorphous light eruption. A seven-year follow-up evaluation of 114 patients. *Arch. Dermatol.* 120 (1984), 862–865.
34 Kämmerer, R., E. Hölzle: Immunophenotypisierung verbessert die Differentialdiagnose zwischen Lupus erythematodes, polymorpher Lichtdermatose und lymphozytärer Infiltration Jessner–Kanof. *Actuel Dermatol.* (1994), 135–137.
35 Karvonen, J., M. Viander, J. Ilonen, C.T. Jansen: PUVA photosensitization in polymorphous light eruption: evaluation of systemic immunological factors. *Acta Derm. Venereol. (Stockh.)* 62 (1982), 497–500.
36 Kaufmann, S.H.E.: Heat shock proteins and the immune response. *Immunol. Today* 11 (1990), 129–136.
37 Kiss, M., S. Husz, A. Dobozy: The occurrence of antinuclear, anti-SSA/Ro and anti-SSB/La antibodies in patients with polymorphous light eruption. *Acta Derm. Venereol. (Stockh.)* 71 (1991), 341–343.
38 Lamp, J.H., B. Shelmire, Z. Cooper, R.J. Horgan, C. Keaty: Solar dermatitis. *Arch. Dermatol. Syph.* 62 (1950), 1–27.
39 Lane, P.R., D.J. Hogan, M.J. Martel, B. Reeder, J. Irving: Actinic prurigo: clinical features and prognoses. *J. Am. Acad. Dermatol.* 26 (1992), 683–692.
40 Lane, P.R., D.P. Sheridan, D.J. Hogan, A. Moerland: HLA typing in polymorphous light eruption. *J. Am. Acad. Dermatol.* 24 (1991), 570–573.
41 Lehmann P., E. Hölzle, P. Kind, G. Goerz, G. Plewig: Experimental reproduction of skin lesions in lupus erythematosus by UVA and UVB radiation. *J. Am. Acad. Dermatol.* 22 (1990), 181–187.
42 Lehmann, P., E. Hölzle, G. Plewig: Vesikobullöse Form der polymorphen Lichtdermatose. *Allergologie* 9 (1986), 32–35.
43 Lindmaier, A., R. Neumann: Der PLD-Patient. *Hautarzt* 42 (1991), 430–433.
44 Magnus, I.A.: Studies with a monochromator in the common idiopathic photodermatoses. *Br. J. Dermatol.* 76 (1964), 245–264.
45 Mattheus, A., C. Rader, H. Heise: Nicotinsäureamid bei polymorphen Lichtdermatosen. *Dermatologische Monatschrift* 174 (1988), 142–146.
46 McFadden, N.: UVA sensitivity and topical photoprotection in polymorphous light eruption. *Photodermatol.* 1 (1984), 76–78.
47 McFadden, J.P., P.G. Norris, G. Orchard, R. Cerio, J.L.M. Hawk, R.D.R. Camp: 65 Kilodalton heat shock protein expression in experimentally induced polymorphic light eruption (Abstract). *Br. J. Dermatol.* 125 (1991), 484.
48 Molin, L., G. Volden: Treatment of polymorphous light eruption with PUVA and prednisolone. *Photodermatol.* 4 (1987), 107–108.
49 Moncada, B., R. Gonzalez-Amaro, M.L. Baranda, C. Coredo, R. Urbina: Immuno-pathology of polymorphous light eruption. *J. Am. Acad. Dermatol.* 10 (1984), 970–973.
50 Morison, W.L., R.S. Stern: Polymorphous light eruption. A common reaction uncommonly recognized. *Acta Dermatovenereol. (Stockh.)* 62 (1982), 237–240.
51 Muhlbauer, J.E., A.K. Bhan, T.J. Harrist, J.D. Bernhard, M.C. Mihm: Papular polymorphic light eruption: an immunoperoxydase study using monoclonal antibodies. *Br. J. Dermatol.* 108 (1983), 153–162.
52 Muhlbauer, J.E., M.C. Mihm, T.J. Harrist: Papular polymorphous light eruption. Fibrin, complement, and immunoglobulin deposition. *Arch. Dermatol.* 120 (1984), 866–868.
53 Murphy, G.M., J.L. Hawk, I.A. Magnus: Hydroxychloroquine in polymorphic light eruption: a controlled trial with drug and visual sensitivity monitoring. *Br. J. Dermatol.* 116 (1987), 379–386.
54 Murphy, G.M., R.A. Logan, C.R. Lovell, R.W. Morris, J.L.M. Hawk, I.A. Magnus: Prophylactic PUVA and UVB therapy in polymorphic light eruption: a controlled trial. *Br. J. Dermatol.* 116 (1987), 531–538.
55 Neumann, R: Polymorphous light eruption and oral contraceptives. *Photodermatol.* 5 (1988), 40–41.
56 Neumann, R., H. Pohl-Markl, R.M. Knobler: Polymorphous light eruption: experimental reproduction of skin lesions by whole-body UVA radiation. *Photodermatol.* 4 (1987), 252–256.
57 Neumann, R., E. Rappold, H. Pohl-Markl: Treatment of polymorphous light eruption with nicotinamide: a pilot study. *Br. J. Dermatol.* 115 (1986), 77–80.
58 Norris, D.A.: Pathomechanisms of photosensitive lupus erythematosus. *J. Invest. Dermatol.* 100 (1993), 58S–68S.
59 Norris, P.G., J.N.W.N. Barker, M.H. Allen, K.M. Leiferman. D.M. McDonald, D.O. Haskard, J.L.M. Hawk: Adhesion molecule expression in polymorphic light eruption. *J. Invest. Dermatol.* 99 (1992), 504–508.
60 Norris, P.G., J.L.M. Hawk: Succesful treatment of severe polymorphous light eruption with azathioprine. *Arch. Dermatol.* 125 (1989), 1377–1379.
61 Norris, P.G., J.L.M. Hawk: Polymorphic light eruption. *Photodermatol.* 5 (1990), 186–191.
62 Norris, P.G., J. Morris, D.M. McGibbon, A.C. Chu, J.L.M. Hawk: Polymorphic light eruption: an

immunopathological study of evolving lesions. *Br. J. Dermatol.* 120 (1989), 173–180.

63 Norris, P.G., S. Thomas, R. Poston, J.L.M. Hawk, D.O. Haskard: Expression of endothelial leucocyte adhesion molecule-1 (ELAM-1) and intercellular adhesion molecule-1 (ICAM-1) in the evolution of cutaneous inflammation: a comparison of ultraviolet-B (UVB) erythema and delayed type hypersensitivity (DTH). *Br. J. Dermatol.* 122 (1990), 274–275.

64 Orr. P.H., A.R. Birt: Hereditary polymorphous light eruption in Canadian Inuit. *Int. J. Dermatol.* 23 (1984), 472–475.

65 Ortel, B., H. Tanew, W. Hönigsmann, K. Wolff: Polymorphous light eruption. Action spectrum and protoprotection. *J. Am. Acad. Dermatol.* 14 (1986), 748–753.

66 Ortel, B., D. Wechdorn, A. Tanew, H. Hönigsmann: Effect of nicotinamide on the phototest reaction in polymorphous light eruption. *Br. J. Dermatol.* 118 (1988), 669–673.

67 Petzelbauer, P., M. Binder, P. Nikolakis, B. Ortel, H. Hönigsmann: Severe sun sensitivity and the presence of antinuclear antibodies in patients with polymorphous light eruption-like lesions. A forme fruste of photosensitive lupus erythematosus? *J. Am. Acad. Dermatol.* 26 (1992), 68–74.

68 Plewig, G., E. Hölzle, P. Lehmann: Phototherapy for photodermatoses. In: *Therapeutic Photomedicine*, (1986), pp. 254–264. Eds.: H. Hönigsmann, G. Stingl, Karger, Basel.

69 Przybilla, B., A. Galosi, M. Heppeler, T. Ruzicka, J. Ring: Polymorphous light eruption: eliciting and inhibiting wavelengths. *Acta Dermatovenereol. (Stockh.)* 68 (1987), 173–176.

70 Przybilla, B., M. Heppeler, T. Ruzicka: Preventive effect of an *E. coli*-filtrate (Colibiogen) in polymorphous light eruption. *Br. J. Dermatol.* 121 (1989), 229–233.

71 Raffle, E.J., T.M. MacLeod, F. Hutchinson: *In vitro* lymphocytic studies in chronic polymorphic light eruption. *Br. J. Dermatol.* 89 (1975), 143–148.

72 Rasch, C: Om et polymorft (erythematost, vesikulost og ekzematoidt) lysudslet. *Hospitalstid* 43 (1900), 478–480.

73 Ruzicka, T., B. Przybilla: Eicosanoid release in polymorphous light eruption; selective UV-A-induced LTB4 generation by peripheral blood leukocytes. *Skin Pharmacol.* 1 (1988), 186–191.

74 Sheridan, D.P., P.R. Lane, J. Irving, M.J. Martel, D.J. Hogan: HLA typing in actinic prurigo *J. Am. Acad. Dermatol.* 22 (1990), 1019–1023.

75 Sontheimer, R.D., J.R. Thomas, J.N. Gilliam: Subacute cutaneous lupus erythematosus: a cutaneous marker for a distinct lupus erythematosus subject. *Arch. Dermatol.* 115 (1989), 1409–1415.

13: Photoallergy

Günter Goerz, Hans F. Merk & Erhard Hölzle

Introduction

From the clinical point of view allergic reactions are divided into four categories: Types I–IV according to Coombs and Gell [3]. The pathology of types I–III is mainly dependent on antibody-dependent reactions. Type IV reactions are induced by specific T lymphocytes and their dermatological manifestations are allergic contact dermatitis or a granulomatous reaction, e.g. tuberculin type reaction [2, 5, 25].

Photoallergic reactions however require not only the allergen – a chemical substance – but also light (visible light or UV radiation) for their manifestation. Consequently, photoallergic reactions involve only the areas of the skin which are uncovered and exposed to light.

The chemical substances to be considered as potential photoallergens must share a special property: absorption of light or UV irradiation at particular wavelengths. These 'photosensitizers' are capable of triggering not only photoallergic but also phototoxic reactions. The latter group of reactions is well known in the field of photochemotherapy. A photosensitizer, e.g. 8-methoxypsoralen, accumulates to a threshold concentration within the skin and can be stimulated by light (UVA) at a critical energy level to induce an inflammatory reaction. Oxygen radicals and metabolites or other mediators are liberated in the course of the phototoxic reaction which is triggered by an O_2-independent energy transfer and direct reaction with DNA. In a second type of phototoxic reaction induced by light-activated porphyrins, O_2 is used to produce oxygen metabolites and other damaging radicals (Plate 13.1, facing p. 180).

When evaluating the literature [4, 6, 8–14, 17, 18, 21–23, 28, 32] it is extraordinarily difficult to categorize photoallergic reactions according to the Coombs–Gell classification of allergic reactions. Among the photoallergic reactions a dominance of Type IV reactions exists whereby the determining factor in most cases is whether the photoallergen is an exogenously applied compound, such as salicylanilide, buclosamide and UV filter or an orally or parenterally administered substance, such as chlorothiazide, phenothiazine or quinidine sulfate. In most of the cases published to date, the reaction was first seen 24 h after photoprovocation and oral or parenteral exposure and must therefore be classified as a delayed type.

These considerations together with findings published in the literature [4, 6, 8–14, 17, 18, 21–23, 28, 32] lead us to conclude that photoallergens are predominantly those chemicals that are capable of eliciting and sustaining a Type IV reaction or a photoallergic contact dermatitis.

In the past, numerous chemical substances have been 'blacklisted' as photoallergens, although judgment has often been based on unproved, anecdotal reports from the literature (Table 13.1). It is thanks to the 'Deutschsprachige Arbeitsgemeinschaft für Photopatch-Testung' (DAPT) – an association of 48 dermatologic centers in Austria, Switzerland and Germany – that these substances have now been critically examined and tested [9, 11]. From the original test series of 32 substances (Table 13.2), 11 have been eliminated on the basis of their lack of relevance to the disorders under discussion (Table 13.3). These are: tiaprofenic acid, musk mix, perfume mix, 6-methyl cumarin, furosemide, cyclamate, saccharine, wood tar, rosin, tolbutamide and thiourea. In contrast, the protective UV filters used in sunscreen agents and for

Table 13.1 Photoallergens. Substances which lead to photoallergic or phototoxic reactions by additional adequate light exposure after (par)enteral or local application

Drugs for the treatment of psychiatric disorders and antihistamines:
Chloropromazine, Promazine, Triflupromazine, Trimeprazine, Levopromazine, Diethazine, Promethazine, Ethopromazine, Mepazine, Thioridazine, Perazine, Prochloroperazine, Thiopropazat, Prothipendyle, Isothipendyle, Chloroprothixen, Imipramin, Karbamazepin, Amitriptylin, Chlorodiazepoxid

Sulfonamids (chemotherapeutics, antidiabetics and diuretics):
Sulfanilamid, Sulfathiazole, Sulfoxazole, Sulfaguanidine, Sulfaphenazole, Sulfamethoxypyridazine, Sulfamonomethoxine, Carbutmide, Tolbutamide, Metabutamide, Chloropropamide, Chlorothiazid, Hydrochlorothiazide, Chlorothalidon, Clopamide, Zyklopenthiazide, Furosemide, Thiabutazide, Quinethazone

Optical clearifier (Stilbenc)

Nalidixine-acid

Artificial sweeteners (Cyclamate, Saccharine)

Psoralen

Laxatives (Triazetyldiphenylisatin)

Halogenic sylicylanilids (disinfectants, antimicrobials, antifunguals):
Tetrachlorsalicylanilids, Tribromsalicylanilids, Dibromsalan, Fenticlor, Buclosamide, Bithionol, Multifungin, Mycanodine

Oral contraceptives (oestrogens)

UV filter:
p-Aminobenzoic acid, Digalloyl-Trioleat, Benzocain,

Griseofulvin

Quinidine

Olaquinalox

other purposes have recently received considerably more attention as contact allergens and also as photoallergens. The DAPT has extended the list for photoallergic testing to cover five light-filtering substances (Table 13.3). To present a better overall view we have differentiated further (Table 13.4) between predominantly or exclusively internally in comparison to primarily externally applied substances – drugs, cosmetics and food ingredients.

For historical reasons we have summarized the substances in Table 1 that have been presented in numerous reviews as photoallergic or phototoxic substances [4, 6, 8–14, 17, 18, 21–23, 28, 32]. Medicines with phototoxic and/or photoallergic effects are rarely encountered nowadays since they have either been replaced by more effective drugs or simply been taken off the market, e.g. the laxatives triacetyldiphenyl-isatine [15, 16] or the larger number of halogenated salicylanilides commonly used in antiseptics or in cosmetics.

Table 13.2 Photopatch test set of the Austro-Swiss-German task force group (DAPT)

Tetrachlorosalicylanilide
Bromsalicylanilide
Tribromsalicylanilide
Buclosamide
Fenticlor
Hexachlorophene
Bithionol
Triclosan
Sulfanilamide
Chloropromazine
Promethazine
Carprofene
Tiaprofenic acid*
Quinidine sulfate
Musk ambrette
Musk mix*
Perfume mix*
6-Methycoumarine*
p-Aminobenzoic acid
Hydrochlorothiazide
Furosemide*
2-Hydroxy-4-methoxybenzophenone
3-(4-Methylbenzylidene)-camphor
4-Isopropyldibenzoylmethane
Cyclamate*
Saccharine*
Wood tar*
Colophony*
Balsam of Peru
Compsitae mix
Tolbutamide*
Thiourea*

*Marked substances have not been considered in the patch tests presented in Table 13.3.

Table 13.3 Photopatch-test reactions to photo sensitizers

	Photopatch test		Results of photopatch testing		
	(%)*	(W/P)†	Photoallergic (%)	Phototoxic (%)	Nonclassifiable (%)
Tetrachlorsalicylanilide	0.1	P	0.5	0.5	2.3
Bromalicylchloranilide	1.0	P	–	1.0	0.5
Tribromsalicylanilide	1.0	P	–	0.6	0.9
Buclosamide	5.0	P	0.1	1.1	0.8
Fenticolor	1.0	W	1.6	3.0	6.5
Hexachlorophene	1.0	P	0.2	3.7	2.0
Bithionol	1.0	P	0.3	0.6	1.8
Triclosan	5.0	W	–	0.4	0.4
Sulfanilamide	0.1	W	0.4	6.0	4.2
Chlorpromazine	1.0	W	0.7	19.8	14.1
Promethazine	5.0	W	1.3	7.4	5.3
Carprofen	1.0	W	0.1	0.5	0.5
Quinidine sulfate	1.0	Ethanol	0.5	0.8	0.9
Musk ambrette	0.5	W	–	0.2	0.5
Hydrochlorothiazide	25.0	P	–	1.4	2.8
Balsam of Peru	6.5	Ethanol	0.1	0.9	0.8
Composite mix	5.0	Ethanol	0.1	0.4	1.1
p-Aminobenzoic acid	2.0	P	0.7	0.7	1.8
2-Hydroxy-4-methoxybenzophenone	5.0	P	–	0.7	0.8
3-(4-Methylbenzylidene)-camphor	5.0	P	0.9	1.9	2.7
4-Isopropyldibenzoylmethane	5.0	P	0.9	1.9	2.7

Additionally five other ointments protecting the skin against light (from which there are still no results) have been tested: 4-tert-butyl-4-methoxydibenzylmethane; 2-ethylhexyl-*p*-dimethylaminobenzoate; 2-ethylhexyl-*p*-methoxyzinamate; phenylbenzinmidazol-sulfonic acid; *p*-methoxycinnamic acid.
* Percentage concentration. † W, water; P, petroleum.

There are a few drugs of minor significance in human medicine which do however play an important role in veterinary medicine or in professional animal breeding (chloropromazine, Olaquinalox). Workers in certain professions, especially agriculture, e.g. farmers, can therefore still become exposed to such drugs and are able to develop the corresponding photodermatitis [26, 27].

Photoallergic reactions

Clinical findings

As already noted, under certain conditions the effects of UV radiation or visible light on photosensitizers induce their transformation into a photoallergen and the formation of specific T lymphocytes or antibodies. On repeated exposure to the photosensitizer and light of appropriate wavelengths an inflammatory reaction occurs which is restricted to uncovered skin areas.

Case report 1. Quinidine sulfate-induced photodermatitis [20, 33]

An 83-year-old white man who had been suffering from arrhythmic cardiac insufficiency was treated for several years with various heart medications, diuretics and antiarrhythmics such as digitoxine, isosorbidinitrate, furosemide, spironolactone and quinidine sulfate. Several months prior to hospitalization, increased sensitivity to UV radiation, accompanied by a dermatitis in all uncovered skin

Table 13.4 From the Austro-Swiss-German task force group [9, 11] recommended series of test arranged substances by 'oral' or 'topical' application

Predominantly orally applied compounds	
Drugs	*Indication*
Hydrochlorothiazide	Diuretics
Quinidinesulfate	Antiarrhythmics
Carprofen	Antiphlogistics (NSAID)
Chlorpromazine	Sedatives (tranquilizers)
Promethazine	Sedatives
Sulfanilamide	Sulfonamide
Bithionol	Chemotherapeutic, anthelmintic
Predominantly topically applied compounds	
Substance	*Indication*
Tetrachlorsalicylanilide	Antiseptic
Monobromsalicylchloranilide	
Tribromsalicylanilide	
Hexachlorophene	
Triclosan	
p-Aminobenzoic acid	
Fenticlor	Antifungals
Buclosamide	
Musk ambrette	Fragrances
Balsam of Peru	
Composite mix	Component of plants
2-Hydroxy-4-methoxybenzophenone	UV filter
3-(4-Methylbenzyliden)-camphor	
4-Isopropyldibenzoylmethane	
4-tert-Butyl-4-methoxydibenzoylmethane	
2-Ethylhexyl-*p*-dimethylaminobenzoate	
2-Ethylhexyl-*p*-methocyzinamate	
Phenylbenzimidazolsulphon acid	
p-Methoxycinnamonacid-isoamylate	

areas (see Plate 13.1, facing p. 180), occurred. Application of 10 J UVA/cm^2 on unexposed areas induced a skin photocontact dermatitis (Plate 13.2, facing p. 180) over 72 h which diminished after a long period of approximately 8 weeks; at the time of testing the patient was still treated with all the drugs cited above. The patient was treated with disopyramide instead of quinidine sulfate, an antiarrhythmic drug chemically unrelated to quinidine. Four weeks later the skin lesions and the photosensitivity disappeared (Plate 13.3, facing p. 180).

In conclusion the patient suffered from a photoallergic dermatitis induced by quinidine sulfate.

Case report 2. Neotri®-(chlorothiazide)-induced photodermitis [1, 24]

A 75-year-old white man who had suffered for 3 years from light sensitivity and dermatitis in all uncovered skin areas (Plate 13.4, facing p. 180). Edemas of the legs were treated for 6 years with Neotri®: 60 mg triamterene and 20 mg Xipamide

(4-chloro[1]-5-sulfamoy1-2′,6′-salicyloxylide). After cessation of the drug, skin lesions and photosensitivity disappeared (Plate 13.5, facing p. 180). Photopatch tests did not reveal a photoallergen. The patient was further treated with Neotri® and test areas on the back were exposed for 1, 3 or 6 h respectively with 5 J UVA/cm^2 after intake of the tablet. Twenty-four hours later in the test area – exposed 3 and 6 h after challenge – he developed a dermatitis (Plate 13.6, facing p. 180). Xipamide – a substituted salicyloxyilide is a well known phototoxic or photoallergic compound and therefore we assume that this constituent in Neotri® was the causal photoallergic substance.

In conclusion our diagnosis was allergic photodermatitis induced by Xipamide (salicyloxylide derivative).

Case report 3. Hydrochlorothiazide-induced photodermatitis [1, 24]

A 90-year-old white man noticed 24 h after sun exposure – nearly 1 to 2 h duration – a diffuse erythema followed by desquamation at all light-exposed areas. The skin (Plate 13.7, facing p. 180) on the face, neck, backs of fingers or hands, and forearms showed an acute dermatitis. For 3 months the patient had been treated with the diuretic Moduretik® (containing 5 mg amiloride, a chloropyrazinecarboxamide derivate, and 50 mg hydrochlorothiazide). Phototesting with UVA and UVB was without pathological findings. In addition photopatch testing with various photosensitizers (see Table 10.2) were negative. The patient was exposed with 50 mg hydrochlorothiazide (6-chloro-3,4-dihydro-2h-1,2,4,-benzothiadiazine-7-sulfonamide-1,1-dioxide) and the test area was irradiated 1, 2 and 5 h later with 10 J UVA/cm^2. Twenty-four hours later a dermatitis could be demonstrated in the test areas.

In conclusion the photo testing after hydrochlorothiazide exposure proved the diagnosis: hydrochlorothiazide-induced photodermatitis.

Table 13.5 Allergic contact dermatitis due to sunscreens in an EPP patient (Case 4). Results of patch testing

Substance	24 h	48 h	72 h
Carrot Sun Milk®	++	++	+
Contralum Ultra® cream	+	++	+
Ilrido Plus 15® cream	++	++	+
Isopropyldibenzoylmethane	+	+	+
Methoxydibenzoylmethane	+	+	+
Methylbenzylidene-camphor	++	++	+

Case report 4. Allergic contact dermatitis to sunscreens in a patient with erythropoietic protoporphyria (EPP) [7, 19, 29–31]

A 25-year-old white patient had been suffering from phototoxic reactions since he was 2 years of age. An increased porphyrin content in red cells of 1812 μg/dl (upper normal limit: 60 μg/dl) revealed the diagnosis of EPP. The patient was treated orally with β-carotene and topically with various sunscreen preparations. In summer 1991 a contact dermatitis developed after application of Carrot Sun Milk® on all sunscreen-treated skin areas. The later application of other preparations such as Contralum Ultra® and/or Ilride Plus 15® induced the same skin lesions (Plate 13.8, facing p. 180). Patch testing (Table 13.5) showed a contact allergic reaction to all three sunscreens which had been used (Carrot Sun Milk®, Contralum Ultra Plus 15® cream (Plate 13.9, facing p. 180) and the corresponding UVB filter: isopropyldibenzoylmethane and methoxydibenzoylmethane). Irradiation with UVA (photopatch test) did not increase the reactions induced from these contact allergens.

In conclusion the patient suffering from EPP since birth developed a contact dermatitis to UVA and UVB filters.

Diagnostic procedures

The diagnosis of allergic reactions, especially of drug-induced allergic reactions, is often problematic. Patch or scratch tests with systemically applied drugs are often disappointing.

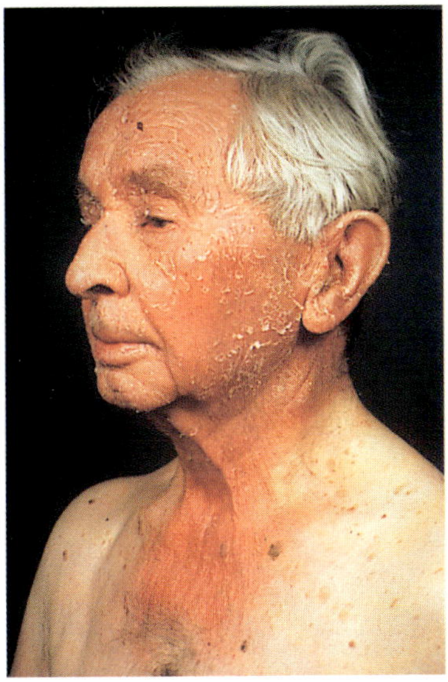

Plate 13.1

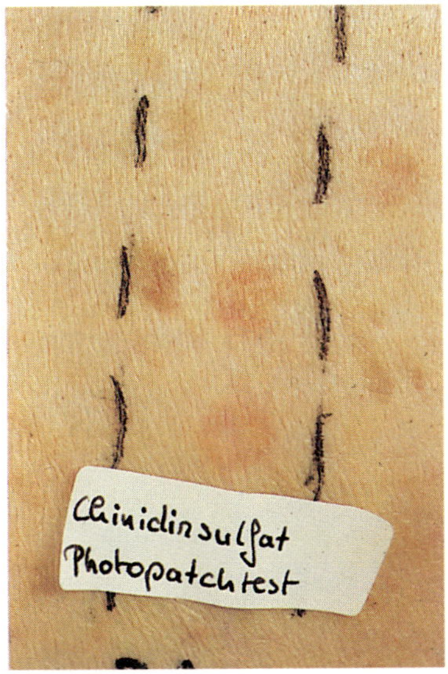

Plate 13.2

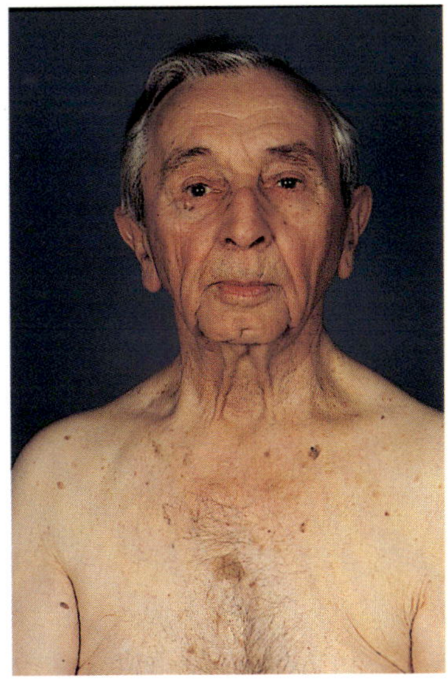

Plate 13.3

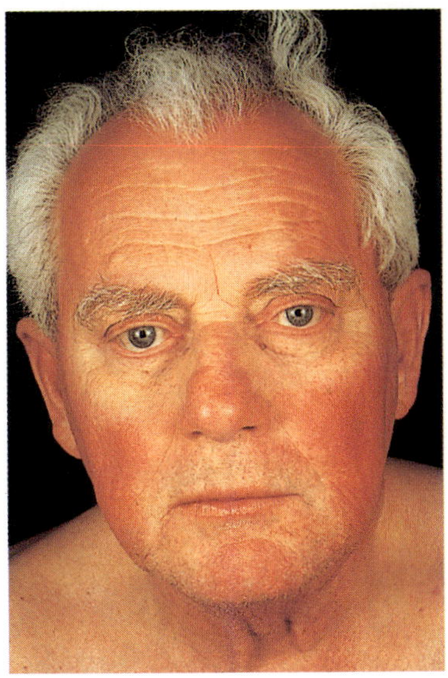

Plate 13.4

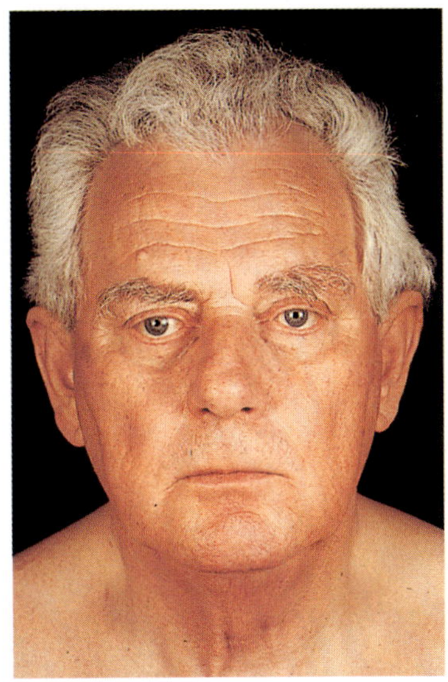

Plate 13.5

Plate 13.1 Quinidine sulfate-induced allergic photodermatitis. Acute dermatitis in uncovered skin areas of face and neck (Case 1).

Plate 13.2 UVA irradiation: 10 J UVA/cm^2 on the test area after intake of quinidine sulfate provoked a photoallergic reaction 24 h later in the exposed skin (Case 1).

Plate 13.3 The same patient as in Plate 13.1: 4 weeks after cessation of the quinidine treatment (Case 1).

Plate 13.4 Neotri® (Xipamide plus Triamterene) induced allergic photodermatitis. Acute dermatitis in chronic light-exposed areas of the patient's face (Case 2).

Plate 13.5 Disappearance of the skin lesions 1 week after cessation of therapy with the diuretic Neotri® (Case 2).

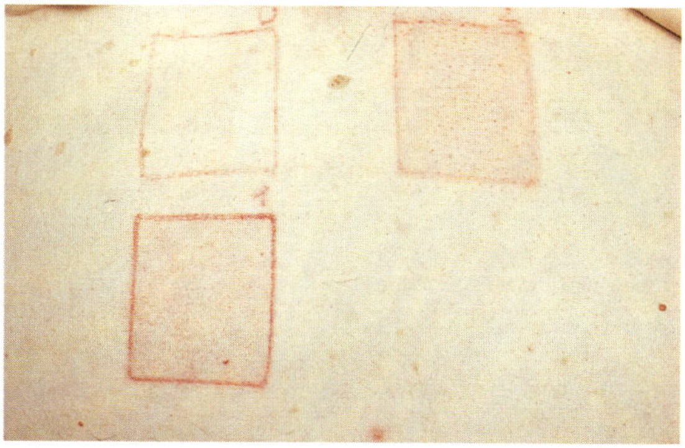

Plate 13.6

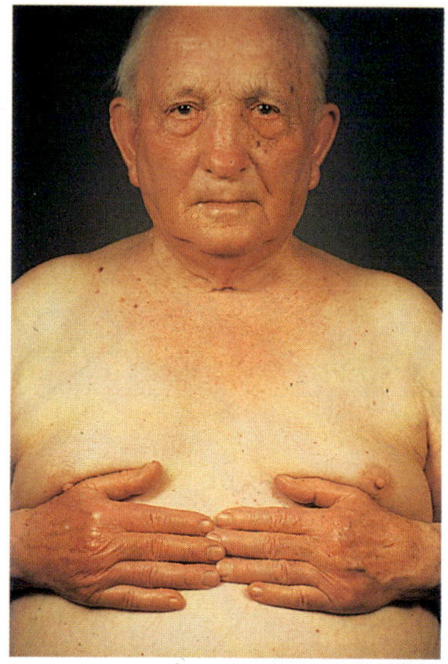

Plate 13.7

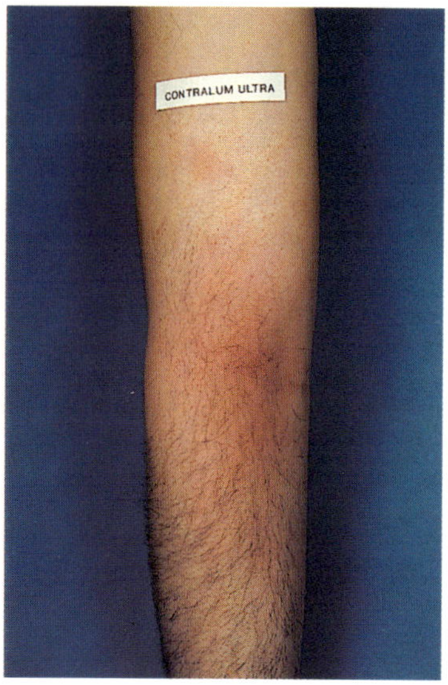

Plate 13.8

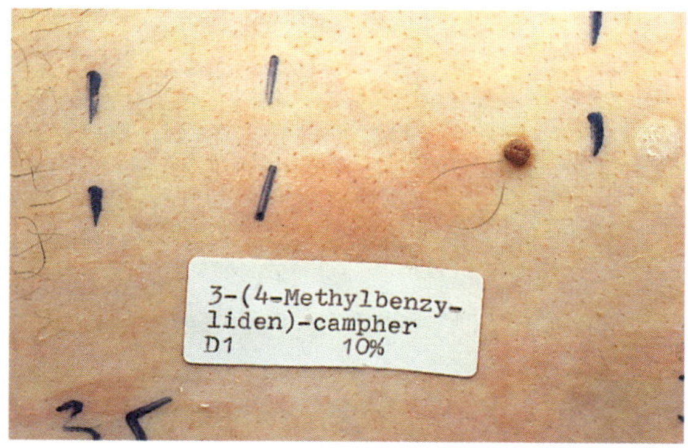

Plate 13.9

Plate 13.6 Phototests. Top left, 5 J UVA/cm^1 1 h after. Top right, 5 J UVA/cm^2 3 h after. Bottom, 5 J UVA/cm^2 6 h after Neotri® intake (Case 2).

Plate 13.7 Hydrochlorothiazide-induced allergic photodermatitis. Chronic dermatitis in light-exposed uncovered skin (Case 3).

Plate 13.8 Allergic contact dermatitis to sunscreens in an EPP patient (Case 4). Contralum Ultra: local application, D1 + reaction.

Plate 13.9 Allergic contact dermatitis to sunscreens in an EPP patient (Case 4). Methylbenzyliden-camphor: patch test, D1 + reaction.

Diagnostic procedures for light-induced toxic or allergic skin diseases are complicated by the necessity of electromagnetic waves to promote the manifestation of these skin reactions. Appropriate wavelengths and adequate energy levels play a crucial role. It is impossible to trigger the skin reactions either by light or the drug alone.

Photoallergic or phototoxic reactions can only be elicited when a photosensitizer is acting in combination with subthreshold light stimulation. Therefore in addition to irradiation either epidermal application as in patch tests, or oral (or parenteral) administration of the drug in question (as in drug challenge tests) must be included in the trials. In the latter case the time schedule of photoprovocation should be based on the pharmacokinetics of the drug under investigation.

Since the pioneer studies of Schulz *et al.* [32], Ippen [13] and Epstein *et al.* [4] a wide variety of test systems has been developed, which differ to a considerable extent. In the USA, in Scandinavia, and also among German-speaking dermatologists (DAPT), working groups have been set up to try to standardize photopatch testing: this is an ongoing process [9, 11, 14, 21, 22].

Principle of photopatch testing

Photocontact allergy (photoallergic contact dermatitis)

As shown in Tables 13.3 and 13.4, the chemical substances are placed in their corresponding vehicles using Finn chambers on the patient's back in the manner employed for patch testing. The patch tests must, however, each be applied twice, on the right and the left side of the back. After 24 h the tests are removed and the results evaluated. One set is followed as a control for 72 h to evaluate plain contact reactions, the second set is subjected to UVA irradiation (320–400 nm) and observed for a further 72 h.

The mode of photopatch testing proposed by DAPT [11], Hölzle *et al.* [9] or Lehmann [18] is particularly useful, yielding reproducible findings, and will be described in detail (Table 13.6).

The results are evaluated in accordance with the characteristics used in patch testing but with minor modifications (see below).

Table 13.6 Methods of photopatch testing

Light sources	Polychromatic UVA, 320–400 nm
Test material	Finn chambers
Test substances	Recommended commercially available chemicals in recommended concentrations and vehicles (Tables 13.3 & 13.4) are applied to unaffected skin on the patient's back
The tests are left on the skin for 24 h, one side serving as a patch test control. The second test series on the other side of the back is exposed after 24 h. Readings are performed immediately, then 24, 48 and 72 h after irradiation	
Light exposure	Each test area is subjected to 10 J UVA/cm^2 using one of the light sources listed in Table 13.3

Photopatch tests must be controlled immediately or 20 min after UVA irradiation. Subsequent readings are taken 24, 48 and 72 h later, and if possible over a longer time period, especially when puzzling or doubtful reactions have been observed. It may be that the patient to be tested is extremely sensitive to light or UVA. In such a case a preliminary test must be performed to establish the UVA threshold dose. For subsequent exposure sub-threshold doses should be used.

Evaluation

According to international agreements concerning the evaluation of patch tests the reactions are categorized as follows:

+	= erythema and flat papules or infiltrations
++	= erythema and papular vesicles
+++	= erythema, papules and blisters

The toxic irritative reaction (IR) must be differentiated from an allergic test reaction. IR is characterized by a sharply demarcated erythema sometimes accompanied by necrosis.

Within the framework of the various photopatch test groups however, the following gradings have been proposed:

+	= erythema
++	= erythema with infiltrations
+++	= erythema with infiltrations, papules and small vesicles
++++	= erythema with blisters or erosions

It is confusing and a little unfortunate that varying measures have had to be used in the tests included in the patch and photopatch test series. However, we believe that it is impossible to simplify or standardize the tests any further without introducing errors. Whether the reading pattern (see below) will produce new approaches and perhaps more compatible results remains to be seen.

Photopatch test reaction pattern

As discussed above, photopatch tests are evaluated in a similar way to patch tests with some modifications (see Table 13.5). Hölzle *et al.* [10] and Neumann *et al.* [22] observed the following peculiarities during the evaluation of tests from 2114 patients included in the DAPT survey: the symptoms – erythema, infiltration, papulovesicles and blisters – represent four parameters to be taken into account in photopatch testing. When the appearance of one particular symptom is noted throughout the observation period a typical pattern can be obtained for a given substance.

First. Decrescendo reaction pattern type. This is comparable to a toxic irritative skin reaction in that it occurs in positive photopatches with furosemide (0.5% in 0.9% NaCl solution). As in the case of toxic irritative patch test reactions, a sharply delineated erythema is seen within the test area. However, since there are no subsequent inflammatory reactions, a pharmacologically induced vascular response could be assumed. Fig. 13.1 demonstrates such a pattern found in a few cases of patients tested with furosemide in a multicenter study [22].

In conclusion, we recommend that furosemide (and perhaps other diuretics, such as thiazides) should be tested preferentially by systemic application combined with a corresponding photoprovocative test, since on one hand these drugs are systemically applied and on the other hand local phototoxic irritations cannot be excluded after topical use of furosemide in phototesting.

Second. Combined reaction pattern type. The pattern consists of an immediate erythema on day 1 followed by infiltrations and vesicles on days 2, 3 and 4. This type of reaction was preferentially found after phototesting of tetrachlorosalicylanilides (0.1% in petroleum) as demonstrated in Fig. 13.2.

According to Neumann *et al.* [22] this pattern is found (apart from salicylanides) in related disinfectants, *p*-aminobenzoic acid, musk ambrette or perfume mixture. The pattern can be interpreted as an initial phototoxic irritation of the skin followed by a photoallergic reaction.

Third. Plateau reaction pattern type. As demonstrated in Fig. 13.3 this type of reaction, consisting of an erythema persisting for 1–4 days, often combined with infiltration but, rarely with vesicles or with bullae, was very frequently found in this multicenter study [22].

This 'plateau type pattern' was seen preferentially after photopatch testing with chlorpromazine or tiaprofenic acid.

These reactions arise frequently without previous exposure, so that an allergic reaction can be excluded in most cases and in additon because the observation periods are too short for a primary sensitization. The marked erythematous reaction can be considered to be a phototoxic irritative or a pharmacological reaction, while the infiltrations which occur to almost the same degree between day 2 and day 4 are extremely difficult to interpret. Owing to the unacceptable proportion of false-positive, probably phototoxic, reactions that occurred during the tiaprofenic acid tests, this non-steroidal antiphlogistic drug has had to be eliminated from the photopatch test series [9, 11, 22].

Fourth. Crescendo reaction pattern type. This

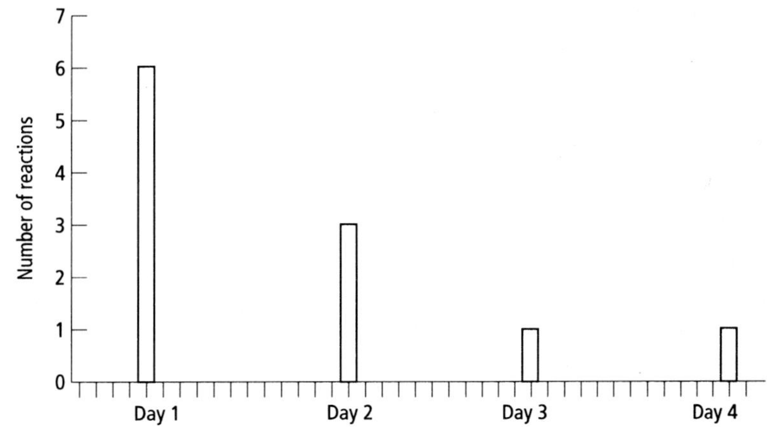

Fig. 13.1 Decrescendo reaction pattern type. Erythema reaction with a maximum at day 1 in patients tested with furosemide (1% in petroleum) and UVA irradiation (for details see text and Neumann *et al*. [22]).

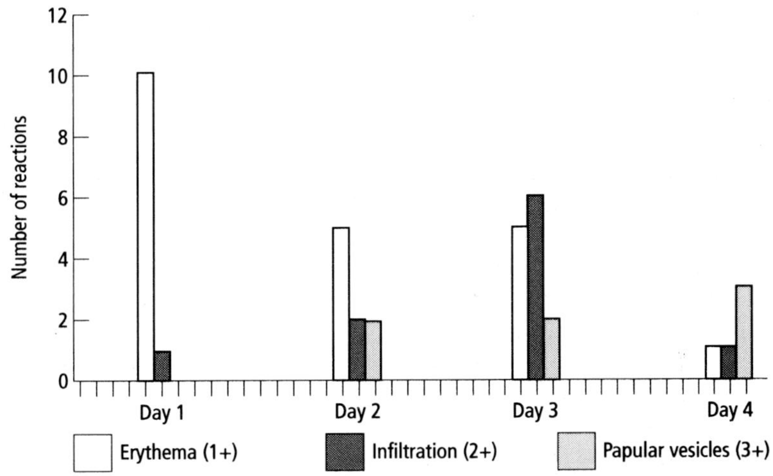

Fig. 13.2 Combined reaction pattern type. Photopatch test with tetrachlorosalicylanide as an example of this combined pattern (for details see text and Neumann *et al*. [22]).

pattern characterizes the typical allergic and photoallergic contact dermatitis (Fig. 13.4). The erythema is not the predominant sign, these are papules and vesicles. In all examples of a photoallergic contact dermatitis there is no evidence of a contact dermatitis on the non-irradiated patch test site. In recent series fentichlor, 2-hydroxy-4-methoxybenzophenone, 4-isopropyldibenzoylmethane and *p*-methoxyisoamylcinnamate have been highlighted as the most frequently found photocontact allergens. Recent studies have shown that many of the sunscreen preparations commercially available are potent contact allergens and in addition potent photocontact allergens.

In conclusion it is assumed from the above that the differentiation of the symptoms in phototesting – erythema, infiltration, papular vesicles and blisters in correlation with the time course (day 1–4) – is very helpful for interpretation or more specifically for differentiation between photoallergicand phototoxic- or pharmacologically induced reactions.

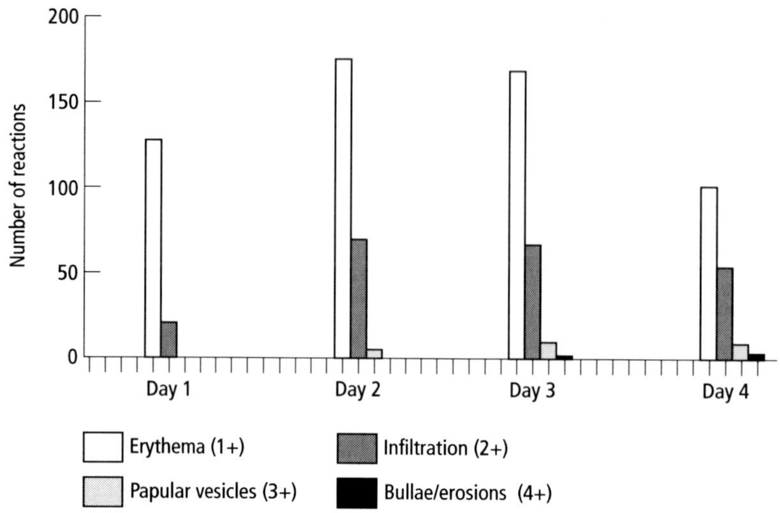

Fig. 13.3 Plateau reaction pattern type of tiaprofenic acid. An erythema persisting for 1–4 days is combined with infiltrations in many patients and with vesicles in some patients (for details see text and Neumann *et al.* [22]).

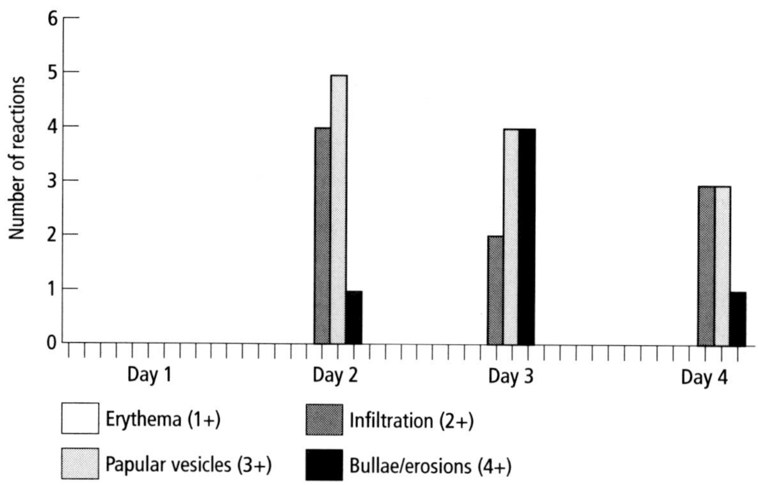

Fig. 13.4 Crescendo reaction pattern type. Typical patterns of an allergic contact or photocontact dermatitis respectively, illustrated in patients with photoallergic contact dermatitis dermatto, the sunscreen constituent *p*-methoxy-isoamylcinnamate (10%) in petrolatum [22].

Systemic photoprovocation

Systemic exposure to the photosensitizer and subsequent light exposure

With regard to systemically administered photoallergens, it is hard to imagine how epidermal application of the allergen could adequately reproduce the photoallergic reaction.

Exposure

The photosensitizer in question – usually a drug – must be administered to the patient either orally or parenterally. In general the amount given is two or three times the single dosage.

Light provocation

Before and after systemic application of the drugs, several 5 × 5 cm test areas on the patient's back are irradiated with 19 J UVA/cm^2 (320–400 nm). The time frame of irradiation depends on the pharmacokinetics of the drug. In general, irradiations before, then 1, 3 and 6 h after medication are performed.

Reading

Immediately and after 24, 48, 72 h or longer, the test reactions are read and evaluated according to the scale cited above. In uncertain cases or for final conformation of a diagnosis it may be necessary to take a biopsy for histopathological analysis.

References

1 Addo, H.A., I. Ferguson: Thiazide-induced photosensitivity. *Br. J. Dermatol.* 116 (1987), 749–760.

2 Bos, J.D., M.L. Kapsenberg: The skin immune system: progress in cutaneous biology *Immunol. Today* 14 (1993), 75–78.

3 Coombs, R.R.A., P.G.H. Gell: Classification of allergic reactions responsible for clinical hypersensitivity and disease. In: *Clinical Aspects of Immunology*, 3rd edn, (1975), pp. 761–781. Eds.: P.G.H. Gell, R.R.A. Coombs, P.J. Lachmann, Blackwell Scientific Publications, Oxford.

4 Epstein, J.H., K.D. Wuepper, H.I. Maibach: Photocontact dermatitis to halogenated salicylanilides and related compounds. *Arch. Derm.* 97 (1967), 236–244.

5 Fisher, A.A.: *Contact Dermatitis* (1986). Lea & Febinger, Philadelphia.

6 Giudichi, P.A., H.C. Maguire: Experimental photoallergy to systemic drugs. *J. Invest. Dermatol.* 85 (1985), 207–217.

7 Goldermann, R., E. Vadarman, N. Neumann, K. Scharfetter-Kochanek, G. Goerz: Contact dermatitis from UVA and UVB filters in a patient with erythropoietic protoporphyria. *Contact Dermatitis* 28, (1993), 300–301.

8 Harber, L.C., H. Harris, R. Baer: Photoallergic contact dermatitis. *Arch. Dermatol.* 94 (1966), 255–262.

9 Hölzle, E., die Mitglieder der Deutschsprachigen Arbeitsgemeinschaft Photopatch-Test: Photopatch Test: Ergebnisse der multizentrischen Studie. *Akt. Dermtol.* 17 (1991), 117–123.

10 Hölzle, E., N. Neumann, G. Goerz: Photoallergie – Mechanismus und Diagnostik. In: *Jahrbuch der Dermatologie 1992/93, Licht und Haut*, (1992), pp. 135–142. Eds.: E. Macher, G. Kolde, E.B. Bröcker, Biermann-Verlag, Zülpich.

11 Hölzle, E., N. Neumann, B. Hausen, B. Przybilla, S. Schauder, H. Hönigsmann, A. Bircher, G. Plewig: Photopatch testing: the 5-year experience of the German, Austrian and Swiss Photopatch Group. *J. Amer. Acad. Dermatol.* 25 (1991), 59–68.

12 Hruza, L.L., A.P. Pentland: Mechanisms of UV-induced inflammation. *J. Invest. Dermatol.* 100 (1993), 35S–41S.

13 Ippen, H.: Arzneimittelausschläge durch Malariamittel, Phenothiazine, Sulfonamide, Tetracycline und andere Medikamente, die zu Photodermatosen führen können. *Zschr. Haut-Geschl.-Krkh.* 42 (1967), 47–52.

14 Jansen, C.T., G. Wennersten, I. Rystedt, P. Thune, H. Brodthagen: The Scandinavian standard photopatch test procedure. *Contact Dermatitis* 8 (1982), 155–158.

15 Jung, E.G.: Photoallergie durch Triacetyldiphenolisatin (TDI) I. Kasuistik. *Arch. Klin. U. Exp. Dermatol.* 229 (1967), 170–173.

16 Jung, E.G.: Photoallergie durch Triacetyldihenolisatin (TDI) II. Photochemische Untresuchungen zur Pathogenes. *Arch. Klin. U. Exp. Dermatol.* 231 (1967), 39–49.

17 Jung, E.G., E. Bohner: Lichtbiologie der Haut. In: *Normale und Pathologische Physiologie der Haut II. Handbuch der Haut- und Geschlechtskrankheiten*, (1979), pp. 459–540. Eds.: E. Schwarz, K.W. Spier, G. Stüttgen, J. Jadassohn, Ergänzungswerk I/4A, Spring-Verlag, Berlin.

18 Lehmann, P.: Photodiagnostische Testverfahren. *Akt. Dermatol.* 20 (1994), 41–46.

19 Lehmann, P., P. Scharfetter, P. Kind, G. Goerz: Erythropoetische Protoporphyrie: Synopsis von 20 Patieten. *Hautarzt* 42 (1991), 570–574.

20 Marx, J.L., B.A. Eisenstat, A.H. Gladstein: Quindine photosensitivity. *Arch. Dermatol.* 119 (1983), 39–43.

21 Menz, J., S.A. Muller, S.M. Connolly: Photopatch testing: a six year experience. *J. Am. Acad. Dermatol.* 18 (1988), 1044–1047.

22 Neumann, N.J., E. Hölzle, P. Lehmann, S. Benediter, B. Tapernoux, G. Plewig: Pattern analysis of photopatch test reactions. *Photodermatol. Photoimmunol. Photomed.* 10 (1994), 65–73.

23 Norris, D.A.: Pathomechanisms of photosensitive lupus erythematosus. *J. Invest. Dermatol.* 100 (1993), 58S–68S.

24 Robinson, H.N., W.L. Morison, H.F. Hood: Thiazide diuretic therapy and chronic photosensitivity. *Arch. Dermatol.* 121 (1985), 522–524.

25 Roitt, I.M.: *Leitfaden der Immunologie* (1993). Blackwell Wissenschaft, Berlin.

26 Satanove, A., J.S. McIntosh: Phototoxic reactions induced by high doses of chlorpromazine and

thioridazine. *JAMA* 200 (1967), 209–212.
27 Schauder, S.: How to avoid phototoxic reactions in photopatch testing with chlorpromazine. *Photodermatologie* 2 (1985), 95–100.
28 Schauder, S.: Photodiagnostik. *Z. Hautkr.* 65 (1990), 247–255.
29 Schauder, S.: *Sonnenschutzmittel: lichtfilterhaltige Hautpflege Präparate in Deutschland. Göttinger Liste 1991/92* (1991). Deutscher Apotheker Verlag, Stuttgart.
30 Schauder, S., H. Ippen: Photoallergic and allergic contact dermatitis from Dibenzoylmethanes *Photodermatol.* 3 (1986), 140–147.
31 Schauder, S. H. Ippen: Photoallergisches und allergisches Kontaktekzem durch Dibenzoylmethan-Verbindungen und andere Lichtschutz-Filter. *Hautarzt* 39 (1988), 435–440.
32 Schulz, K.H., A. Wiskemann, K. Wolf: Klinische und experimentelle Untersuchungen über die photodynamische Wirksamkeit von Phenothiazinderivaten, insbesondere von Megaphen. *Arch. Klin. Exp. Dermatol.* 202 (1956), 285–298.
33 Schürer, N.Y., E. Hölzle, G. Plewig, P. Lehmann: Photosensitivity induced by quinidine sulfate: experimental reproduction of skin lesions. *Photodermatol. Photoimmunol. Photomed.* 9 (1992), 78–82.

14: Chronic actinic dermatitis

H. du P. Menagé, Stephen M. Breathnach & John L.M. Hawk

Chronic actinic dermatitis (CAD) is an uncommon photodermatosis affecting middle-aged and elderly individuals, predominantly males. The clinical features are of a persistent, confluent eczematous or pseudolymphomatous eruption, predominantly of light-exposed sites, although in more severely affected individuals the condition may spread to affect covered sites. The eruption is induced by ultraviolet and occasionally short wavelength visible irradiation and may persist throughout the year, although exacerbations tend to occur during spring and summer. It is a chronic disease, although on occasion slow spontaneous remission may be observed. Although the pathogenesis is not yet fully understood, the immunophenotype of the infiltrate and the therapeutic response to immunomodulatory drugs suggest an immune pathogenesis, compatible with delayed-type hypersensitivity, conceivably to a cutaneous photoallergen. Many patients have associated contact and photocontact cutaneous allergies which may pre-date the onset of photosensitivity, and some have had preceding endogenous eczema; these conditions may therefore predispose to the development of the photosensitivity.

A unifying terminology

Chronic actinic dermatitis is synonymous with the *photosensitivity dermatitis/actinic reticuloid (PD/AR) syndrome*, a term coined by Frain-Bell and colleagues [33]. Understanding of the disease has been hampered by the proliferation of other diagnostic terms describing different variants of the condition; these are therefore now unified under the label 'chronic actinic dermatitis' [45], or less commonly the PD/AR syndrome. Originally proposed by Hawk and Magnus [45] the term CAD encompasses the conditions previously known as photosensitive eczema, photosensitivity dermatitis, actinic reticuloid, and persistent light reactivity [45, 85].

The phenomenon of *persistent light reactivity* was first recognised by Wilkinson in 1962 [101] and can be described as CAD developing from localized photocontact dermatitis. Sufferers gradually develop eczema on all light-exposed sites despite avoidance of the original photocontact allergen and irrespective of whether or not the affected sites have previously been exposed to allergen. Whereas photocontact dermatitis is usually UVA-induced (315–400 nm), as the condition evolves into CAD the action spectrum alters to include the UVB (280–315 nm) wavelengths as well. Topical photosensitizers reported to predispose to induce CAD include: topical sulphonamides, used extensively in the Second World War to treat infected skin conditions and withdrawn shortly thereafter; halogenated phenols, such as tetrachlorsalicylanilide, tribromosalicylanilide, fentichlor and bithionol, used as antibacterial agents in soaps, antiseptics and toiletries in the 1960s and early 1970s and since withdrawn; the fragrance musk ambrette, a synthetic musk used widely in after-shave and other preparations and still available in certain parts of the world; quinoxaline-N-dioxide in animal foodstuffs; phenothiazines; pitch; zinc pyrithione; bleaching agents [14, 15, 20, 53, 54, 101, 103, 104]; and more recently, chemical sunscreens (authors' unpublished observations). Localized persistent reactivity, which differs from true CAD, has also arisen following sensitization to epoxy resins, and the halogenated phenols multifungin and buclosomide [4, 16, 97]. It is less clear if systemic drug photosensitivity may predispose to the development of CAD. Although thiazide diuretics [87] and quinidine [64] have been reported to lead to persistent light reactivity following withdrawal of the drug, intake of medication, including potential photosensitizers, is common in the age group affected by CAD, and the phenomenon of eczematous photoallergy caused by systemic medication thus remains a disputed phenomenon.

Actinic reticuloid was the term used by Ive *et al.* in 1969 [51], who described a chronic photodermatosis of elderly men induced by UVB, UVA and occasionally visible wavelengths. It was characterized by infiltrated erythematous plaques arising on eczematous or normal skin, with histopathological features suggestive of cutaneous T-cell lymphoma and negative photopatch tests, although clinical similarity was noted to severe cases of photoallergic contact dermatitis. *Photosensitive eczema*, described by Ramsay and Kobza-Black in 1973 [86], refers to an eczematous eruption with features of actinic reticuloid in some patients, again with no preceding photoallergy; it differs from actinic reticuloid in the action spectrum, which is in the UVB region alone. *Photosensitivity dermatitis* describes a similar eczematous eruption but with photosensitivity often extending to include also the UVA wavelengths [33].

Transition has been described between the eczematous and pseudolymphomatous variants of CAD and vice versa [33, 44], and from persistent light reactivity to actinic reticuloid and vice versa [48], while other patients have been described with the histological features of eczema and the photobiological abnormalities associated with actinic reticuloid [45]. Thus it appears certain that these conditions are all variants of the same syndrome, now unified under the term CAD.

Clinical aspects

Clinical features

CAD is an uncommon condition with an incidence of only 1:6000 in the Tayside region of Scotland [29]. Males are more often affected, only 10% [29] to 22% [24] of sufferers being women, although female prevalence appears to be increasing [24]. Patients are generally middle-aged to elderly, with an average age of 65 [24]; sufferers under 50 are exceptional. No entirely satisfactory explanation has been proposed to explain this age incidence, although barrier function against antigen penetration and subsequent antigen clearance may be impaired in aged skin [76]. All races are affected, the condition having been reported in Japanese [48], Afro-Caribbeans [59] and Asians [24] in addition to the more commonly described Caucasians. Dark-skinned individuals comprised up to 10% of 86 patients in one study from a city centre hospital [24], their profuse melanin thus not affording effective protection in the presence of such extreme sensitivity to ultraviolet radiation. Familial incidence is not reported. Outdoor workers, particularly gardeners, appear more likely to be affected, possibly because they are exposed at length to sunlight and to exacerbating allergens, although the condition is not necessarily associated with any particular occupation or hobby [29], the majority of patients being retired at presentation [24].

Affected patients present with chronic or subacute, often lichenified eczema, predominantly on light-exposed sites, notably the face, back and sides of neck, upper chest, scalp, and the backs of hands, characteristically with cut-off where clothing begins (Plates 14.1 & 14.2, facing p. 204). The upper eyelids, submental area, skin protected by the earlobes (the so called 'Wilkinson's triangle'; Plate 14.3, facing p. 204), skin creases and finger webs are often spared. However, in patients with severe light sensitivity, the eruption frequently spreads to affect covered areas as well, and episodes of erythroderma may develop, the trigger for such episodes not always being evident. In severe cases, erythematous, infiltrated plaques, occasionally confluent, and smooth-topped papules (Fig. 3) may develop on erythematous, eczematous or normal skin, mimicking cutaneous lymphoma [51]. Palmar and occasionally plantar eczema may also be a feature. Areas of hyper- or hypopigmentation may develop during the course of the disease, the latter sometimes closely resembling widespread vitiligo and attributed to the cytotoxic destruction of melanocytes [99]. Eyebrow and scalp hair may be lost or become stubbly or prematurely grey (authors' unpublished observations). Although the condition overall is generally worse in the summer months, many patients and unwary physicians may initially be unaware of the relationship to ultraviolet exposure, as the eruption may be delayed for hours to days following exposure, may persist into the winter months, and may be complicated by the features of associated contact dermatitis.

There is a suggestion that geographic variation in the incidence of CAD may occur, with peak incidence in more temperate climates, although no formal study of this has been undertaken. If this is so, a possible explanation is that suppression of the immune response may be occurring in areas of higher ambient ultraviolet intensity. In support of this is the fact that induction of photoallergy in guinea-pigs is a dose-related phenomenon, with decreased responses following larger induction doses of UVA, possibly due to loss of antigenic properties or to UV-mediated immune suppression [65]. Further support is lent by evidence that preteatment with UVB prior to sensitization suppresses induction of photocontact allergy [40].

Criteria for diagnosis

The diagnosis of chronic actinic dermatitis is suggested by clinical findings supported by the histological features of eczema (Plate 14.4, facing p. 204), sometimes pseudolymphomatous in severe cases (Plate 14.5, facing p. 204). Phototesting of unaffected skin to broadband or monochromatic irradiation is necessary to confirm the diagnosis and is always abnormal: there is a reduction in the minimum dose required to induce a response at 24 h (the minimal response dose), and the response is often of abnormal morphology, with eczematous lesions characteristic of the disease itself [33]. Very rarely, such phototest abnormalities may not become manifest for some months after onset of the disease [76]. All patients react abnormally to the UVB wavelengths, and 90% to the UVA wavelengths as well; in 10–24% of patients short wavelength visible abnormalities also occur [24, 33]. Circulating antinuclear antibodies, anti-SSA(Ro) and anti-SSB(La) titres and blood, urine and faecal porphyrin concentrations are normal.

Association with contact and photocontact allergy and endogenous eczema

Patch and photopatch testing are essential ancillary investigations in the diagnosis and management of CAD, positive responses to one or more allergens being found consistently in about 75% of patients [24, 31, 43]; sensitivity to two or more allergens being found in 65% [24]. CAD evolves in some instances from photoallergic contact dermatitis (so-called persistent light reactivity), while photoallergic contact dermatitis may also develop during the course of the disease [41]. Contact allergy may also precede the development of photosensitivity [71] and frequently coexists [29]. In our recent study of 86 patients, 9% had both contact and photocontact allergy, 2% had associated photocontact allergy alone and 64% had associated contact allergy alone. The photocontact allergens implicated were musk ambrette and sunscreen chemicals. Contact allergies were as follows: 36% were sensitive to Compositae plants (historically this had been noted with testing to plant extracts [3, 32] and was confirmed in our patients by testing to a mixture of sesquiterpene lactones, the main antigenic constituents of Compositae), 21% to fragrance compounds, 20% to colophony, 20% to metals and 14% to rubber compounds. In addition, 5% were sensitive to epoxy resins and 5% to phosphorus sesquisulfide, while occasional allergies were noted to medicaments, preservatives and vehicle bases. Similar findings have been reported by others [1, 3, 24, 32, 43, 95]. These reactions are persistent and reproducible and do not simply represent increased cutaneous reactivity [2]. A proportion of patients, including those with associated contact and photocontact allergies, have a history of endogenous eczema such as atopic, palmar or plantar eczema [86]. In our experience, 16% had a preceding history of endogenous eczema; 12% of patients had neither associated contact nor photocontact allergies, nor associated eczema [24].

Differential diagnosis

Contact dermatitis to airborne allergens and photocontact dermatitis may closely mimic CAD, being distinguished by normal cutaneous irradiation tests in the absence of antigen exposure. However, both may also occur in association with CAD. Light-exacerbated atopic or seborrhoeic eczema can generally be distinguished by the distribution of the eruption, or in more difficult situations by negative cutaneous irradiation tests. Erythrodermic chronic actinic dermatitis may be distinguished from other erythrodermas by irradiation skin testing once clearance of the eruption has been achieved by confinement of the patient in a light-protected room; in some instances such patients may have circulating Sézary cells in the peripheral blood [73]. Systemic drug-induced photosensitivity is associated with normal cutaneous irradiation tests or abnormalities in the UVA range only, and withdrawal of the drug is considered to result in resolution of the clinical and phototest abnormalities within 6 months [11]. Also, cutaneous T-cell lymphoma may occasionally be clinically and histologically indistinguishable from severe CAD. In addition photosensitivity is occasionally demonstrated on formal testing in cutaneous T-cell lymphoma, but this is usually minimal and the cutaneous irradiation tests are only mildly abnormal, principally to the UVA wavelengths [98]. The ratio of $CD4^+$/$CD8^+$ circulating lymphocytes may be helpful in distinguishing the two conditions in difficult cases, with $CD8^+$ predominant in CAD, particularly severe variants [27] and $CD4^+$ cells predominant in cutaneous T-cell lymphoma [18]. Although there have also been isolated reports of the onset of dermal reticuloses in patients with CAD [5, 52, 68, 93, 94], and it has been suggested that persistent abnormal cellular response to antigen may predispose to these conditions [92], neither clinical evidence [29] nor experimental evidence support this as a regular phenomenon. A flow cytometric study showed no evidence of DNA aneuploidy in the cutaneous infiltrate of patients with actinic reticuloid [78], and T-cell receptor and immunoglobulin gene rearrangement studies have demonstrated no evidence of a clonal lymphoid population in CAD [27].

Management

Ultraviolet avoidance, through minimizing outdoor exposure when irradiance is high, and use of protective clothing in conjunction with high sun protection factor sunscreens, is the cornerstone of management. Non-irritant reflectant sunscreen preparations are generally preferable in view of the risk of contact allergy to chemical preparations. Ultraviolet screening blinds in the home may be helpful to limit exposure to long wavelength ultraviolet radiation. Careful avoidance of presumed inducing and exacerbating allergens is also essential, while apparent worsening of the condition may be due to the development of superimposed contact or photocontact allergy, notably to sunscreens [41]. Acute flares of the disease may require nursing in light-protected cubicles, which is highly effective but time consuming and costly. Emollients and topical or short-term oral corticosteroids are helpful adjunctive therapy. Care should be taken to stop corticosteroid therapy several days before phototesting to avoid inhibition of the response [66]. For persistent or frequently relapsing cases, PUVA or immunosuppressive therapy may also be required. Some patients respond to PUVA therapy, although this may be poorly tolerated in the early phase of treatment even under systemic steroid cover, and maintenance therapy, possibly restricted to light-exposed sites, is generally also required [46, 47]. The mode of action of PUVA is unknown but it may involve modification of the immune response, in addition perhaps to inducing protective skin thickening and melanogenesis. Azathioprine is generally effective if tolerated [70] – exerting a purine antimetabolite effect on activated lymphocytes, which require purines during the proliferative phase – as is cyclosporin A [75], which prevents clonal expansion of activated T cells through inhibition of interleukins 1 and 2. Danazol has also been reported to be effective in a single patient with associated α1 antitrypsin deficiency, but this finding needs confirmation [49].

Although CAD is generally persistent and disabling, occasionally resolution occurs; in those patients who continue to suffer, careful management can usually allow the patient to maintain a satisfactory life-style.

Pathogenesis

The clinical response to T-cell immunomodulatory drugs, and a number of immunohistochemical studies of lesional skin, support an immunological basis for the disease, which is thought to be compatible with delayed-type hypersensitivity; cellular hypersensitivity to ultraviolet radiation has also been proposed as an alternative pathogenic basis for the disease. Humoral immunity appears to be normal [67], and although others have found occasional raised peripheral IgA, IgE and IgG or eosinophilia, such disturbances are unlikely to be primary phenomena [28, 56].

The skin immune system

Advances in our understanding of the skin immune system are of relevance to the investigation of cellular events in CAD. The skin contains the elements of an intrinsic immune system [13], comprising lymphocytes, resident antigen-presenting CD1+HLA-DR+ Langerhans cells, which circulate between the epidermis and the regional lymph nodes, dermal antigen-presenting cells (including perivascular factor XIIIa+ dermal dendritic cells and lymph node based RFD-1+ interdigitating reticulum cells) and keratinocytes, which secrete a wide range of proinflammatory cytokines with chemotactic and/or immunoregulatory properties. These include the interleukins IL-1, IL-6, IL-7, IL-8 and IL-10, as well as tumour necrosis factor α (TNFα), the colony stimulating factors (CSF) IL-3, granulocyte macrophage CSF, and macrophage CSF, together with growth factors, including transforming growth factors α and β, platelet-derived growth factor and basic fibroblast growth factor, as well as α-melanocyte stimulating hormone (αMSH), and a specific inhibitor of IL-1, IL-1 receptor antagonist (IL-1Ra). In general, IL-1, IL-6, IL-7, IL-8 and TNFα have proinflammatory properties, while IL-10, TGFβ, αMSH and IL-1Ra downregulate cutaneous inflammation. Recruitment of inflammatory cells into the extravascular spaces of the dermis is dependent upon their attachment to blood vessel endothelium, which in turn depends on the interaction between sets of adhesion molecules on the inflammatory cells, such as lymphocyte function-associated antigen 1 (LFA-1), and on endothelial cells, such as intercellular adhesion molecule 1 (ICAM-1), endothelial leucocyte adhesion molecule-1 (E-selectin) and vascular cell adhesion molecule 1 (VCAM-1). Since expression of these adhesion molecules is modulated by the action of cytokines, it has been proposed that keratinocyte production of IL-1 and TNFα following various noxious stimuli is sufficient to initiate cutaneous inflammation in an antigen-independent fashion; amplification and persistence of inflammation is then dependent on antigen-dependent mechanisms involving T cell – Langerhans cell or –dermal macrophage interactions. Interferon γ (IFN-γ)-induced keratinocyte expression of ICAM-1 may be important in determining the exocytosis of leukocytes into the epidermis.

Evidence for a delayed hypersensitivity reaction

Clinically, histologically and experimentally, lesions of CAD are compatible with a process of cell-mediated hypersensitivity. Clinically, the lesions resemble those of allergic contact dermatitis, which differs from the tuberculin response, the traditional cutaneous model of delayed-type hypersensitivity, in that the antigen is applied cutaneously and not intradermally. Histologically, the changes are very similar to those in allergic contact dermatitis, with epidermal spongiosis and a variable degree of acanthosis, and a lymphohistiocytic infiltrate with occasional plasma cells and eosinophils, predominantly in the upper dermis in a perivascular distribution and extending into the epidermis (Plate 14.4, facing p. 204). In the pseudolymphoma or actinic reticuloid stage, the infiltrate is much heavier (Plate 14.5, facing p. 204) with atypical lymphocytes with convoluted nuclei (Sézary-like cells) and Pautrier-like microabcesses in the epidermis. Similar changes have been reported in certain forms of chronic contact dermatitis, so-called lymphomatoid contact dermatitis [80], suggesting that the pseudolymphomatous lesions represent a severe stage in the spectrum of chronic actinic dermatitis, presumably occurring after persistent antigenic stimulation.

A number of immunohistochemical studies have been performed to characterize the dermal infiltrate in lesions of chronic actinic dermatitis. This consists predominantly of T cells, both in fully evolved [12, 77, 84, 91, 96], and in induced lesions [26, 35] (Plates 14.6 & 14.7, facing p. 204), together with Langerhans cells, interdigitating reticulum cells, and monocyte macrophages including CD36 (OKM5$^+$), factor XIIIa$^+$, CD11b$^+$, CD11c$^+$ and CD14$^+$ cells [26, 35]. Leukcocyte epidermotropism is observed, and keratinocytes express MHC class II antigens in induced lesions [26, 35]. The relative proportion of CD4$^+$ (T-helper/inducer) and CD8$^+$ (T-suppressor/cytotoxic) cells in the dermal infiltrate has varied in different studies; in some a predominance of CD4$^+$ cells has been reported [26, 84], while in others equal ratios [84, 91], or a predominance of CD8$^+$ cells was found [35, 96], particularly in lesions with more florid histological changes [77]. Most studies agree that CD4$^+$ T cells predominate over CD8$^+$ T cells in the early stages of allergic contact dermatitis and the tuberculin response [17, 19, 36, 37, 57, 82, 83, 88], with an increase in CD8$^+$ cells occurring in the late stage [35, 57]. The kinetics of the inflammatory cell infiltrate in induced lesions of CAD is consistent with a delayed-type hypersensitivity response [19, 36, 81–83, 88], with a peak in the intensity of infiltrate, including dermal and epidermal activated T cells, epidermal Langerhans cells and monocyte macrophages, at 24 to 48 h [26].

Analysis of circulating lymphocyte subsets in CAD has also been variable in the literature. In some patients they have been normal [61, 77] while in others a reduction in the CD4$^+$/CD8$^+$ ratio has been observed [7, 18, 35, 77, 91], generally due to a predominance of CD8$^+$ circulating lymphocytes, which was most marked in patients with the actinic reticuloid variant of CAD [27]. The latter finding correlates with the reported increase in CD8$^+$ cells in the cutaneous infiltrate of more florid cases of CAD [77].

Specific changes in the kinetics and pattern of cell surface adhesion molecule expression have been reported in the delayed-type hypersensitivity response. Following PPD injection, prolonged expression of endothelial cell ELAM-1, expression of VCAM-1 by perivascular and dermal interstitial cells, and early ICAM-1 expression by keratinocytes have all been reported [74]. Similar changes have been found in allergic contact dermatitis: keratinocyte ICAM-1 expression occurs early in the induction of lesions and upregulation of ICAM-1, VCAM-1 and ELAM-1 is observed on dermal perivascular cells [42]. Time course biopsies following eight to ten minimal response doses of solar-simulated irradiation (less than 1 minimal erythema dose in normals) in five patients with CAD revealed upregulation of perivascular ICAM-1, VCAM-1 and ELAM-1, and of dermal interstitial VCAM-1 and ICAM-1, from 1 to 5 h on, persisting at 5–7 days (Plates 14.8–14.10, facing p. 204). There was also increased expression of ICAM-1 by basal keratinocytes and focally throughout the epidermis at 1 h, persisting at 5–7 days [25] (Plate 14.8, facing p. 204). These changes are not simply a function of ultraviolet radiation; following UVB irradiation, expression of ELAM-1 is less prolonged, and induction of VCAM-1 and of keratinocyte ICAM-1 is not observed, and following combined UVB and UVA irradiation, expression of ICAM-1 on keratinocytes *in vitro* is delayed by 48 h [74, 79]. Thus the pattern of adhesion molecule expression observed in CAD is likely to be of significance in the development of lesions, and suggests delayed-type hypersensitivity.

Cytokine expression

Cytokine production from keratinocytes plays an

important complementary role to Langerhans and other antigen-presenting cell activity in the initiation and propagation of allergic contact dermatitis [9]. Studies on the role of cytokines in the pathogenesis of the lesions of CAD are limited. IL-1α has been reported to be proinflammatory *in vitro* [21]. We therefore investigated the distribution of IL-1α, its receptor (IL-R) and IL-1Ra in CAD by immunohistochemical staining [23]. Epidermal and perivascular IL-1α expression was upregulated at 5 h, peaked between 24 and 48 h, and had waned by 6 days after ultraviolet irradiation; similar upregulation of IL-R and IL-1Ra was seen which, however, persisted at 6 days. IL-1Ra expression was also seen in association with dermal inflammatory cells [23]. These findings support a proinflamatory role for IL-1α, mediated by its receptor, and regulated by IL-1Ra, in CAD.

Cellular hypersensitivity to UVA

It has been suggested that fibroblast cell lines from patients with CAD are abnormally sensitive to UVA irradiation, with resultant inhibition of RNA synthesis and cytopathological changes, including an increase in DNA strand breaks [38]. Excision repair, as assessed by quantitative and action spectrum studies for unscheduled DNA synthesis, is normal in CAD [50] and no increase in UVR-induced skin cancer has been noted (authors' unpublished observations). The observed cellular hypersensitivity may be corrected by coculture with normal cell lines [34], or by culture with an antioxidant analogue [62], suggesting that the sensitivity is due to deficiency of scavengers of active oxygen species. These free radicals may be responsible for induction of the inflammatory response [38], conceivably through the generation of photoallergens [69]. However, further studies have shown wide variation in strand breaks and cellular hypersensitivity in normal cells [55], and either failed to detect UVA sensitivity in CAD cells [55] or demonstrated it only in certain strains [39], making this theory increasingly difficult to support.

Nature of the antigen

The antigen(s) triggering the postulated delayed-type hypersensitivity response in CAD is (are) unknown. It has been suggested that a photo-contact allergen may persist in the skin. Alternatively, a cutaneous neoantigen may be induced following ultraviolet irradiation, initially with involvement of an intermediary chemical, either exogenous or endogenous (or conceivably in the absence of such an initiating chemical), and subsequently in response to ultraviolet irradiation alone. Such theories must take into consideration the role of associated contact sensitivities.

Persistence of allergen in the skin

Some patients with chronic actinic dermatitis have a preceding photocontact dermatitis (so-called persistent light reactors). Willis and Kligman in 1968 induced local photosensitivity to halogenated salicylanilides and related compounds, and by studying the UV absorption spectra of skin extracts from these patients showed persistence of the original photosensitizer in the skin for months, and occasionally for a year or more [102]. They therefore proposed that small persistent amounts of exogenous photoallergen were triggeing the photosensitivity. *In vitro* studies with tetrachlor-salicylanilide show that it binds to human serum albumin [60], and binding to epidermal proteins of guinea-pigs has been demonstrated [72]. Thus exogenous photoallergens could conceivably persist in the skin and give rise to the disease. However, this theory does not explain the development of generalized reactivity, unless the photosensitizing chemical were somehow distributed throughout the skin; nor does it explain spread beyond the original UVA elicitation spectrum to include UVB sensitivity in CAD; nor the pathogenesis of CAD arising in the absence of documented exogenous photoallergy.

Induction of neoantigen

In 1964 Baer and Kopf [8] proposed that a carrier protein, albumin or another skin protein, which binds a photoallergen might alter during the process of photocontact dermatitis to become a neoantigen capable of stimulating the immune system. Support for this theory was lent by Kochevar and Harber in 1977 [60], who showed that, *in vitro*, tetrachlorsalicylanilide binds strongly to human albumin and promotes the oxidation of the histidine moiety of albumin, rendering it weakly antigenic; subsequent absorption of UV radiation by albumin alone in the absence of the initiating exogenous photosensitizer then produced oxidized protein. The same phenomenon could occur *in vivo*, and, as albumin is widely distributed throughout the skin, this could explain the potential for a delayed-type hypersensitivity response at all skin sites in the absence of the initial photosensitizer. The initiating photosensitizer might be an exogenous chemical such as a phototoxic or photoallergic contact allergen, or alternatively, an endogenous phototoxic chemical such as a porphyrin or kynurenic acid. An association has been reported between CAD and porphyria cutanea tarda [89], but this association has not been upheld, and porphyrin metabolism is normal in CAD. Kynurenic acid is an endogenous protein which is a metabolic byproduct of tryptophan metabolism, and has been shown to cause phototoxic oxidation of histidine *in vitro* [90, 97]. Levels of kynurenic acid were found to be increased in the skin of one patient with CAD [90], while another study revealed abnormal tryptophan metabolism with accumulation of kynurenic acid [10]. However, these findings have not proved to be reproducible [100]. Thus no endogenous photosensitizer has yet been identified. It is also conceivable, however, that the generation of antigenic oxidized protein occurs in the absence of initiating endogenous or exogenous chemicals, as has been described in guinea-pigs with adjuvant and UVA alone [58]; this would account for patients who develop CAD in the absence of preceding contact or photocontact dermatitis.

Role of contact allergens

The high prevalence of associated allergic contact dermatitis may be of relevance to the pathogenesis of the condition. Patients with allergic contact dermatitis may develop superimposed photosensitivity [97] as has been shown with chromium [30] and Compositae dermatitis [71]. The precise mechanism for the development of photosensitivity is unexplained, although it may involve phototoxic oxidation of endogenous protein, which occurs subsequently in the absence of the initiating contact antigen. The more prevalent contact allergens associated with the disease, namely Compositae oleoresins, fragrance materials, and colophony [1, 32, 63], do have phototoxic potential. Missed allergens, presumably ubiquitous, could conceivably be implicated in those cases where no formal contact allergy has been demonstrated, although there is no evidence for this. However, chronic actinic dermatitis has been reported to develop in other conditions producing persistent T-cell stimulation, notably long-standing endogenous eczema [24, 86] and rarely long-standing PLE, both of which have been postulated to be delayed-type hypersensitivity responses. It is therefore probable that contact allergens play a facilitative rather than a causal role in the pathogenesis of the condition, acting as an adjuvant [58] and predisposing to chronic actinic dermatitis in affected patients by enhancing cutaneous immune function and facilitating generation or recognition of weak endogenous antigen.

Failure of immune modulation or the development of autoreactivity

Ultraviolet irradiation generally suppresses cutaneous immune responses including the induction phase of delayed-type hypersensitivity reactions [6]. It is conceivable that CAD may result from dysregulation of such suppressor mechanisms, enabling the development of hypersensitivity to UV-induced neoantigens. Indirect support for the pathogenic role of such neoantigens comes from the action spectrum (wavelengths most effective for eliciting the response) for CAD; the shape of the action spectrum for induction of lesions resembles the action spectrum for induction of UV erythema ([22], authors' unpublished observations). This implies that the chromophores (or chemical absorbers) responsible for the changes of CAD are found in normal UV-irradiated skin. This, in turn, implies that the antigen responsible for CAD is an altered self protein.

Summary

CAD is a rare disease characterized by exquisite sensitivity to ultraviolet radiation, which initiates an inflammatory process resembling a delayed-type hypersensitivity response. It is possible, although there is as yet no direct support for this, that a weak endogenous photoallergen is responsible for this process. A state of heightened reactivity of the skin immune system, due to contact allergic or endogenous eczema, may favour the development of the disease by as yet undefined mechanisms. Alternatively, an alteration of immune modulation, or a process of autoreactivity, may occur. Further work is needed to study the mechanisms involved in local T-cell stimulation in this disease.

References

1 Addo, H.A., J. Ferguson, B.E. Johnson, W. Frain-Bell: The relationship between exposure to fragrance materials and persistent light reaction in the photosensitivity dermatitis with actinic reticuloid syndrome. *Br. J. Dermatol.* 107 (1982), 261–274.

2 Addo, H.A., W. Frain-Bell: Persistence of allergic contact sensitivities in subjects with the photosensitive dermatitis and actinic reticuloid syndrome. *Br. J. Dermatol.* 117 (1987), 555–559.

3 Addo, H.A., S.C. Sharma, J. Ferguson, B.E. Johnson, W. Frain-Bell: A study of Compositae plant extract reactions in photosensitivity dermatitis. *Photodermatology* 2 (1985), 68–79.

4 Allen, H., K. Kaibey: Persistent photosensitivity following occupational exposure to epoxy resins. *Arch. Dermatol.* 115 (1979), 1307–1310.

5 Ashinoff, R., M.R. Buchness, H.W. Lim: Lymphoma in a black patient with actinic reticuloid treated with PUVA: possible aetiologic considerations. *J. Am. Acad. Dermatol.* 21 (1989), 1134–1137.

6 Baadsgaard, O.: *In vivo* ultraviolet irradiation causes profound perturbation of the immune system. *Arch. Dermatol.* 127 (1991), 99.

7 Baadsgaard, O., C. Geisler, T. Plesner, K. Thestrup-Pedersen: Lymphocyte subsets in patients with Compositae olleoresin dermatitis and increased UVA sensitivity during treatment with azathioprine. *Arch. Dermatol. Res.* 277 (1985), 304–306.

8 Baer, R.L., A.W. Kopf: Editor's comment In: *Yearbook of Dermatology, 1963–64 series,* (1964), p. 133. Eds.: R.L. Baer, A.W. Kopf, Yearbook Medical Publishers, Chicago.

9 Barker, J.: Role of keratinocytes in allergic contact dermaitis. *Contact Dermatitis* 26 (1992), 145–148.

10 Binazzi, M., P. Calandra: Actinic reticuloid pathogenic aspects. *Arch. Derm. Forsch.* 241 (1971), 391–395.

11 Bisland, D., J. Ferguson: Management of chronic photosensitive eczema (Letter). *Arch. Dermatol.* 127 (1991), 1065–1066.

12 Braathen, L.R., Ø. Førre, J.B. Natvig: An anti-human T-lymphocyte antiserum: *in situ* identification of T cells in the skin of delayed type hypersensitivity reactions, chronic photosensitivity dermatitis, and mycosis fungoides. *Clin. Immunol. Immunopath.* 13 (1979), 211–219.

13 Breathnach, S.M.: The skin immune system and psoriasis. *Clin. Exp. Immunol.* 91 (1993), 342–345.

14 Burckhardt, W.: Photoallergic eczema due to blankophores (optic brightening agents). *Hautartz* 8 (1957), 486–488.

15 Burry, J.N.: Photoallergies to fentichlor and Multifungin. *Arch. Dermatol.* 95 (1967), 287–291.

16 Burry, J.N.: Persistent light reaction from buclosamide. *Arch. Dermatol.* 101 (1970), 95–97.

17 Carr, M.M., P.A. Botham, D.J. Gawkrodger, E. McVittie, J.A. Ross, J.C. Stewart, J.A. Hunter: Early cellular reactions induced by dinitrochlorobenzene in sensitized human skin. *Br. J. Dermatol.* 110 (1984), 637–641.

18 Chu, A.C., D. Robinson, J.L.M. Hawk, R. Meacham, M.F. Spittle, N.P. Smith: Immunologic differentiation for the Sézary syndrome due to cutaneous T-cell lymphoma and chronic actinic dermatitis. *J. Invest. Dermatol.* 98 (1986), 134–137.

19 Chu, C.Q., M. Field, E. Andrew, D. Haskard, M. Feldmann, R.N. Mainz: Detection of cytokines at the site of tuberculin-induced delayed-type hypersensitivity in man. *Clin. Exp. Immunol.* 90 (1992), 522–529.

20 Crow, K.D., E. Alexander, W.H.I. Buck, B.E. Johnson, J.A. Magnus, A.D. Porter: Photosensitivity due to pitch. *Br. J. Dermatol.* 73 (1961), 220–232.

21 Dowd, P.M., R.D.R. Camp, M.W. Greaves: Human recombinant interleukin-1 alpha is pro-inflamatory in normal human skin. *Skin Pharmacol.* 1 (1988), 30–37.

22 du P. Menagé, H., J.L.M. Hawk: The action spectrum in chronic actinic dermatitis (Abstract). *Br. J. Dermatol.* 129 (1993), 496.

23 du P. Menagé, H., M. Kristensen, C.Q. Chu, F.M. Brennan, J.L.M. Hawk, S.M. Breathnach: Upregulation of interleukin 1α, its receptor and interleukin 1 receptor antagonist levels in induced lesions of chronic actinic dermatitis. *Br. J. Dermatol.* 127 (Suppl) (1992), 429.

24 du P. Menagé, H., J. Ross, J.L.M. Hawk, I.R. White: Contact and photocontact sensitisation in chronic actinic dermatitis sensitivity: sesquiterpene lactone mix is an important allergen. *Br. J. Dermatol.* (1995), (in press).

25 du P. Menagé, H., N. Sattar, D.O. Haskard, J.L.M. Hawk, S.M. Breathnach: The kinetics and pattern of adhesion molecule expression in induced lesions of chronic actinic dermatitis suggest delayed type hypersensitivity (Abstract). *J. Invest. Dermatol.* 100 (1993), 498.

26 du P. Menagé, H., N. Sattar, J.L.M.H. Hawk, S.M. Breathnach: Immunophenotyping of the inflammatory cell infiltrate during the evolution of induced lesions of chronic actinic dermatitis (Abstract). *J. Invest. Dermatol.* 100 (1993), 482.

27 du P. Menagé, H., S.J. Whittaker, Y.L. Ng, N.P. Smith, J.L.M. Hawk: Analysis of T-cell receptor genes in chronic actinic dermatitis: no evidence of clonality (Abstract). *J. Invest. Dermatol.* 98 (1992), 546.

28 Ehlers, G., K. Scharr: Aktinisches Reticuloid. *Z. Hautkr.* 55 (1979), 1443–1464.

29 Ferguson, J.: Photosensitivity dermatitis and actinic reticuloid syndrome (chronic actinic dermatitis). *Semin. Dermatol.* 9 (1990), 47–54.

30 Frain-Bell, W.: Dowling oration 1978. What is that thing called light? *Clin. Exp. Dermatol.* 4 (1979), 1–29.

31 Frain-Bell, W.: Photosensitivity and Compositae dermatitis. *Clinics Dermatol.* 4 (1986), 122–126.

32 Frain-Bell, W., A. Hetherington, B.E. Johnson: Contact allergic sensitivity to chrysanthemum and the photosensitivity dermatitis and actinic reticuloid syndrome. *Br. J. Dermatol.* 101 (1979), 491–501.

33 Frain-Bell, W., T. Lakshmipathi, J. Rogers, J. Willock: The syndrome of chronic photosensitivity dermatitis and actinic reticuloid. *Br. J. Dermatol.* 91 (1974), 617–634.

34 Francis, A.J., F. Gianelli: Cooperation between human cells sensitive to UVA radiations: a clue to the mechanism of cellular hypersensitivity associated with different clinical conditions. *Experiment. Cell Research* 195 (1991), 47–52.

35 Fujita, M., Y. Miyachi, T. Horio, S. Imamura: Immunohistochemical comparison of actinic reticuloid with contact dermatitis. *J. Dermatol. Sci.* 1 (1990), 289–296.

36 Fullmer, M.A., J.-Y. Shen, R.L. Modlin, T.H. Rea: Immunohistological evidence of lymphokine production and lymphocyte activation antigens in tuberculin reactions. *Clin. Exp. Immunol.* 67 (1987), 383–390.

37 Gawkrodger, D.J., E. McVittie, M.M. Carr, J.A. Hunter: Phenotypic characterisation of the early cellular responses in allergic and irritant contact dermatitis. *Clin. Exp. Immunol.* 66 (1986), 590–598.

38 Gianelli, F., P.K. Botcherby, B. Marimo, I.A. Magnus: Cellular hypersensitivity to UVA: a clue to the aetiology of actinic reticuloid? *Lancet* i (1983), 88–91.

39 Gibbs, N.K., P.K. Botcherby, R.W. Morris, A.R. Young, F. Gianelli: UVA dose response of normal and actinic reticuloid cell fibroblasts: a statistical analysis (Abstract). *Br. J. Dermatol.* 118 (1988), 289.

40 Grantstein, R.D., W.L. Morison, M.L. Kripke: The role of UVB radiation in the induction and elicitation of photocontact hypersensitivity to TCSA in the mouse. *J. Invest. Dermatol.* 80 (1983), 158–162.

41 Green, C., M. Catterall, J.L.M. Hawk: Chronic actinic dermatitis and sunscreen allergy. *Clin. Exp. Dermatol.* 16 (1991), 70–71.

42 Griffiths, C.E.M., J.N.W.N. Barker, S. Kunkel, B.J. Nickoloff: Modulation of leucocyte adhesion molecules, a T-cell chemotactin (IL-8), and a regulatory cytokine (TNF-α) in allergic contact dermatitis (rhus dermatitis). *Br. J. Dermatol.* 124 (1991), 519–526.

43 Hannuksela, M., R. Suhonen, L. Forstrom: Delayed contact allergies in patients with photosensitivity dermatitis. *Arch. Dermatovener. (Stockh.)* 61 (1981), 303–306.

44 Hawk, J.L.M., I.A. Magnus: Resolution of actinic reticuloid with transition to photosensitive eczema. *J. R. Soc. Med.* 71 (1978), 608–609.

45 Hawk, J.L.M., I.A. Magnus: Chronic actinic dermatitis – an idiopathic photosensitivity syndrome including actinic reticuloid and photosensitive eczema. *Br. J. Dermatol.* 101 (Suppl 17) (1979), 24.

46 Hindson, C., A. Downey, S. Sinclair, B. Cominos: PUVA therapy of chronic actinic dermatitis: a 5 year follow-up. *Br. J. Dermatol.* 123 (1990), 264–277.

47 Hindson, C., J. Spiro, A. Downey: PUVA therapy of chronic actinic dermatitis. *Br. J. Dermatol.* 113 (1984), 156–160.

48 Horio, T.: Actinic reticuloid via persistent light reactivity from photoallergic contact dermatitis. *Arch. Dermatol.* 118 (1982), 339–342.

49 Humbert, P., C. Drobacheff, M. Vigan, E. Quencez, R. Laurent, P. Agache: Chronic actinic dermatitis responding to danazol. *Br. J. Dermatol.* 124 (1991), 195–197.

50 Ichihashi, M., C.A. Ramsay: Excision repair of DNA in some photodermatoses. *Br. J. Dermatol.* 95 (1976), 13–17.

51 Ive, F.A., I.A. Magnus, R.P. Warin, E. Wilson Jones:

'Actinic reticuloid'; a chronic dermatosis associated with severe photosensitivity and the histological resemblance to lymphoma. *Br. J. Dermatol.* 81 (1969), 469–485.

52 Jensen, N.E., I.B. Sneddon: Actinic reticuloid with lymphoma. *Br. J. Dermatol.* 82 (1970), 287–291.

53 Jillson, O.F., R.D. Baughman: Contact photodermatitis from bithionol. *Arch. Dermatol.* 88 (1963), 103–112.

54 Johnson, B.: Cellular mechanisms of photosensitivity. *Proc. R. Soc. Med.* 67 (1974), 871–873.

55 Johnson, B.E., E.M. Walker, J. Ferguson, W. Frain-Bell: Cellular sensitivity to UVA in photosensitivity dermatitis/actinic reticuloid (PD/AR). *Br. J. Dermatol.* 118 (1988), 286.

56 Johnson, S.C., D.J. Cripps, D.H. Norbach: Actinic reticuloid, a clinical, pathological and action spectrum study. *Arch. Dermatol.* 115 (1979), 1078–1083.

57 Kanerva, L., A. Ranki, J. Lauharanta: Lymphocytes and Langerhans cells in patch tests: an immunohistochemical and electron microscopic study. *Contact Derm.* 11 (1984), 150–155.

58 Katsumura, Y., B.S.J. Tanaka, H. Ichikawa, S. Kato, T. Kobayashi, T. Horio: Persistent light reaction: induction in the guinea pig. *J. Invest. Dermatol.* 87 (1986), 330–333.

59 Kingston, T.P., N.J. Lowe, H.L. Sofen, D.P. Weingarten: Actinic reticuloid in a black man: successful treatment with azathioprine. *J. Am. Acad. Dermatol.* 16 (1987), 1079–1083.

60 Kochevar, I.E., L.C. Harber: Photoreactions of 3′,3′,4′,5-tetrachlorosalicylanilide with proteins. *J. Invest. Dermatol.* 68 (1977), 151–156.

61 Kofoed, M.L., B. Munch-Petersen, J.K. Larsen, G. Wantzin: Non-replicative DNA synthesis detected in peripheral lymphocytes from a patient with actinic reticuloid. *Photodermatology* 3 (1986), 158–163.

62 Kralli, A., S.H. Moss: The sensitivity of an actinic reticuloid cell strain to near ultraviolet radiation and its modification by Trolox-C, a vitamin E analogue. *Br. J. Dermatol.* 116 (1987), 761–772.

63 Krutmann, J., B. Rzany, E. Schopf, A. Kapp: Airborne contact dermatitis from colophony: phototoxic reaction? *Contact Dermatitis* 21 (1989), 275–276.

64 Lim, H.W., M.R. Buchness, R. Ashinoff, N.A. Soter: Chronic actinic dermatitis: study of the spectrum of chronic photosensitivity in 12 patients. *Arch. Dermatol.* 126 (1990), 317–323.

65 Lovell, W.W., D.J. Sanders: Dose–response study of ultraviolet radiation for induction of photoallergy to tetrachlorsalicylanilide in guinea pigs. *Photodermatol. Photoimmunol. Photomed.* 7 (1990), 192–197.

66 Lowe, J.G., J. Ferguson: A double blind control study to assess the effect of pretreatment with a potent topical steroid on the phototest response of photosensitivity dermatitis and actinic reticuloid syndrome (PD/AR). *Scot. Med. J.* 34 (1989), 509–514.

67 Mentor, M.A., R.A. McKerron, H.E. Amos: Actinic reticuloid: an immunological investigation providing evidence of basement membrane damage. *Br. J. Dermatol.* 90 (1974), 507–515.

68 Meynadier, J., J.L. Peyron, G. Barneon, B. Guillot, A. Habib, J.J. Guilhou: Hodgkin's disease complicating actinic reticulosis. *Ann. Dermatol. Venereol.* 111 (1984), 999–1003.

69 Miyachi, Y., S. Imamura, Y. Niwa, V. Tokura, M. Takig: Mechanisms of contact photosensitivity in mice. Oxygen intermediaries are involved in contact photosensitisation but not in ordinary contact sensitisation. *J. Invest. Dermatol.* 86 (1986), 26–28.

70 Murphy, G.M., P.M. Maurice, P.G. Norris *et al.*: Azathioprine in the treatment of chronic actinic dermatitis: a double-blind controlled trial with monitoring of exposure to ultraviolet radiation. *Br. J. Dermatol.* 121 (1989), 639–646.

71 Murphy, G.M., I.R. White, J.L.M. Hawk: Allergic airborne dermatitis to Compositae with photosenstivity-CAD in evolution. *Photodermatol. Photoimmunol. Photomed.* 7 (1990), 38–39.

72 Nakayama, Y., M.A. Pathak: Photoconjugation of 3′4′5′-tetrachlorsalicylanilide to proteins of guinea-pig skin. *J. Invest. Dermatol.* 66 (1976), 273–274.

73 Neild, V.S., J.L.M. Hawk, R.A.J. Eady, J. Cream: Actinic reticuloid with Sézary cells. *Clin. Exp. Dermatol.* 7 (1982), 143–148.

74 Norris, D.A., M. Bradley-Lyons, M. Middleton, J.J. Yohn, M. Koshiharagawani: Ultraviolet radiation can either suppress or induce expression of intercellular adhesion molecule 1 on the surface of cultured human keratinocytes. *J. Invest. Dermatol.* 95 (1990), 132–138.

75 Norris, P.G., R.D.R. Camp, J.L.M. Hawk: Actinic reticuloid: response to cyclosporin. *J. Am. Acad. Dermatol.* 21 (1989), 307–309.

76 Norris, P.G., J.L. Hawk: Chronic actinic dermatitis: a unifying concept. *Arch. Dermatol.* 126 (1990), 376–378.

77 Norris, P.G., J. Morris, N.P. Smith, A.C. Chu, J.L.M. Hawk: Chronic actinic dermatitis: an immunohistologic and photobiologic study. *J. Am. Acad. Dermatol.* 21 (1989), 966–971.

78 Norris, P.G., J.A. Newton, R.S. Camplejohn *et al.*: A flow cytometric study of actinic reticuloid. *Clin. Exp. Dermatol.* 14 (1989), 128–131.

79 Norris, P., R.N. Poston, D.S. Thomas, M. Thornhill, J.L.M. Hawk, D.O. Haskard: The expression of endothelial leukocyte adhesion molecule-1 (ELAM-1), intercellular adhesion molecule-1 (ICAM-1), and vascular cell adhesion molecule-1 (VCAM-1) in experimental cutaneous inflammation: a comparison of ultraviolet B erythema and delayed type hyper-

sensitivity. *J. Invest. Dermatol.* 96 (1991), 763–770.

80 Orbeneja, J.G., L.I. Diez, J.L.S. Lozano, L.C. Salazar: Lymphomatoid contact dermatitis: a syndrome produced by epicutaneous hypersensitivity with clinical features and a histopathologic picture similar to that of mycosis fungoides. *Contact Dermatitis* 2 (1976), 139–143.

81 Platt, J.L., B.W. Grant, A.A. Eddy, A.F. Michael: Immune cell populations in cutaneous delayed-type hypersensitivity. *J. Exp. Med.* 158 (1983), 1227–1242.

82 Poulter, L.W., G.J. Seymour, O. Duke, G. Janossy, G. Panayi: Immunohistological analysis of delayed type hypersensitivity in man. *Cell Immunol.* 74 (1982), 358–369.

83 Ralfkiaer, E., G.L. Wantzin: *In situ* immunological characterisation of the infiltrating cells in positive patch tests. *Br. J. Dermatol.* 111 (1984), 13–22.

84 Ralfkiaer, E., G.L. Wantzin, H. Stein, D.Y. Mason: Photosensitive dermatitis with actinic reticuloid syndrome: an immunohistological study of the cutaneous infiltrate. *Br. J. Dermatol.* 114 (1986), 47–56.

85 Ramsay, C.A.: The enigma of chronic photosensitivity. *Clin. Exp. Dermatol.* 6 (1981), 665–671.

86 Ramsay, C.A., A. Kobza-Black: Photosensitive eczema. *Trans. St John's Hospital Dermatol. Soc.* 59 (1973), 152–158.

87 Robinson, H.N., W.L. Morison, A.F. Hood: Thiazide diuretic therapy and chronic photosensitivity. *Arch. Dermatol.* 121 (1985), 522–524.

88 Scheynius, A., T. Fischer, U. Forsum, L. Klareskog: Phenotypic characterisation *in situ* of inflammatory cells in allergic and irritant contact dermatitis in man. *Clin. Exp. Immunol.* 55 (1984), 81–90.

89 Schnitzler, L., J.L. Verret, B. Schubert, B. Belperron: Actinoréticulose survenant au décours d'une porphyrie cutanée tardive – étude optique et ultrastructurelle d'un cas. *Arch. Anat. Pathol. (Paris)* 23 (1975), 145–150.

90 Swanbeck, G., G. Wennersten: Evidence for kynurenic acid as a potential photosensitiser in actinic reticuloid. *Acta. Derm. Venereol. (Stockh.)* 53 (1973), 109–113.

91 Takigawa, M., Y. Tokura, S. Shirahama, H. Sagimoto, M. Yamada: Actinic reticuloid: an immunohistochemical study (Letter). *Arch. Dermatol.* 123 (1987), 296–297.

92 Tan, R.S.-H., C.M. Butterworth, H. McLaughlin, S. Malka, R.D. Samman: Mycosis fungoides – a disease of antigen persistence. *Brit. J. Dermatol.* 91 (1974), 607–616.

93 Thestrup-Pedersen, K., C. Zachariae, K. Kaltoft, G. Pallesen, H. Sogaard: Development of cutaneous pseudolymphoma following cyclosporin therapy of actinic reticuloid. *Dermatologica* 177 (1988), 376.

94 Thomsen, K.: The development of Hodgkins disease in a patient with actinic reticuloid. *Clin. Exp. Dermatol.* 2 (1977), 109–113.

95 Thune, P.: Allergy to lichens with photosensitivity. *Contact Derm.* 3 (1977), 213–214.

96 Toonstra, H., S.C.J. van der Putte, D.F. van Wichen, H. van Weelden, C.J.M. Herquet, W.A. Van Vloten: Actinic reticuloid: immunohistochemical analysis of the cutaneous infiltrate in 13 patients. *Br. J. Dermatol.* 120 (1989), 779–789.

97 Vandermaesen, J., R. Roelandts, H. Degreef: Light on the persistent light reaction–photosensitivity dermatitis–actinic reticuloid syndrome. *J. Am. Acad. Dermatol.* 15 (1986), 685–692.

98 Volden, G., P.O. Thune: Light sensitivity in mycosis fungoides. *Br. J. Dermatol.* 97 (1977), 279–284.

99 Von Den Driesch, P., M. Fartasch, O.P. Hornstein: Chronic actinic dermatitis with vitilligo-like depigmentation. *Clin. Exp. Dermatol.* 17 (1992), 38–43.

100 Vonderheid, E.C., E.L. Sobel, R.D. Hoeldtke, G.J. Faerber, V.S. Sardi: Kynurenic acid and xanthurenic acid excretion after tryptophan loading in actinic reticuloid. *Int. J. Dermatol.* 26 (1987), 33–41.

101 Wilkinson, D.S.: Patch test reactions to certain halogenated salicylanalides. *Br. J. Dermatol.* 74 (1962), 302–306.

102 Willis, I., A.M. Kligman: The mechanism of the persistent light reactor. *J. Invest. Dermatol.* 51 (1968), 385–394.

103 Yates, V.M., O.A. Finn: Contact allergic sensitivity to zinc pirithione followed by the photosensitivity dermatitis and actinic reticuloid syndrome. *Contact Derm.* 6 (1980), 349–350.

104 Zaynoun, S., B.E. Johnson, W. Frain-Bell: The investigation of quindoxin photosensitivity. *Contact Derm.* 2 (1976), 343–352.

15: Solar urticaria

Erhard Hölzle

Summary

Solar urticaria is a rare but incapacitating photodermatosis in afflicted patients. Generally, immediate-type hypersensitivity is assumed as the pathomechanism inducing itching, erythema and whealing after exposure to small amounts of sunlight or artificial electromagnetic radiation. Sometimes wheals are confined to distinct circumscribed skin sites. This variety may by termed fixed solar urticaria. The action spectrum is frequently within a broad portion of the electromagnetic spectrum, mostly involving UVA and visible light. The diversity of the action spectrum is probably due to a variety of chromophores activated by radiation and thus creating the photoallergen. Rarely, certain portions of the electromagnetic spectrum inhibit solar urticaria. Photoinactivation of the photoallergen is probably the underlying mechanism. In most patients the presence of a putative photoallergen in serum or plasma is demonstrated by *in vitro* activation experiments and passive and reverse passive transfer tests. Development of transient skin tolerance by repeated exposures to eliciting wavelengths is a characteristic phenomenon of solar urticaria. During the state of tolerance, lasting for only a few days, binding sites of IgE on mast cells probably remain occupied by the photoallergen. A working classification of two types of solar urticaria has been proposed. In type I, IgE-mediated hypersensitivity to a specific photoallergen, which is generated only in patients with this type of solar urticaria, represents the pathomechanism. In type II, specific IgE is formed against an unspecific photoallergen which is generated in patients with solar urticaria as well as in normal subjects. Although histamine probably represents the major mediator, antihistamines are of limited therapeutic value. Phototherapy and, as the treatment of choice, photochemotherapy offer practical means. Plasmapheresis seems to be helpful in some patients in whom a circulating photoallergen is present.

Introduction

Solar urticaria, first described in 1904 by Merklen [33], is a rare photosensitivity disorder of unknown etiology. It belongs to the urticarias elicited by physical agents – in solar urticaria it is electromagnetic radiation. The incidence of solar urticaria is rare, however, afflicted patients are highly incapacitated owing to an extreme photosensitivity. Despite the elusive cause, an immunological basis of solar urticaria is generally accepted. It is assumed that a putative photoallergen is formed by irradiation in skin, and in many patients also in serum or plasma [17, 21, 26]. The photoallergen elicits a type I immunoreaction. According to studies on action spectra, results of passive transfer tests and reverse passive transfer tests Harber *et al.* [13] proposed a classification of six types (Table 15.1). More recently. Leenutaphong and coworkers reviewed all available reports on experimental studies in patients with solar urticaria and came up with a working classification distinguishing just two types [29]. Although there is general agreement as to the formal pathomechanism of solar urticaria, which is parallel with other types of immediate immunoreactions, more questions have been raised than have been answered regarding the pathogenesis of solar urticaria [18]. There are several reasons for lack of

Table 15.1 Classification of solar urticaria according to Harber *et al.* [13]

Type	Action spectrum	Passive transfer	Reverse passive transfer
I	285–320 nm	Positive	Positive
II	320–400 nm	Negative	Negative
III	400–500 nm	Negative	Negative
IV	400–500 nm	Positive	Negative
V	280–500 nm	Negative	Negative
VI	400 nm	n.d.	n.d.

n.d., no data

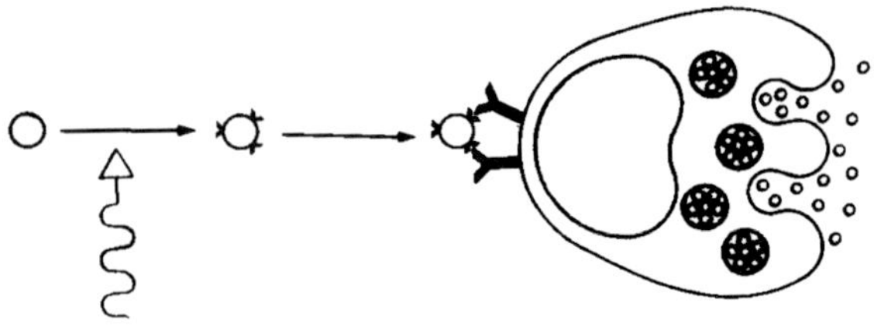

Fig. 15.1 Pathogenesis of solar urticaria. A precursor (chromophore) is activated by electromagnetic irradiation, thus forming a photoproduct (photoallergen). Specific IgE directed against the photoallergen is bound to the surface of mast cells. Bridging of IgE molecules by the photoallergen leads to release of histamine and probably also of other mediators.

progress in understanding the mechanisms. Action spectra, which could provide clues about chromophores, are usually broad and lack specific absorption maxima. For ethical reasons it is no longer feasible to perform passive and reverse passive transfer tests. In addition, an animal model is not available.

Pathomechanism and working classification of solar urticaria

In 1942, Rajka [43] passively transferred solar urticaria to normal subjects with the patient's serum. This passive transfer test (Prausnitz–Küster test) provides evidence for immediate hypersensitivity as the basic pathogenesis of solar urticaria. In addition, Epstein showed in 1949 [40] that reverse passive transfer of solar urticaria was also possible. In such experiments skin of normal subjects was first irradiated and subsequently the patient's serum was intradermally injected into these sites. Inconsistent results of passive and reverse passive transfer studies were obtained from different investigators, indicating that the mechanisms involved were not uniform in all cases. This and the awareness of different action spectra led to the classification developed by Harber *et al.* [13]. Subsequently, experimental evidence was produced revealing that this classification might not be appropriate. Type six (according to Harber) belongs to erythropoietic protoporphyria. Also, during the course of solar urticaria a broadening of the action spectrum was observed, indicating a transition from one type to another [37].

It is generally assumed that a precursor in the skin of patients acts as a chromophore and absorbs electromagnetic radiation of the action spectrum. The resulting photoproduct represents the putative photoallergen. Consecutively, specific IgE directed against the photoallergen is generated and bound to mast cells in the skin. If in such a sensitized patient the allergen is produced by further exposure to eliciting radiation, an allergic reaction of immediate type with whealing and flare is started as a result of histamine release from mast cells (Fig. 15.1).

Although the proposed pathogenesis is an immediate-type hypersensitivity reaction in all patients, the precursors (chromophores) seem to differ among patients. The precursor may be an abnormal endogenous substance generated only in the patient, or a normal factor present in both affected patients and normal subjects. On the basis of this concept two types of solar urticaria have been postulated [29]. The nature of the precursor can be evaluated by performing *in vitro* activation studies involving the injection of activated serum or plasma into the patient's skin or into the skin of normal subjects. In addition, the activated serum or plasma may be derived either from solar urticaria patients or from normal controls. The results of such experiments determine type I or type II of solar urticaria.

Type I

If an urticarial response develops in the patient's skin after injection of his or her own irradiated serum or plasma, but not to irradiated samples from normal control subjects, he or she is classified as having type I solar urticaria. In this case the precursor would be an abnormal endogenous substance occurring only in the patient. After irradiation, the specific precursor is converted into a photoproduct, the specific photoallergen. In type I solar urticaria the principal pathologic process is thus the presence of an individually specific precursor resulting in a specific photoallergen. Consecutively, specific IgE is formed.

Type II

If the patient reacts not only to his or her own irradiated serum or plasma but also to irradiated samples from normal subjects, he or she is classified as having type II solar urticaria. In this case the precursor would be a normal skin constituent present in both the patient and normal subjects. Irradiation converts this non-specific precursor into a photoproduct. It is only the patient who develops specific IgE against this photoproduct. Thus the pathologic process in type II solar urticaria is the development of an immediate-type hypersensitivity to a non-specific photoproduct acting as the photoallergen.

Results of *in vitro* activation studies as well as passive and reverse passive transfer tests support this hypothesis. A synopsis of experimental findings is presented in Table 15.2; further explanations are given in the section on chromophores and action spectra.

Table 15.2 Results of *in vitro* activation studies, passive and reverse passive transfer tests according to type of solar urticaria

Experimental procedure	Type I	Type II
In vitro activation studies		
Reaction of patient to irradiated serum or plasma of patient	Urticae	Urticae
Reaction of normal subject to irradiated serum or plasma of patient	No effect	No effect
Reaction of patient to irradiated serum or plasma of normal subject	No effect	Urticae
Reaction of normal subject to irradiated serum or plasma of normal subject	No effect	No effect
Passive transfer test	Variable	Positive
Reverse passive transfer test	Negative	Variable

Clinical features

Solar urticaria is a rare disease, exact figures on incidence are not available. Owing to the exquisite photosensitivity, afflicted patients are incapacitated in their daily lives, professionally and privately. Small amounts of sunlight, usually of a few minutes duration, or even artificial lighting in patients sensitive to visible irradiation, suffice to elicit the discomfortable eruption. The disorder appears throughout the world without specific geographic limitations. It has been reported in all races and ethnic groups [47]. The age of onset varies, ranging from 1 year to the eighth decade [11]; it seems to be common between the ages of 20 and 40. No predisposing genetic factors are known; only one familial case has been reported [42]. Solar urticaria occurs almost exclusively in otherwise healthy individuals. No relation to other disorders is known except for some other types of physical urticaria such as cold contact urticaria [46], dermographism [46], heat urticaria [55] and pressure urticaria (author's personal observation).

Within minutes after exposure to sunlight or appropriate artificial electromagnetic radiation, the typical wheal and flare reaction accompanied by intensive itching is induced. In some instances a delayed onset of several hours has been reported [35]. Wheals fade spontaneously without residues within a few hours. If urticarial lesions develop slowly and persist for days, an urticarial type of leukocytoclastic vasculitis induced by irradiation has to be considered [2]. All body areas including mucous membranes [23] are sensitive. Chronically sun-exposed skin areas, such as face and dorsa of the hands, are less sensitive owing to a hardening phenomenon following repeated sun exposure.

After exposure of large skin areas, shock fragments or collapse may occur owing to massive mediator release, probably mainly histamine [15, 16, 24, 52]. In addition, we have observed in one patient who was extremely light-sensitive, dizziness, sleepiness and muscular aches. The natural course of solar urticaria is not well documented. In most patients a sudden onset is reported. It generally persists for years [11]; spontaneous remissions seem to be rare.

Recently a variant of solar urticaria was described in which whealing occurred in a patchy pattern strictly localized to the same circumscribed skin areas [48]. In three patients wheals were reproducible in their pattern and location by eliciting radiation; the action spectra ranged from 320 to 700 nm, 320 to 585 nm and 400 to 560 nm. Intradermal injections of the patient's plasma activated by *in vitro* irradiation induced wheals only in the affected skin sites. For this subset of solar urticaria the term *fixed solar urticaria* was proposed. It is hypothesized that differences in the mast cell population of afflicted and non-afflicted skin sites account for the peculiar distribution pattern of lesions. Mast cells in affected skin sites contained numerous lipid bodies within the cytoplasm and were conspicuously devoid of association to fibrocytes as revealed by electron microscope studies.

The fixed variant of solar urticaria seems to be associated with only a moderate degree of photosensitivity. The patients are able to manage their disease by themselves. Graduated exposures to natural sunlight induce hardening. Systemic symptoms are absent, probably due to the relatively small skin areas involved and the resulting small amount of mediators released. Thus the patients are not dependent on medical assistance and this variant may be largely underreported.

Histopathology and ultrastructural findings

Histopathology of wheals in solar urticaria basically reveals features of other types of urticaria with some edema in the dermis and a scant inflammatory infiltrate consisting of mononuclear cells with an admixture of neutrophils and eosinophils. Studying the evolution of solar urticaria histologically, Norris and coworkers [38] found a dose-dependent increase, predominantly perivascular, in upper dermal neutrophil and eosinophil numbers at 5 min and 2 h after eliciting irradiation. At 24 h, mononuclear cells were increased in number. The appearance of eosinophils in the dermis was accompanied by tissue deposition of the eosinophil major basic protein, as a result of eosinophil degranulation in the course of solar urticaria [32]. It is still a matter of debate whether vasculitis is associated with solar urticaria. There are some case reports suggesting vasculitis as a feature of solar urticaria [2]; in some patients it has been documented by immunoelectron microscope investigations [41].

Sequential ultrastructural analysis demonstrated marginalization and activation of platelets, formation of interendothelial clefts, and alteration of nerve fibers as primary events in solar urticaria preceding mast cell degranulation [4]. Mediators other than histamine are most probably involved in the early stage of solar urticaria. Once the wheal is formed, mast cell degranulation is evident by dissolution of granular matrix, fusion of perigranular membranes, including labyrinth formation, and opening to the extracellular space. This is followed by extravasation of eosinophils, erythrocytes, and neutrophils. Nerve fibers, which initially demonstrate partial swelling, then show edema of endoneurium [4].

Histamine and other mediators

After local challenge of the skin various investigators found increased levels of serum histamine in the draining venous blood [15, 24, 52]. Also mast cell degranulation was demonstrated by electron microscope studies [4, 15, 24] as mentioned above. It is, however, one of the enigmas of solar urticaria that classical H1-blocking antihistamines are of little therapeutic value in patients with solar urticaria. At best, minimal urticaria doses (MUDs) are increased by a factor of two to three [34]. New H1-receptor antagonists with little sedative effect, like terfenadine and astemizole, were found to suppress the development of wheals in the majority of patients [6, 11, 36] if administered in high doses. In many patients erythema was not diminished. This indicates that mediators other than histamine are involved in the erythemal response. This is confirmed by the electron microscope finding that formation of interendothelial clefts in the dermal blood vessels precede the degranulation of mast cells in the dermis [4]. Candidates as mediators for immediate erythema are neurotransmitters like substance P or vasoactive intestinal protein. Also leukotrienes might be involved [1]. The lipid bodies found in mast cells in the so-called fixed solar urticaria might contain such lipid mediators.

Action spectra, inhibition spectra and chromophores

The action spectrum varies greatly among patients. It is usually found within a broad portion of optical radiation. It may reach from UVC, UVB and UVA to visible light [13, 45]. In most patients, however, the spectrum is confined to UVA and/or visible light (Table 15.3). Also, such patients have been reported from Japan [14, 17, 26]. Since patients are highly photosensitive, the minimal urticaria doses (MUD) are usually small, comparable to sun exposure for a few minutes.

A peculiar feature of some solar urticaria patients is the identification of inhibiting wavelengths. In most of these patients the activating wavelengths are in the UVA or short visible region and the inhibiting radiation is within the longer visible range [14, 21, 22]. We have, however, observed two patients with inhibiting spectra of shorter wavelengths than the eliciting irradiation [28, 48] (Table 15.4). In most patients inhibition is only possible when the skin is first activated by irradiation and subsequently or simultaneously exposed to inhibitory wavelengths [14, 21, 22, 28]. Exceptionally, inhibition by preirradiation has been reported [22, 53]. Inhibitory effects can also be exerted by *in vitro* irradiation of serum or plasma and subsequent injection into the patient's skin [28]. It seems likely that the photoallergen must first be generated before inhibition, probably by photodestruction of the photoallergen, can occur. Experiments suggest inactivation of the photoallergen rather than interaction with the precursor [21, 28]. The mechanism of inhibition by preirradiation remains unknown. Also an enhancement of solar urticaria by wavelengths other than the action spectrum was described. Solar urticaria confined to the visible range was enhanced by UVA. In this case the existence of an augmentation spectrum was postulated [19].

From the assumed pathomechanism of solar urticaria the existence of a chromophore in the skin is derived. Horio was the first to show that the chromophore or precursor is also present in serum or plasma of patients and can be activated by *in vitro* irradiation, thus forming the photoallergen [20]. In fact, in many patients the so-called serum factor has been identified by various investigators [8, 17, 26, 30, 51]. It is, however, not found in all patients [27]. Kojima *et al.* [26] as well as our group [30] observed patients whose precursor was present in plasma, but not in serum. In addition, we investigated a patient in whom a precursor was present neither in serum nor in plasma. The patient reacted to eluates of irradiated isolated epidermis from his own skin; unirradiated samples failed to induce wheals. This provided evidence that in this patient the precursor was probably only present in the epidermis [27].

Kojima *et al.* [26] investigated the nature of the photoallergen. They separated the *in vitro* activated serum or plasma by gel filtration and found the molecular weight of the photoallergen to be between 25 and 45 kD in three patients whose action spectrum was in the short visible range (400–500 nm). In other patients with a wide action spectrum the molecular weight of the photoallergen ranged between 300 and 1000 kD. It was concluded that the diversity of the action spectrum among different patients was due to a difference in chromophores.

Table 15.3 Synopsis of phototest results in solar urticaria

Sex	Age	Action spectrum (nm)	MUD (J/cm^2)	Inhibition spectrum (nm)	Plasma/serum factor	Distribution of wheals
F	30	Vis 400–800	5.0	n.d.	n.d.	Generalized
F	69	Vis 400–520	40.0	–	Positive	Generalized
M	22	Vis 400–500	0.8	–	Positive	Generalized
M	17	Vis 400–500	3.0	320–400	Positive	Generalized
M	47	Vis 400–430 UVA 330–400	20.0 6.0	–	Negative	Generalized
F	21	Vis 400–560 UVA 330–400	n.d. 2.0	–	n.d.	Fixed
M	55	Vis 400–700 UVA 330–400	7.0 5.0	280–320	Positive	Fixed
F	51	Vis 400–585 UVA 330–400	10.0 3.0	n.d.	Positive	Fixed
M	36	Vis 580 ± 10 UVA 360 ± 10 UVB 300 ± 5 UVC 250 ± 5	0.4 0.1 0.01 0.01	–	Negative	Generalized
F	40	UVA 330–400 UVB 280–320	0.5 0.0025	–	Positive	Generalized
F	41	UVA 360 ± 5 UVB 300 ± 5 UVC 250 ± 5	1.0 0.0015 0.01	n.d.	n.d.	Generalized
M	25	UVA 360 ± 5 UVB 300 ± 5 UVC 250 ± 5	0.025 0.001 0.01	n.d.	n.d.	Generalized

n.d., no data; MUD, minimal urticaria dose; vis, visible light.

Table 15.4 Action and inhibition spectra

Author, year, reference	Action spectrum (nm)	Inhibition spectrum (nm)
Hasei & Ichihashi 1982 [14]	400–500	>503
Torinuki *et al.* 1983 [53]	400–500	>660
Horio *et al.* 1984 [21]	400–500	>530
Leenutaphong *et al.* 1988 [28]	400–500	320–400
Reinauer *et al.* 1993 [48]	320–700	280–320

Passive transfer involves the injection of the patient's serum into normal skin followed by irradiation of the skin site. If skin irradiation is performed prior to injection of the patient's serum the procedure is referred to as reverse passive transfer testing. Sams [50] studied the passively transferred factor in two patients and demonstrated that it was neither IgG nor IgM. He suggested that it would probably be IgE. Solar urticaria that could not be transferred passively to normal subjects had been considered to be of a non-immunologic nature, probably phototoxic. Kojima *et al.* [26] found that irradiated serum from such patients yielded positive skin reactions when intradermally

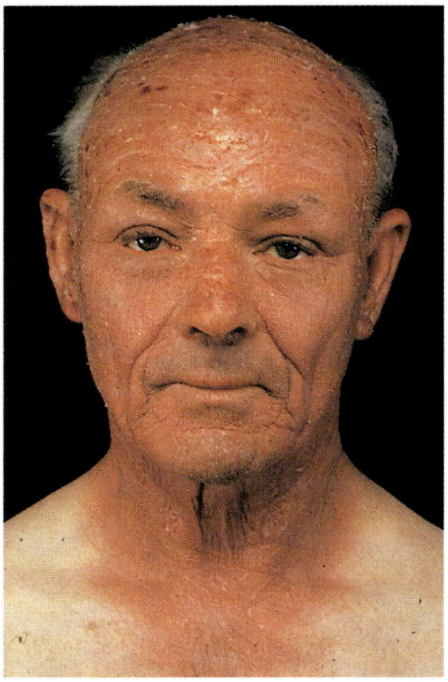

Plate 14.1

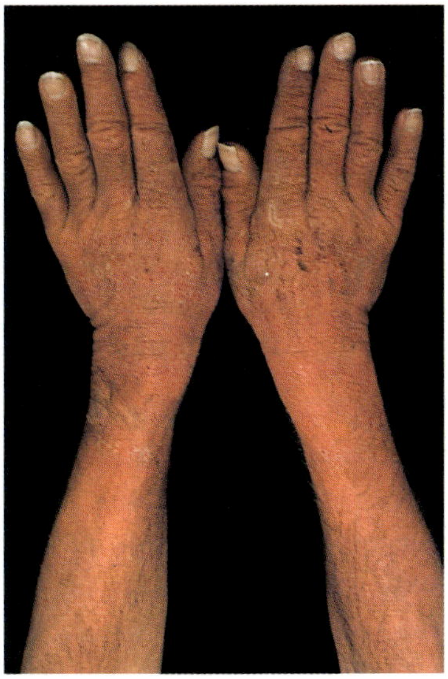

Plate 14.2

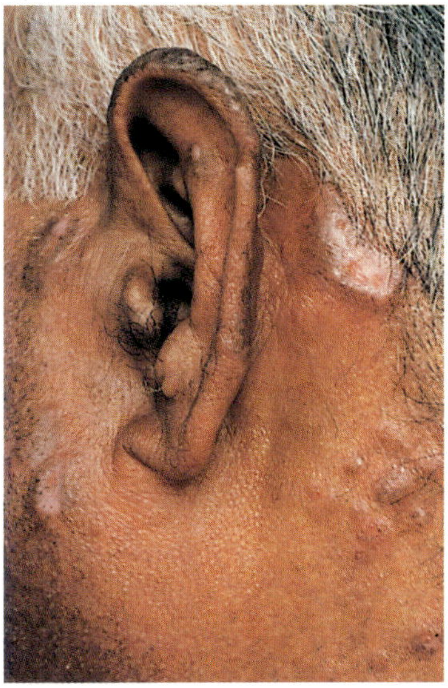

Plate 14.3

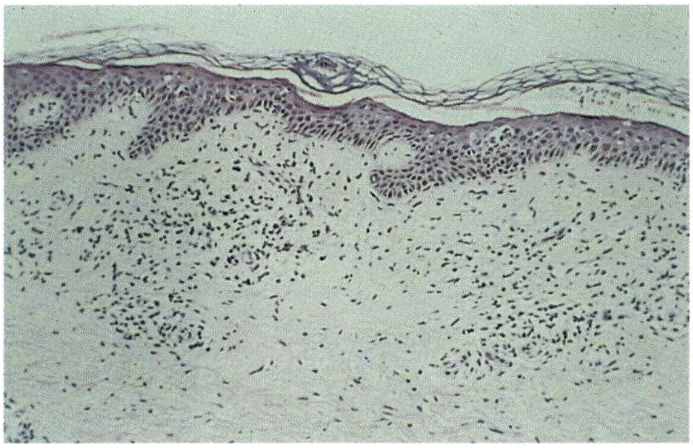

Plate 14.4

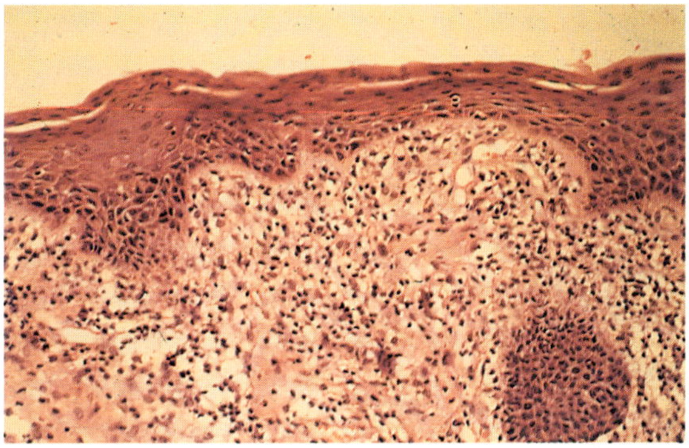

Plate 14.5

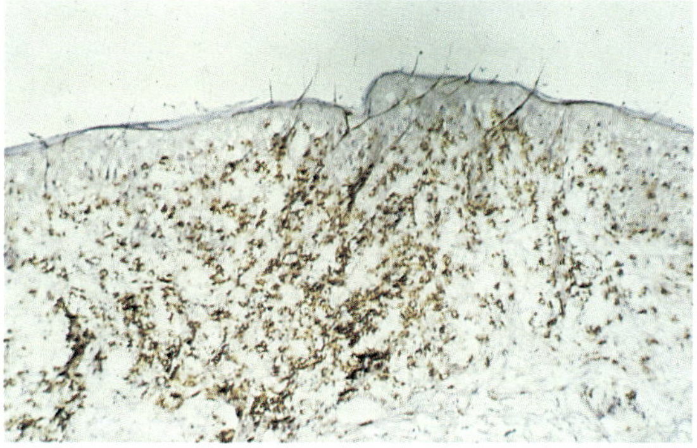

Plate 14.6

Plate 14.1 Eczematous change of CAD on light-exposed sites of the face and upper chest.

Plate 14.2 Eczematous change on the dorsum of hands and forearms.

Plate 14.3 Papular lesions in a dark-skinned patient with CAD, with sparing of the light-protected site behind the ear.

Plate 14.4 (H & E staining.) Lesions of CAD induced with solar-simulated irradiation and biopsied at 24 h, showing uper dermal lymphohistioytic infiltrate and epidermotropism.

Plate 14.5 (H & E staining.) Pseudolymphomatous change in CAD, with florid lymphohistiocytic infiltrate.

Plate 14.6 (Indirect immunoperoxidase stained with anti-CD 3.) T-cell infiltrate in the upper dermis extending into the epidermis in induced lesion of CAD biopsied at 24 h.

[*facing page 204*]

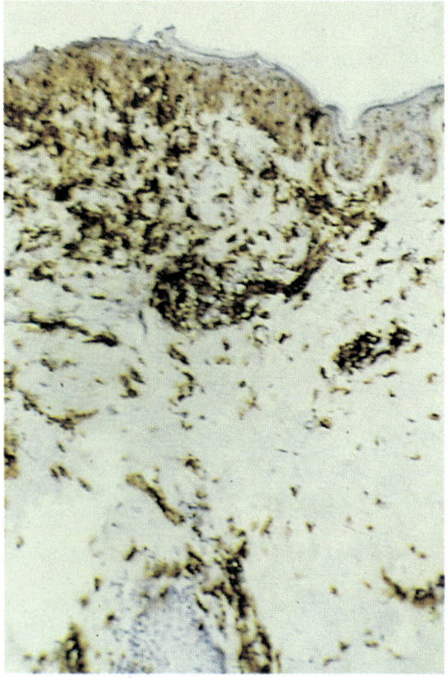

Plate 14.7

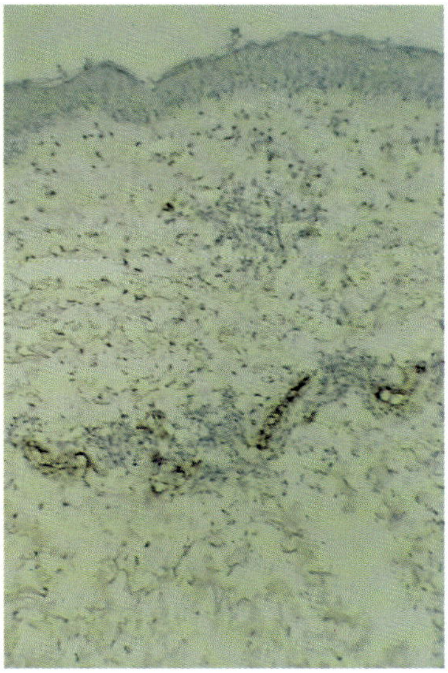

Plate 14.9

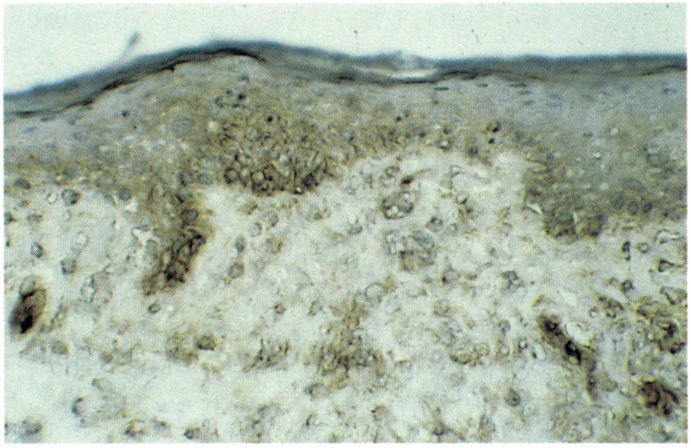

Plate 14.8

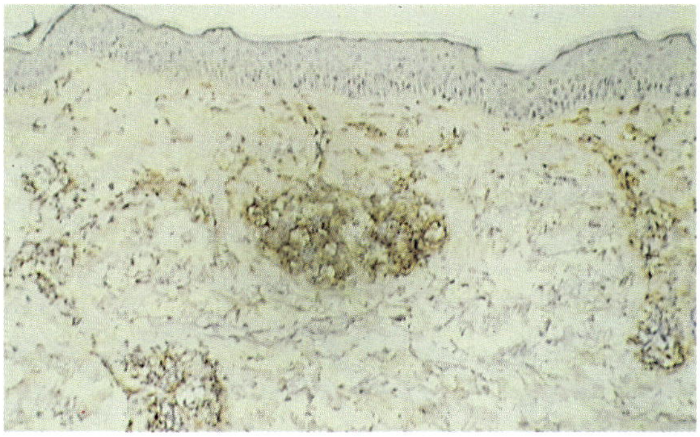

Plate 14.10

Plate 14.7 (Indirect immunoperoxidase stanied with anti-HLA DR.) HLA DR expression on keratinocytes, in a perivascular location, and on infiltrating cells, in induced lesion of CAD biopsied at 24 h.

Plate 14.8 (Indirect immunoperoxidase.) Induced lesions of CAD showing persistent upregulation of ICAM-1 on keratinocytes and in a perivascular distribution at 5 days.

Plate 14.9 (Indirect immunoperoxidase.) Persistent E-selectin expression in a perivascular location at 5 days.

Plate 14.10 (Indirect immunoperoxidase.) Upregulation of VCAM-1 in a perivascular location and on dermal interstitial cells at 24 h.

injected in the same patients but not in normal subjects. Thus, solar urticaria with negative passive transfer test results might also be based on an allergic mechanism. It was also Kojima and his group who pointed out the importance of a time lag between intradermal injection of serum and consecutive irradiation of the injected site.

To interpret passive and reverse passive transfer tests, several factors must be considered. First, samples of serum or plasma from the patient contain the precursor of the photoallergen as well as some specific IgE directed against the photoallergen. Second, the lag period between injection of serum and irradiation is required, since during this peroid IgE transferred with the serum adheres to the surface of mast cells. Kojima *et al.* [26] reported that results obtained in passive transfer tests 30 min after injection of serum were negative, but were positive when irradiation was postponed for 2–6 h. Third, in type I solar urticaria in which the pathologic agent is the precursor formed only in the patient, the precursor concomitantly transferred with the IgE antibody has to remain in the skin until IgE has bound to mast cells. If the precursor is already eliminated when irradiation with eliciting wavelengths is performed, the photoallergen cannot be generated and whealing is prevented. Therefore, results of passive transfer tests may be positive in some [17, 54] and negative in other patients [26] with type I solar urticaria. Fourth, in type II solar urticaria, results of passive transfer tests are always positive [26]. In this case, IgE, which binds to mast cells, is transferred, and histamine release is induced by formation of a non-specific photoallergen upon irradiation with eliciting wavelengths. The photoallergen is derived from a precursor that represents a natural factor and exists in the skin of everyone. As mentioned before, however, a lag period between injection of serum and irradiation is essential.

Epstein [10] first showed that reverse passive transfer of solar urticaria is also possible. Results of this test are negative in all patients with type I solar urticaria [17, 20, 54] because the precursor for type I solar urticaria is absent in normal human skin. In contrast, the photoallergen in type II solar urticaria is always produced by irradiation in human skin. If it remains at the test site until IgE transferred by the patient's serum adheres to the surface of mast cells, the result of reverse passive transfer is positive. If it is eliminated before IgE binds to mast cells, the test would be negative [26] (see Table 15.2).

Mechanisms of tolerance

In most types of physical urticaria, including solar urticaria, it is possible to induce tolerance by repeated exposure to the inducing stimulus [44]. The mechanism by which tolerance develops is not fully understood; a simple explanation would be tachyphylaxis to histamine or complete degranulation of mast cells, resulting in the exhaustion of mediator reserves. This, however, seems not to be the case. If histamine or a histamine-releasing agent (substance 48/80, codeine) are injected into tolerant skin, wheal and flare responses are unaltered compared to normal skin of the same patient [24, 30]. It was also shown by ultrastructural studies [4, 24] that mast cells in skin tolerant to further irradiation contain intact histamine granules. These findings indicate that mast cell mediator depletion or histamine tachyphylaxis are not responsible for the inhibition of solar urticaria. In further experiments [30] *in vitro* irradiated activated plasma or serum was used. This activated plasma or serum contained the photoallergen and was found to produce an urticarial reaction in the unexposed skin of the patients, but failed to produce a reaction at sites tolerant to radiation, proving that tolerance was not due to exhaustion of the photoallergen in skin. In addition, it was demonstrated that repeated injections of serum or plasma activated by *in vitro* irradiation also induced tolerance to eliciting radiation, but again, wheal and flare responses to intradermal codeine or substance 48/80 were unaltered. This demonstrates that elevation of the mast cell degranulation threshold as proposed by Keahey *et al.* [24] due to direct exposure of mast cells to UV irradiaton can be ruled out as the main underlying mechanism of tolerance. It is likely that the binding sites of IgE on mast cells remain occupied by the photoallergen during the state of tolerance and IgE-mediated histamine release from mast cells is blocked until new IgE is generated.

Table 15.5 Experimental investigations on mechanism of tolerance

Mode of inducing tolerance	Type of challenge	Result
Eliciting radiation	Histamine	Urticae
Eliciting radiation	Codeine	Urticae
Eliciting radiation	*In vitro* irradiated patient's plasma*	No response
None (control)	*In vitro* irradiated patient's plasma*	Urticae
In vitro irradiated patient's plasma*	*In vitro* irradiated patient's plasma*	No response
In vitro irradiated patient's plasma*	Eliciting radiation	No response
In vitro irradiated patient's plasma*	Codeine	Urticae

*In most patients, serum can be used also.

In Table 15.5 a set of experiments confirming this notion is presented.

Treatment of solar urticaria

Photoprotection by clothing and sunscreens

Sun protection by clothing or physical sunscreens (opaque preparations) provides protection, but these are not always feasible. Chemically absorbing sunscreens are of no help in view of the broad action spectrum, reaching frequently into the visible light, and exquisite photosensitivity.

Systemic medication

Reports on the effect of systemic medication are equivocal. There are sporadic observations of patients benefiting from treatment with antimalarial drugs [9]. Indomethacin and beta-carotene have been disappointing [17, 25]. Antihistamines provide protection to a variable degree. As mentioned above, the protection is only modest [34] and the use of this type of therapy on a regular basis is not recommended because of daytime sedative effects. Diffey and Farr [6], Ferguson [11] and our group [44] studied the effect of terfenadine, a high potency H1-receptor antagonist, and found that the majority of patients benefited from this regimen. It was also mentioned [6] that the wheal and flare reaction in solar urticaria, but not the immediate erythema, was suppressed by terfenadine. In some patients astemizole, another highly potent H1-receptor antagonist, was also shown to be successful in suppressing immediate erythema [36, 48]. Further studies are needed to determine the therapeutic value of new generation antihistamines. In our experience they fail in highly photosensitive patients given as monotherapy, but may be used in combination with other therapeutic regimens.

Phototherapy and photochemotherapy

Light hardening is a well-known feature of solar urticaria [7]. Induction of tolerance by phototherapy offers some practical means of treatment. The therapeutic effect is, however, short lived, lasting only for some days, usually less than 72 h [5, 44].

Photochemotherapy using psoralen and sunlight [44] or UVA irradiation [16, 39] is superior to the light hardening procedure, since it provides protection lasting for weeks [16, 40]. It is the treatment of choice. We adhere to a modified regimen originally proposed by Roelandts [49], employing a light-hardening procedure before PUVA therapy is commenced. Multiple partial or subthreshold whole body exposures induce complete tolerance before photochemotherapy is started, and then carried out following the usual PUVA protocol. This procedure eliminates the cumbersome initial phase and shortens the course of PUVA therapy. The mechanism by which PUVA therapy exerts its long-lasting effect is unknown. Since patients reacting to visible light alone fail to respond to UVA phototherapy [5], but do respond to photochemotherapy [40], and since the protective effect of PUVA lasts markedly longer than that of UVA alone, other mechanisms than induction of tolerance seem to be involved. It is conceivable that in addition to photoprotection by pigment formation and thickening of the stratum corneum, immunosuppression, perhaps downregulation of

IgE production, plays an important role. It was shown in atopic patients that UVB irradiation decreases plasma levels of IgE [3]. On the other hand, UVA and UVB irradiation were found to decrease the releasability of histamine from normal mast cells *in vivo* when challenged with codeine [12]. Perhaps such direct effects on mast cells should also to be considered.

Plasmapheresis

The rationale for employing plasmapheresis is elimination of the putative photoallergen from the patient's plasma. Experimental experience is scarce. Few patients have been reported [8, 31]. It was found that plasmapheresis was only effective in patients demonstrating a serum factor. Most patients experienced a partial transient remission. Only two patients remained completely free of symptoms.

References

1 Armstrong, R.B.: Solar urticaria. *Dermatologic Clinics* 4 (1986), 253–259.

2 Armstrong, R.B., D.B. Horan, D.N. Silver: Leukocytoclastic vasculitis in urticaria induced by ultraviolet irradiation. *Arch. Dermatol.* 121 (1985), 1145–1148.

3 Barlag, K., E. Hölzle, N. Schürer, G. Goerz: UVB-phototherapy reduces total serum IgE levels in patients with atopic dermatitis. *Arch. Dermatol. Res.* 283 (1991), 23–24.

4 Behrendt, H., P. Lehmann, V. Leenutaphong, E. Hölzle, G. Plewig: Sequential ultrastructural analysis of solar urticaria: inflammatory cells, blood vessels, and nerve fibers (Abstract). *J. Invest. Dermatol.* 92 (1989), 400.

5 Bernhard, J.D., K. Jaenicke, K. Momtaz-T, J.A. Parrish: Ultraviolet A phototherapy in the prophylaxis of solar urticaria. *J. Am. Acad. Dermatol.* 10 (1984), 29–33.

6 Diffey, B.L., P.M. Farr: Treatment of solar urticaria with terfenadine. *Photodermatol.* 5 (1988), 25–29.

7 Duke, W.W.: Urticaria caused by light. *JAMA* 80 (1923), 235–238.

8 Duschet, P., P. Leyen, T. Schwarz, P. Höcker, J. Greiter, F. Gschnait: Solar urticaria: effective treatment by plasmapheresis. *Clin. Exp. Dermatol.* 12 (1987), 185–188.

9 Epstein, J.H., J.J. Vandenberg, W.L. Wright: Solar urticaria. *Arch. Dermatol.* 88 (1963), 135–141.

10 Epstein, S.: Urticaria photogenica: report of two cases, one of them associated with purpura photogenica. *Ann. Allergy* 7 (1949), 443–457.

11 Ferguson, J.: Idiopathic solar urticaria: natural history and response to non-sedative antihistamine therapy. A study of 26 cases (Abstract). *Br. J. Dermatol.* 119 (suppl 33) (1988), 16.

12 Gollhausen, R., K. Kaidbey, N. Schechter: UV suppression of mast cell-mediated wealing in human skin. *Photodermatol.* 2 (1985), 58–67.

13 Harber, L.C., R.M. Holloway, V.R. Wheatley, R.L. Baer: Immunologic and biophysical studies in solar urticaria. *J. Invest. Dermatol.* 41 (1963), 439–443.

14 Hasei, K., M. Ichihashi: Solar urticaria: determinations of action and inhibition spectra. *Arch. Dermatol.* 118 (1982), 346–350.

15 Hawk, J.L.M., R.A.J. Eady, A.V.J. Challoner, A. Kobza-Black, T.M. Keahey, M.W. Greaves: Elevated blood histamine levels and mast cell degranulation in solar urticaria. *Br. J. Clin. Pharmacol.* 9 (1980), 183–186.

16 Hölzle, E., C. Hofmann, G. Plewig: PUVA-treatment for solar urticaria and persistent light reaction. *Arch. Dermatol. Res.* 269 (1980), 87–91.

17 Horio, T.: Photoallergic urticaria induced by visible light: additional cases and further studies. *Arch. Dermatol.* 114 (1978), 1761–1764.

18 Horio, T: Solar urticaria—sun, skin and serum. *Photodermatol.* 4 (1987), 115–117.

19 Horio, T., K. Fujigaki: Augmentation spectrum in solar urticaria. *J. Am. Acad. Dermatol.* 18 (1988), 1189–1193.

20 Horio, T., K, Minami: Solar urticaria: photoallergen in a patient's serum. *Arch. Dermatol.* 113 (1977), 157–160.

21 Horio, T., A. Yoshioka, H. Okamoto: Production and inhibition of solar urticaria by light exposure. *J. Am. Acad. Dermatol.* 11 (1984), 1094–1099.

22 Ichihashi, M., K. Hasei, K. Hayashibe: Solar urticaria: further studies on the role of inhibition spectra. *Arch. Dermatol.* 112 (1985), 503–507.

23 Illig, I., Born W.: Untersuchungen zur Pathogenese der Lichturticaria. *Arch. Dermatol. Res.* 220 (1964), 19–37.

24 Keahey, T.M., R.M. Lavker, K.H. Kaidbey, P.C. Atkins, B. Zweiman: studies on the mechanism of clinical tolerance in solar urticaria. *Br. J. Dermatol.* 110 (1984), 327–338.

25 Kobza, A., C.A. Ramsay, I.A. Magnus: Oral beta-carotene therapy in actinic reticuloid and solar urticaria. Failure to demonstrate a photoprotective effect against long wave ultraviolet and visible radiation. *Br. J. Dermatol.* 88 (1973), 157–166.

26 Kojima, M., T. Horiko, Y. Nakamura, T. Aoki: Solar urticaria: relationship of photoallergen and action spectrum. *Arch. Dermatol.* 122 (1986), 550–555.

27 Leenutaphong, V., E. Hölzle, G. Plewig: Evidence for an epidermal photoallergen in solar urticaria (Abstract). *J. Invest. Dermatol.* 91 (1988), 408.
28 Leenutaphong, V., E. Hölzle, G. Plewig: Solar urticaria induced by visible light and inhibition by UVA. *Photodermatol.* 5 (1988), 170–174.
29 Leenutaphong, V., E. Hölzle, G. Plewig: Pathomechanism and classification of solar urticaria: a new concept. *J. Am. Acad. Dermatol.* 21 (1989), 237–240.
30 Leenutaphong, V., E. Hölzle, G. Plewig: Solar urticaria: studies on mechanisms of tolerance. *Br. J. Dermatol.* 122 (1990), 601–606.
31 Leenutaphong, V., E. Hölzle, G. Plewig, B. Kutkuhn, B. Grabensee: Plasmapheresis in solar urticaria. *Dermatologica* 182 (1991), 35–38.
32 Leiferman, K.M., P.G. Norris, G.M. Murphy, J.L. Hawk, R.K. Winkelmann: Evidence for eosinophil degranulation with deposition of granule major basic protein in solar urticaria. *J. Am. Acad. Dermatol.* 21 (1989), 75–80.
33 Merhlen, P.: Urticae. In: *La Practique Dermatologique*, (1904), pp. 728–771. Eds.: E. Besnier, L. Brocq, L. Jacquet, Masson & Cie, Paris.
34 Michell, P., J.L.M. Hawk, A. Shafir, M.F. Corbett, I.A. Magnus: Assessing the treatment of solar urticaria. The dose–response as a quantifying approach. *Dermatologica* 160 (1980), 198–207.
35 Monfrecola, G., P. Nappa, D. Pini: Solar urticaria with delayed onset: a case report. *Photodermatol.* 5 (1988), 103–104.
36 Monfrecola, G., P. Nappa, D. Pini: Solar urticaria in the visible spectrum successfully treated with astemizole. *Dermatologica* 180 (1990), 154–156.
37 Murphy, G.M., Hawk, J.L.M.: Broadening of action spectrum in a patient with solar urticaria. *Clin. Exper. Dermatol.* 12 (1987), 455–456.
38 Norris, P.G., G.M. Murphy, J.L. Hawk, R.K. Winkelmann: A histological study of the evolution of solar urticaria. *Arch. Dermatol.* 124 (1988), 80–83.
39 Parrish, J.A., K.F. Jaenicke, W.L. Morison, K. Momtaz, C. Shea: Solar urticaria: treatment with PUVA and mediator inhibitors. *Br. J. Dermatol.* 106 (1982), 575–580.
40 Plewig, G., E. Hölzle, P. Lehmann: Phototherapy for photodermatoses. In: *Current Problems in Dermatology Band 15. Therapeutic Photomedicine*, (1986), pp. 254–264. Eds.: H. Hönigsmann, G. Stingl, Karger, Basel.
41 Plewig, G., H.H. Wolff, E. Hölzle: Solar urticaria: immunoelectron microscopic study of leukocystoclastic vasculitis (Abstract). *Arch. Dermatol. Res.* 267 (1980), 209.
42 Polano: Urticaria solaris. *Dermatologica* 97 (1948), 327–328.
43 Rajka, E.: Passive transfer in light urticaria. *J. Allergy Clin. Immunol.* 13 (1942), 327–345.
44 Ramsay, C.A.: Solar urticaria treatment by inducing tolerance to artificial radiation and natural light. *Arch. Dermatol.* 113 (1977), 1222–1225.
45 Ramsay, C.A., R.J. Scrimenti, D.J. Cripps: Ultraviolet and visible action spectrum in a case of solar urticaria. *Arch. Dermatol.* 101 (1970), 520–523.
46 Rantanen, T., R. Suhonen. Solar urticaria: a case with increased skin mast cells and good therapeutic response to an antihistimine. *Acta Dermatovenereol. (Stockh.)* 60 (1980), 363–365.
47 Ravits, M., R.B. Armstrong, L.C. Harber: Solar urticaria: clinical features and wavelength dependence. *Arch. Dermatol.* 118 (1982), 228–231.
48 Reinauer, S., V. Leenutaphong, E. Hölzle: Fixed solar urticaria. *J. Am. Acad. Dermatol.* (1993), 161–165.
49 Roelandts, R.: Pre-PUVA UVA desensitization for solar urticaria. *Photodermatol.* 2 (1985), 174–176.
50 Sams, W.M.: Solar urticaria: studies of the active serum factor. *J. Allerg.* 45 (1970), 295–301.
51 Sams, W.M., J.H. Epstein, R.K. Winkelmann: Solar urticaria. Investigation of pathogenetic mechanisms. *Arch. Dermatol.* 99 (1969), 390–397.
52 Soter, N.A., S.L. Wassermann, M.A. Pathak, J.A. Parrish, K.F. Austen: Solar urticaria: release of mast cell mediators into the circulation after experimental challenge (Abstract). *J. Invest. Dermatol.* 72 (1979), 282.
53 Torinuki, W., N. Kumai, T. Miura: Solar urticaria inhibited by visible light. *Dermatologica* 166 (1983), 151–155.
54 Torinuki, W., H. Tagami: Solar urticaria without inhibitory spectrum: demonstration of both circulating photoallergen and reaginic antibodies. *Dermatologica* 173 (1986), 116–119.
55 Willis, I., J.H. Epstein.: Solar- vs heat-induced urticaria. *Arch. Dermatol.* 110 (1974), 389–392.

16: Photoimmunology of lupus erythematosus

David A. Norris

Introduction

Lupus erythematosus is the most familiar photosensitivity disorder known to physicians in all fields of medicine. Even the word lupus refers to something dreaded and vicious invoking the appearance of skin lesions resembling the bite of a wolf. For as long as physicians have appreciated the role of sunlight in exacerbating cutaneous and systemic disease in lupus erythematosus, they have debated the mechanisms by which UVR activates clinical disease in lupus. This review will consider many of the important aspects of UVR as it pertains to the syndrome of lupus erythematosus.

Clinical spectrum of photosensitive lupus

Lupus erythematosus is a spectrum of overlapping clinical syndromes that encompass many different patterns of autoimmune damage to numerous organ systems [106]. The production of autoantibodies is a hallmark of most forms of lupus. These autoantibodies are usually specific for sequestered nuclear and/or cytoplasmic antigens [107, 123]. When these antigens are released or translocated, they bind autoantibodies to form circulating immune complexes or tissue complexes which activate immunologic effector mechanisms that produce organ or cellular damage [2, 95, 100].

Clinical involvement of the skin in lupus is seen in a diverse spectrum of clinical syndromes. Photosensitivity is one of the major diagnostic criteria of systemic lupus erythematosus (SLE), and is seen in other types of LE with more limited organ involvement such as subacute cutaneous lupus erythematosus (SCLE), neonatal lupus erythematosus (NLE) and chronic cutaneous lupus erythematosus (CCLE).

Patients with SLE may develop an acute cutaneous lupus erythematosus (ACLE) syndrome with the distinctive photosensitive malar ('butterfly') eruption in association with palmar erythema, oral erosions, cheilitis, urticarial plaques, and an interphalangeal eruption on the dorsal aspects of the fingers. A number of autoantibodies including anti-SSA/Ro may be found [93, 107].

CCLE or discoid lupus is a very distinctive papulosquamous eruption producing characteristic histologic features and clinical features of dyspigmentation, follicular plugging, induration and scarring [25, 112]. Although it is commonly exacerbated by sun exposure, it can also be found in photoprotected areas [112]. It is usually seen with limited systemic disease [111, 112]. Although many CCLE patients have no significant autoantibodies [110, 112] some have anti-SSA/Ro or anti single-stranded DNA [68, 107].

SCLE and NLE are two highly photosensitive syndromes with similar papulosquamous eruptions characterized by a diffuse non-indurated erythema [25, 75], a sparse lymphocytic infiltrate [9] and intraepidermal deposits of IgG in a particulate pattern [75]. SCLE is commonly associated with arthritis, but usually not with cerebritis or nephritis. It is highly associated with antibodies to SSA/Ro and to SSB/La [27, 100, 118, 120]. NLE presents with isolated congenital heart block (ICHB), with a photosensitive papulosquamous eruption or with both. The association with antibodies to SSA/Ro or SSB/La is nearly 100%. The skin disease rapidly fades coincident with the catabolism in the infant of these maternally transmitted autoantibodies.

Photosensitivity and phototesting in lupus erythematosus

It was suspected for many years that light was a precipitating factor in lupus erythematosus. Case reports of apparent induction and spread of lupus lesions by artificial light sources or natural sunburn brought attention in the United States to the potential risk associated with light exposure [14, 33]. The specificity of sun exposure in inducing lupus was questioned by Kestin and Slatkin, who proposed that induction of lupus was a Koebner-like reaction [60]. In 1965, three groups clearly demonstrated the ability of UVR to induce abnormal skin reactions in lupus patients [8, 31, 32]. These studies demonstrated that involved skin was more likely to show abnormal responses to light, and that sequential small doses of radiation were necessary to induce lesions that were histologically and clinically similar to lupus. In multiple studies of photosensitive SLE and CCLE, the action spectrum was in the UVB range [23, 35, 64].

Recently, interest in UVA in the induction of lupus lesions has grown. Hölzle summarized extensive studies of UVR induction of lesions in patients with CCLE, SCLE and SLE. SCLE patients were the most sensitive, with 90% showing sensitivity to sequential UVR administered to large areas of skin; 40% of DLE and 10% of SLE patients were also susceptible. Although UVB was necessary for induction, UVA also proved to be quite effective [52]. In a recent extension of these studies, Kind and his colleagues showed that 53% of SCLE patients reacted to a combination of UVA plus UVB, 33% reacted to UVB alone, and 14% to UVA alone [62, 81]. Kind also showed that the presence of anti-Ro was not necessary for development of photosensitivity in these SCLE patients. However, anti-Ro antibodies were highly correlated with the epidermal changes seen in one subclass of the SCLE patients. Anti-Ro was not associated with positive phototesting in CCLE or in lupus tumidus [62]. However, anti-Ro was associated with photosensitivity in photosensitive SLE, as originally described by Rothfield [91].

These modern phototesting protocols underline a number of principles important in understanding the mechanisms of induction of lupus lesions by UVR.

1 Both UVA and UVB are involved in inducing clinical lupus lesions, implicating dermal effects of UVA on mast cells and endothelial cells as well as UVB effects on the epidermis in inducing lupus skin lesions.

2 Photosensitivity in lupus is heterogeneous even in apparently similar patient populations.

3 Induction of skin lesions by UV light alone is difficult, requiring multiple exposures of large areas of skin over extended periods of time.

4 Antibodies to SSA/Ro are highly associated with photosensitivity in SLE, SCLE and NLE, but are not universal requirements for photosensitivity in lupus.

5 An exaggerated sunburn response in lupus patients is probably not the only component of UVR induction of LE skin lesions.

Proposed mechanisms of photosensitivity in lupus erythematosus

By what mechanism(s) does UVR induce the development of lupus skin lesions? Is UVR a specific trigger of the immunologic mechanisms that produce lupus skin lesions? Does UVR induce photosensitive lupus through an exaggerated release of mediators involved in sunburn? Does UVR trigger both specific and non-specific pathways which produce clinical lupus? The answers to these questions have not yet been resolved.

Initial studies of UVR effects in lupus found that cells from lupus patients showed enhanced cytotoxicity when exposed to UVR, and attempted to link this enhanced cytotoxicity to increased DNA damage. Zamansky and colleagues demonstrated that fibroblasts from SLE patients or from mouse models of lupus show enhanced cellular cytotoxicity following UVB and UVC radiation [39, 141, 142]. Other investigators showed spontaneous increases in DNA synthesis in murine SLE [37]. It was proposed that such cytotoxic damage was due to diminished repair of DNA-photoadducts, as observed in xeroderma pigmentosum [139]. However, cells from SLE patients were found to repair UVA-, UVB- and UVC-induced DNA damage normally [39, 114, 138]. The enhanced cytotoxic damage seen when testing cells from lupus patients has not yet been adequately explained.

Effects of UV light on autoantibody production and tissue binding of autoantibodies, and effects of the epidermal and dermal immunologic environment, have demonstrated many possible sites of action. Both UVB and UVA can enhance DNA release and induce anti-DNA antibodies in animal models [2, 37]. Irradiation of skin in mice immunized to DNA can induce tissue binding of anti-DNA [93]. However, the histology of the lesions in these models is not similar to those seen in photosensitive LE.

A number of important UV effects on keratinocytes have been described that may contribute to UVR induction of lupus skin lesions. Ultraviolet light can induce translocation of antigens recognized by autoantibodies associated with photosensitive cutaneous lupus [36, 38, 77, 80]. Antibodies bound to translocated antigen may be a factor in induction of epidermal damage in photosensitive SCLE and NLE. UVB and UVC may also directly disrupt intermediate filament structure in keratinocytes [140]. UVB also triggers release of cytokines from keratinocytes [115].

In some early phototesting protocols, prolonged UV erythema was described in lupus patients, suggesting an abnormal level of UVR-induced mediators or an enhanced response to UVR-induced mediators. UVB and UVA induce a number of important inflammatory mediators necessary for activation and mobilization of leukocytes involved in LE skin lesions [53]. UVB and UVA produce striking changes in the immunologic environment of the epidermis which promote the specific immunologic reactions which are believed to be involved in different types of lupus skin lesions. It is likely that photosensitive lupus results from both a concurrence of multiple components, involving both exaggerated response to UVR-induced mediators, and the induction or enhancement of immunologic cytotoxic mechanisms. The purpose of the remainder of this review will be to consider the complex UVR effects that appear to contribute to the induction of photosensitive LE.

Determinants of clinical spectrum

It is not clear what defines the differences among the photosensitive lupus syndromes. The unifying characteristic of lupus syndromes in humans and animals is the production of autoantibodies to sequestered nuclear or cytoplasmic antigens [107, 123]. These autoantibodies mediate the specific organ damage which produces different patterns of clinical disease. There are many arguments which link antibodies to SSA/Ro, or SSB/La directly to keratinocyte cytotoxicity in photosensitive LE, especially in SCLE and NLE [99, 100, 120]. The specificity and role of autoantibodies in photosensitivity in SLE is less clear, and in CCLE other immunologic effector mechanisms such as autoreactive T cells may be critical.

Keratinocyte cytotoxic damage is a common feature in photosensitive lupus syndromes [125], but in some patients dermal targets appear to be more important. Ultraviolet light produces many important effects in photosensitive LE including release or translocation of sequestered antigens [6, 36, 37, 80] and release of proinflammatory mediators and cytokines [3, 135]. The mechanisms of photosensitive lupus are likely to involve specific immunologic factors and less specific target organ features.

I have proposed that photosensitivity in lupus is determined by three possibly interrelated characteristics [99].

1 Susceptibility to UVR-induced release of epidermal and dermal cytokines.

2 Susceptibility to UVR-induced release or translocation of sequestered antigens in the epidermis or dermis.

3 Different specific immunologic effector mechanisms, activated by cytokines and directed against discrete dermal or epidermal targets.

There are significant differences in the patterns of autoantibody and/or genetic associations among the different photosensitive lupus syndromes. Therefore it seems quite logical to propose that the differences in the photosensitive lupus syndromes are determined in some measure by the nature of the specific cytotoxic effector mechanisms which mediate keratinocyte and/or endothelial cell injury. Genetically determined differences in cytokine release following UVR may be one important determinant of which lupus patients develop photosensitive skin disease, and of the racial and ethnic differences in clinical presentations of different cutaneous lupus syndromes. On the other hand, it may be found that some UVR effects such as cytokine or inflammatory mediator release, or antigen translocation may be quite similar among the different photosensitive lupus syndromes.

Ultraviolet light-induced mediators

The mediators of UVR-induced erythema are likely to be an important factor in the induction of photosensitive LE lesions. The mediators involved in induction of UV-induced erythema are complex, including eicosanoids, vasoactive mediators such as histamine, and possibly cytokines. They are released from keratinocytes, mast cells, endothelial cells, and perhaps fibroblasts and Langerhans cells. UVB induced erythema peaks at 24–48 h after irradiation, and is mediated by release of arachidonic acid, PGE_2 and $PGF_{2\alpha}$ from keratinocytes, and by histamine from mast cells [13, 42, 48]. Eicosanoids such as PGD, LTC4 and LTD4 may also be released from mast cells by UVR, but this has not been proven experimentally. UVA-induced erythema is also mediated by increased arachidonic acid, PGE_2 and $PGF_{2\alpha}$, prostacyclin from endothelial cells and histamine [47]. UVB and UVA both increase phospholipase activity in keratinocyte cell membranes, increasing arachidonic acid [20], which is a substrate for both cyclooxygenase and lipoxygenase pathways. PG synthesis is increased early by increased phospholipase activity which increases the AA substrate, and later by direct increases in enzymatic cyclooxygenase activity [44]. The fine control of PG synthesis following UVB appears to be quite complex, having been attributed to histamine [105], to combined influence by TNFα and IL-1 on keratinocytes [43] or to tyrosine kinase receptor activation, possibly through the EGF receptor [104].

The immunoregulatory cytokines IL-1, IL-6 and TNFα are also released by UVR, and are likely to be necessary for immunological activation of the epidermis, dermis and leukocytes necessary for induction of lupus lesions. UVB radiation induces transcription of the gene for IL-1 [70], and induces circulating levels of IL-1-like activity in irradiated humans [41]. However, keratinocyte IL-1 is an intracellular molecule without a signal sequence necessary for secretion. IL-1α can be expressed on the cell membrane, but is released from monocytes and macrophages by induction of cell death. Triggering monocyte necrosis causes release of processed and unprocessed IL-1α and unprocessed IL-1β. Induction of apoptosis in the same cells results in release of processed IL-1α and IL-1β [50]. Release of such molecules by UVR probably requires keratinocyte lysis [51]. In addition, biologically active keratinocyte IL-1α occurs along with a 10–20-fold excess of its antagonist IL-1ra, which is also an intracellular molecule without a signal sequence, probably released by lysis [12, 46]. At present it is not clear whether ambient UVR induces sufficient keratinocyte lysis to induce significant IL-1 release, or whether the accompanying icIL-1ra significantly blunts the biologic results of this response.

Keratinocytes synthesize and release IL-6. UVR is one trigger of IL-6 release [63]. Measurable serum levels of IL-6 can be found in human subjects following UVR [127]. However, in a transgenic model of IL-6 overexpression in the epidermis of mice, limited cellular infiltrates were seen, bringing into question the proinflammatory aspects of IL-6 alone in the epidermis [126] without concomitant leukocyte activation, or without concomitant action of IL-1 or TNFα.

TNFα is released from human keratinocytes by UV radiation [65]. This effect is UVB-, not UVA-dependent, and is greatest in more differentiated keratinocytes [90]. There is strong circumstantial evidence that the release of TNFα by UVR is dependent on the isomerization of *trans*-UCA to *cis*-UCA [72]. UVB induces local suppression of contact hypersensitivity through effects on epidermal Langerhans cells [130], which are significantly dependent on the release of TNFα [72, 136, 137]. UVR also induces local release of IL-10 which can suppress delayed-type hypersensitivity in the skin [113].

IL-1 and TNFα are 'primary' cytokines with profound effects on many cellular functions [3]. IL-1 is a major costimulatory signal for T-cell activation by conventional or superantigens; it is a pyrogen, stimulates acute phase reactants, and stimulates collagenase production; it stimulates Langerhans cell function in the epidermis; it induces the secondary release of IL-1, GM-CSF and IL-8 [28, 69]. TNFα stimulates monocytes, B and T lymphocytes, LAK cells, monocytes and neutrophils; it also stimulates Langerhans cell emigration from the epidermis or inactivation and immobilization of resident Langerhans cells within the epidermis; it can be a direct cytotoxic factor; it stimulates the secondary release of numerous cytokines and mediators (IL-1, IL-6, IL-8, IFN-β, GM-CSF, PAF and PG) from various cellular sources [3, 19, 54, 82].

Ultraviolet light induction of adhesion molecules

UVR induces the release of cytokines which act alone or in synergy with cytokines released by leukocytes to stimulate the expression of adhesion molecules on endothelial cells, leukocytes or parenchymal targets of inflammation. ICAM-1 (CD54) is the prototype of adhesion molecules crucial to normal cellular function in the immune response. ICAM-1 is a ligand for the B2 integrin LFA-1 [86], found on most leukocytes. The ICAM-1/LFA-1 interaction is necessary to initiate many of the important immunologic functions which require leukocyte/leukocyte, leukocyte/endothelial cell, or leukocyte target interaction [14, 74, 83]: specific antigen recognition by B and T lymphocytes [24, 73, 83], superantigen stimulation of T lymphocytes [34], leukocyte migration into tissue [117] and leukocyte-mediated cytotoxicity [84, 98]. Unlike endothelial cells, fibroblasts and inflammatory cells, keratinocytes have very low constitutive expression of cell surface ICAM-1 [29]. It has been proposed that such differences in ICAM-1 expression are important in the functional differences among skin cells during immune responses [97, 98].

Modulation of expression of ICAM-1 on the surface of epidermal keratinocytes by UVR is likely to be of central importance in those types of photosensitive lupus in which keratinocyte damage is seen. Epidermal keratinocytes and melanocytes express low levels of cell surface ICAM-1 [29, 134]. The low constitutive expression of ICAM-1 in keratinocytes is determined by a promoter region which suppresses ICAM-1 gene transcription [22]. Figure 16.1 indicates the major cytokines which influence keratinocyte cytokine expression. Keratinocytes express IFN-γ receptors [91], IL-1 type I receptors [71] and the 55 kd TNFα receptors [124]. Ligand binding to the IL-1 receptor can be inhibited by the IL-1 receptor antagonist (IL-1ra). In monocytes IL-1ra is found as a secreted form released by cellular activation; in keratinocytes IL-1ra is found in 20-fold excess over IL-1α as a non-secreted intracellular form [12]. TNFα or -β binding to the 55-kd receptor can be blocked by the soluble TNF–INF

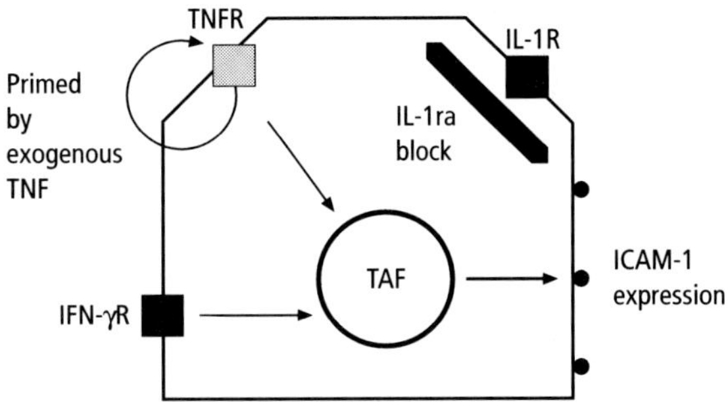

Fig. 16.1 Cytokine control of keratinocyte ICAM-1 production. Three major receptor pathways for induction of ICAM-1 expression in keratinocytes are illustrated schematically. Ligand binding and activation of IFN-γR, TNFαR or IL-1R induces transcription of ICAM-1 mRNA through responsive elements in the ICAM-1 promoter region. The number of TNF receptors is increased by TNFα release or exogenous cytokine. The activation of IL-1R and subsequent ICAM-1 induction appears to be blocked by the ic form of IL-1ra.

which is a proteolytic fragment of the receptor [116]. IFN-γ and TNF synergistically activate transcription of the ICAM-1 gene [10]; both IFN and TNF responsive elements are present in the promoter region of the ICAM-1 gene [18, 22]. An apparent IL-1 responsive element is also present, but is not inducible when the gene is transfected into keratinocytes [22]. Experimental evidence supports the hypothesis that activation of the ICAM-1 gene by IL-1 binding to the IL-1 receptor is blocked by the large excess levels of the intracellular form of IL-1ra found in epidermal keratinocytes. Affinity purified icIL-1ra blocks ICAM-1 induction by IL-1 in responsive cell lines [89]. Keratinocyte cell lines which respond to IL-1 with increased ICAM-1 expression are those with low levels of icIL-1ra; those which do not respond to IL-1 have high levels of icIL-1ra [89]. The blockade of IL-1 effects by IL-1ra in keratinocytes appears to be selective, since GM-CSF and IL-8 production are induced by IL-1 even in the presence of high levels of IL-1ra [89].

Ultraviolet light has a biphasic effect on ICAM-1 expression. In the first 24 h following UVR, ICAM-1 induction by gamma interferon is inhibited. After 24 h, ICAM-1 expression is induced by UVR, and IFN-γ-induced ICAM-1 expression is significantly enhanced by UVR [67, 101]. In the first 24 h following UVR, ICAM-1 induction is inhibited by interference with transcription through disruption of TAF [66]. The late effects of UVR on ICAM-1 appear to be mediated through TNFα release and IL-1 in different keratinocyte-derived cell lines. In the transformed KB cell line, both IL-1 and TNFα appear to be involved in UVR induction of ICAM-1 [43]. In non-transformed early passage human keratinocyte cultures, TNFα appears to be the principal mediator of ICAM-1 induction. TNFα-induced ICAM-1 expression can be blocked with anti-TNF in some strains of UVR-treated human keratinocytes [89]. IL-1 induction of ICAM-1 expression may be naturally blocked by the high levels of icIL-1ra in keratinocytes [89]. Exogenous purified icIL-1ra blocks respond to IL-1 in induction of ICAM-1 in cultured human keratinocytes [89]. Finally, there are significant individual differences in the level of ICAM-1 expression induced by IFN-γ, TNFα or UVR. In a large study of over 50 individual strains of keratinocytes, TNFα and UVR responses are closely linked in individual strains [89]. None of the human keratinocyte strains tested showed significant induction of ICAM-1 expression by exogenous IL-1.

Thus UVR produces profound effects on the epidermis, inducing release of inflammatory mediators, induction of adhesion molecules, and migration of cell populations, as illustrated in Fig. 16.2. UV-irradiated keratinocytes release IL-1 and TNFα, the latter through UVR production of *cis*-UCA. Irradiated epidermis also releases IL-10, most likely from keratinocyte sources [30]. IL-1α may be released by keratinocyte lysis, but whether significant biologic effects are produced may depend on how much icIL-1ra is also released. The relative excess of IL-1ra is increased in more differentiated keratinocytes, the cells which receive the greatest UVR. Other cytokines such as GM-CSF, IL-6 and IL-8 are released directly or secondarily through IL-1 or TNFα. Local production and release of these cytokines, plus prostaglandins, prostacyclin and histamine may produce important local or systemic effects. Release of TNFα (and perhaps IL-1) and histamine induce ICAM-1 expression throughout the epidermis but especially in basal layer keratinocytes. The effects of UVR on cytokine release and on ICAM-1 expression in the epidermis are stratified. The greatest cytokine release is in the more differentiated superficial epidermis, while the greatest ICAM-1 expression is induced in the basal layer. UVB decreases the number of epidermal Langerhans cells. Streilein proposed that UVB immobilizes Langerhans cells in the epidermis through UVB effect [130, 136], while others have shown that UVB acting through TNFα induces Langerhans cells to emigrate from the epidermis to regional lymph nodes [61, 92]. Melanocyte proliferation and melanogenesis may be inhibited by IL-1, IL-6 and TNF [122]. Inflammatory mediators released in the dermis may also participate in induction of keratinocyte adhesion molecules which promote leukocyte keratinocyte attach-

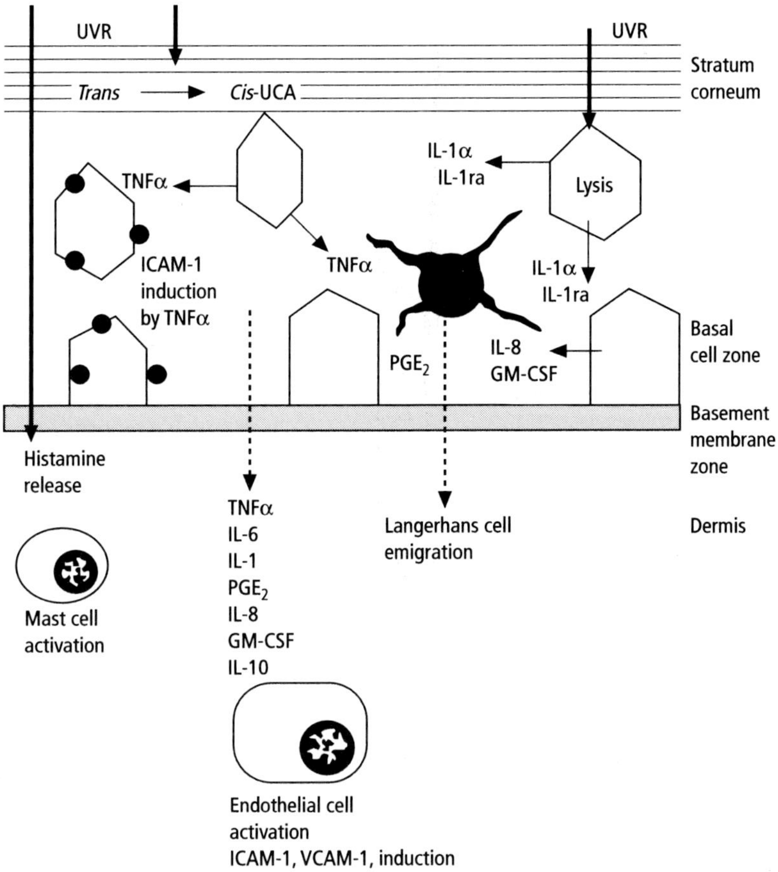

Fig. 16.2 UVR effects on epidermal cell populations. A complex array of cytokines and inflammatory mediators are released in the epidermal environment by UVR. UVR effects cytokine release through formation of *cis*-UCA, through keratinocyte lysis, and secondarily through the effects of released IL-1 and TNFα. A complex mix of cytokines (TNFα, IL-6, IL-1, PGE_2, IL-8, GM-CSF and IL-10) are all liberated by UVR and may have local or systemic effects. Mast cell degranulation and endothelial cell activation, plus upregulation of keratinocyte ICAM-1 can be induced by UVR. There is still some debate as to whether cytokines released by UVR induce Langerhans cells to be immobilized or stimulated to leave the epidermis.

ment preceding immunologic keratinocyte lysis.

Endothelial cell activation is also important in induction of photosensitive lupus. Injection of IL-1, TNFα or IL-6 into skin induces mixed leukocyte infiltrates comprised of macrophages, lymphocytes and neutrophils [128]. Migration of leukocytes into the skin requires sequential expression of adhesion molecules on leukocytes and endothelial cells, gradients of chemotactic factors, and epidermal, dermal or adnexal sources of attractants. Transient attachment of leukocytes to regionally activated endothelial cells is mediated by receptors termed selectins. Activation of loosely adherent leukocytes upregulates integrin receptors which bind to induced endothelial receptors, causing enhanced attachment and then directed emigration into tissue. Induction of E-selectin or P-selectin on endothelial cells induces attachment of lymphocytes including memory T cells and skin-homing T cells to the endothelial cells in the cutaneous

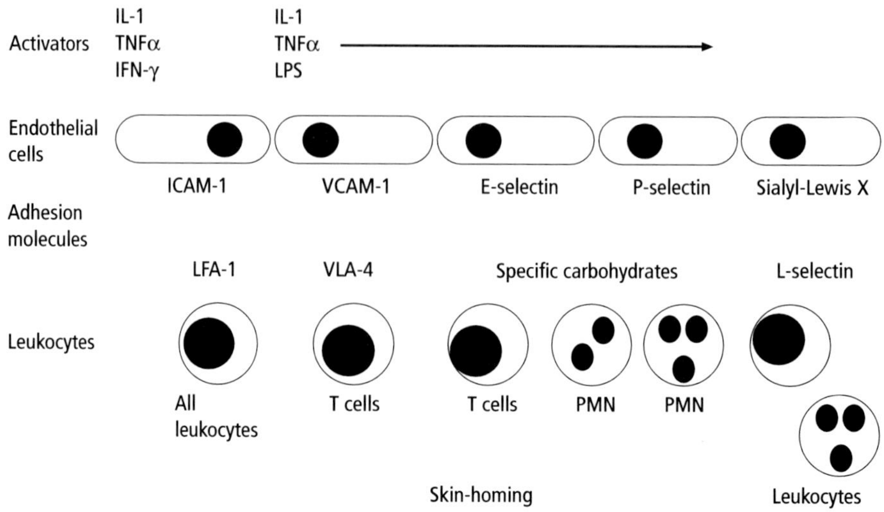

Fig. 16.3 The attachment of leukocytes to endothelial cells is mediated by several families of adhesion structures. The expression of adhesion structures on endothelial cells is stimulated by IL-1, TNFα, IFN-γ or LPS as summarized in the figure. The adhesion pairs on leukocytes can also be stimulated by cytokines. Changes in pattern of these adhesion molecules are felt to be crucial in directing leukocytes into different tissue sites. Reprinted with permission from [99].

microvasculature [88, 99, 121]. Increase in endothelial VCAM-1 promotes binding of VLA4$^+$ T cells, and increase in ICAM-1 enhances leukocyte binding through LFA-1. Expression of all of these adhesion molecules is influenced by cytokines produced by inflammatory cells, mast cells, endothelial cells or keratinocytes, as illustrated in Fig. 16.3. Ultraviolet light selectively induces E-selectin, but not VCAM-1 in dermal endothelial cells [102]; it also induces increased endothelial cell ICAM-1 expression [102].

UV effects in the stratified immunologic environment of the epidermis which might promote photosensitive lupus

The induction of ICAM-1 throughout the epidermis is not uniform. I propose that the epidermis is stratified with respect to induction of ICAM-1 expression (Fig. 16.4). Both IFN-γ and TNFα induce greater ICAM-1 expression in cultured keratinocytes with the most undifferentiated phenotype than in more differentiated keratinocytes [58]. Dermal mediators such as histamine many synergistically combine with TNFα to influence ICAM-1 expression [92]; if relevant *in vivo*, this would mainly influence basal keratinocytes. IL-1 responses are relatively dampened by high levels of IL-1ra; the ratio of IL-1ra/IL-1 increases in more differentiated keratinocytes, which are the cells exposed to the greatest UVR levels *in vivo*. In irradiated epidermis, diffuse ICAM-1 expression is seen with spotty accentuation in the basal cell layer.

In biopsies of skin lesions from SCLE patients, diffuse ICAM-1 expression is found in the epidermis, with some enhancement in the basal cell region [11, 96]. This pattern is consistent with low level induction throughout the epidermis by UVR-released mediators, and with enhanced response to mediators in the basal layer, augmented by dermal mediators such as histamine and leukocyte derived gamma interferon and TNFβ.

Recently, the importance of genetic polymorphisms in both TNF and IL-1ra genes has been

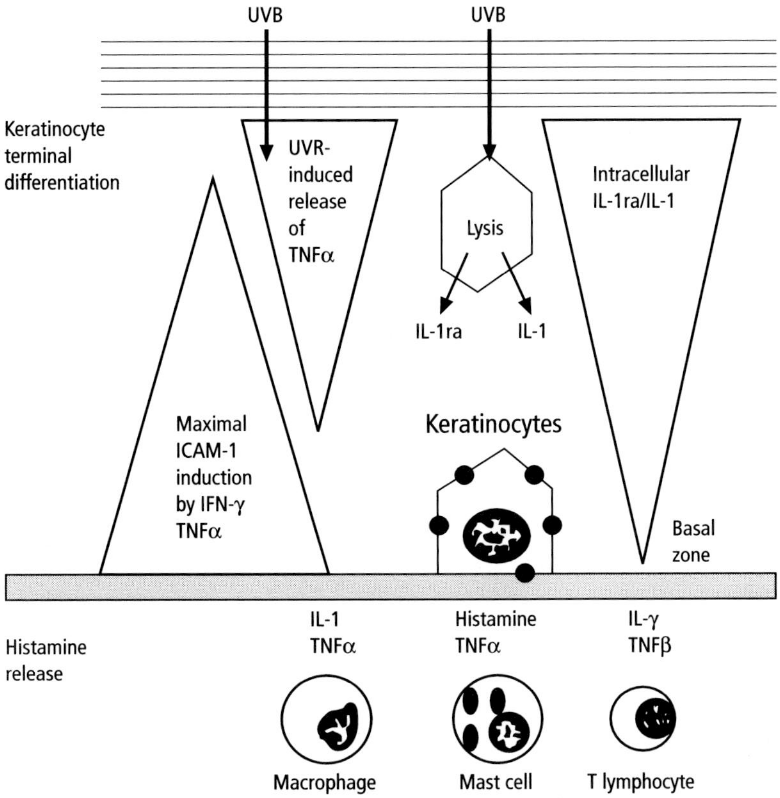

Fig. 16.4 Control of ICAM-1 expression in the epidermis by cytokine and UVR is shown. Triangles indicate regions of stratified ICAM-1 induction by TNFα and IFN-γ, stratified release of TNFα by UVR and stratified ratio of IL-1ra to IL-1α. The upper epidermis has greater UVR-induced TNFα release, and greater UVB-induced release of IL-1 and IL-1rα through keratinocyte lysis, but the relative levels of inhibitory IL-1rα are also greater. The basal keratinocytes can be stimulated by cytokines released by UVR in the upper epidermis and by cytokines and mediators released in the dermis and dermo-epidermal junction from inflammatory cells and mast cells.

appreciated. High TNFα response is linked to the DR3 gene [133], associated with autoantibody responses in a number of diseases, including many patients with photosensitive lupus [76, 120]. One IL-1rα polymorphism is strongly correlated with photosensitive CCLE [21]. The hypothesis that particular polymorphisms in TNFα or IL-1ra may cause enhanced ICAM-1 induction by UVR and serve as important risk factors in the development of photosensitive LE is now being evaluated.

Possible combinations of control of ICAM-1 in different individuals are shown in Fig. 16.5. High TNFα releasers and responders would exhibit predominantly TNFα-driven ICAM-1 expression. Patients with polymorphisms in icIL-1rα might have enhanced IL-1 responses. Patients with dense infiltrates of activated lymphocytes such as in CCLE would likely have TNFβ- and IFN-γ-driven ICAM-1 expression. It will be informative to see if such differences may determine the unique characteristics of SCLE, CCLE and ACLE photosensitive lesions.

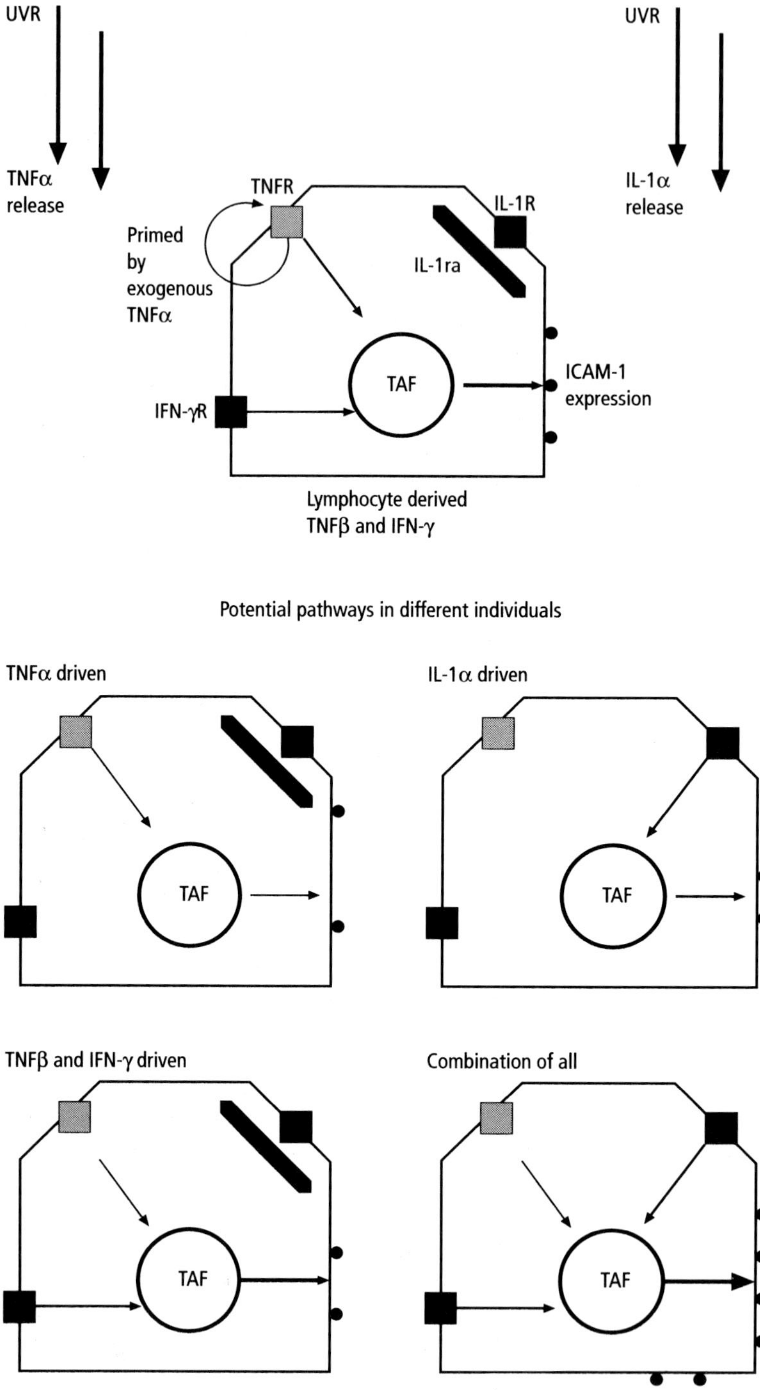

Fig. 16.5 The control of induction of ICAM-1 expression in the keratinocyte is shown. In photosensitive lupus, UVR released TNFα and IL-1 plus TNF and IFN-γ released by inflammatory cells in the dermal infiltrate are likely mediators of induction of ICAM-1 expression. We propose that in certain individuals, individual cytokine-driven pathways may be more active, based on genetic polymorphisms in TNFα release, in IL-1ra levels or function, or in gamma interferon responses.

Autoantibodies and photosensitive lupus

Antibodies to SSA/Ro and/or SSB/La are highly associated with photosensitivity in SCLE and NLE, and in photosensitive SLE patients [59, 118, 120, 132], although there are some clinical studies that dispute this evidence [17]. There is strong circumstantial evidence that these autoantibodies might participate in the epidermal and dermal damage seen in these forms of lupus.

The Ro antigen is a molecular complex containing a 60 kD and 52 kD protein, complexed to specific hY mRNAs. There is controversy as to whether the 47 kD calreticulin protein is also a component of this complex [87]. Antibodies with anti-Ro specificity react with confirmation determinants in the 60 kD protein, with linear determinants in the 52 kD protein, and in some patients with conformational determinants in 47 kD calreticulin not seen in the recombinant molecule [87].

In NLE, nearly 100% of patients have high titers of anti-Ro or anti-La antibodies [16, 106, 132]. In NLE the occurrence of anti-Ro with anti-La is nearly concordant [16, 45]. Buyon reported an increased incidence of anti-52 kD Ro and anti-48 kD La in 20 NLE mothers [16], but this result might be dependent on differential sensitivities in the immunoblotting techniques used. Lee found increases in 50 kD La antigen in neonatal tissue compared to adult tissues [79]. However, to date there are no absolutely convincing data that explain differential organ involvement in NLE.

The association with anti-SSA/Ro and La is not so strong in SCLE as in NLE, but this may be partially due to criteria for diagnosis. Most clinical studies report a strong association of anti-SSA/Ro and SCLE [118], but there continue to be studies which report SCLE patients with low prevalence of anti-SSA/Ro [17]. In a recent study by Lee and colleagues [78], 17 SCLE patients with high titers of precipitating antibodies to Ro were described. All had particulate intraepidermal IgG. All had high titers to 60 kD Ro by ELISA and 15/17 had high titers to 52 kD by ELISA. Only four had anti-La. In the SCLE patients described by this group, circulating anti-Ro antibodies and particulate intraepidermal IgG are an almost universal finding.

In CCLE (discoid LE) the association with autoantibodies is less clear. Lee reported that 13/14 DLE patients had granular immunoreactants in the DEJ but no particulate IgG deposits in the epidermis as are seen in SCLE patients [77]. Although the DLE patients did have anti-Ro antibodies, the response was qualitatively different than in SCLE. The DLE patients lacked precipitating antibodies, and the titers to 60 and 52 kD Ro were low. Interestingly, 12/14 of the DLE patients had other autoantibodies detected by immunoblotting.

Antibodies to SSA/Ro are seen in a number of connective tissue disease syndromes such as Sjogren's syndrome and in Sjogren's/SLE overlap syndromes, often without clinical skin disease or photosensitivity [1, 49, 108, 109]. What are the other factors that are necessary for induction of clinical photosensitivity in anti-Ro positive individuals?

Antigen translocation and antibody binding in photosensitive lupus

Initially, the associations of antibodies to Ro and La with NLE and SCLE were viewed as 'markers' of disease activity, and not as functional participants in the induction of disease. The temporal association of these autoantibodies with disease activity in NLE argued in favour of functional significance for these antibodies. In 1985 LeFeber *et al.* demonstrated that antibodies to the extractable nuclear antigens Ro, RNP and Sm would bind to the surface of cultured human keratinocyte that had been UVR irradiated [80]. It was proposed that this represented translocation of the Ro antigen complex to the cell surface following UVR since this was an energy-dependent–protein synthesis-dependent process and not just cell death and antigen 'leakage' [36]. These results were verified *in vitro* in cell culture using FACS analysis, and have subsequently been validated by a number of different groups [38, 55, 56].

Detailed ultrastructural evaluation of La translocation to the plasma membrane by UVR was provided by Bachmann [6]. Similar translocation of La to the cell membrane was induced by a number of investigators [4, 5, 7].

In SCLE and NLE, discrete epidermal cytotoxicity is observed histologically [9, 125]. IgG from anti-Ro positive serum binds to intact human epidermis in a number of model systems, and in biopsies from patients with SCLE and NLE [36, 57, 77]. Human skin grafted onto SCID or nude mice showed preferential deposition of IgG within the epidermis following injection of antisera containing anti-Ro, anti-La, anti-RNP or anti-Sm, all of which are antibodies to extractable nuclear antigens which can be released from cells by saline extraction [57, 77]. Affinity purified anti-Ro binds to the epidermis in this model, producing cell surface, cytoplasmic and nuclear patterns of staining. In this model, binding of anti-Ro occurs in grafted human epidermis even without UVR, but the intensity of antibody binding is increased by UVR. This same pattern of staining, a faint particulate epidermal staining, can be seen in lesional skin of SCLE [77]. Nieboer *et al.* have reported similar IgG deposition as 'faint dust-like particles' (DLP) in SCLE patients [96]. In their series, the association with anti-Ro antibodies is not as strong as reported by Lee. Without more definite histologic or biologic markers of the SCLE clinical phenotype, it is unclear whether the patients included in these studies are the same.

Golan *et al.* have recently shown that binding of anti-Ro, anti-La, and anti-RNP to cultured human keratinocytes is very dramatic following UVR radiation, especially if the keratinocytes are from photosensitive SLE patients [38]. Using autologous sera from the patients themselves also enhanced antibody binding following irradiation. This raises the intriguing possibility that the degree of UVR-induced translocation of extractable nuclear antigens may be a genetic trait, and that the antibodies themselves may enhance this effect.

It has been proposed that the binding of anti-Ro to Ro antigen on the surface of cells can induce antibody-dependent cellular cytotoxicity [100, 103] or other antibody-mediated tissue damage. Indeed IgG from anti-Ro and anti-La positive sera bind to the surface of irradiated keratinocytes, and are then internalized to the cytoplasm and nucleus [36]. Similar findings were seen with anti-RNP, perhaps explaining the positive ANA seen in direct immunofluorescence examinations of some skin biopsies from anti-RNP positive MCTD patients. Elkon has shown that internalization of autoantibodies by keratinocytes can occur with a number of antibody specificities [40]. There are not yet any direct functional studies which determine the significance of internalization of anti-Ro in photosensitive SCLE.

Blood vessels and mast cells

The importance of UVA in experimental induction of SCLE lesions [81] suggests that mast cell and blood vessel activation are important in the development of photosensitive lupus lesions. In a guinea-pig model of photosensitivity, Sontheimer showed that UVR-induced blood flow in the skin was augmented by SCLE patient's sera containing anti-Ro positive antibodies [26].

UVB erythema is dependent on the release of arachidonic acid and PGE_2 and $F_{2\alpha}$ from the epidermis, and also on mast cell-derived factors [85, 105]. UVR selectively increases endothelial expression of E-selectin, but not VCAM-1 in dermal endothelial cells [102]. It is unknown whether local cytokine release in photosensitive lupus patients may occur, promoting leukocyte emigration. Individual variability in release of, and response to, TNFα might be an important factor in endothelial activation in lupus.

T-lymphocyte activation in cutaneous lupus

The effects of UVR on T-cell activation or on suppression of T-cell responses in contact hypersensitivity or delayed-type hypersensitivity are complex, and are not necessarily germane to the lymphocyte activation seen in the lesions of photosensitive lupus. The cellular infiltrates of SCLE and DLE contain macrophages and $CD4^+$ and $CD8^+$ lymphocytes [129]. Keratinocytes in these lesions are HLA class II^+ and leukocytes within the dermal infiltrate show signs of activation such as class I antigen expression [129]. This suggests T-

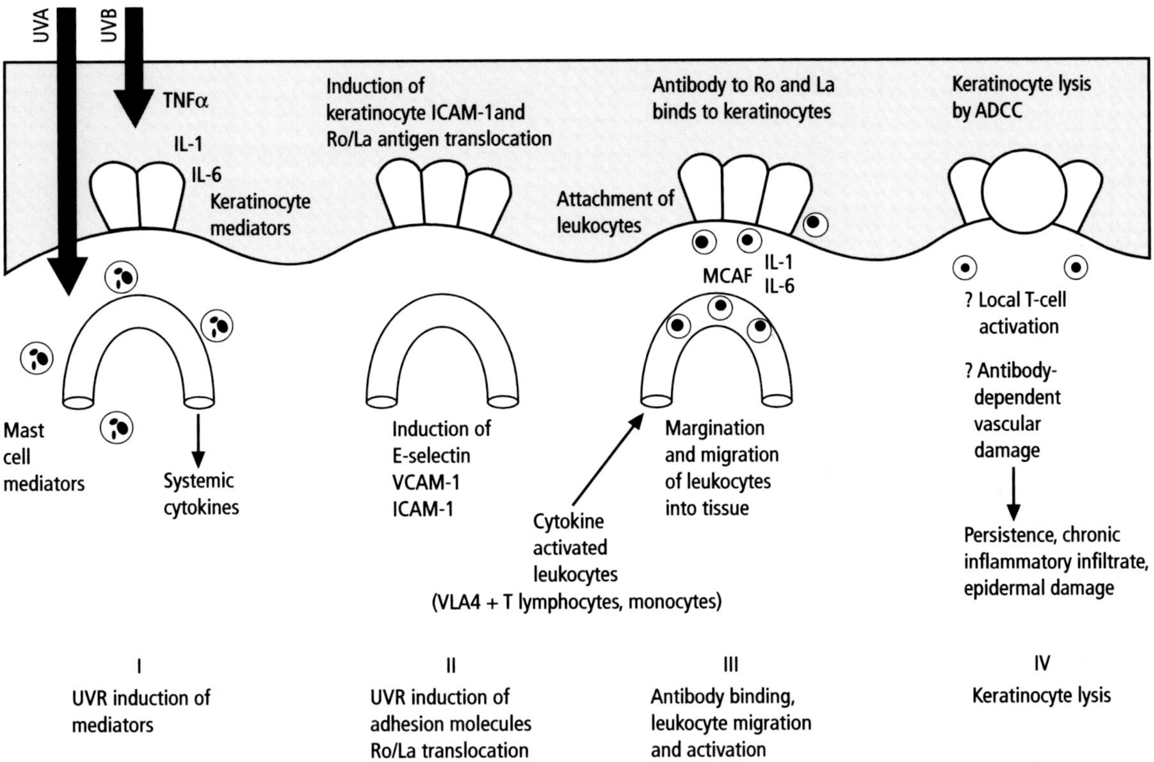

Fig. 16.6 A proposed mechanism for induction of skin lesions in photosensitive subacute cutaneous lupus erythematosus. The individual components of this proposed pathway are discussed throughout this review. Reprinted with permission from [99].

cell activation and gamma interferon release. The epidermis of photosensitive lupus is relatively depleted of Langerhans cells [119, 129], suggesting UVR-induced TNFα release. In two studies, this decrease in Langerhans cells contrasted to normal or elevated levels in PMLE or lichenoid dermatitis [119, 129].

Proposed mechanisms for SCLE

The best evidence for the mechanisms of photosensitive lupus has been produced in the study of SCLE and NLE. The basic components of the mechanism of tissue damage in SCLE are proposed in Fig. 16.6. Four stages of disease are proposed: UVR induction of mediator release, UVR induction of adhesion molecules and antigen translocation, antibody binding and leukocyte migration and activation, and keratinocyte lysis. These steps are not discrete, but are probably overlapping. Although there is a definite sequence resulting in keratinocyte lysis, continued UVR radiation should produce a persistent environment of mediator release, induction of cytokines and adhesion molecules, and immunologic damage. Although the keratinocyte is proposed as the principal target of immunologic damage, damage may also occur to blood vessels, innocent bystander melanocytes and other cells. In this model, ADCC mediated via Ro-specific antibodies is proposed as the primary effector of keratinocyte damage. However, cytotoxic T lymphocytes or cytokines from T lymphocytes may also be potent mediators of damage.

This model can be modified to apply to CCLE and ACLE. In CCLE, cytotoxic gamma/delta T cells may be the effectors of specific keratinocyte lysis, as proposed by the recent findings of Volc Platzer *et al*. [131]. In ACLE, vascular damage may be more prominent, and keratinocyte damage may be more through T lymphocyte-released cytokines than through ADCC or specific antibody. However, in both of these photosensitive syndromes, similar patterns of UVR-released inflammatory mediators and cytokines may be necessary for induction of clinical disease, even if specific immunologic effector mechanisms are somewhat different.

Conclusion

This review has presented a broad view of the factors involved in induction of skin lesions in lupus by ultraviolet light. It is clear that most aspects of the relationship of UVR to the development of lupus skin lesions are incompletely understood. The approach presented here to understanding the mechanisms of photosensitive lupus is a combination of established facts and hypotheses based on the basic tenents of photoimmunology. When these hypotheses are tested, it is likely that we will have a better understanding of the determinants of the different photosensitive lupus syndromes.

References

1 Alexander, E.L., J. McNicholl, R.M. Watson, W. Bias, M. Reichlin, T.T. Provost: The immunogenetic relationship between anti-Ro (SS-A)/La (SS-B) antibody positive Sjogren's/lupus erythematosus overlap syndrome and the neonatal lupus syndrome. *J. Invest. Dermatol.* 93 (1989), 751–756.

2 Ansel, J.C., J. Mountz, A.D. Steinberg, E. DeFabo, I. Green: Effects of UV radiation on autoimmune strains of mice: increased mortality and accelerated autoimmunity in BXSB male mice. *J. Invest. Dermatol.* 85 (1985), 181–186.

3 Ansel, J., P. Perry, J. Brown, D. Damm, T. Phan, C. Hart, T. Luger, S. Hefeneider: Cytokine modulation of keratinocyte cytokines. *J. Invest. Dermatol.* 94 (1990), 101S–107S.

4 Baboonian, C., P.J. Venables, J. Booth, D.G. Williams, L.M. Roffe, R.N. Maini: Virus infection induces redistribution and membrane localization of the nuclear antigen La (SS-B): a possible mechanism for autoimmunity. *Clin. Exp. Immunol.* 78 (1989), 454–459.

5 Bachmann, M., H. Althoff, H. Troster, C. Selenka, D. Falke, W.E. Muller: Translocation of the nuclear autoantigen La to the cell surface of herpes simplex virus type 1 infected cells. *Autoimmunity* 12 (1992), 37–45.

6 Bachmann, M., S. Chang, H. Slor, J. Kukulies, W.E. Muller: Shuttling of the autoantigen La between nucleus and cell surface after UV irradiation of human keratinocytes. *Exp. Cell Res.* 191 (1990), 171–180.

7 Bachmann, M., D. Falke, H.C. Schroder, W.E. Muller: Intracellular distribution of the La antigen in CV-1 cells after herpes simplex virus type 1 infection compared with the localization of U small nuclear ribonucleoprotein particles. *J. Gen. Virol.* 70 (1989), 881–891.

8 Baer, R., L.C. Harber: Photobiology of lupus erythematosus. *Arch. Dermatol.* 92 (1965), 124–128.

9 Bangert, J.L., R.G. Freeman, R.D. Sontheimer, J.N. Gilliam: Subacute cutaneous lupus erythematosus and discoid lupus erythematosus. Comparative histopathologic findings. *Arch. Dermatol.* 120 (1984), 332–337.

10 Barker, J.N.W.N., V. Sarma, V. Dixit, R.S. Mitra, B.J. Nickoloff, Marked synergism between tumor necrosis factor alpha and interferon-gamma in regulation of keratinocyte-derived chemotactic factors and adhesion molecules. *J. Clin. Invest.* 95 (1990), 605–608.

11 Bennion, S.D., M.H. Middleton, K.M. David Bajar, S. Brice, D.A. Norris: Varied patterns of epidermal interleukin adhesion molecule-1 (ICAM-1) in three diseases indicate different triggers of disease. *J. Invest. Dermatol.* (manuscript submitted).

12 Bigler, C.F., D.A. Norris, W.L. Weston, W.P. Arend: Interleukin-1 receptor antagonist production by human keratinocytes. *J. Invest. Dermatol.* 98 (1992), 38–44.

13 Black, A.K., N. Fincham, M.W. Greaves, C.N. Hensby: Time course changes in levels of arachidonic acid and prostaglandins D2, E2, F2 alpha in human skin following ultraviolet B irradiation. *Br. J. Clin. Pharmacol.* 10 (1980), 453–457.

14 Boyd, A.W., S.O. Wawryk, G.F. Burns, J.V. Fecondo: Intercellular adhesion molecule 1 (ICAM-1) has a central role in cell–cell contact dependent mediated immune mechanism. *Proc. Natl Acad. Sci. USA* 85 (1988), 3095–3099.

15 Brian, R.T.: Lupus erythematosus replacing the lesions of erythema. *Proc. Roy. Soc. Med.* 26 (1993), 784.

16 Buyon, J.P., E.E. Ben Chetrit, S. Karp, R.A. Roubey, L. Pompeo, W.H. Reeves, E.M. Tan, R. Winchester: Acquired congenital heart block. Pattern of maternal antibody response to biochemically defined antigens of the SSA/Ro-SSB/La system in neonatal lupus. *J. Clin. Invest.* 84 (1989), 627–634.

17 Callen, J.P., J. Klein: Subacute cutaneous lupus erythematosus. Clinical, serologic, immunogenetic, and therapeutic considerations in seventy-two patients. *Arthritis Rheum.* 31 (1988), 1007–1013.
18 Caughman, S.W., L.J. Li, K. Degitz: Human intercellular adhesion molecule-1 gene and its expression in the skin. *J. Invest. Dermatol.* 98 (1992), 61S–65S.
19 Chouaib, S., D. Branellec, W.A. Buurman: More insights into the complex physiology of TNF. *Immunol. Today* 12 (1991), 141–142.
20 Cohen, D., V.A. DeLeo: Ultraviolet radiation-induced phospholipase A2 activation occurs in mammalian cell membrane preparations. *Photochem. Photobiol.* 57 (1993), 383–390.
21 Cork, M.J., J.K. Tarlow, A.J.F. Blakemore, A.J.G. McDonagh, A.G. Messenger, S.S. Bleehan, G.W. Duff: Genetics of interleukin I receptor antagonist in inflammatory skin diseases. *J. Invest. Dermatol.* 100 (1993), 522.
22 Cornelius, L.A., J.T. Taylor, K. Degitz, L.J. Li, T.J. Lawley, S.W. Caughman: A 5′ portion of the ICAM-1 gene confers tissue-specific differential expression levels and cytokine responsiveness. *J. Invest. Dermatol.* 100 (1993), 753–758.
23 Cripps, D.J., J. Rankin: Action spectra of lupus erythematosus and experimental immunofluorescence. *Arch. Dermatol.* 107 (1973), 563–567.
24 Damle, N.K., K. Klussman, P.S. Linsley, A. Aruffo: Differential costimulatory effects of adhesion molecules B7, ICAM-1, LFA-3, and VCAM-1 on resting and antigen-primed CD4+ T lymphocytes. *J. Immunol.* 148 (1992), 1985–1992.
25 David Bajar, K.M., S.D. Bennion, J.D. DeSpain, L.E. Golitz, L.A. Lee: Clinical, histologic, and immunofluorescent distinctions between subacute cutaneous lupus erythematosus and discoid lupus erythematosus. *J. Invest. Dermatol.* 99 (1992), 251–257.
26 Davis, T.L., C.B. Lyde, B.M. Davis, R.D. Sontheimer: Perturbation of experimental ultraviolet light-induced erythema by passive transfer of serum from subacute cutaneous lupus erythematosus patients. *J. Invest. Dermatol.* 92 (1989), 573–577.
27 Deng, J.S., R.D. Sontheimer, J.N. Gilliam: Relationships between antinuclear and anti-Ro/SS-A antibodies in subacute cutaneous lupus erythematosus. *J. Am. Acad. Dermatol.* 11 (1984), 494–499.
28 diGiovine, F.S., G.W. Duff: Interleukin 1: the first interleukin. *Immunol. Today* 11 (1990), 13–20.
29 Dustin, M.L., K.H. Singer, D.T. Tuck, T.A. Springer: Adhesion of T lymphoblasts to epidermal keratinocytes is regulated by interferon-gamma and is mediated by intercellular adhesion molecule-1 (ICAM-1). *J. Exp. Med.* 167 (1988), 1323–1340.
30 Enk, A.H., S.I. Katz: Identification and induction of keratinocyte-derived IL-10. *J. Immunol.* 149 (1992), 92–95.
31 Epstein, J.H., D. Tuffanelli, E.L. Dubois: Light sensitivity and lupus erythematosus. *Arch. Dermatol.* 91 (1965), 483–485.
32 Everett, M.A., R.L. Olson: Response of cutaneous lupus erythematosus to ultraviolet light. *J. Invest. Dermatol.* 44 (1993), 133–139.
33 Fiet, H.: Light as an exciting agent in lupus erythematosus and other dermatoses. *J. Med. Soc. NJ* 24 (1927), 226–229.
34 Fischer, H., A. Gjorloff, G. Hedlund, H. Hedman, E. Lundgren, T. Kalland, H.O. Sjogren, M. Dohlsten: Stimulation of human naive and memory T helper cells with bacterial superantigen. Naive CD4+ 45RA+ T cells require a costimulatory signal mediated through the LFA-1/ICAM-1 pathway. *J. Immunol.* 148 (1992), 1993–1998.
35 Freeman, R.G., J.M. Knox: Cutaneous lesions of lupus erythematosus induced by monochromatic light. *Arch. Dermatol.* 100 (1969), 677–682.
36 Furukawa, F., M. Kashihara Sawami, M.B. Lyons, D.A. Norris: Binding of antibodies to the extractable nuclear antigens SS-A/Ro and SS-B/La is induced on the surface of human keratinocytes by ultraviolet light (UVL): implications for the pathogenesis of photosensitive cutaneous lupus. *J. Invest. Dermatol.* 94 (1990), 77–85.
37 Golan, T.D., Y. Borel: Spontaneous increase of DNA turnover in murine systemic lupus erythematosus. *Eur. J. Immunol.* 13 (1983), 430–433.
38 Golan, T.D., K.B. Elkon, A.E. Gharavi, J.G. Krueger: Enhanced membrane binding of autoantibodies to cultured keratinocytes of systemic lupus erythematosus patients after ultraviolet B/ultraviolet A radiation. *J. Clin. Invest.* 90 (1992), 1067–1076.
39 Golan, T.D., V. Foltyn, A. Roueff: Increased susceptibility to *in vitro* ultraviolet B radiation in fibroblasts and lymphocytes cultured from systemic lupus erythematosus patients. *Clin. Immunol. Immunopathol.* 58 (1991), 289–304.
40 Golan, T.D., A.E. Gharavi, K.B. Elkon: Penetration of autoantibodies into living epithelial cells. *J. Invest. Dermatol.* 100 (1993), 316–322.
41 Granstein, R.D., D.N. Sauder: Whole-body exposure to ultraviolet radiation results in increased serum interleukin-1 activity in humans. *Lymphokine Res.* 6 (1987), 187–193.
42 Greaves, M.W., J. Sondergaard: Pharmacologic agents released in ultraviolet inflammation studied by continuous skin perfusion. *J. Invest. Dermatol.* 54 (1970), 365–367.
43 Grewe, M., U. Trefzer, A. Ballhorn, K. Gyufko, H. Henniger, J. Krutmann: Analysis of the mechanism of ultraviolet (UV) B radiation-induced prostaglandin E synthesis by human epidermal carcinoma cells. *J. Invest. Dermatol.* 101 (1993), 528–531.
44 Hanson, D.L., V.A. DeLeo: Long wave ultraviolet

radiation stimulates arachidonic acid release and cyclooxygenase activity in mammalian cells in culture. *Photochem. Photobiol.* 49 (1989), 423–430.

45 Harley, J.B., H. Yamagata, M. Reichlin: Anti-La/SSB antibody is present in some normal sera and is coincident with anti-Ro/SSA precipitins in systemic lupus erythematosus. *J. Rheumatol.* 11 (1984), 309–314.

46 Haskill, S., G. Martin, L. VanLe, J. Morris, A. Peace, C.F. Bigler, G.J. Jaffe, C. Hammerberg, S.A. Sporn, S. Fong *et al.*: cDNA cloning of an intracellular form of the human interleukin 1 receptor antagonist associated with epithelium. *Proc. Natl. Acad. Sci. USA* 88 (1991), 3681–3685.

47 Hawk, J.L., A.K. Black, K.F. Jaenicke, R.M. Barr, N.A. Soter, A.I. Mallett, B.A. Gilchrest, C.N. Hensby, J.A. Parrish, M.W. Greaves: Increased concentrations of arachidonic acid, prostaglandins E2, D2 and 6-oxo-F1 alpha, and histamine in human skin following UVA irradiation. *J. Invest. Dermatol.* 80 (1983), 496–499.

48 Hensby, C.N., N.A. Plummer, A.K. Black, N. Fincham, M.W. Greaves: Time-course of arachidonic acid, prostaglandins E2 and F2 alpha production in human abdominal skin, following irradiation with ultraviolet wavelengths (290–320 nm). *Adv. Prostaglandin. Thromboxane Res.* 7 (1980), 857–860.

49 Hochberg, M.C., R.E. Boyd, J.M. Ahearn, F.C. Arnett, W.B. Bias, T.T. Provost, M.B. Stevens: Systemic lupus erythematosus: a review of clinicolaboratory features and immunogenetic markers in 150 patients with emphasis on demographic subsets. *Medicine (Baltimore)* 64 (1985), 285–295.

50 Hogquist, K.A., M.A. Nett, E.R. Unanue, D.D. Chaplin: Interleukin 1 is processed and released during apoptosis. *Proc. Natl. Acad. Sci. USA* 88 (1991), 8485–8489.

51 Hogquist, K.A., E.R. Unanue, D.D. Chaplin: Release of IL-1 from mononuclear phagocytes. *J. Immunol.* 147 (1991), 2181–2186.

52 Hölzle, E.: Photodermatoses – diagnostic procedures and their interpretation. *Photodermatol.* 4 (1987), 109–114.

53 Hruza, L.L., A.P. Pentland: Mechanisms of UV-induced inflammation. *J. Invest. Dermatol.* 100 (1993), 35S–41S.

54 Jacob, C.O.: Studies on the role of tumor necrosis factor in murine and human autoimmunity. *J. Autoimmun.* 5 (1992), 133–143.

55 Jones, S.K.: Ultraviolet radiation (UVR) induces cell-surface Ro/SSA antigen expression by human keratinocytes *in vitro*: a possible mechanism for the UVR induction of cutaneous lupus lesions. *Br. J. Dermatol.* 126 (1992), 546–553.

56 Jones, S.K.: The effects of hormonal and other stimuli on cell-surface Ro/SSA antigen expression by human keratinocytes *in vitro*: their possible role in the induction of cutaneous lupus lesions. *Br. J. Dermatol.* 126 (1992), 554–560.

57 Jones, S.K., S. Coulter, C. Harmon, D.A. Norris, M. Rosandich, L.A. Lee: Ro/SSA antigen in human epidermis. *Br. J. Dermatol.* 118 (1988), 363–367.

58 Kashihara Sawami, M., D.A. Norris: The state of differentiation of cultured human keratinocytes determines the level of intercellular adhesion molecule-1 (ICAM-1) expression induced by gamma interferon. *J. Invest. Dermatol.* 98 (1992), 741–747.

59 Kephart, D.C., A.F. Hood, T.T. Provost: Neonatal lupus erythematosus: new serologic findings. *J. Invest. Dermatol.* 77 (1981), 331–333.

60 Kesten, B., M. Slatkin: Diseases related to light sensitivity. *Arch. Dermatol.* 67 (1953), 284–301.

61 Kimber, I., M. Cumberbatch: Stimulation of Langerhans cell migration by tumor necrosis factor alpha (TNF-α). *J. Invest. Dermatol.* 99 (1992), 48S–50S.

62 Kind, P., P. Lehmann, G. Plewig: Phototesting in lupus erythematosus. *J. Invest. Dermatol.* 100 (1993), 53S–57S.

63 Kirnbauer, R., A. Kock, T. Schwarz, A. Urbanski, J. Krutmann, W. Borth, D. Damm, G. Shipley, J.C. Ansel, T.A. Luger: IFN-beta 2, B cell differentiation factor 2, or hybridoma growth factor (IL-6) is expressed and released by human epidermal cells and epidermoid carcinoma cell lines. *J. Immunol.* 142 (1989), 1922–1928.

64 Kochevar, I.: Action spectrum and mechanism of UV-radiation-induced injury in lupus erythematosus. *J. Invest. Dermatol.* 85 (1985), 140S–143S.

65 Kock, A., T. Schwarz, R. Kirnbauer, A. Urbanski, P. Perry, J.C. Ansel, T.A. Luger: Human keratinocytes are a source for tumor necrosis factor alpha: evidence for synthesis and release upon stimulation with endotoxin or ultraviolet light. *J. Exp. Med.* 172 (1990), 1609–1614.

66 Krutmann, J., W. Czech, F. Parlow, U. Trefzer, A. Kapp, E. Schopf, T.A. Luger: Ultraviolet radiation effects on human keratinocyte ICAM-1 expression: UV-induced inhibition of cytokine-induced ICAM-1 mRNA expression is transient, differentially restored for IFN gamma versus TNF alpha, and followed by ICAM-1 induction via a TNF alpha-like pathway. *J. Invest. Dermatol.* 98 (1992), 923–928.

67 Krutmann, J., A. Kock, E. Schauer, F. Parlo, A. Moller, A. Kapp, E. Forster, E. Schopf, T.A. Luger: Tumor necrosis factor beta and ultraviolet radiation are potent regulators of human keratinocyte ICAM-1 expression. *J. Invest. Dermatol.* 95 (1990), 127–131.

68 Kulick, K.B., T.T. Provost, M. Reichlin: Antibodies to single-stranded DNA in patients with discoid lupus erythematosus. *Arthritis Rheum.* 25 (1982), 639–646.

69 Kupper, T.S.: Mechanisms of cutaneous inflam-

mation. *Arch. Dermatol.* 125 (1989), 1406–1412.

70 Kupper, T.S., A.O. Chua, P. Flood, J. McGuire, U. Gubler: Interleukin 1 gene expression in cultured human keratinocytes is augmented by ultraviolet irradiation. *J. Clin. Invest.* 80 (1987), 430–436.

71 Kupper, T.S., F. Lee, N. Birchall, S. Clark, S. Dower: Interleukin 1 binds to specific receptors on human keratinocytes and induces granulocyte macrophage colony-stimulating factor mRNA and protein. A potential autocrine role for interleukin 1 in epidermis. *J. Clin. Invest.* 82 (1988), 1787–1792.

72 Kurimoto, I., J.W. Streilein: cis-Urocanic acid suppression of contact hypersensitivity induction is mediated via tumor necrosis factor-alpha. *J. Immunol.* 148 (1992), 3072–3078.

73 Lane, P.J., F.M. McConnell, E.A. Clark, E. Mellins: Rapid signaling to B cells by antigen-specific T cells requires CD18/CD54 interaction. *J. Immunol.* 147 (1991), 4103–4108.

74 Lanier, L.L., J.J. Ruitenberg, J.H. Phillips: Functional and biochemical analysis of CD16 antigen on natural killer cells and granulocytes. *J. Immunol.* 141 (1988), 3478–3485.

75 Lee, L.A.: Neonatal lupus erythematosus. *J. Invest. Dermatol.* 100 (1993), 9S–13S.

76 Lee, L.A., W.B. Bias, F.C. Arnett Jr, J.C. Huff, D.A. Norris, C. Harmon, T.T. Provost, W.L. Weston: Immunogenetics of the neonatal lupus syndrome. *Ann. Intern. Med.* 99 (1983), 592–596.

77 Lee, L.A., K.K. Gaither, S.N. Coulter, D.A. Norris, J.B. Harley: Pattern of cutaneous immunoglobulin G deposition in subacute cutaneous lupus erythematosus is reproduced by infusing purified anti-Ro (SSA) autoantibodies into human skin-grafted mice. *J. Clin. Invest.* 83 (1989), 1556–1562.

78 Lee, L.A., C.M. Roberts, B. Frank, V. McCubbin, D. Rice, M. Reichlin: Autoantibody markers of cutaneous lupus erythematosus. *J. Invest. Dermatol.* 100 (1993), 573.

79 Lee, L.A., K.B. Rundell: The 50 kD La protein is present in greater amounts in neonatal versus adult tissues (Abstract). *Clin. Res.* 38 (1990), 605A.

80 LeFeber, W.P., D.A. Norris, S.R. Ryan, J.C. Huff, L.A. Lee, M. Kubo, S.T. Boyce, B.L. Kotzin, W.L. Weston: Ultraviolet light induces binding of antibodies to selected nuclear antigens on cultured human keratinocytes. *J. Clin. Invest.* 74 (1984), 1545–1551.

81 Lehmann, P., E. Hölzle, P. Kind, G. Goerz, G. Plewig: Experimental reproduction of skin lesions in lupus erythematosus by UVA and UVB radiation. *J. Am. Acad. Dermatol.* 22 (1990), 181–187.

82 Luger, T.A., T. Schwarz: Evidence for an epidermal cytokine network. *J. Invest. Dermatol.* 95 (1991), 1005–1041.

83 Makgoba, M.W., M.E. Sanders, G.E. Ginther Luce, M.L. Dustin, T.A. Springer, E.A. Clark, P. Mannori, S. Shaw: ICAM-1 a ligand for LFA-1-dependent adhesion by B, T and myeloid cells. *Nature* 331 (1988), 86–91.

84 Makgoba, M.W., M.E. Sanders, G.E. Ginther Luce, E.A. Gugel, M.L. Dustin, T.A. Springer, S. Shaw: Functional evidence that intercellular adhesion molecule 1 (ICAM-1) is a ligand for LFA-1-dependent adhesion in T cell-mediated cytotoxicity. *Eur. J. Immunol.* 18 (1988), 637–640.

85 Mandel, M.J., H.W. Lim: Effect of ultraviolet radiation and PUVA on mediator release from purified mast cells. *Photodermatol.* 5 (1988), 211–217.

86 Marlin, S.D., T.A. Springer: Purified intercellular adhesion molecule (ICAM-1) is a ligand for lymphocyte function associated antigen. *Cell* 51 (1987), 813–819.

87 McCauliffe, D.P., R.D. Sontheimer: Molecular characterization of the Ro/SS-A autoantigens. *J. Invest. Dermatol.* 100 (1993), 73S–79S.

88 McEver, R.P.: Leukocyte–endothelial cell interactions. *Curr. Opin. Cell Biol.* 4 (1993), 840–849.

89 Middleton, M., W. Arend, D.A. Norris: High intracellular interleukin-1 receptor antagonist (icIL1ra) levels are correlated with low expression of intracellular adhesion molecule-1 sideroblasts and fibroblasts. *J. Invest. Dermatol.* 102 (1994), 524.

90 Middleton, M.H., D.A. Norris: Cytokine-induced ICAM-1 expression in human keratinocyte strains from different donors. *J. Invest. Dermatol.* (1995, in press).

91 Mitra, R.S., B.J. Nickoloff: Epidermal growth factor and transforming growth factor-alpha decrease gamma interferon receptors and induction of intercellular adhesion molecule (ICAM-1) on cultured keratinocytes. *J. Cell. Physiol.* 150 (1992), 264–268.

92 Mitra, R.S., Y.J. Shimuzu, B.J. Nickoloff: Histamine and cis-urocanic acid augmented tumor necrosis factor-alpha mediated induction of keratinocyte intercellular adhesion molecule-1 expression. *J. Cell. Physiol.* 156 (1993), 348–357.

93 Mond, C.B., M.G. Peterson, N.F. Rothfield: Correlation of anti-Ro antibody with photosensitivity rash in systemic lupus erythematosus patients. *Arthritis Rheum.* 32 (1989), 202–204.

94 Moodycliffe, A.M., I. Kimber, M. Norval: The effect of ultraviolet B irradiation and urocanic acid isomers on dendritic cell migration. *Immunol.* 77 (1992), 394–399.

95 Natali, P.G., E.M. Tan: Experimental skin lesions in mice resembling systemic lupus erythematosus. *Arthritis Rheum.* 16 (1973), 579–589.

96 Nieboer, C., Z. Tak Diamand, H.E. VanLeeuwen Wallau: Dust-like particles: a specific direct immunofluorescence pattern in sub-acute cutaneous lupus erythematosus. *Br. J. Dermatol.* 118 (1988), 725–729.

97 Norris, D.A.: Immunological cytotoxicity of cutaneous cellular targets. In: *Immune Mechanisms in Cutaneous Disease*, (1989), pp. 123–166. Ed.: D.A. Norris, Marcel Dekker, Inc., New York.
98 Norris, D.A.: Cytokine modulation of adhesion molecules in the regulation of immunologic cytotoxicity of epidermal targets. *J. Invest. Dermatol.* 95 (1990), 111S–120S.
99 Norris, D.A.: Pathomechanisms of photosensitive lupus erythematosus. *J. Invest. Dermatol.* 100 (1993), 58S–68S.
100 Norris, D.A., L.A. Lee: Antibody-dependent cellular cytotoxicity and skin disease. *J. Invest. Dermatol.* 85: (1985), 165S–175S.
101 Norris, D.A., M.B. Lyons, M.H. Middleton, J.J. Yohn, M. Kashihara Sawami: Ultraviolet radiation can either suppress or induce expression of intercellular adhesion molecule 1 (ICAM-1) on the surface of cultured human keratinocytes. *J. Invest. Dermatol.* 95 (1990), 132–138.
102 Norris, P., R.N. Poston, D.S. Thomas, M. Thornhill, J. Hawk, D.D. Haskard: The expression of endothelial leukocyte adhesion molecule-1 (ELAM-1), intercellular adhesion molecule-1 (ICAM-1), and vascular cell adhesion molecule-1 (VCAM-1) in experimental cutaneous inflammation: a comparison of ultraviolet B erythema and delayed hypersensitivity. *J. Invest. Dermatol.* 96 (1991), 763–770.
103 Norris, D.A., S.R. Ryan, K.A. Fritz, M. Kubo, E.M. Tan, J.S. Deng, W.L. Weston: The role of RNP, Sm, and SS-A/Ro-specific antisera from patients with lupus erythematosus in inducing antibody-dependent cellular cytotoxicity (ADCC) of targets coated with nonhistone nuclear antigens. *Clin. Immunol. Immunopathol.* 31 (1984), 311–320.
104 Pentland, A.P., P. Hale, C.C. Miller: PGE synthesis by UVB-irradiated human keratinocyte cultures is tyrosine-kinase dependent. *J. Invest. Dermatol.* 100 (1993), 514,
105 Pentland, A.P., M. Mahoney, S.C. Jacobs, M.J. Holtzman: Enhanced prostaglandin synthesis after ultraviolet injury is mediated by endogenous histamine stimulation. A mechanism for irradiation erythema. *J. Clin. Invest.* 86 (1990), 566–574.
106 Provost, T.T.: Subsets in systemic lupus erythematosus. *J. Invest. Dermatol.* 72 (1979), 110–113.
107 Provost, T.T., R. Herrera-Esparaza, L.A. Diaz: Nucleoprotein autoantibodies in lupus erythematosus. *J. Invest. Dermatol.* 85 (Suppl) (1985), 133S–139S.
108 Provost, T.T., N. Talal, W. Bias, J.B. Harley, M. Reichlin, E.L. Alexander: Ro (SS-A) positive Sjogren's/lupus erythematosus (SC/LE) overlap patients are associated with the HLA-DR3 and/or DRw6 phenotypes. *J. Invest. Dermatol.* 91 (1988), 369–371.
109 Provost, T.T., R. Watson: Anti-Ro (SS-A) HLA-DR3-positive women: the interrelationship between some ANA negative, SS, SCLE, and NLE mothers and SS/LE overlap female patients. *J. Invest. Dermatol.* 100 (1993), 14S–20S.
110 Prystowsky, S.D., J.N. Gilliam: Discoid lupus erythematosus as part of a larger disease spectrum. Correlation of clinical features with laboratory findings in lupus erythematosus. *Arch. Dermatol.* 111 (1975), 1448–1452.
111 Prystowsky, S.D., J.N. Gilliam: Antinuclear antibody studies in chronic cutaneous discoid lupus erythematosus. *Arch. Dermatol.* 113 (1977), 183–186.
112 Prystowsky, S.D., J.H. Herndon Jr, J.N. Gilliam: Chronic cutaneous lupus erythematosus (CLE) – a clinical and laboratory investigation of 80 patients. *Medicine (Baltimore)* 55 (1976), 183–191.
113 Rivas, J.M., S.E. Ullrich: Systemic suppression of delayed-type hypersensitivity by supernatants from UV-irradiated keratinocytes. An essential role for keratinocyte-derived IL-10. *J. Immunol.* 149 (1992), 3865–3871.
114 Rosenstein, B.S., R.B. Rosenstein, G.B. Zamansky: Repair of DNA damage induced in systemic lupus erythematosus skin fibroblasts by simulated sunlight. *J. Invest. Dermatol.* 98 (1992), 469–474.
115 Sauder, D.N., D. Wong, C. Laskin: Epidermal cytokines in murine lupus. *J. Invest. Dermatol.* 100 (1993), 42S–46S.
116 Seckinger, P., J.H. Zhang, B. Hauptmann, J.M. Dayer: Characterization of a tumor necrosis factor alpha (TNF-alpha) inhibitor: evidence of immunological cross-reactivity with the TNF receptor. *Proc. Natl. Acad. Sci. USA* 87 (1990), 5188–5192.
117 Smith, C.W., S.D. Marlin, R. Rothlein, C. Toman, D.C. Anderson: Cooperative interactions of LFA-1 and MAC-1 with intercellular adhesion molecule 1 in facilitating adherence and transendothelial migration of human neutrophils *in vitro*. *J. Clin. Invest.* 83 (1989), 2008–2017.
118 Sontheimer, R.D.: Questions pertaining to the true frequencies with which anti-Ro/SS-A autoantibody and the HLA-DR3 phenotype occur in subacute cutaneous lupus erythematosus patients. *J. Am. Acad. Dermatol.* 16 (1987), 130–134.
119 Sontheimer, R.D., P.R. Bergstresser: Epidermal Langerhans cell involvement in cutaneous lupus erythematosus. *J. Invest. Dermatol.* 79 (1982), 237–243.
120 Sontheimer, R.D., P.J. Maddison, M. Reichlin, R.E. Jordon, P. Stastny, J.N. Gilliam: Serologic and HLA associations in subacute cutaneous lupus erythematosus, a clinical subset of lupus erythematosus. *Ann. Intern. Med.* 97 (1982), 664–671.
121 Swerlick, R.A., T.J. Lawley: Role of microvascular endothelial cells in inflammation. *J. Invest. Dermatol.*

100 (1993), 111S–115S.

122 Swope, V.B., Z. Abdel Malek, L.M. Kassem, J.J. Nordlund: Interleukins 1 alpha and 6 and tumor necrosis factor-alpha are paracrine inhibitors of human melanocyte proliferation and melanogenesis. *J. Invest. Dermatol.* 96 (1991), 180–185.

123 Tan, E.M.: Autoantibodies to nuclear antigens (ANA): their immunobiology and medicine. *Adv. Immunol.* 33 (1982), 167–240.

124 Trefzer, U., M. Brockhaus, M. Loetscher, F. Parlow, A. Kapp, E. Schopf, J. Krutmann: 55-kd Tumor necrosis factor receptor is expressed by human keratinocytes and plays a pivotal role in regulation of human keratinocyte ICAM-1 expression. *J. Invest. Dermatol.* 97 (1991), 911–916.

125 Tuffanelli, D.L., D. Kay, K. Fukuyama: Dermal–epidermal junction in lupus erythematosus. *Arch. Dermatol.* 65 (1969), 143–153.

126 Turksen, K., T. Kupper, L. Degenstein, I. Williams, E. Fuchse: Interleukin 6: insights to its function in skin by overexpression in transgenic mice. *Proc. Natl. Acad. Sci. USA* 89 (1992), 5068–5072.

127 Urbanski, A., T. Schwarz, P. Neuner, J. Krutmann, K. Kirnbauer, K. Kock, T.A. Luger: Ultraviolet light induces increased circulating interleukin-6 in humans. *J. Invest. Dermatol.* 94 (1990), 808–811.

128 Vassalli, P.: Subcutaneous perfusion of tumor necrosis factor induces local proliferation of fibroblasts, capillaries, and epidermal cells, or massive tissue necrosis. *Am. J. Pathol.* 136 (1990), 103–110.

129 Velthuis, P.J., H. vanWeelden, D. vanWichen, H. BaartdelaFaille: Immunohistopathology of light-induced skin lesions in lupus erythematosus. *Acta Derm. Venereol. (Stockh.)* 70 (1990), 93–98.

130 Vermeer, M., J.W. Streilein: Ultraviolet B light-induced alterations in epidermal Langerhans cells are mediated in part by tumor necrosis factor-alpha. *Photodermatol. Photoimmunol. Photomed.* 7 (1990), 258–265.

131 Volc Platzer, B., B. Anegg, S. Milota, W. Pickl, G. Fischer: Accumulation of gamma delta T cells in chronic cutaneous lupus erythematosus. *J. Invest. Dermatol.* 100 (1993), 84S–91S.

132 Weston, W.L., C. Harmon, C. Peebles, D. Manchester, H.L. Franco, J.C. Huff, D.A. Norris: A serological marker for neonatal lupus erythematosus. *Br. J. Dermatol.* 107 (1982), 377–382.

133 Wilson, A.G., N. deVries, F. Pociot, F.S. diGiovine, L.B. vanderPutte, G.W. Duff: An allelic polymorphism within the human tumor necrosis factor alpha promoter region is strongly associated with HLA A1, B8, and DR3 alleles. *J. Exp. Med.* 177 (1993), 557–560.

134 Yohn, J.J., M. Critelli, M.B. Lyons, D.A. Norris: Modulation of melanocyte intercellular adhesion molecule-1 by immune cytokines. *J. Invest. Dermatol.* 95 (1990), 233–237.

135 Yohn, J.J., J.G. Morelli, S.J. Walchak, K.B. Rundell, D.A. Norris, M.R. Zamora: Cultured human keratinocytes synthesize and secrete endothelin-1. *J. Invest. Dermatol.* 100 (1993), 23–28.

136 Yoshikawa, T., I. Kurimoto, J.W. Streilein: Tumour necrosis factor-alpha mediates ultraviolet light B-enhanced expression of contact hypersensitivity. *Immunol.* 76 (1992), 264–271.

137 Yoshikawa, T., J.W. Streilein: Tumor necrosis factor-alpha and ultraviolet B light have similar effects on contact hypersensitivity in mice. *Reg. Immunol.* 91 (1990), 139–144.

138 Zamansky, G.B.: Recovery from UV-induced potentially lethal damage in systemic lupus erythematosus skin fibroblasts. *Int. J. Radiat. Biol. Relat. Stud. Phys. Chem. Med.* 50 (1986), 305–314.

139 Zamansky, G.B.: Varying sensitivity of human skin fibroblasts to polychromatic ultraviolet light. *Mutat. Res.* 160 (1986), 55–60.

140 Zamansky, G.B., I.N. Chou: Disruption of keratin intermediate filaments by ultraviolet radiation in cultured human keratinocytes. *Photochem. Photobiol.* 52 (1990), 903–906.

141 Zamansky, G.B., L.F. Kleinman, J.C. Kaplan, P.H. Black: Effect of ultraviolet light irradiation on the survival of New Zealand black mouse cells. *Arthritis Rheum.* 23 (1980), 866–867.

142 Zamansky, G.B., D.F. Minka, C.L. Deal, K. Hendricks: The *in vitro* photosensitivity of systemic lupus erythematosus skin fibroblasts. *J. Immunol.* 134 (1985), 1571–1576.

PART 4
THERAPEUTIC PHOTOIMMUNOLOGY

17: Extracorporeal photochemotherapy

Francis P. Gasparro & Richard L. Edelson

Introduction

Cutaneous T-cell lymphoma, also known as mycosis fungoides, is a malignant clonal proliferative disease of helper T cells which is usually divided into three phases – the early eczematous stage, the plaque stage and the tumor stage [13]. The first stage, where the pathology is localized to the skin, can last for months or years before developing into plaques and then tumors. In the early stage the disease is often misdiagnosed. Treatment regimens for this stage have included topical steroids and PUVA (psoralen plus UVA) therapy [21]. Response to PUVA is often dramatic and can be confirmed histologically. Later, malignant cells may disseminate to involve lymph nodes and internal organs. Until recently there had been no effective treatment for the later stages of CTCL. Thus it is not surprising that many different therapies have been employed (X-rays, electron beam radiation and chemotherapy, including topical nitrogen mustard [29]). In 1981 an extracorporeal photochemotherapy with 8-methoxypsoralen and long-wavelength ultraviolet radiation (UVA, 320–400 nm) was developed. In this process, which is also called photopheresis, a portion of the patient's T cells are directly exposed to UVA [14]. In this report we survey clinical results for the treatment of CTCL and several other diseases with photopheresis. The follow-up analysis on the CTCL patients from the original clinical trial is also summarized.

The molecular events involving 8-MOP and UVA in cells are also described. Some new aspects of psoralen photochemistry and photobiology (visible activation and photoadduct distribution) are also presented. When activated with UVA, 8-MOP forms photoadducts with DNA which inhibit DNA synthesis. In the hyperproliferative keratinocytes characteristic of psoriatic lesions, it is likely that these photoadducts inhibit cell replication. Although the mechanism of action of photopheresis has not been fully elucidated, photoinduced damage in the target cells has been well characterized and evidence for immune-augmenting effects has been documented. When the treated cells are reinfused into the patient there appears to be an autovaccination against the pathogenic T-cell clones. This conclusion is based on two kinds of observations. First, several diseases not characterized by hyperproliferation (e.g. rheumatoid arthritis [35] and progressive systemic scleroderma [46]) respond to photopheresis. Second, animal models in which immune responses are assayed directly confirm an immune-modulating effect [5, 42, 48].*

Therapeutic procedure

The photopheresis treatment involves oral administration of the drug (0.6 mg/kg) followed by the exposure of the leukapheresed blood fraction to UVA [14]. The principal advantage of this system is that the drug is inert until exposed to UVA. Because the UVA activation of 8-MOP occurs extracorporeally, only the target cells are affected and the vast majority of the patient's cells are spared the cytotoxic effects of the activated drug. Red blood cells, platelets and plasma proteins are not significantly affected [33]. Sufficient 8-MOP is administered by oral ingestion (typically 0.6 mg/kg) to achieve a minimum plasma level of 50 ng/ml

* For a current review see Perez *et al.* [41].

1–2 h after ingestion. Heparinized venous blood is centrifuged, separating the red blood cell, white blood cell and plasma fractions. The latter two fractions are combined with each other (240 ml and 300 ml respectively) and then with 200 ml normal saline used to prime the system. It is essential to remove a significant fraction of the red blood cells (up to 95–97%) because they shield the psoralen-containing lymphocytes from the UVA. Six leukapheresis cycles are performed in each photopheresis treatment. The total volume of 740 ml is recirculated through an irradiation chamber (as a 1 mm film). Approximately 100 ml at a time is exposed to UVA radiation over a 3 h period. In this time period the leukocytes complete 10–15 excursions through the UVA field. The average lymphocyte is exposed to a UVA dose of ~2 J/cm^2 [14]. Approximately 10–15% of a patient's circulating lymphocytes are treated; however, only a fraction of these are malignant cells [23]. The entire treated volume and the red blood cells are returned to the patient. Typically a patient is treated on two consecutive days at monthly intervals. Although this schedule has been altered in a few selected cases, there is no evidence that any other schedule (more or less frequent) may be more efficacious in the initial treatment of patients.

Tapering of therapy

Patients who show signs of response, e.g. improved skin scores and CD4/CD8 ratios, are tapered off therapy by gradually lengthening the interval between treatments. In 3 month increments 1 week at a time is added to the standard 4 week interval as long as the patient continues to improve. If a complete response in maintained, the 1 week increment is added until an 8 week interval is reached. After three cycles at 8 weeks, if there is no recurrence, the therapy is discontinued. If any relapse occurs during the tapering process the 4 week schedule is resumed.

Technical details

Apparatus

Two major modifications to the apparatus have been made since the first clinical trial when treatment took an entire day (up to 10 h). In the original clinical device the blood was first centrifuged using a plasmapheresis instrument. The collected cells were then transfered to the irradiation apparatus which consisted of a set of twelve 6 foot long UVA bulbs. The cells were pumped through a series of

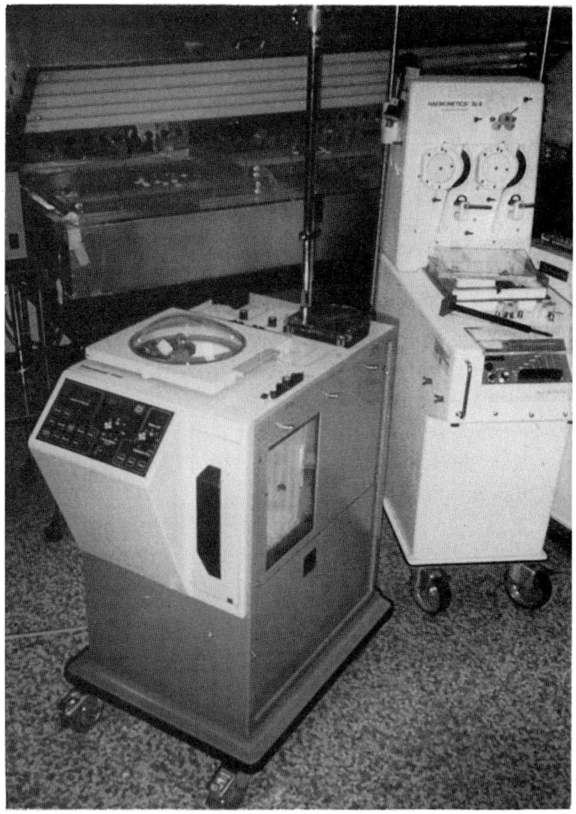

Fig. 17.1(a) Instruments used for photopheresis. Shown in the foreground, the current instrument; the blood plate (c) is behind the window on the right side of the instrument. In the right background: the Haemonetics leukapheresis device which was used originally to separate white blood cells; and behind to the left, the original UVA irradiator.

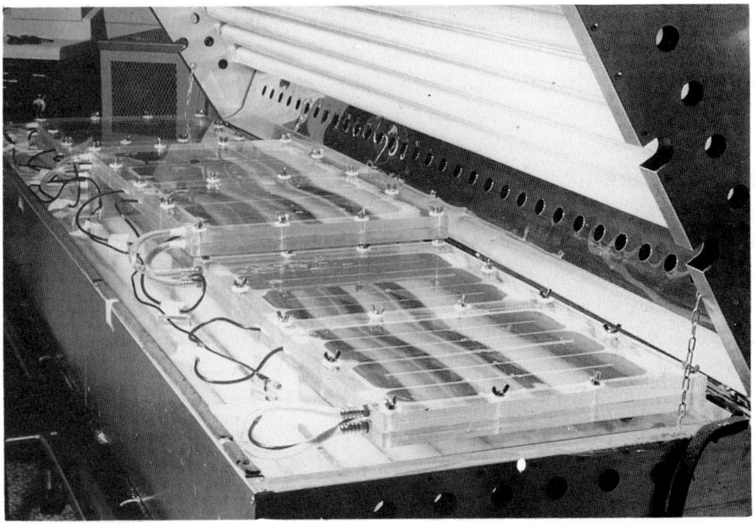

Fig. 17.1(b) Close-up of blood cell bags sandwiched between water cooling plates on the original irradiator.

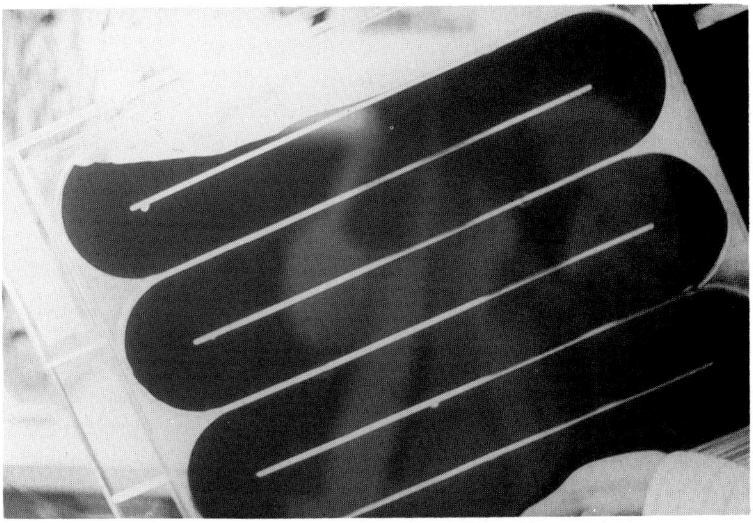

Fig. 17.1(c) Close-up of blood plate used in current device – this plate is situated between two banks of UVA lamps behind the plate on the right side of the instrument.

six plastic bags sandwiched between water cooling plates with the UVA lights above and below (Fig. 17.1). In the second generation instrument, more efficient techniques of cell separation and irradiation were devised and the entire procedure starting with drug ingestion is performed by one device and now takes less than 6 h. A new device currently undergoing initial testing for efficacy treats 150 ml of whole blood at a time. This volume of blood is centrifuged in a bonnet-like bowl. The centrifugally induced layer of leukocytes is then exposed to UVA radiation using circular, concentrically fitted lamps. This procedure is repeated on nine more 150 ml whole blood fractions.

8-MOP plasma levels

In PUVA, MED testing is performed to determine initial UVA doses as well as subsequent maintenance doses. In photopheresis, the clinical device delivers a fixed UVA dose. The one variable over which there is little control is the amount of 8-MOP in patient tissue at the time of UVA exposure. A streamlined solid phase extraction method was developed [18] to replace the time-consuming multistep organic solvent extraction procedure previously used (10 minutes as compared to 2 h). The extracted sample is then analyzed by HPLC. After oral administration, 8-MOP is absorbed unpredictably, primarily due to its low aqueous solubility (~30 μg/ml) [24]. The two available forms of the drug (crystalline and liquid or 'ultra') have very similar bioavailability characteristics as demonstrated by studies in our laboratory in a random series of patients [17]. There was no statistical difference between the average 8-MOP levels (124 ± 98 for the crystalline and 114 ± 101 for the liquid). Data such as these highlight why it is crucial to measure 8-MOP blood levels. Studies purporting to show the superiority of one form of 8-MOP or another are often performed under idealized conditions that do not represent the clinical setting. In a study of photopheresis patients, we measured the 8-MOP (crystalline and liquid) concentration (ng/ml) over a 4 h period (Table 17.1). These data illustrate a high level of interindividual variation for both the crystal and liquid forms of 8-MOP. Similar variations are seen for a single patient who ingested 8-MOP (crystal, $n = 4$): 1 h, 88 ± 39; 2 h, 115 ± 68; 3 h, 86 ± 54; and 4 h, 52 ± 27.

In another patient who ingested 8-MOP (liquid, $n = 92$), we found a great variation in 2 h 8-MOP levels and an unexpected dependence on the dosage: 30 mg ($n = 9$), 73 ± 62; 40 mg (15), 142 ± 80; 50 mg (28), 77 ± 60; and 60 mg (40), 103 ± 79. The highest average 8-MOP level was obtained when the patient ingested 40 mg of the drug; in addition, the 40 mg dose showed the smallest variation. These results highlight the importance of determining 8-MOP levels and reinforce the findings by other investigators (see below).

A monoclonal antibody which recognizes 8-MOP has been prepared by immunizing mice with 8-MOP conjugated to bovine serum albumin with 1-ethyl-3-(3-dimethylaminopropyl) carbodiimide [4]. After characterization by competitive ELISA, we showed as little as 1 ng/ml of 8-MOP could be detected in 50 μl plasma. This antibody has only been used experimentally but it has the potential of being incorporated into a colorimetric dipstick or biosensing electronic probe, enabling rapid assays which would permit administration of augmenting doses of 8-MOP prior to or during photopheresis to achieve optimal plasma drug levels.

Improvements in 8-MOP formulation

As discussed in the preceding section, the oral formulations of the drug are absorbed unpredictably. In 1979 Wagner *et al.* demonstrated that poorly responding PUVA patients had consistently low plasma levels [60]. Deviations from the expected 2 h peak level were observed in 50% of problem cases and only 14% of control patients (responders). *No correlation between ingested doses and 8-MOP plasma levels was observed for either group of patients.* Three of fourteen patients showed improved therapeutic response when the time of UVA exposure was adjusted to coincide with time of optimal plasma levels. In 1991 Walther and Haustein found abnormally low or 'deviated' serum levels in seven of eleven PUVA problem patients [61]. They independently reproduced the findings of Wagner *et al.* In addition, however, they also

Table 17.1 Absorption kinetics for two formulations of 8-MOP (ng/ml)

	0.5 h	1 h	2 h	3 h	4 h
Crystal	–	73 ± 55	121 ± 72	106 ± 80	85 ± 53
Liquid	135 ± 152	146 ± 118	112 ± 131	50 ± 33	32 ±28

noted that 8-MOP formulation affected plasma levels. These two studies are cited because they indicate that plasma levels play an important role in the non-life-threatening disease of psoriasis in which a facile determination of patient response can be made. It is also likely that sub-optimal plasma levels may affect the outcome of CTCL patients being treated with photopheresis. However, the relatively slow response to CTCL and other diseases treated with photopheresis makes it difficult to attribute lack of response to low 8-MOP levels.

The ideal 8-MOP formulation would be a sterile aqueous solution which could be injected directly into the leukocyte/plasma fraction just prior to UVA exposure. Considerably smaller amounts of drug would be required (100–200 μg in 740 ml as compared to an average oral dose of ~40 mg). An *ex vivo* formulation of this type has been described by Knobler *et al.* who have demonstrated its efficacy in a series of 16 patients being treated with photopheresis [30]. In these patients oral ingestion of 8-MOP (0.6 mg/kg) led to an average 8-MOP plasma level of 99 ± 49 ng/ml (91 treatments; 4–224 ng/ml range). After leukapheresis of the blood the expected 8-MOP would be ~60 ng/ml during the irradiation phase. Six levels were below the minimum considered effective. When the *ex vivo* formulation was used in the same series of patients, the average plasma level was 126 ± 24 ng/ml (108 treatments; 66–178 ng/ml range). None were below 60 ng/ml. In the latter cases 135 μg of 8-MOP was added *ex vivo*. Thus the maximum plasma level should have been 182 ng/ml. The target value is not reached because the plastic bags and tubing which are essential for the manipulation of the blood absorb a significant fraction of the *ex vivo* administered 8-MOP (F.P. Gasparro, unpublished results). In spite of this absorption, an *ex vivo* method would still be preferred because the administered dose can be adjusted directly to compensate for this loss. On the other hand, when 8-MOP is administered orally, there is no control over the ultimate plasma level and increased dosing will not have the desired effect (see above). Another *ex vivo* formulation has recently been patented in the United States [3].

CTCL clinical results

In addition to the original clinical trial for photopheresis, three other trials have been performed. The results of these studies have been published [1, 4, 22, 23] and hence are only summarized here in Table 17.2.

Survival statistics on patients from original clinical trial

Median survival time after CTCL diagnosis is 31–33 months [32, 62]. The patient group from the original multicenter clinical trial have a median survival to more than 60 months (from the date of diagnosis) and 48 months (from the start of photopheresis) [23]. Note that this clinical trial included several patients who had failed previous therapies. Four of six complete responders remain alive and in continued remission, whereas only one of the five non-responders remains alive. In a comparison of complete responders and non-responders, a correlation with near normal CD4/CD8 ratio prior to photopheresis was related to clinical improvement at a statistically significant level. It was also shown that the interval between diagnonsis of erythrodermic CTCL and the initiation of therapy was correlated with therapeutic response – complete responders having an interval of ~12 months and non-responders ~28 months.

Combination modalities

Because some patients do not respond to photopheresis or show a relapse after an initial response, several groups have established adjunctive therapy protocols. Heald *et al.* demonstrated an enhanced response in five of eight patients who had been unresponsive to photopheresis alone by administering low doses of methotrexate [22]. Rook *et al.* and Vonderheid *et al.*, in anecdotal reports, described the use of interferon-α-2b (5 million units, subcutaneously) [47, 59]. Alternative therapies for CTCL have been reviewed recently by Holloway *et al.* [27] who also summarized some new experimental therapies (e.g. radiolabeled

Table 17.2 Survey of photopheresis clinical trials in CTCL

Trial	Patients	Responders	Comments
Original clinical trial [14]	37	27	25% Reduction in skin score for 4 weeks
		9	>75% Improvement; patients with Sezary syndrome more likely to respond than those with plaques or tumors
YNH [22]	32	22	Of these erythrodermic patients, 25% went in remission; half of these had marked skin improvement but retained CTCL lesions; in 25% of the patients there was no significant improvement
Morristown [1]	8	5 (erythrodermic)	4/5 – At least partial remission
		1 (extensive)	Partial remisison
		1 (patch/plaque)	Partial remission
		2 (tumors)	One with complete remission of all disease at 20 months; the other partial remission at 5 months
Follow-up on [14] [23]	29	6	Complete remission (one patient up to 7 years)
		6	>50% Improvement
		10	Variable improvement in skin score
		5	No change or deterioration
		2	Lost to follow-up

monoclonal antibodies, autologous bone marrow transplatation and cyclosporine).

Photopheresis and other diseases

Small clinical trials have been conducted in several other diseases [7, 12, 35, 46, 48, 63] and promising results have been obtained (Table 17.3).

8-MOP photophysics, photochemistry and photobiology

Psoralens are furocoumarins with a well-characterized photosensitizing potential [11, 16]. 8-Methoxypsoralen (or methoxsalen), the most commonly used natural product analog, is found in a variety of fruits and vegetables (e.g. limes and celery). Although several water-soluble psoralen derivatives have been synthesized [25], only the naturally occurring compounds have been used clinically. The properties of these compounds as a family (UV absorption, DNA binding and photoreactivity) are very similar to those for 8-MOP, which are summarized in this section.

Table 17.3 Photopheresis and other diseases

Disease	Patients	Responders	Comments
Rheumatoid arthritis [35]	7	4	Significant improvement in joint scores
Psoriatic arthritis [63]	5	4	Slight to moderate improvement over 5–15 months. Need for large contolled study highlighted
PSS [46]	31	21 (ECP)	Statistically significant but small advantage for photopheresis
	25	8 (D-pen)	as compared to D-penicillamine
ARC [7]	5	4	Enhanced immune functions observed

Physical properties of psoralens

Psoralens are tricyclic aromatic compounds (furocoumarins) in which the 2,3 furan bond is fused to the 6,7 bond of the double ringed coumarin moiety (Fig. 17.2). The extended aromaticity of the resulting structure is responsible for its ability to absorb ultraviolet radiation with strong absorption bands near 250 and 300 nm [16] (Fig. 17.3). These latter wavelengths do *not* correspond to the regions which photoactivate the compound because these wavelengths are energetic enough to cause photodegradation [54] and photoreversal [55]. Thus longer ultraviolet wavelengths (320–400 nm) are used for the activation of 8-MOP. Recently we showed that the shorter wavelength portion of the visible spectrum could be used to activate 8-MOP [20]. When irradiated with UVA radiation or visible light, psoralen is photoactivated and can photobind directly to cellular macromolecules [37]. Other reactions may also occur indirectly as oxygen-dependent events with the production of singlet oxygen.

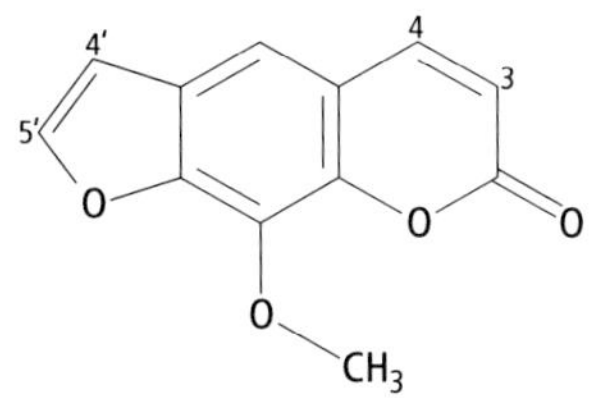

Fig. 17.2 Molecular structure of 8-MOP – the photoreactive sites are labeled: 4′,5′ and 3,4.

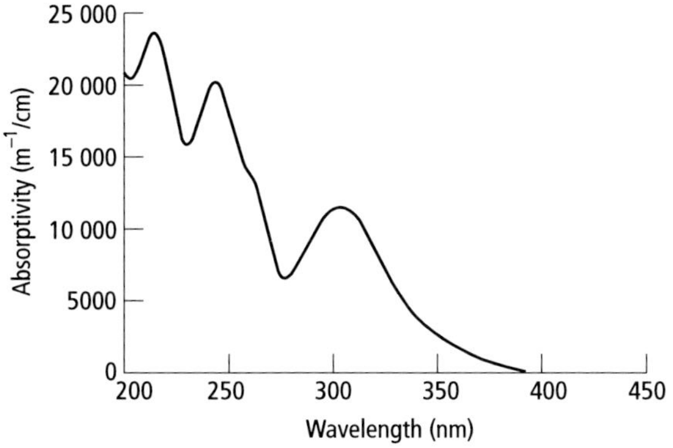

Fig. 17.3 Ultraviolet absorption spectrum of 8-MOP. The absorption spectrum of 8-MOP dissolved (10 mg/ml in ethanol) is shown. Absorptivity in $m^{-1}cm^{-1}$ for the following wavelengths: 300 nm, 11 500; 350 nm, 2200; 400 nm, 51; 420 nm, 3.5; 440 nm, 1.3 (see [20] for details).

DNA photoadducts

The photochemical reactions of psoralens with nucleic acids are the most well characterized, having been studied for nearly three decades [11]. Psoralen molecules form complexes with DNA hydrogen-bonded base pairs in the dark. Following irradiation with UVA radiation or visible light, photoadducts can form at the 4′,5′ double bond and perhaps the 3,4 double bond (see below) of the psoralen molecule. Each of these adducts (Fig. 17.4) has a cyclobutyl bond involving the 5,6-double bond of a pyrimidine. The extent of photoadduct formation depends on the suitability (base sequence) and accessibility (protein contacts, DNA bending) of intercalation sites between DNA base pairs. The 4′,5′ adducts (furanside) can absorb a second photon and form an interstrand crosslink between two thymines from the adjacent base pairs in a T–A site. These photoreactions have been completely characterized *in vitro* using either calf thymus DNA or various synthetic polynucleotides. Sage and Moustacchi used short segments of natural DNA to conclusively demonstrate that repetitive runs of adenine and thymine were most suitable for photoadduct formation under *in vitro* conditions [50]. The determination of these same parameters *in vivo* has been hindered by the lack of techniques with sufficient sensitivity to detect the photoadducts at the part per million level.

Photoadduct formation is wavelength dependent [55]. Monochromatic 400 nm (bandpass 10 nm) leads to a preponderance of 4′,5′-monoadducts (>80%) while irradiation at 350 nm produces more crosslinks (~40%) and many fewer 4′,5′-monoadducts [6]. We have found that 3,4-monoadducts are a very low yield photoadduct. In synthetic polynucleotides a *split-dose* irradiation protocol (400 nm followed by 350 nm) also results in the conversion of 4′,5′-monoadducts to 3,4-monoadducts [39]. In recent studies with lymphocytes

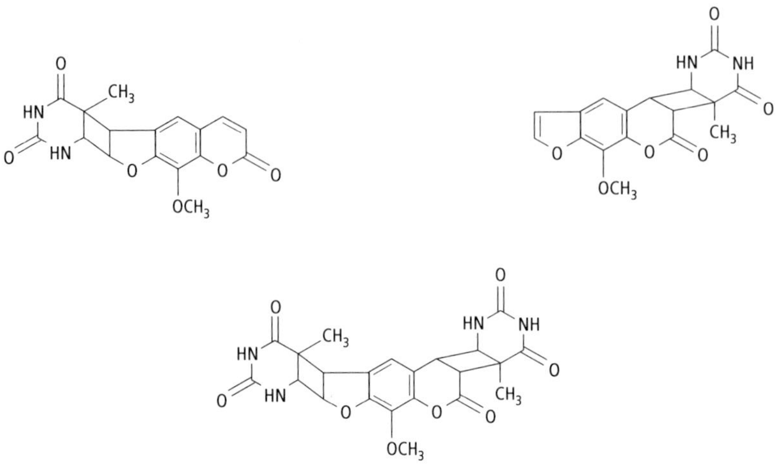

Fig. 17.4 Psoralen–DNA photoadducts: upper left, 4′,5′-monoadduct; upper right, 3,4-monoadduct; below, crosslink.

treated in this manner, the subsequent exposure to 350 nm radiation led to the formation of 3,4 adducts as well as the expected crosslinks. This is consistent with the results of Tessman *et al.* who demonstrated that 4′,5′-monoadducts can undergo intrahelical isomerization upon absorption of additional photons [55].

UVA wavelengths less than 320 nm may cause photoreversal of previously formed adducts and degradation of non-intercalated psoralen molecules [54]. This degradation process is very efficient at 300 nm. The rate of destruction is proportional to the ability of 8-MOP to absorb a particular wavelength. Studies in which different wavelengths of ultraviolet and visible radiation were used to activate 8-MOP have shown that 3,4-monoadducts do not appear to be the result of primary photoreactions. Rather they appear to form as a result of the absorption of photons by previously formed 4′,5′-monoadducts [39]. A survey of the literature shows no direct evidence for the formation of 3,4-monoadducts as primary photoadducts.

Experiments carried out on various cells (e.g. human lymphocytes [19], murine keratinocytes [56] and fibroblasts and bovine aorta smooth muscle cells [20]) as well as synthetic DNA have enabled the quantification and characterization of the photoadducts following the extraction and enzymatic hydrolysis of DNA. The extent of photoadduct formation is reported as the number of photoadducts per million base pairs (mbp). Increasing doses of 8-MOP and UVA lead to increased yields of photoadducts – with a correlation coefficient of 1.0 [56]. We have previously reported [19] that the number of photoadducts formed is directly related to the product of 8-MOP concentration in ng/ml and the UVA dose in J/cm^2 (8-MOP concentrations over the range 10–100 ng/ml and UVA doses over the range 1–10 J/cm^2). At 10 ng/ml 8-MOP and 1 J/cm^2 UVA, 0.4 adducts/mbp are formed, and with 100 ng/ml 8-MOP and 1 J/cm^2 UVA, ~4 adducts/mbp result. It has been shown that in lymphocytes treated with 100 ng/ml 8-MOP, 80% of the photoreaction occurs at 4′,5′ site resulting in the formation of either 4′,5′-MA or cross-links. Using 10 ng/ml 8-MOP the 4′,5′ site does not appear to be as favored and a twofold increase in the yield of 3,4-MA is seen [6]. This may be a result of more photons being absorbed by 4′,5′-MA (see above) because fewer free 8-MOP molecules are intercalated between the DNA base pairs. In keratinocytes treated with 50 ng/ml 8-MOP and 3 J/cm^2 UVA, 3,4-monoadducts also formed to a greater extent than expected [56].

Lipid photoadducts

Although nuclear DNA has been the primary subject of psoralen photobiological research, psoralens also modify other macromolecules (e.g. lipids and proteins) under both *in vitro* and *in vivo* conditions [37]. Lipid photoadditions and/or oxidations may take place, the former resulting in covalent adducts with unsaturated fatty acids (e.g. linoleic and linolenic acids). These photoadducts appear to form with quantum efficiencies comparable to those for DNA photomodification. Cell membrane phospholipids may also be modified, profoundly affecting regulation of cell processes. Inhibition of lipid hydrolysis necessary for the activation of protein kinase may also be a result of lipid photoadditions.

In a different approach, two reports demonstrated that exogenously formed peroxidation of low density lipoproteins (LDL) was cytotoxic to HUT-78 helper T cells and non-toxic towards normal lymphocytes [15, 28]. These cytotoxic effects were induced with 8-MOP/UVA, hydrogen peroxide or peroxidase treatments. In this vein it is noteworthy to mention that Heck *et al.* detected the production of hydrogen peroxide during photopheresis [26]. While this latter study was performed *in vivo*, it is important to note that Fossell used *supraphysiological doses* of 8-MOP and UVA *in vitro* [15].

Protein photoadducts

The biological impact of proteins photomodified by psoralens has not been elucidated. More than 13 years ago it was shown that 8-MOP/UVA induced modifications, including covalent binding, subunit crosslinking, and photooxidation, between furocoumarins and proteins such as lysozyme, serum albumin, histone and ribonuclease *in vitro* [64]. UVA-induced psoralen crosslinking of glutamate dehydrogenase, alcohol dehydrogenase and catalase was also observed [58]. However no amino acid psoralen photoadduct has ever been characterized. This gap in our knowledge is a direct result of our inability to systematically disassemble proteins into their amino acid components with gentle enzymatic reactions. There are enzymes which digest proteins into smaller oligopeptides, but there is no protein equivalent to an endonuclease such as phosphodiesterase which would yield monomeric amino acids.

Proteins located on cell membranes may bind with psoralen in the presence of UVA and affect antigenicity. Covalent binding of 8-MOP to specific amino acids has been observed, although a direct role in altered protein function has not been determined. Using copolymers consisting of amino acids known to be photosensitive towards 8-MOP, Olack and Gasparro has shown that tyrosine and tryptophan can be modified with a quantum efficiency comparable to that for nucleic acid components [38].

The antigenic signals at the surface of cells are mediated by oligopeptide binding proteins. Endogenously processed proteins are displayed by class I MHC molecules and exogenously processed proteins by class II molecules [49]. These oligopeptides are small fragments of larger proteins which have been enzymatically hydrolyzed by proteosomes. Psoralen–amino acid photoadducts at crucial enzymatic sites might alter the type of oligopeptide produced during proteosome processing and hence different antigens might be bound to the MHC molecules. Alternatively, psoralen-modified amino acids in an oligopeptide might enhance the binding of a previously less-favored oligopeptide. Either of these events could lead to a new immune signal and enhanced antigenicity and immunogenicity.

The impact of photoadducts on cells

DNA synthesis

The inhibition of DNA synthesis is usually attributed to the presence of DNA photoadducts. As discussed above the degree of inhibition of DNA synthesis in human lymphocytes *in vitro* is related to the product of the 8-MOP dose (ng/ml) and UVA (J/cm^2), as demonstrated by reduced levels of tritiated thymidine incorporation after mitogen stimulation [19]. Nuclear processes appear to be completely suppressed when the product of the 8-MOP and UVA doses is greater than 50. An alternate explanation for these results is that these relatively low doses of 8-MOP and UVA affect the ability of the cells to bind PHA and hence reduce the usual stimulatory effects of PHA on the cells. However, another study supports the previous interpretation. The ability of cells to repair 8-MOP adducts follows a similar titration curve. At combined doses >100 ng/ml no repair occurs. At doses greater than 300 ng/ml the ability of the cells to exclude trypan blue decreases significantly as the damage to cell membranes reaches a critical level. Whether photodamage to cell membranes at lower concentrations could induce more specific, subtle antigenic effects has not been determined.

In metabolically active keratinocytes inhibition of DNA synthesis or decreased viability was observed in cells containing fewer than two adducts per million bases (produced by a combined 8-MOP/UVA dose of 60). At a combined dose of 30, 0.9 photoadducts were formed and no inhibition of DNA synthesis or decreased viability could be detected [56]. Photoadduct formation was unaltered by level of differentiation or source of keratinocytes (murine or human). At sublethal doses of 8-MOP and UVA, other cell processes may also be affected. For example, interleukin-1 production is reduced [57]. In hyperproliferative epidermis characteristic of psoriatic skin, it is assumed that these photoadducts inhibit cell replication. Although it is fashionable to attribute this blockage to DNA crosslinks, other psoralens (angular psoralens or angelicins), which cannot form crosslinks, also block DNA synthesis [45] and lead to therapeutic effects. Thus 8-MOP monoadducts may also play some role in the therapeutic efficacy of PUVA.

DNA repair

Extensive repair studies have been carried out in bacterial systems as well as in mammalian cells [51]. Several repair pathways have been described and include excision repair (the most common), postreplication recombination repair, and photoreactivation. 8-MOP photoadducts are known to be repaired, although studies on the repair of discrete adducts have been limited. It has often been assumed that crosslinks would not be repaired. However, an excision–recombination mechanism has been proposed to account for crosslink repair in bacteria [10]. Whether such a mechanism operates in mammalian cells is not known at present.

Freshly isolated human lymphocytes treated with 100 ng/ml 8-MOP and 1 J/cm^2 UVA show no repair after 48 h. However, cells treated with lower 8-MOP doses (10–20 ng/ml) are capable of removing photoadducts [19]. The removal rate is 25% in 48 h for cells treated with 10 ng/ml 8-MOP and 1 J/cm^2 UVA. The removal of these adducts is also associated with recovery of the proliferative activity as evidenced by the increased levels of tritiated thymidine incorporation after varying repair periods (see above).

Keratinocyte photoadduct removal has been found to be 54% at 20 h after the treatment of cells with 10 ng/ml 8-MOP and 1 J/cm^2 UVA but reduced to 12% at 20 h when 100 ng/ml 8-MOP and 1 J/cm^2 UVA were used. The extent of repair has also been shown to be dependent on calcium concentration – the rate being decreased in low calcium media [56]. Keratinocyte repair is therefore more efficient than lymphocyte repair *in vitro*. This may be due to the increased metabolic activity of keratinocytes.

Apoptosis

The cytotoxic effects of 8-MOP and UVA have been well characterized. Because the intended effect of photopheresis is immune augmentation, we have not focused our attention on its cytoreductive aspects. Recently however, Marks and Fox suggested that the treatment of cells with 8-MOP and UVA could lead to apoptosis, a process in which the cell actively participates in its own destruction [36]. In this process enzymes are produced that lead to chromatin condensation, the inhibition of cell–cell interactions and cytoskeleton disruption. Apoptosis culminates in the phagocytosis of the apoptotic cells by neighboring cells. A hallmark of this process is the production of multiple 180 base pair units by a nuclear endonuclease. Marks and Fox [36] showed that 300 ng/ml 8-MOP and 10 J/cm^2 UVA induced the formation of apoptotic cells (~30%).

Apoptosis has been the focus of new interest among researchers in diverse areas of biological research [44]. The induction of apoptosis by cells treated with 8-MOP and UVA could lead to programmed cell death of a selected group of cells, which in the process of disassembling could release their proteins, which could then lead to the production of new oligopeptide fragments which may be displayed in surface MHC molecules. These events could result in a higher level of antigenicity.

Mutagenicity

Any agent which modifies DNA has the potential of being mutagenic. UVA-induced psoralen mutagenesis has been reported in viruses, and in prokaryotic and eukaryotic cells. In V79 Chinese hamster cells [2] it has been shown that the exposure of 'preformed' 8-MOP 4′,5′-monoadducts to additional UVA radiation (in the absence of free 8-MOP), leads to the formation of crosslinks and an increase in the number of mutants. This was interpreted as an indication that crosslink repair occurs and leads to a greater mutation rate as compared to monoadducts. This irradiation regimen for crosslink induction may, however, be flawed. In poly(dA-dT) Tessman *et al.* have shown that the exposure of 'preformed' 4′,5′-monoadducts to additional UVA radiation causes not only crosslink formation but also an intrahelical isomerization to 3,4-monoadducts [55]. Therefore it is also possible that 3,4-monoadducts could also be responsible, at least in part, for the increase in mutagenic activity after the second irradiation step. Olack *et al.* have shown that this photoisomerization also occurs in lymphocytes treated with the *split-dose* protocol (see above) [39].

For the reasons cited above we have begun mutagenesis studies in which the extent and distribution of photoadducts are always determined. In order to limit crosslink formation during the first step of split-dose studies we have employed visible radiation to activate 8-MOP [20]. Using longer wavelengths to activate 8-MOP it is possible to reduce the extent of crosslink formation by nearly an order of magnitude (from ~20% at 400 nm to 2% at 419 nm).

The repair of photoadducts while the DNA is undergoing replication is known as recombination repair and may be error-prone and therefore lead to mutations. The base substitutions found are thymine to cytosine and vice versa; thymine or cytosine to adenine or guanine or vice versa. This is of clinical relevance as psoriatic patients using PUVA may require treatment over many years. It has now been demonstrated that there is an association between cumulative exposure to PUVA and increased risk of squamous cell carcinoma [52].

A study comparing the mutagenicity of two psoralens (8-MOP and 5-MOP) provides an interesting illustration of the impact of monoadducts and crosslinks [40]. Papadopoulo and Averbeck showed that the treatment of cells with the stronger DNA binding agent, 5-MOP, led to a greater number of photoadducts (monoadducts and crosslinks) as compared to 8-MOP [40]. However, on a peradduct basis, 8-MOP was more mutagenic. The basis of this finding can be understood in terms of the physical chemistry of the DNA-binding phenomenon. The *more weakly binding 8-MOP* will occupy the *most ideal binding sites* (TA) and thus when exposed to UVA photons have a high probability of forming crosslinks. When more *strongly binding 5-MOP* is used, the additional binding to DNA is reflected in the occupation of *less ideal sites* and even though more adducts may be formed, *fewer of them will be crosslinks*. Although the therapeutic role of monoadducts or crosslinks is unknown, it is interesting to note that 5-MOP is not as effective as 8-MOP in the treatment of psoriasis [9].

Do mutagenic events occur during PUVA photochemotherapy? Albertini and Strauss demonstrated that lymphocytes obtained from patients receving PUVA for either psoriasis or vitiligo had a significant increase in the number of 6-thioguanine resistant cells [53] which could be attributed to a mutation at the *hprt* locus. These studies were performed on lymphocytes obtained from patients who were receving PUVA therapy and thus the lymphocytes were only exposed indirectly. Studies are in progress to determine the extent of 6-thioguanine resistance in directly irradiated lymphocytes.

In the 10-year history of photopheresis, the most severe side-effect of the therapy has been nausea. In spite of some fears that the well-known immunosuppressive effects of 8-MOP and UVA might lead to an enhanced susceptibility to opportunistic infections this has not occurred. Recently Beer and Zmudzka have shown that only *supraphysiological* doses of 8-MOP and UVA [65] could lead to enhanced HIV promoter activity. They have yet to determine the impact of the cumulative effect of repeated incremental doses of 8-MOP and UVA.

Antigenicity

When lymphocytes are treated with moderate doses of 8-MOP (20–200 ng/ml) and UVA the cell membrane remains intact but nuclear processes are inhibited. However, low levels of highly specific cell membrane photomodifications resulting in the formation of new antigenic moieties may have occurred [31]. During photopheresis unfractionated populations of autologous mononuclear cells, containing an expanded clone(s) of pathogenic T lymphocytes, can immunize against the activity of the same clone if extracorporeally altered in an appropriate fashion. This has been demonstrated in animal studies using photoactivated effector lymphocytes which were shown to temporarily suppress graft rejection [43]. It is difficult to envisage how these modifications alone can lead to an enhanced immunogenic effect in untreated cells. However, they may stimulate the immune system in such a way that crossrecognition of other antigens also occurs.

Tum system

Boon has shown that the chemical mutagen, N-methyl-*N,N′*-nitronitrosoguanidine or N-MNNG – a DNA adduct-forming agent, can dramatically alter the antigenicity and subsequently the immunogenicity of a variety of tumor cell lines [8]. Mutagenization of P1 cells, a normally easily transplantable and highly tumorigenic line, led to growth in the immunosuppressed but not in normal syngeneic recipient mice. Preimmunization with some of the cells from the normal syngeneic recipients provided significant protection against challenge with the original tumorigenic cell line. In one such system a single point mutation, which caused one amino acid difference in an oligopeptide that is bound to MHC class I molecules, enhanced the antigenicity of the original cell line. The tumor cell is then recognized by the immune system and a response generated. This may be relevant in the treatment of CTCL by photopheresis where the photoactivated psoralen may cause a gene mutation(s) leading to a single amino acid change in the oligopeptides that are displayed by MHC molecules. This in turn would alter antigens on the surface of tumor T cells and serve as a signal which primes a cytotoxic T-cell response against the tumor cells.

To determine whether 8-MOP and UVA could enhance immunogenicity by a similar mechanism, P1 cells were treated with 8-MOP (10 ng/ml) and UVA (1 J/cm^2). More than 99% of the cells were destroyed in each of three successive treatments [34]. Outgrowth of residual cells and subsequent cloning led to 72 new clones of which 71 were tumorigenic in normal DBA/2 mice. One clone, which was tumorigenic in immunosuppressed mice, failed to demonstrate detectable growth in normal mice during a six-week period. Challenge of these resistant mice with the highly tumorigenic P1 parent clone resulted in the survival of 40% of the mice at 16 weeks. In a second series of experiments in which the tumor cells were treated with 10 rounds of 8-MOP and UVA, 102 clones were obtained, seven of which were non-tumorigenic. These new clones are currently being tested in crossprotection and adoptive transfer experiments. The results with 8-MOP and UVA are consistent with Boon's findings with N-MNNG and strongly suggest that 8-MOP/UVA-induced mutations enhance the recognition of both treated and untreated cells. Adoptive transfer experiments with the original tumor negative clone support the involvement of cytotoxic cells, although we have yet to conclusively characterize any CD8$^+$ cells.

Summary

Extracorporeal photochemotherapy is a direct method of targeting cells responsible for CTCL. It is most effective in the erythrodermic stages of the disease but also has a place in the treatment of later stages. It has also been used in the treatment of other T-cell mediated diseases. Psoralen plus UVA is known to interact with cellular DNA, lipids and proteins. Although the mechanism underlying clinical responses in diseases treated with photopheresis is not completely understood, it appears that the cell surface changes, induced directly or indirectly by 8-MOP and UVA, render cells susceptible to immune surveillance and destruction. In addition, changes in antigen presentation due to altered processing of cellular proteins, or to enhanced expression of class I MHC, may also contribute to therapeutic responses.

Acknowledgments

The authors thank Drs Gerry Olack, Paola Gattolin, Michelle Malane and Peter Glazer for permission to use preliminary data in this manuscript. These photobiology studies were supported by gifts from Therakos, Inc and the New York Cardiac Center and grants from the NIH: AR37629 (FPG), CA43058 (RLE), and AR07016 (RLE).

References

1 Armus, S., B. Keyes, C. Cahill, C. Berger, D. Scarborough, A. Klainer: Photopheresis for the treatment of cutaneous T-cell lymphoma. *J. Am. Acad. Dermatol.* 23 (1990), 898–902.

2 Babudri, N., B. Pani, S. Venturini, M. Tamaro, C. Monti-Bradagin, F. Bordin: Mutation induction and killing of V79 Chinese hamster cells by 8-methoxypsoralen plus near-ultraviolet light: relative effects of monoadducts and crosslinks. *Mut. Res.* 91 (1981), 391–394.

3 Bachynsky, M.O., M.H. Infeld, R.L. Margolic, D.A. Perla: Psoralen formulation for UV-A photopheresis. Eur. Pat. Appl. EP. (1990) 392,429; US Appl. (1990) 336,179.

4 Baoyun, Y., F. Gasparro, P. Bevilacqua, R.M. Santella: Quantitation of plasma levels of 8-methoxypsoralen by competitive enzyme-linked immunosorbent assay. *J. Invest. Dermatol.* 97 (1991), 1001–1004.

5 Berger, C.L.: Experimental murine and primate models for dissection of immunosuppressive potential of photochemotherapy in autoimmune disease and transplantation. *Yale J. Biol. Med.* 62 (1989), 611–620.

6 Bevilacqua, P.M., R.L. Edelson, F.P. Gasparro: High performance liquid chromatography analysis of 8-methoxypsoralen monoadducts and crosslinks in lymphocytes and keratinocytes. *J. Invest. Dermatol.* 97 (1991), 151–155.

7 Bisaccia, E., C. Berger, A. Klainer: Extracorporeal photopheresis in the treatment of AIDS-related complex: a pilot study. *Ann. Int. Med.* 113 (1990), 270–275.

8 Boon, T.: Towards a genetic analysis of tumor rejection antigens. *Adv. Cancer Res.* 58 (1992), 177–210.

9 Calzavar-Pinton, P., B. Ortel, A. Carlino, H. Honigsmann, G. DePanfilis: A reappraisal of the use of 5-methoxypsoralen in the therapy of psoriasis. *Exp. Dermatol.* 1 (1992), 46–51.

10 Cortes, F., W.F. Morgan, E.R. Varcarel, J.E. Cleaver, S. Wolff: Both cross-links and monoadducts induced in DNA by psoralen can lead to sister chromatid exchange formation. *Expt. Cell. Res.* 196 (1991), 127–130.

11 Dall'Acqua, F., S. Marciani, L. Ciavatta, G. Rodighiero: Formation of inter-strand cross linkings in the photoreactions between furocoumarins and DNA. *Z. Naturforsch.* 26b (1971), 561–569.

12 Deeg, H., K. Erickson, R. Storb *et al.*: Ultraviolet (UV) manipulation of donor/host interactions: effects on sensitization and graft-versus-host disease (GVHD). *Photochem. Photobiol.* 53 (1991), 57S–58S.

13 Edelson, R.L.: Cutaneous T-cell lymphoma – perspective. *Ann. Intern. Med.* 83 (1975), 548–552.

14 Edelson, R., C. Berger, F. Gasparro *et al.*: Treatment of cutaneous T-cell lymphoma by extracorporeal photochemotherapy. *N. Engl. J. Med.* 316 (1987), 297–303.

15 Fossel, E.T., J.C. Fletcher, J. McDonagh, K.K.S. Hui: Selective cytotoxicity of low density lipo-proteins to helper T-cells of cutaneous T-cell lymphoma after photoperoxidation. *J. Natl. Can. Inst.* 83 (1991), 1316–1321.

16 Gasparro, F.P.: Psoralen–DNA interactions: thermodynamics and photochemistry. In: *Psoralen–DNA Photobiology*, vol. I, (1988), pp. 5–36. Ed.: F. Gasparro, CRC Press, Boca Raton, FL.

17 Gasparro, F.P., L.A. Amici, R.M. Knobler: Practical photochemotherapy: 8-methoxypsoralen levels (Abstract). *J. Invest. Dermatol.* 98 (1992), 654A.

18 Gasparro, F.P., J. Battista, J. Song *et al.*: Rapid and sensitive analysis of 8-methoxypsoralen in plasma. *J. Invest. Dermatol.* 90 (1988), 234–236.

19 Gasparro, F.P., P.M. Bevilacqua, D. Goldminz *et al.*: Repair of 8-MOP photoadducts in human lymphocytes. In: *DNA Damage and Repair in Human Tissues* (1990), pp. 137–151. Eds.: B.M. Sutherland, A.D. Woodhead, Plenum Press, New York.

20 Gasparro, F.P., P. Gattolin, B.E. Sumpio, L.I. Deckelbaum: Excitation of 8-methoxypsoralen with visible light. *Photochem. Photobiol.* 57 (1993), 1007–1010.

21 Gilchrest, B.A.: Methoxsalen photochemotherapy for

mycosis fungoides. *Cancer Treat. Rep.* 63 (1979), 663–667.

22 Heald, P., M. Perez, I. Christensen *et al.*: Photopheresis therapy of cutaneous T cell lymphoma: the Yale–New Haven Hospital experience. *Yale J. Biol. Med.* 62 (1989),629–638.

23 Heald, P.W., A. Rook, M. Perez, B. Wintroub, R. Knobler, B. Jegasothy, F. Gasparro, C. Berger, R. Edelson: Treatment of erythrodermic cutaneous T-cell lymphoma with extracorporeal photochemotherapy. *J. Am. Acad. Dermatol.* 27 (1992), 427–433.

24 Hearst, J.: Psoralen photochemistry and nucleic acid structure. *J. Invest. Dermatol.* 77 (1981), 39–44.

25 Hearst, J.E.: Psoralen photochemistry. *Ann. Rev. Biophys. Bioeng.* 10 (1981), 69–86.

26 Heck, D.E., E. Bisaccia, S. Armus, J.D. Laskin: Production of hydrogen peroxide by cutaneous T-cell lymphoma after photopheresis with psoralen and ultraviolet light. *Can. Chemotherap. Pharmacol.* 28 (1991), 344–350.

27 Holloway, K.B., F.P. Flowers, F.A. Ramos-Cara: Therapeutic alternatives in cutaneous T-cell lymphoma. *J. Am. Acad. Dermatol.* 27 (1992), 367–378.

28 Hornicek, F.J., G.I. Malinin, T.I. Malinin, H.K. Lo: Turnover of phospholipids in HUT 102 lymphoblasts and chromatographic characterization of purified lecithins after their exposure to UV light, psoralen, and UV light and psoralen. *J. Photochem. Photobiol.* B12 (1992), 359–374.

29 Kaye, F.J., P.A. Bunn, S.M. Steinberg *et al.*: A randomized trial comparing combination electron beam radiation and chemotherapy with topical therapy in advance cutaneous T-cell lymphoma. *N. Engl. J. Med.* 321 (1989), 1784–1790.

30 Knobler, R., C. Gruenwald, F. Trautinger, W. Ramer, R. Neumann, W. Graninger, P. Neuner: Extracorporeal application of 8-methoxypsoralen leads to predictable and reproducible levels without side-effects (Abstract). *J. Invest. Dermatol.* 96 (1991), 586A.

31 Krylenkov, V.A., D.I. Roshshupkin, A.Y. Potapenko, A.M. Maygin: Changes in the expression of lymphocyte surface antigens under UV-irradiation: effect of antioxidants. *Photobiochem. Photobiophys.* 8 (1984), 19–24.

32 Lamberg, S.I., S.B. Green, D.P. Byar *et al.*: Status report of 376 mycosis fungoides patients at 4 years: mycosis fungoides cooperative group. *Cancer Treat. Rep.* 63 (1979), 701–707.

33 Lee, K., J. Garro: Engineering aspects of extracorporeal photochemotherapy. *Yale J. Biol. Med.* 62 (1989) 621–628.

34 Malane, M.S., R.E. Tigelaar, F.P. Gasparro: Treatment with 8-methoxypsoralen and ultraviolet A radiation enhances immune recognition in a murine mastocytoma model. *J. Invest. Dermatol.* 98 (1992), 554A.

35 Malawista, S.E., D.H. Trock, R.L. Edelson: Treatment of rheumatoid arthritis by extracorporeal photochemotherapy – a pilot study. *Arthr. Rheum.* 34 (1991) 646–654.

36 Marks, D.I., R.M. Fox: Mechanism of photochemotherapy-induced apoptotic cell death in lymphoid cells. *Biochem. Cell. Biol.* 69 (1991), 754–760.

37 Midden, W.R.: Chemical mechanisms of the bioeffects of furocoumarins. In: *Psoralen–DNA Photobiology*, vol. II, (1988), pp. 1–49. Ed.: F. Gasparro, CRC Press, Boca Raton, FL.

38 Olack, G.A., F.P. Gasparro: UVA induced binding of 8-methoxypsoralen to polypeptides. American Society for Photobiology, 19th Annual Meeting June 22–27, 1991. Pergamon Press, New York.

39 Olack, G.A., P. Gattolin, F.P. Gasparro: Improved high performance liquid chromatographic analysis of 8-methoxypsoralen monoadducts and crosslinks in polynucleotide, DNA and cellular systems: analysis of split dose protocols. *Photochem. Photobiol.* 57 (1993), 941–949

40 Papodopoulu, D., D. Averbeck: Gentotoxic effects and DNA photoadducts induced in Chinese hamster V79 cells by 5-methoxypsoralen and 8-methoxypsoralen. *Mut. Res.* 151 (1985), 281–291.

41 Perez, M.I.: Effects of UV light. In: *Mechanisms of Regulation of Immunity*. Ed.: R. Granstein, Karger, Basel (in press).

42 Perez, M.I., C.L. Berger, Y. Yamane, L. John, L. Laroche, R.L. Edelson: Inhibition of anti-skin allograft immunity induced by infusions with photoinactivated effector T lymphocytes (PET cells). The congenic model. *Transplantation* 51 (1991) 1283–1289.

43 Perez, M., R.L. Edelson, L. Laroche, C. Berger: Inhibition of antiskin allograft immunity by infusions with syngeneic photoinactivated effector lymphocytes. *J. Invest. Dermatol.* 92 (1989), 669–676.

44 Raff, M.C.: Social controls on cell survival and cell death. *Nat* 356 (1992), 397–400.

45 Rodighiero, G., F. Dall'Acqua, D. Averbeck: New psoralen and angelicin derivatives. In: *Psoralen–DNA Photobiology*, vol. 1, (1988), pp. 37–133. Ed.: F. Gasparro, CRC Press, Boca Raton, FL.

46 Rook, A.H., B. Freundlich, B.V. Jegasothy, M.I. Perez, W.G. Barr, S.A. Jimenez: Treatment of systemic sclerosis with extracorporeal photochemotherapy: results of a multicenter trial. *Arch. Dermatol.* 128 (1992), 337–346.

47 Rook, A.H., M.B. Prystowsky, M. Cassin, M. Boufal, S.R. Lessin: Combined therapy of the Sezary syndrome with extracorporeal photochemotherapy and low dose interferon alpha: clinical, molecular and immunologic observations. *Arch. Dermatol.* 127 (1991), 1535–1549.

48 Rose, E.A., M.L. Barr, H.-Xu, P. Pepino, M.P. Murphy, M.A. McGovern, A.J. Ratner, J.F. Watkins,

C.C. Marboe, C.L. Berger: Photochemotherapy in human heart transplant recipients at high risk for fatal rejection. *J. Heart Lung Transplant.* 11 (1992), 746–750.

49 Rotzschke, O., K. Falk: Naturally occurring peptide antigens derived from MHC class I-restricted processing pathway. *Immunol. Today.* 12 (1991), 447–455.

50 Sage, E., E. Moustacchi: Sequence context effects on 8-methoxypsoralen photobinding to defined DNA fragments. *Biochem.* 26 (1987), 3307–3314.

51 Sladek, F.M., M.M. Munn, W.D. Rupp, P. Howard-Flanders: *In vitro* repair of psoralen–DNA crosslinks by RecA, uvrABC, and the 5′-exonuclease of DNA polymerase I. *J. Biol. Chem.* 264 (1989), 6755–6765.

52 Stern, R.S., R. Lange: Members of the Photochemotherapy Follow-up Study: Non-melanoma skin cancer occurring in patients treated with PUVA five to ten years after first treatment. *J. Invest. Dermatol.* 91 (1988), 120–124.

53 Strauss, G.H., R.J. Albertini, P.A. Krusinskie, R.D. Baughman: 6-Thioguanine resistant peripheral blood lymphocytes in humans following long-wave ultraviolet light (PUVA) therapy. *J. Invest. Dermatol.* 73 (1979), 211–216.

54 Taylor, A., F.P. Gasparro: Extracorporeal photochemotherapy for cutaneous T-cell lymphoma and other diseases. *Sem. Hematol.* 29 (1992), 132–141.

55 Tessman, J.W., S.I. Isaacs, J.E. Hearst: Photochemistry of furan-side 8-methoxypsoralenthymidine monoadduct inside the DNA helix. Conversion to diadduct and to pyrone-side monoadduct. *Biochem.* 24 (1985), 1669–1676.

56 Tokura, Y., R.L. Edelson, F.P. Gasparro: Formation and removal of 8-MOP–DNA photoadducts in keratinocytes: effects of calcium concentration and retinoids. *J. Invest. Dermatol.* 96 (1991), 942–949.

57 Tokura, Y., J. Yagi, R.L. Edelson *et al.*: Inhibitory effect of 8-MOP plus UVA on IL-1 production by murine keratinocytes. *Photochem. Photobiol.* 53 (1991), 517–523.

58 Veronese, F., O. Schiavon, R. Bevilacqua: Photoinactivation of enzymes by linear and angular furocoumarins. *Photochem. Photobiol.* 36 (1982), 25–30.

59 Vonderheid, E.C., R. Thompson, K.A. Smiles, A. Lattanand: Recombinant interferon-alpha-2b in plaque-phase mycosis fungoides. Intralesional and low-dose intramuscular therapy. *Arch. Dermatol.* 123 (1987), 757–763.

60 Wagner, G., C. Hofmann, U. Busch, J. Schmid, G. Plewig: 8-MOP plasma levels in PUVA problem cases in psoriasis. *Br. J. Dermatol.* 101 (1979), 285–292.

61 Walther, F., U.-F. Haustein: 8-Methoxypsoralen serum levels and poor response to photochemotherapy. *Pharmacol. Therap.* 30 (1991), 516–518.

62 Weinstock, M.A., J.W. Horm: Population-based estimate of survival and determinants of prognosis with mycosis fungoides. *Cancer* 62 (1988), 1658–1661.

63 Wilfert, H., H. Honigsmann, G. Steiner *et al.*: Treatment of psoriatic arthritis by extracorporeal photochemotherapy. *Br. J. Dermatol.* 122 (1990), 225–232.

64 Yoshikawa, K., N. Mori, S. Sakakibara *et al.*: Photoconjugation of 8-methoxypsoralen with proteins. *Photochem. Photobiol.* 29 (1979), 1123–1127.

65 Zmudzka, B.Z.: Activation of HIV by UVB radiation and PUVA treatment *in vitro*: an evaluation of the safety of medical procedures and cosmetic applications (Abstract). *Photochem. Photobiol.* 55 (1992), 89S.

18: Ultraviolet A1 radiation-induced immunomodulation: high-dose UVA1 therapy of atopic dermatitis

Jean Krutmann

Introduction

According to spectral regions, ultraviolet (UV) radiation may be divided into UVC (<290 nm), UVB (290–315 nm), UVA2 (315–340 nm) and UVA1 (340–400 nm) radiation (see Chapter 2). The capacity to exert immunomodulatory effects is well established for UVBR [25, 28]. Ultraviolet A radiation may also modulate immune responses, if UVAR is combined with photosensitizers such as psoralens, as is the case for photochemotherapy (e.g. PUVA) [28]. Recent phototherapeutic developments indicate that UVA1R, in the absence of exogenous photosensitizers, but applied in relatively high doses (high-dose UVA1 therapy), may be efficiently used to treat patients with atopic dermatitis [29]. Since the pathogenetic background of this inflammatory skin disease is thought to be immunologic, the development of high-dose UVA1 therapy prompted us to address the issue of UVA1-induced immunomodulation more closely. From these studies it appears that UVA1R, similarly to UVBR or PUVA, is capable of affecting the skin immune system [26], and that the photobiological mechanisms responsible for UVA1R- versus UVBR-induced immunomodulation differ from each other. The present chapter provides an overview of the current clinical status of high-dose UVA1 therapy, as well as of recently identified immunomodulatory effects associated with UVA1R.

UVA1R as a therapeutic tool: high-dose UVA1 therapy

High-dose UVA1 therapy was originally designed to treat patients with atopic dermatitis, and this chapter will therefore focus mainly on the current status of this modality for the treatment of this common inflammatory skin disease. Increasing knowledge about photoimmunological effects associated with high-dose UVA1 therapy has led us to examine its therapeutic effectiveness in other skin diseases, e.g. urticaria pigmentosa, and a short summary of these preliminary studies will be included.

High-dose UVA1 therapy of atopic dermatitis

Atopic dermatitis (AD) is a chronic inflammatory skin disease that is characterized clinically by eczematous skin lesions in typical localization, e.g. flexural skin of the elbows and knees [17, 31]. Chronic atopic dermatitis may cause significant morbidity including hospitalization for control of skin symptoms, and may result in severe disturbance of the patient's physical, social, and psychological well-being. There is currently no known curative therapy for atopic dermatitis. Treatment decisions are directed towards symptomatic relief and depend on the cutaneous symptoms at a given time point. The standard therapy for severe exacerbation of atopic dermatitis is topical and/ or systemic treatment with glucocorticosteroids. Since the long-term use of glucocorticosteroids is associated with a variety of unwanted side-effects, attempts have been made to develop alternative therapeutic strategies.

Exposure to sunlight has been known for years to be of benefit for the majority of patients with atopic dermatitis, and dermatologists have therefore been interested in employing UV phototherapy in the management of this inflammatory skin disease [19]. From these studies it appears that conventional UVB phototherapy is of limited effectiveness in the management of atopic dermatitis and is not suitable for the treatment of acute exacerbation of this disease [20]. Instead its use is mainly restricted to chronic disease stages, for which it is used as an adjuvant modality in combination with topical glucocorticosteroids. Systemic photochemotherapy (PUVA) is, in contrast, more effective. The use of PUVA for the treatment of atopic dermatitis, however, is limited by a number of disadavantages [32]. For example, cessation of PUVA therapy is associated with the occurrence of rebound phenomena in a high percentage of patients if photochemotherapy is not combined with systemic glucocorticosteroid therapy. In addition, recent reports indicate that long-term use of PUVA may be associated with an increased risk of developing skin cancer [44]. Conventional UVB phototherapy or photochemotherapy are thus of limited use for the management of patients with AD and do not represent equivalent alternatives to glucocorticosteroids in the treatment of severe exacerbation of atopic dermatitis.

It was therefore of great interest to learn that UVA1R (340–400 nm), if applied in relatively high doses (high-dose UVA1 therapy), may, in marked contrast to conventional phototherapeutic approaches, be used effectively as a monotherapy to treat exacerbation of atopic dermatitis [29]. The rationale for UVA1R in the treatment of AD was based on the observation that the therapeutic effectiveness of conventional UVB therapy may be improved by increasing the UVA portion of the action spectrum (UVAB therapy) [19]. In addition, *in vivo* exposure of human skin to relatively high doses of UVA1R was found to functionally affect epidermal Langerhans cells [3], a cell population which appears to be of pathogenetic relevance for atopic dermatitis [5].

In a pilot study conducted at the Department of Dermatology at the University of Freiburg, the therapeutic effectiveness of high-dose UVA1R was compared to that of a conventional UVAB therapy in the treatment of patients with severe exacerbation of chronic AD [27]. In this study the two different modalities were employed as monotherapies, that is additional therapy was restricted in both groups to the use of emollients. For high-dose UVA1 therapy, patients were exposed fifteen times to a single daily dose of 130 J/cm^2 UVA1R, whereas patients undergoing UVAB therapy were irradiated fifteen times in an MED-dependent fashion. Therapeutic effectiveness was assessed by two means: by an established clinical scoring system that had been developed by Costa *et al.* [7], and by monitoring serum levels of eosinophil cationic protein. Serum levels of eosinophil cationic protein were previously shown to represent a laboratory parameter, which could be objectively measured, and which would reflect disease activity in atopic dermatitis [8]. Assessment of clinical scores revealed that high-dose UVA1 irradiation was highly efficient in improving clinical symptoms in patients with atopic dermatitis, and that in comparison to conventional UVAB therapy, significant differences in favour of high-dose UVA1 therapy were observed [27]. These clinical results were corroborated and extended by corresponding laboratory findings. Elevated serum levels of eosinophil cationic protein in atopic dermatitis patients undergoing high-dose UVA1 therapy were significantly decreased, whereas no such decrease could be observed in patients who were treated with conventional UVABR [27].

These results indicated that high-dose UVA1 irradiation could represent a novel phototherapeutic approach to effectively treat patients with severe exacerbation of atopic dermatitis. As with most pilot studies there are a number of open questions which have to be addressed before high-dose UVA1 therapy may be regarded as an established phototherapeutic modality. In particular, the design of the pilot study was open, the number of patients enrolled in this study was limited to 25, and only a single center had been involved in conducting this study. We therefore initiated a clinical trial which was designed to provide a direct comparison between high-dose UVA1 therapy (first branch) versus conventional UVAB therapy (second branch) versus a topical glucocorticosteroid treatment (third branch). In this randomized study, enrolment and therapy of patients was provided by three separate dermatological departments, whereas a fourth, independent department was in charge of randomization and data analysis. The practical phase of this trial has been finished very recently, and a preliminary analysis of therapeutic effectiveness based on clinical scores appears to confirm the therapeutic effectiveness of high-dose UVA1 therapy (J. Krutmann, E. Schöpf *et al.*, manuscript submitted for review). Results from this trial indicate that a proportion of patients with atopic dermatitis (approximately 20%) do not improve under high-dose UVA1 therapy. These non-responders are not characterized by a specific skin type, since they include both fair-skinned individuals (skin types 1 and 2) and individuals with skin types 3 and 4. Photoimmunological studies on the molecular mechanisms relevant for UVA1-induced immunomodulation indicate an important role for the generation of free radical species in UVA1R-induced immunomodulation [7, 36]. In this regard it is of interest that the capacity of a given cell to respond to UVA1R critically depends on its level of free radical quenchers, e.g. its glutathione content [36]. Since different individuals are known to exhibit great variety with regards to their dermal and epidermal free radical quencher systems, it is conceivable to speculate that these interindividual differences may at least partially determine whether a given patient will or will not respond to high-dose UVA1 therapy.

Increased UVA1 sensitivity, which may have resulted in an exacerbation of UVA1-sensitive skin diseases, was excluded prior to high-dose UVA1 therapy by phototesting with UVA1R. Due to the limited experiences with this new modality, one may only speculate on potential long-term side-effects. Animal studies indicate the exposure of skin to high doses of UVA1R may contribute to photoaging of skin [23], and more importantly, may increase the risk of developing skin cancer [43]. Accordingly, exposure of hairless, albino Skh-hr1 mice to high doses of UVA1R was reported to induce the appearance of squamous cell carcinoma. In this regard it is of interest that the dose of 130 J/cm^2 UVA1R which is currently employed in high-dose UVA1 therapy was selected on the basis of photoimmunological studies indicating that this dose is sufficient to functionally inhibit epidermal Langerhans cells in human skin [3]. This does not necessarily imply that optimal therapeutic effectiveness may not be achieved with smaller UVA1R doses, and further clinical trials are therefore necessary to define precisely the lowest UVA1R dose sufficient to achieve a maximal therapeutic effect. It should, however, be noted that a fivefold lower UVA1R dose appears to be significantly less effective than UVAB therapy in the treatment of patients with atopic dermatitis [21].

Cessation of high-dose UVA1 therapy was not associated with the occurrence of rebound phenomena [27]. Although no controlled studies have been performed to determine how long the therapeutic effect achieved by high-dose UVA1 therapy will last, it may be assumed that at least for a limited period of time atopic dermatitis may not reexacerbate, even upon reexposure of patients' skin to relevant inhalant allergens. This assumption is based on studies employing inhalant allergen patch testing as a model for the initiation phase of atopic dermatitis [38]. From these studies it appears that in approximately 47% of atopic dermatitis patients with known sensitivity to house dust mite allergens, positive patch test lesions may be provoked after 24 to 48 h patch testing with DP allergens on uninvolved, unirradiated skin. In marked contrast, in none of these patients could a positive patch test reaction be elicited if patch testing was performed in a skin area that prior to testing has been exposed to high-dose UVA1R [46]. This suppressive effect appears to be immunologic in nature, since skin reactions to irritants were not affected by high-dose UVA1R, and could be due to UVA1R-induced tolerance towards inhalant allergens (S. Walter, J. Krutmann, unpublished observations).

High-dose UVA1 therapy of other skin diseases

There are currently no controlled studies assessing the therapeutic effectiveness of high-dose UVA1 therapy for skin diseases other than atopic dermatitis. Based on photoimmunological observations it may be speculated, however, that high-dose UVA1 therapy could be of benefit for several other skin diseases. For example, immunohistochemical studies indicate that dermal and epidermal mast cells represent target cells for UVA1R [11]. This observation prompted us to determine whether UVA1R may be used to treat patients with urticaria pigmentosa. To this end, two patients with urticaria pigmentosa have been irradiated with 130 J/cm^2 UVA1R on a daily basis for 4 consecutive weeks. In both patients an almost immediate relief of itching could be observed, which was associated with a significant decrease of the major histamine metabolite 1-methyl-4-imidazoleacetic acid in the urine (H. Stege, J. Krutmann, manuscript submitted for review). Similar therapeutic effects have been reported in urticaria pigmentosa patients undergoing PUVA therapy, but in these patients a reversal of all beneficial effects was reported to occur 5 to 8 months after cessation of photochemotherapy [12, 24]. However, even if a similar loss of therapeutic effectiveness may be observed after high-dose UVA1R, this modality may nevertheless represent a valuable alternative to PUVA, since it is easier and safer to perform.

In addition to affecting mast cells, high-dose UVA1 therapy was found to be associated with a reduction of the inflammatory T-cell infiltrate in patients with atopic dermatitis (R. Block, J. Krutmann, unpublished observations). Based on these observations it is of interest to speculate that high-dose UVA1R may be of benefit for the treatment of other T-cell mediated skin diseases as well. Accordingly, we have recently initiated studies to examine the therapeutic effectiveness of high-dose UVA1R in the treatment of patients with alopecia areata.

UVA1 radiation-induced immunomodulation

The clinical observation that UVA1R, even in the absence of exogenously added photosensitizers, could be used to treat patients with atopic dermatitis effectively, prompted us to analyse the immunomodulatory effects which are associated with UVA1R more closely. From these studies it appears that both epidermal and dermal cells represent potential cellular targets for UVA1R. These include keratinocytes [6, 36], Langerhans cells [3, 42], T lymphocytes [14, 16], fibroblasts [47], mast cells [11] and dermal endothelial cells [18]. Some of these modulatory effects have previously been observed for UVBR-induced immunomodulation. Interestingly, studies on the underlying photobiological mechanisms indicate that UVA1R and UVBR may exert similar or even identical immunomodulatory effects, for example on keratinocytes, which are, however, mediated via different photobiologic mechanisms.

UVA1R effects on keratinocytes

Epidermal keratinocytes represent a well-recognized cellular target for UVBR-induced immunomodulation [25, 28]. Accordingly, *in vitro* as well as *in vivo* UVBR exposure of murine and human keratinocytes was found to profoundly affect their capacity to produce and secrete soluble mediators including cytokines, eicosanoids and neuropeptides. Recent studies indicate that UVBR exposure may also modulate the expression of immunologically relevant surface molecules in human keratinocytes, e.g. expression of MHC class II molecules or the adhesion molecule intercellular adhesion molecule-1 (ICAM-1) [22, 30]. The ICAM-1 molecule on the surface of human keratinocytes plays a crucial role in the generation and maintenance of an epidermal inflammatory infiltrate by serving as a matrix to which LFA-1 positive leukocytes are capable of binding. This conclusion was initially based on the *in vitro* observation that the major pathway by which leukocytes are capable of adhering to cultured keratinocytes is mediated by keratinocyte ICAM-1 molecules [9]. In support of this view is the *in vivo* observation that in contrast to normal human skin, which is practically devoid of ICAM-1 expressing keratinocytes, strong keratinocyte ICAM-1 expression may be observed in inflammatory skin diseases including atopic dermatitis [16, 41, 45]. Immunohistochemical studies have revealed that high-dose

UVA1 therapy of atopic dermatitis was associated with a reduction of the inflammatory T-cell infiltrate (R. Block, J. Krutmann, unpublished observations). This anti-inflammatory effect could be the result of UVA1R-induced downregulation of human keratinocyte ICAM-1 expression, since clinical improvement achieved by high-dose UVA1R was found to be associated with an almost complete reduction of keratinocyte surface expression of ICAM-1 molecules [16].

This *in vivo* observation prompted us to examine whether UVA1R, similarly to UVBR, is capable of directly affecting the regulation of ICAM-1 expression in human keratinocytes. *In vitro* exposure of long-term cultured normal human keratinocytes to sublethal doses of UVA1R was found to inhibit cytokineinduced upregulation of ICAM-1 surface expression, if keratinocytes were UVA1 irradiated prior to cytokine stimulation [6]. Further studies revealed that UVA1R inhibited human keratinocyte ICAM-1 expression by a pretranslational mechanism, which was associated with a UVA1R dose-dependent decrease in steady state levels of ICAM-1 specific mRNA. Similarly to UVBR-induced ICAM-1 expression, UVA1R-induced ICAM-1 expression was found to be biphasic [30]. Accordingly, in addition to suppressing cytokine-mediated ICAM-1 upregulation early after UVA1R exposure, UVA1R was found to induce constitutive ICAM-1 expression at later time points (24–48 h) [6, 36]. Ultraviolet B radiation-induced upregulation of human keratinocyte ICAM-1 expression was found to be mediated via UVBR-induced, endogenously produced IL-1α, which, in an autocrine manner, induced keratinocyte ICAM-1 expression via a two step pathway, consisting of upregulation of IL-1 receptor type I expression, and subsequent induction of ICAM-1 mRNA and surface expression [30]. Ultraviolet A1 radiation-induced ICAM-1 expression may also be mediated in an autocrine manner via keratinocyte-derived cytokines, since recent studies from this laboratory indicate that UVA1R, similarly to UVBR, is capable of inducing cytokine production by cultured human keratinocytes [15]. However, further studies are required to determine whether different or identical autocrine mechanisms account for UVA1R-induced human keratinocyte ICAM-1 expression. In this regard it is of interest that Norris *et al.* were able to block UVR-induced keratinocyte ICAM-1 expression by incubating irradiated cells in the presence of neutralizing anti-TNFα antibodies, indicating that TNFα rather than IL-1α may mediate UVR-induced ICAM-1 expression [35]. Since the irradiation device used by Norris was a solar simulator, which emits considerably more UVAR than the FS20 sun lamps which were used in the present studies, these divergent results may indicate that different wavelengths induce keratinocyte ICAM-1 expression via different autocrine pathways. In support of this assumption is the observation that UVA1R modulated human keratinocyte ICAM-1 expression in non-transformed human keratinocytes, but failed to do so in transformed cells (KB, A431), whereas UVBR was capable of regulating ICAM-1 expression in all of these cell types [6]. These studies provide strong evidence that UVA1R and UVBR modulate human keratinocyte ICAM-1 by different mechanisms (see also the section on molecular mechanisms below).

UVA1 radiation effects on T-cell-derived cytokine expression

The basis of atopic dermatitis is immunologic in nature, and there is increasing evidence that a dysregulated, cytokine-mediated response of the immune system to environmental allergens is an important factor contributing to the pathogenesis of this disease [4, 31]. Histologically, eczematous skin lesions in atopic dermatitis patients contain an inflammatory infiltrate mainly consisting of CD4-positive T cells. In animals, and possibly also in humans, CD-4 positive T cells or helper T cells may be grouped into two subclasses, depending on their cytokine secretion pattern [33]. The T_{h1}-like subtype is characterized by predominant production of γ-interferon, whereas the T_{h2}-like subtype typically synthesizes IL-4, IL-5 and IL-10. By employing a highly sensitive RT-PCR method to semiquantitatively assess *in situ* cytokine expression in lesional skin of patients with chronic atopic dermatitis, we have recently observed that in the majority (85%) of atopic dermatitis lesions assessed, increased levels of IFN-γ mRNA are present [15]. In contrast, only one third (25–30%) of patients expressed increased levels of IL-4 mRNA in lesional skin. Most interestingly, clinical improvement achieved by high-dose UVA1 therapy was associated with a reduction of increased IFN-γ mRNA levels to background, whereas increased IL-4 mRNA levels remained essentially unaltered [16]. These studies identify the T_{h1}-cell-derived cytokine IFN-γ as a potential target for UVA1R. Since IFN-γ is among the most potent inducers of human keratinocyte expression of the adhesion molecule ICAM-1 [9], UVA1R-induced downregulation of *in situ* expression of IFN-γ may represent an alternative or additional mechanism responsible for UVA1R-induced downregulation of keratinocyte ICAM-1 expression in lesional atopic skin. It is not currently known whether inhibition of *in situ* expression of IFN-γ mRNA expression is caused by direct effects of UVA1R on dermal and epidermal T cells, or whether an indirect mechanism, possibly involving keratinocyte-derived cytokines, may be operative in this system. In support of the latter possibility is our recent observation that *in vitro* exposure of cultured human keratinocytes to UVA1R is associated with a strong induction of interleukin 10 (IL-10), mRNA expression and protein secretion [15]. Interleukin 10 is known to be produced by human keratinocytes *in vivo* [10] and to function as a potent inhibitor of IFN-γ production by T_{h1}-like T cells. Ultraviolet A1 radiation-induced keratinocyte-derived IL-10 may thus act on epidermal and dermal T cells by inhibiting IFN-γ production. Alternatively, UVA1R may exert direct effects on T_{h1}-like cells. Accordingly, UVA1R at relatively low doses was recently shown to affect normal human peripheral blood T lymphocytes functionally by inducing tyrosine phosphorylation and calcium signaling [39]. Further studies will be needed to assess whether different T cell subsets exhibit selective susceptibility towards immunomodulatory effects induced by UVA1R.

UVA1 radiation effects on epidermal Langerhans cells

Langerhans cells, in addition to keratinocytes, represent a crucial target cell population in UVBR-induced immunomodulation, but little is currently known about the effects of UVA1R on these important epidermal cells. *In vivo* studies indicate that a single exposure of human skin to 130 J/cm^2 UVA1R leads to a depletion of epidermal Langerhans cell surface markers, as was demonstrated by immunohistochemical studies assessing the number of CD1$^+$ DR$^+$ dendritic epidermal cells [3]. These morphological alterations were associated with a UVA1R-induced abrogation of Langerhans cell function to activate alloreactive T cells. This inhibitory effect was detectable if Langerhans cells were harvested from UVA1-irradiated human skin immediately after irradiation. In contrast, if epidermal cells were harvested 3 days after UVA1R exposure, a complete recovery of alloantigen-presenting cell function could be observed [3]. It should be noted that this immunosuppressive effect on epidermal antigen-presenting cell function is specific for UVA1R, since after UVBR, initial inhibition of Langerhans cell antigen-presenting cell function is not followed by Langerhans cell function recovery. Instead, 3 days after UVBR exposure the appearance of CD1$^-$ DR$^+$ OKM5$^+$ cells was noted, and these cells were capable of preferentially activating T-suppressor cells [2]. As a consequence, UVBR appears to lead to a more sustained immunosuppressive effect, whereas UVA1R-induced immunosuppression, because of its failure to induce the appearance of CD1$^-$ DR$^+$ antigen-presenting cells, and because of the rapid recovery of epidermal Langerhans cell morphology and function, appears to be more transient in nature. It is not currently known whether recovery of Langerhans cell function in UVA1-irradiated human skin is due to the immigration of new Langerhans cells into the UVA1-irradiated skin area, or whether a change in the epidermal milieu, e.g. the cytokine profile released by UVA1-irradiated keratinocytes, would allow UVA1-irradiated Langerhans cells to reexpress CD1 and

DR surface antigens as well as to restore their antigen-presenting cell function. In favor of the first possibility is an immunohistochemical observation made in patients with atopic dermatitis undergoing high-dose UVA1 therapy. Assessment of Langerhans cell surface marker expression in biopsies taken sequentially from these patients during phototherapy revealed that after five to ten consecutive exposures of human skin to a single dose of 130 J/cm^2 UVA1R a significant increase in the number of CD1a$^+$ dendritic epidermal cells could be observed [11, 42]. These cells lacked DR-antigen surface expression, and upon electron microscope examination did not contain Birbeck granules (G. Kolde, J. Krutmann, unpublished observations). This cell population may therefore represent indeterminate cells, which have been migrating into the epidermis from a dermal compartment, possibly following a chemotactic gradient caused by keratinocyte-derived cytokines which were released upon UVA1R exposure. Taken together, these studies indicate that UVA1R, in addition to UVBR, has the capacity to modulate human epidermal Langerhans cell morphology and function. Antigen presentation by dendritic cells is a highly complex process involving antigen processing, presentation of the processed antigen in context with MHC-class II molecules, and the provision of a number of costimulatory signals including selected adhesion molecules and possibly also cytokines [37]. Further studies are therefore needed to define precisely the level at which UVA1R is capable of inhibiting epidermal Langerhans cell antigen-presenting cell function. These studies should provide a better understanding of the clinical relevance of UVA1R-induced modulation of Langerhans cell function, e.g. for high-dose UVA1R therapy.

Molecular mechanisms of UVA1-induced immunomodulation

The identification of UVA1R-resistant versus UVA1R-sensitive keratinocyte populations allowed us to study the molecular mechanisms involved in UVA1R-induced immunomodulation more closely [6]. When transformed human keratinocytes (KB cells) were treated prior to UVA1R exposure with DL-buthionine-[S,R]-sulfoximine (BSO), a specific inhibitor of *de novo* glutathione synthesis, glutathione content of KB cells, which usually exceeds that of normal human keratinocytes by a factor of 4, was reduced to levels below those observed in normal human keratinocytes [36]. Interestingly, lowering of glutathione levels was found to render UVA1-resistant KB cells sensitive to UVA1R-induced modulation of ICAM-1 expression. These studies imply that the capacity of a given keratinocyte to respond to UVA1R critically depends on the intracellular level of free radical quenchers and also indicate that UVA1R-induced modulation of human keratinocyte ICAM-1 expression is mediated via the generation of free radicals. Accordingly, UVA1R-induced ICAM-1 expression in UVA1R-sensitive keratinocytes could be effectively blocked by four different radical quenchers (reduced glutathione, superoxide dismutase, sodiumacide, vitamin E) [36]. Addition of these substances did not inhibit UVBR-induced or cytokine-stimulated ICAM-1 expression. Sodiumacide is known to specifically quench singlet oxygen, indicating a major role of singlet oxygen generation in UVA1R-induced ICAM-1 upregulation. This assumption was further supported by the observation that UVA1R-induced ICAM-1 expression could be significantly enhanced if keratinocytes were irradiated in the presence of deuterium oxide, which increases the survival time of singlet oxygen (S. Olaizola-Horn, J. Krutmann, unpublished observations). Based on these studies we have proposed that the generation of free radical species, and in particular of singlet oxygen, may lead either directly or via subsequent lipid peroxidation to the transduction of regulatory signals, which will eventually cause increased transcription of ICAM-1 (Fig. 18.1; J. Krutmann, M. Grewe, manuscript submitted for review). The potential role of lipid peroxidation in this system is mainly suggested by experiments, in which both UVA1R-induced ICAM-1 expression and UVA1R-induced lipid peroxidation in human keratinocytes could be completely blocked by pretreatment of cells with the membrane-localizing

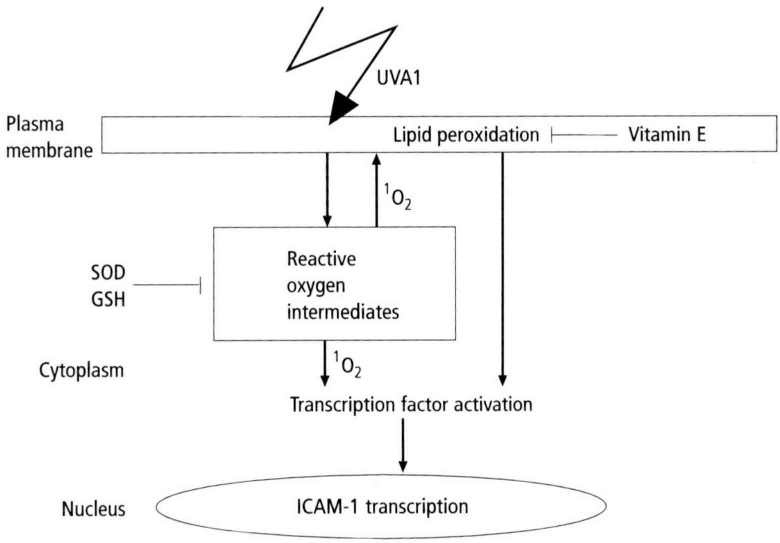

Fig. 18.1 A model showing the presumed involvement of reactive oxygen intermediates and lipid peroxidation products in the ultraviolet A1 radiation-induced regulation of the transcription of the human ICAM-1 gene.

antioxidant vitamin E (S. Olaizola-Horn, J. Krutmann, unpublished observations).

It is not currently known how singlet oxygen and/or lipid peroxides induce human keratinocyte gene expression. In this regard it is of interest that free radical species were previously found to represent important inducers for the activation of transcription factors such as NFkB [40]. Recent studies from this laboratory [13] have demonstrated that exposure of human keratinocytes to doses of UVA1R, which induce keratinocyte ICAM-1 expression, are highly effective in activating the transcription factors NFkB, AP-1, AP-2, p53, and a recently described upstream stimulating factor-like activity, which was previously shown to be activated upon broadband UVA radiation, and to bind to a USF binding region of the hemeoxygenase 1 gene [34]. Most interestingly, time kinetic studies on transcription factor activation in UVB-irradiated versus UVA1-irradiated human keratinocytes revealed different activation patterns [13]. A single exposure of human keratinocytes to UVA1R caused a biphasic activation of transcription factors, that is a first peak 1/2 to 2 h after UVA1R exposure with a maximum after 1 h, and this first activation peak was followed by a second, more sustained activation of transcription factors, which could be detected 8 to 48 h after irradiation. In contrast, a single exposure of human keratinocytes to UVBR was followed by a monophasic response. Activation of transcription factors could be detected 1/2 to 4 h after irradiation, with a maximum at 2 h post-exposure. These studies indicate that both UVBR and UVA1R are capable of activating transcription factors, but by different mechanisms. Accordingly, UVBR and broadband UVAR were recently reported to induce the expression of distinct sets of immediate early genes [1]. Further studies will be needed to establish if UVA1R-induced activation of any of these transcription factors is of functional relevance for the upregulation of human keratinocyte ICAM-1 expression.

Conclusion

The development of high-intensity radiation devices emitting UVA1R, and their widespread use in medicine and cosmetics has recently raised increasing interest in the effects of such radiation on human health. From the work reviewed in this chapter the concept has emerged that UVA1R, similarly to UVBR and/or PUVA, is capable of profoundly affecting the human skin immune system. Further studies are required to define these immunomodulatory effects more closely.

The identification of the immunological key molecule ICAM-1 as a target structure in UVA1R-induced immunomodulation has provided us with a well-defined photoimmunological model system, which will allow direct comparisons of the basic molecular mechanisms involved in UVA1R-versus UVBR-induced immunomodulation. These studies should ultimately lead to a better understanding of how UVA1R may be of therapeutic benefit, but could also provide the basis for a more precise evaluation of the potential deleterious effects that UVA1R may exert on human health.

Acknowledgments

For the work done in my laboratory discussed in this review, I thank my coworkers who have carried out the studies cited, in particular Rüdiger Block, Helmut Christoph, Susanne Grether-Beck, Markus Grewe, Karin Gyufko, Michaela Klammer, Sylvia Olaizola-Horn, Frauke Parlow, Stefanie Sinn, Helger Stege, Uwe Trefzer, and Stephan Walter. The continuous support of Prof. Dr. E. Schöpf, Department of Dermatology, University of Freiburg, is greatly appreciated. This work was supported in part by a grant from the Deutsche Forschungsgemeinschaft Kr 871/3-3.

References

1 Ariizumi, K., P.R. Bergstresser, A. Takashima: Wavelength-specific induction of immediate early genes in keratinocytes by ultraviolet radiation (Abstract). *J. Invest. Dermatol.* 101 (1993), 406.

2 Baadsgard, O., D.A. Fox, K.D. Cooper: Human epidermal cells from ultraviolet light-exposed skin preferentially activate autoreactive $CD4^{+}2H4^{+}$ suppressor–inducer lymphocytes and $CD8^{+}$ suppressor/cytotoxic lymphocytes. *J. Immunol.* 140 (1988), 1738–1744.

3 Baadsgard, O., S. Lisby, G. Lange-Wantzin, H.C. Wulf, K.D. Cooper: Rapid recovery of Langerhans cell alloreactivity, without induction of autoreactivity, after *in vivo* ultraviolet A, but not ultraviolet B exposure of human skin. *J. Immunol.* 142 (1989), 4213–4217.

4 Bos, J.D., E.A. Wierenga, J.H.S. Smitt, F.L. van der Heijden, M.L. Kapsenberg: Immune dysregulation in atopic eczema. *Arch. Dermatol. Res.* 128 (1992), 1509–1514.

5 Bruynzeel-Koomen, C.: IgE on Langerhans cells: new insights into the pathogenesis of atopic dermatitis. *Dermatologica* 172 (1986), 181–184.

6 Christoph, H., F. Parlow, A. Budnik, R. Block, E. Schöpf, J. Krutmann: Ultraviolet A1-induced immunomodulation: UVA1- and UVB-radiation both affect human keratinocyte expression of intercellular adhesion molecule-1 (ICAM-1), although by different mechanisms (Abstract). *Arch. Dermatol. Res.* 285 (1993), 56.

7 Costa, C., A. Rillet, M. Nicolet, J.H. Saurat: Scoring atopic dermatitis: the simpler the better? *Acta Derm. Venereol. (Stockh.)* 69 (1989), 41–47.

8 Czech, W., J. Krutmann, E. Schöpf, A. Kapp: Serum eosinophil cationic protein is a sensitive measure for disease activity in atopic dermatitis. *Br. J. Dermatol.* 126 (1992), 351–355.

9 Dustin, M., K. Singer, D. Tuck, T.A. Springer: Adherence of T-lymphoblasts to epidermal keratinocytes is regulated by IFN gamma and mediated by ICAM-1. *J. Exp. Med.* 167 (1988), 1323–1329.

10 Enk, C.D., S.I. Katz: Ultraviolet B exposure induces interleukin-10 expression in human keratinocytes (Abstract). *J. Invest. Dermatol.* 101 (1993), 400.

11 Grabbe, J., P. Welker, S. Sinn, M. Grewe, E. Schöpf, B.M. Czarnetzki, J. Krutmann: Effects of high-dose UVA1– and UVA–UVB-treatment on IgE binding cells in severe atopic dermatitis (Abstract). *Proceedings of the 3rd Congress of the European Academy of Dermatology and Venerology* (1993), 90.

12 Granerus, G., G. Roupe, G. Swanbeck: Decreased urinary histamine metabolite after successful PUVA treatment of urticaria pigmentosa. *J. Invest. Dermatol.* 76 (1981), 1–4.

13 Grether-Beck, S., M. Klammer, M. Grewe, K. Gyufko, S. Olaizola-Horn, A. Budnik, J. Krutmann: Activation of transcription factors in human keratinocytes by ultraviolet B (280–320 nm) and ultraviolet A1 (340–400 nm) radiation (Abstract). *Arch. Dermatol. Res.* (1993), 38.

14 Grewe, M., K. Gyufko, R. Block, F. Parlow, E. Schöpf, J. Krutmann: Transcript levels of the T-cell derived cytokines interferon γ and interleukin 4 in lesional atopic skin in patients undergoing high-dose UVA1 therapy (Abstract). *Arch. Dermatol. Res.* 285 (1993), 40.

15 Grewe, M., K. Gyufko, J. Krutmann: Interleukin-10 production by cultured human keratinocytes: regulation by ultraviolet B and A1 radiation. *J. Invest. Dermatol.* 104 (1995), 3–6.

16 Grewe, M., K. Gyufko, E. Schöpf, J. Krutmann: *In-situ* expression of interferon γ in chronic atopic eczema. *Lancet* 343 (1994), 25–26.

17 Hanifin, J.M.: Atopic dermatitis. *J. Allergy Clin. Immunol.* 73 (1984), 211–218.

18 Heckmann, M., B. Eberlein-König, A. Wollenberg, B.

Przybilla, G. Plewig: UV-A induces adhesion molecule expression on human dermal microvascular endothelial cells. *Br. J. Dermatol.* 131 (1994), 311–318.

19 Jekler, J.: *Phototherapy of Atopic Dermatitis with Ultraviolet Radiation* (1992). Graphic Systems AB, Göteborg.

20 Jekler, J., O. Larkö: Combined UV-A–UV-B versus UVB phototherapy for atopic dermatitis: a paired-comparison study. *J. Am. Acad. Dermatol.* 22 (1990), 49–53.

21 Jekler, J., O. Larkö: Phototherapy for atopic dermatitis with ultraviolet A (UVA), low-dose UVB and combined UVA and UVB: two paired-comparison studies. *Photodermatol. Photoimmunol. Photomed.* 8 (1991), 151–156.

22 Kan, I.U., K.D. Boehm, C.A. Elmets: Modulation of interferon-gamma-induced HLA-DR expression on the human keratinocyte cell line SCC-13 by ultraviolet radiation. *Photochem. Photobiol.* 57 (1993), 285–289.

23 Kligman, L.H., M. Gebre: Biochemical changes in hairless mouse skin collagen after chronic exposure to ultraviolet-A radiation. *Photochem. Photobiol.* 54 (1991), 233–238.

24 Kolde, G., G.F. Frosch, B.M. Czarnetzki: Response of cutaneous mast cells to PUVA in patients with urticaria pigmentosa: histomorphometric, ultrastructural, and biochemical investigations. *J. Invest. Dermatol.* 83 (1984), 175–180.

25 Kripke, M.L.: Immunological unresponsiveness induced by ultraviolet radiation. *Immunol. Rev.* 80 (1984), 87–95.

26 Krutmann, J.: Ultraviolet A1-induced immunomodulation. In: *Biologic Effects of Light*, (1994), pp. 611–616. Eds.: M.F. Holick, E.G. Jung, Walter de Gruyter, New York.

27 Krutmann, J., W. Czech, T. Diepgen, R. Niedner, A. Kapp, E. Schöpf: High-dose UVA1 therapy in the treatment of patients with atopic dermatitis. *J. Am. Acad. Dermatol.* 26 (1992), 225–230.

28 Krutmann, J., C.A. Elmets: Recent mechanism studies in photoimmunology. Yearly review. *Photochem. Photobiol.* 48 (1988), 787–798.

29 Krutmann, J., E. Schöpf: High-dose-UVA1 phototherapy: a novel and highly effective approach for the treatment of patients with acute exacerbation of atopic dermatitis. *Acta Derm. Venereol. (Stockh.)* 176 (1992), 120–122.

30 Krutmann, J., U. Trefzer: Modulation of the expression of intercellular adhesion molecule-1 (ICAM-1) in human keratinocytes by ultraviolet (UV) radiation. *Springer Semin. Immunopathol.* 13 (1992), 333–344.

31 Leung, D.Y.M.: Immunopathology of atopic dermatitis. *Springer Semin. Immunopathol.* 13 (1992), 427–432.

32 Morison, W.L., J.A. Parrish, T.B. Fitzpatrick: Oral photochemotherapy of atopic eczema. *Br. J. Dermatol.* 98 (1978), 25–28.

33 Mossmann, T.R., R.L. Coffman: Th_1 and Th_2 cells: different patterns of lymphokine secretion lead to different functional properties. *Ann. Rev. Immunol.* 7 (1989), 145–149.

34 Nascimento, A.L.T.O., P. Luscher, R.M. Tyrrell: Ultraviolet A (320–380 nm) radiation causes an alteration in the binding of a specific protein/protein complex to a short region of the promotor of the human heme oxygenase 1 gene. *Nucleic Acids Res.* 21 (1993), 1103–1107.

35 Norris, D.A., S.D. Bennion, M.H. Middleton: Further definition of the effects of ultraviolet light on ICAM-1 (CD54) expression by human epidermal keratinocytes (Abstract). *J. Invest. Dermatol.* 94 (1990), 560.

36 Olaizola-Horn, S., H. Christoph, A. Budnik, M. Grewe, S. Grether-Beck, K. Gyufko, P. Luscher, R.M. Tyrrell, J. Krutmann: Ultraviolet A1 radiation-induced upregulation of keratinocyte expression of intercellular adhesion molecule-1 (ICAM-1) is mediated via the generation of free radicals (Abstract). *Arch. Dermatol. Res.* (1993), 38.

37 Romani, N., G. Schuler: The immunologic properties of epidermal Langerhans cells as a part of the dendritic cell system. *Springer Semin. Immunopathol.* 13 (1992), 265–275.

38 Sager, N., A. Feldmann, G. Schilling, P. Kreitsch, C. Neumann: House dust mite-specific T cells in the skin of subjects with atopic dermatitis: frequency and lymphokine profile in the allergen patch test. *J. Allergy Clin. Immunol.* 89 (1992), 801–805.

39 Schieven, G.L., J.M. Kirihara, L.K. Gilliland, F.M. Uckun, J.A. Ledbetter: Ultraviolet radiation rapidly induces tyrosine phosphorylation and calcium signaling in lymphocytes. *Mol. Biol. Cell* 4 (1993), 523–526.

40 Schreck, R., P. Rieber, P.A. Bäuerle: Reactive oxygen intermediates as apparently widely used messengers in the activation of the NF-kB transcription factor and HIV-1. *EMBO J.* 10 (1991), 2247–2251.

41 Singer, K.H., D.T. Tuck, H.A. Sampson, R.P. Hall: Epidermal keratinocytes express the adhesion molecule intercellular adhesion molecule-1 in inflammatory dermatoses. *J. Invest. Dermatol.* 92 (1989), 746–750.

42 Sinn, S., W. Czech, T. Diepgen, J.C. Simon, A. Kapp, E. Schöpf, J. Krutmann: High-dose UVA1 therapy of patients with atopic dermatitis: modulation of epidermal Langerhans cell surface markers – an immunohistochemical study (Abstract). *Arch. Dermatol. Res.* 284 (1992), 47.

43 Sterenbroigh, H.C.J.M., J.C. van der Leun: Tumorigenesis by a long wavelength UV-A source. *Photochem. Photobiol.* 51 (1990), 325–330.

44 Stern, R.S., Members of the Photochemotherapy Follow-Up Study: Genital tumors among men with psoriasis exposed to psoralen and ultraviolet A

radiation (PUVA) and ultraviolet B radiation. *N. Engl. J. Med.* 322 (1990), 1093–1096.

45 Vejlsgaard, G.L., F. Ralfkiaer, C. Avnstorp, M. Czajkowski, S.D. Marlin, R. Rothlein: Kinetics and characterization of intercellular adhesion molecule-1 (ICAM-1) expression on keratinocytes in various inflammatory skin lesions and malignant cutaneous lymphomas. *J. Am. Acad. Dermatol.* 20 (1989), 782–788.

46 Walter S., M. Grewe, K. Gyufko, W. Czech, A. Kapp, H. Stege, E. Schöpf, J. Krutmann: Inhalant allergen patch tests as a model for the induction of atopic dermatitis: analysis of the *in situ* cytokine expression pattern and modulation by UVA1 irradiation (Abstract). *Arch. Dermatol. Res.* (1994), 36.

47 Wlaschek, M., K. Blosen, G. Herrmann, A. Schwarz, F. Wilmroth, P.C. Heinrich, G. Goerz, K. Scharfetter-Kochanek: UVA-induced autocrine stimulation of fibroblast-derived-collagenase by IL-6: a possible mechanism in dermal photo-damage? *J. Invest. Dermatol.* 101 (1993), 164–168.

19: Photoimmunology of photodynamic therapy

Hasan Mukhtar, Syed I.A. Zaidi, David R. Bickers & Craig A. Elmets

Photodynamic therapy

Photodynamic therapy (PDT) of cancer is a treatment modality which involves the systemic administration of a photosensitizer to a tumor-bearing host and, following its uptake in tumor tissue, local illumination of the lesion by a high intensity visible light source, typically a tunable argon-dye laser [8, 16, 18, 26, 29, 43]. This treatment usually results in gradual ablation of the tumor mass over a period of days. For human use Photofrin-II® (Pf-II), a complex mixture of porphyrins, is currently in Phase III clinical trials in the United States, and has been approved for use in Canada. While Pf-II shows great promise as an agent for photodynamic tumor ablation, a number of undesirable characteristics, including prolonged cutaneous photosensitivity, immunosuppression and a relatively poor penetration of radiant energy to deeper levels of thicker tumors [16, 18, 24, 29, 52], has led to a search for new classes of photosensitizers [16]. Among these phthalocyanines, verdins, chlorins, and purpurins, to name a few, have shown promise both *in vitro* cellular systems and *in vivo* tumor model systems [16]. In our laboratory a series of metallo-phthalocyanines are being examined for their PDT effects using *in vitro* and *in vivo* systems [46, 54].

The mechanism of PDT-induced tumor destruction *in vivo* is very complex, but involves the generation of reactive oxygen intermediates, in particular singlet oxygen [1]. A direct effect of these agents is partially responsible for tumor cytolysis. Tumor necrosis also results from widespread damage to the tumor vasculature which occurs following PDT treatment [29]. Finally, exposure of photosensitizer-treated animals to light nearly always produces a vigorous inflammatory reaction within hours of PDT treatment [13, 57]. Some believe that this inflammatory reaction represents a third component that contributes to tumor destruction [57].

Recent studies indicate that concomitant with its tumor ablative effects, PDT profoundly influences the inflammatory response and the immune system [29, 57]. The overall significance of these alterations to curative or toxic responses associated with PDT is not clear at this time. However, it is essential that a detailed analysis of such consequences of PDT treatment be considered prior to its clinical use. The purpose of this review is to summarize current knowledge of the effects of PDT on immune and inflammatory responses.

Direct effects of PDT on tumors

It is believed that the overall tumor response to PDT depends on photodynamic action in surrounding cells or within the tumor vasculature and subsequent biochemical reactions in neighboring as well as target cells [8, 16, 18, 29, 43]. Particularly in the case of porphyrins it is believed that PDT causes vascular constriction and blood flow stasis, tumor hypoxia, and nutrient depletion [8, 16]. These events alone or in combination lead to vascular occlusion which results in cell death [27, 49, 57]. For more information about this aspect of PDT the reader is referred to recent reviews [8, 16, 18, 29, 43, 57].

Microvascular effects of PDT

Vascular effects of PDT have been known for some time [57] and is antedated by observations that patients with a type of human porphyria known as erythropoietic protoporphyria had damage to their microvascular endothelium following exposure to sunlight [22]. In detailed experiments using transparent window chambers, Star *et al.* [50] demonstrated vessel constriction and blood flow stasis within 1 min following PDT treatment. In other studies, Selman *et al.* [49] using a radioactive microsphere method were able to demonstrate that there was a major reduction in blood flow to transplanted urothelial tumors in rats after porphyrin PDT treatment. It was thus postulated that the tumor cure occurring during PDT was dependent, at least part, on vascular damage rather than on direct tumor cell death. Henderson *et al.* [30] then showed that using the RIF and EMT-6 murine tumor model systems, tumor cell clonogenicity of both of these tumor cell types was unaffected by PDT, if the tumors were excised immediately after irradiation. However, if the tumors were allowed to remain in place for 4 h after the PDT procedure and were then excised, a substantial reduction in tumor cell clonogenicity occurred. This corresponded closely with observations in tumors subjected to oxygen deprivation, suggesting that damage to tumor microcirculation was responsible for the destructive effects of PDT.

The vascular endothelium appears to be a preferential target for PDT. In all probability this relates to the high concentrations of photosensitizer present within the vascular system and to the very high degree of oxygenation to which the endothelium is exposed which makes it especially sensitive to the photodynamic action of PDT.

Photosensitized damage to the vascular endothelium results in the release of endothelial-derived clotting factors into the circulation [3] and exposes sites on the vascular wall to which platelets can adhere. These effects lead to hemostasis and the development of microvascular thrombi comprised of circulating platelets and erythrocytes which can easily be demonstrated histologically and ultrastructurally. The thrombi that develop are responsible for physicochemical changes in the surrounding tissue that are similar, if not identical, to those found during tissue hypoxia [57]. It should also be noted that the permeability of endothelial cells is also increased following PDT treatment. This is caused by photosensitizer-mediated release of vasoactive amines (see below) which may lead to escape of serum proteins and other fluids from the blood resulting in edema around the injured vessels [50]. These two factors together, i.e. the rapid restriction of the blood supply, as well as the onset of edema and hemorrhage in the tumor, lead to hypoxia or even anoxia of the photoinjured neoplastic cells, which eventually undergo necrosis [57].

Inflammation and PDT

The inflammation that accompanies PDT may play an important role in enhancing tumor regression. For example, the inflammatory exudate may deliver protein-bound photosensitizers to deeper areas of the tumor which are less accessible to the delivery of light. It is also possible that the inflammatory response may increase the local PO_2 values in less oxygenated areas of the tumor. Finally, fibrin thrombi within the tumor, resulting from clotting of extravasated fibrinogen, may provide further binding sites for the photosensitizers. Freitas [13] has suggested that the therapeutic efficacy of this inflammatory response might be enhanced by delivering a second dose of photosensitizer at a point when the inflammatory response from the first dose is at a maximum. He postulated that such a regimen would result in: (i) better tissue distribution of the drug due to transport by exudate proteins; (ii) higher cellular uptake of drug due to the presence of a highly proliferative population of tumor cells; and (iii) improved tissue oxygenation brought about by the photosensitizer present within the serous exudate.

Inflammatory mediators and the PDT response

Vasoactive amines

PDT treatment results in the non-specific release of a number of inflammatory mediators including vasoactive amines such as histamine and serotonin, mast cell and neutrophil proteases, products of arachidonic acid metabolism, and cytokines such as tumor necrosis factor alpha (TNFα). Many of them have been implicated both in the destructive effects of PDT on tumors and/or in the prolonged cutaneous photosensitivity that accompanies porphyrin PDT treatment [10, 28, 38].

Both mast cells and platelets are thought to be the sources of the vasoactive amines released following PDT treatment [37]. Inhibitors of vasoactive amines have been shown to inhibit cutaneous photosensitivity in animals treated with porphyrins and light. Similar results have been obtained when animals genetically deficient in mast cells were treated with porphyrin PDT, implicating mast cells as one source of these mediators. Heltianu *et al.* [25] have demonstrated that postcapillary venules have a relatively high density of histamine-binding sites compared to veins and arteries, whereas capillaries and endothelial cells of the aorta have the lowest values. This may provide a rationale for the increased sensitivity of postcapillary venules to PDT-induced damage.

Eicosanoids

Eicosanoids are potent biological mediators that are known to be involved in the development of the inflammatory response [51, 58]. These mediators are generated when cell membranes are disturbed, resulting in mobilization of membrane-bound precursor free fatty acids, the rate limiting step in their biosynthesis [20, 32]. The plasma membrane and membranes of cytoplasmic organelles have been recognized as major targets for photodamage by PDT [9, 14]. Thus it was not surprising to find that the biosynthesis of these biological mediators is greatly augmented by PDT-induced tumor ablation [14, 20, 32]. Henderson and Donovan [28] reported that the PDT greatly enhances the formation and release of PGE_2 from RIF tumor cells and peritoneal exudate macrophages *in vitro*. The release of eicosanoid was found to occur almost immediately following light treatment and returned nearly to background levels within 1 h posttreatment. Its production could be blocked almost entirely by treatment with non-steroidal anti-inflammatory agents such as indomethacin and meclophenamate – which are potent inhibitors of the cyclooxygenase pathway [28]. The corticosteroids hydrocortisone and dexamethasone, which are known to inhibit phospholipase A_2 activity, were also found to inhibit PDT-induced production of PGE_2 [12, 31]. It has been proposed that cell membrane damage leads to increases in Ca^{2+} ion influx into the cell. This then stimulates phospholipase A_2 activity. Phospholipase A_2 is believed to catalyse the liberation of free fatty acids from membrane phospholipids [20]. Other mechanisms by which PDT could influence prostaglandin synthesis might be related to the generation of reactive oxygen species during the photodynamic process. Singlet oxygen, peroxide molecules and other reactive oxygen intermediates which arise during ionizing and UV irradiation, as well as during PDT, are known to result in the peroxidation of membrane lipids which may effect the activity of phospholipase A_2 [14, 41].

Eicosanoid products released during PDT could contribute to the biological effects and side-effects of PDT. Some constituents, such as the unstable prostaglandin endoperoxides PGG_2 and PGH_2, act on vascular smooth muscle and cause transient vasoconstriction before being converted to the vasodilating PGE_2 [58]. Among the immediate effects of PDT are transient vasoconstriction as well as aggregation and agglutination of blood cells, leading to vascular stasis with vasodilation, vessel congestion and hemorrhage. Prostaglandin E_2 is also immunosuppressive. Its generation during the PDT response could also be responsible for suppression of cell-mediated immune responses observed following PDT treatment in animals. In addition, Gomer *et al.* [17] have suggested that the transient decrease of splenic natural killer cells that is observed following PDT in mice might be related to release of eicosanoids into the system.

Effect of TNFα

PDT has also been shown to be an important stimulus for TNFα. When murine thioglycollate-elicited peritoneal exudate macrophages were treated with Pf-II and light, a dose-dependent increase in TNFα production was observed [11]. Furthermore, increased levels of TNFα have been identified in the urine of humans undergoing PDT for bladder cancer [45]. Moreover, the histological features of the vascular effects of TNFα are strikingly similar to those described for PDT [50] and, like PDT, TNFα has a high specific activity for malignant cells [21]. Based on these findings, it has been suggested that TNFα might be a mediator of tumor regression in PDT treated mice, although no studies have directly addressed this issue.

Monocytes, macrophages and PDT

Macrophages and monocytes preferentially accumulate Pf-II and other photosensitizers within their cytoplasm, and as part of the host's natural defense against malignant neoplasms they invade and attempt to destroy neoplastic cells. Thus there has been particular interest in determining the effects of PDT on cells of the macrophage/monocyte lineage. In the earlier studies examining this issue it was shown that there were particularly high levels of hematoporphyrin derivative (a crude preparation of a mixture of porphyrins from which Pf-II is derived) in macrophages scattered throughout transplanted mouse tumors [5]. Similar observations have been made by Henderson and Bellnier [27]. The uptake of Pf-II by murine peritoneal macrophages *in vivo* and *in vitro* has also been reported by Korbelik *et al.* [34]. In other studies, Chan *et al.* [6] have demonstrated that chloroaluminum phthalocyanine tetrasulfonate is present in macrophages extracted from mouse colorectal carcinomas. Moreover, when single cell suspensions were prepared from such tumors, macrophages were among the cells with the highest photosensitizer content. Presumably, circulating monocytes phagocytize the photosensitizers and then later infiltrate into tumor tissue [33].

Based on the fact that monocytes and macrophages preferentially accumulate photosensitizers [5], attempts to enhance the tumor: tissue ratio of photosensitizer by preloading macrophages with the compound and then injecting them into the animal may be a useful therapeutic approach. Photosensitizers that are loaded in micro- or nanospheres and are conjugated with antibodies or other proteins may also have a better chance of delivery by this route [7, 36].

Examination of the functional capacity of PDT-treated macrophages yielded varying results. Pf-II at very low concentrations together with light enhances the phagocytic activity of activated inflammatory murine peritoneal macrophages. Yamamoto *et al.* [54] reported that peritoneal macrophages treated *in vivo* with porphyrins exhibited significant enhancement of Fc-receptor mediated phagocytic activity. They observed that *in vitro* exposure of peritoneal exudate cells (containing macrophages and B- and T-lymphocytes) to a low dose of Pf-II and red light produced an enhanced level of macrophage phagocytic activity. A singlet oxygen quencher, 1,4-diazabicyclo[2,2,2] octane, inhibited this effect. Photodynamic treatment of macrophages alone did not activate them, this occurred only when macrophages were mixed with photodynamically treated non-adherent cells (B- and T-lymphocytes). This study shows that the activation of macrophage is a consequence of peroxidation of lymphocyte membrane lipids by singlet oxygen generated in PDT. Higher doses, however, have been found to suppress phagocytic activity [53]. The latter findings are more consistent with the observations of Krutmann *et al.* [35] who found that treatment of human monocytes with Pf-II and light resulted in a selective and profound inhibition of the high affinity Fc receptor for IgG molecules, cell surface molecules important for phagocytosis and tumor cell killing. This effect of Pf-II-PDT was highly specific and was found to be mediated by the production of reactive oxygen intermediates.

Alterations in cell-mediated immune responses during PDT

Elmets and Bowen [10] were the first to show the immunosuppressive effects of PDT. Using murine contact hypersensitivity as a model of cell-mediated immunity they showed that pretreatment with hematoporphyrin derivative inhibited the contact hypersensitivity response in C3H mice by at least

50%. This porphyrin-induced inhibition of the contact hypersensitivity response was a sustained phenomenon that was still present two weeks after PDT treatment. Photoactivation of the porphyrin molecule was necessary for inhibition of the contact hypersensitivity response since animals treated with photosensitizer alone or light alone responded normally. Immunological suppression could be adoptively transferred to naive syngeneic recipients, thus indicating that the immunosuppression was associated with the development of suppressor cells. Further characterization of the suppressor cells generated in mice by peritoneal PDT was carried out by Lynch *et al.* [39] who showed that the suppressor cells present in the spleen of PDT-treated animals were macrophages. The suppressor cells were not induced in donor mice by PDT treatment but required epicutaneous application of the contact sensitizer as well. Further, they showed that the ability of spleen cells from PDT-treated mice to generate cytotoxic T-lymphocytes after allogeneic stimulation *in vitro* was comparable to that of spleen cells from control mice [39]. However, the ability of spleen cells from PDT-treated mice to act as antigen-presenting cells in the mixed lymphocyte culture was dramatically reduced. These observations provide additional evidence that the primary cell type associated with PDT-associated immunosuppression is the macrophage.

There has been considerable speculation that the mechanism(s) for immunosuppression by PDT-treated macrophages relate to the production by this cell type of soluble cytokines and other mediators of inflammation. Lynch *et al.* [39] showed that peritoneal PDT-treatment may induce resident macrophages to produce and secrete cytokines such as interleukin-1 (IL-1) and prostaglandins, which in other systems are known to inhibit the generation of an effective contact hypersensitivity response [4, 48].

It should be noted that the immunosuppressive effects of PDT are not restricted to porphyrins [44]. Similar observations have been made in our laboratory with a newly developed phthalocyanine [55, 56], termed Pc 4 (C.A. Elmets, S.I.A. Zaidi and H. Mukhtar, unpublished obervations).

Advantages of PDT-induced immunosuppression

In selected situations, the immunosuppressive effects of PDT may be highly desirable. Experiments have been conducted to determine if organ graft survival might be prolonged with PDT [19]. Since it is possible that the site of action of PDT within a tumor mass is the capillary endothelium, foreign antigens present on organ graft capillary endothelial cells could be responsible for the initiation of graft rejection. Therefore, the immunosuppressive effects of PDT may have potential as a useful adjunct in organ transplantation.

Although PDT often leads to selective tumor necrosis, complete responses are not always obtained, particularly in bulky tumors [40]. TNFα given systemically or locally in conjunction with PDT may increase the likelihood of a complete tumor response. It is possible that this approach may be used to reduce either the dose of photosensitizer or of light, which may help in reducing the cutaneous toxicity and normal tissue response respectively. It may also help in reducing the total exposure time and could be advantageous in endoscopic PDT. Combination therapies are therefore desirable to enhance the therapeutic effect and ameliorate the toxicities associated with each modality. In one study, Bellnier [2] used recombinant human TNFα (rhTNFα) and PDT on a mouse mammary adenocarcinoma and found that the response to PDT was significantly enhanced by the addition of a rhTNFα. The mechanism by which these two modalities interact during combination therapy of tumors is unknown. One possibility is a direct effect on tumor cells or the tumor parenchyma since both PDT and rhTNFα are effective *in vitro*. Alternatively, the interaction may take place on the tumor vasculature by both modalities.

Immunoconjugation of photosensitizers as an approach to drug delivery and targeting

Because of the selective localization of photosensitizers and the spatial control of illumination,

PDT is a highly selective modality for cancer treatment. The selectivity of this modality can be further enhanced using immunoconjugates of photosensitizers with tumor specific monoclonal antibodies for targeting their delivery to the tissue. It has been suggested that the photosensitizers conjugated with antibodies or other proteins are more likely to target tumors [23].

Mew *et al.* [42] conjugated hematoporphyrin with monoclonal and polyclonal antibodies directed against tumor-associated antigens and showed increased photodamage with PDT. Oseroff *et al.* [47] observed selective photodestruction of human T-cell leukemia cells using monoclonal antibody-chlorin e6 conjugates. They showed that photodamage was limited to the targeted population bound to monoclonal antibody and no or minimal damage occurred in the unbound population. Goff *et al.* [15] also used an immunoconjugate synthesized from a chlorin derivative, e_6-monoethylenediamine monoamide, as the photosensitizer and an anti-ovarian carcinoma monoclonal antibody OC125. They found that photoimmunoconjugates retain significant antigen-binding specificity and affinity. Ovarian cancer cells exhibited high cytotoxicity while non-ovarian cancer cells under identical conditions exhibited little or no toxicity as assessed by ^{3}H-thymidine incorporation. However, the clinical applicability of this approach is limited because: (i) the large size of immunoconjugates allows them to be easily cleared by the reticuloendothelial system; (ii) the repetitive use of non-host derived immune complexes usually elicits an immune response leading to anti-idiotypes; (iii) antigen targets are often presented on the cell surface in only limited numbers; and (iv) the internalization process for antigen–antibody complexes is frequently inefficient.

References

1 Athar, M., C.A. Elmets, D.R. Bickers, H. Mukhtar: A novel mechanism for the generation of superoxide anions in hematoporphyrin derivative-mediated cutaneous photosensitization: activation of xanthine oxidase pathway. *J. Clin. Invest.* 83 (1989), 1137–1143.

2 Bellnier, D.A.: Potentiation of photodynamic therapy in mice with recombinant human necrosis factor-α. *J. Photchem. Photbiol. B: Biol.* 8 (1991), 203–210.

3 Ben-Hur, E., E. Heldman, S.W. Crane, I. Rosenthal: Release of clotting factor(s) from cultured endothelial cells photosensitized by photodynamic therapy (Abstract) *Laser Med. Sci.* (Abstract Issue) (1988), 107.

4 Benson, M.D., M.J. Aldo-Benson: Effect of purified protein SAA on immune response *in vitro*: mechanism of suppression. *J. Immunol.* 122 (1979), 2077–2082.

5 Bugelski, P.J., C.W. Porter, T.J. Dougherty: Autoradiographic distribution of hematoporphyrin derivative in normal and tumor tissue of the mouse. *Cancer Res.* 41 (1981), 4606–4612.

6 Chan, W-S., J.F. Marchall, G.Y.F. Lam, I.R. Hart: Tissue uptake, distribution, and potency of the photoactivatable dye chloroaluminum sulfonated phthalocyanine in mice bearing transplantable tumors. *Cancer Res.* 48 (1988), 3040–3044.

7 Couvreur, P., B. Kante, V. Lenaerts, M. Ronald, P. Speiser: Tissue distribution and antitumor drugs associated with polyalklycynoacrylate nanoparticles. *J. Pharm. Sci.* 69 (1980), 199–202.

8 Dougherty, T.J.: Photodynamic therapy – new approaches. *Semn. Surg. Oncol.* 5 (1989), 6–16.

9 Dubbelman, T.M.A.R.: Porphyrin-photosensitized modification of subcellular structures. In: *Photodynamic Therapy of Tumors and Other Diseases*, (1985), pp. 93–99. Eds.: G. Jori, C. Perria, Edizioni Libreria Progetto, Padua.

10 Elmets, C.A., K.D. Bowen: Immunological suppression in mice treated with hematoporphyrin derivative photoradiation. *Cancer Res.* 46 (1986), 1608–1611.

11 Evans, S., W. Mathews, R. Perry, D. Fraker, J. Norton, H.I. Pass: Effect of photodynamic therapy on tumor necrosis factor production by murine macrophages. *J. Natl Cancer Ins.* 82 (1990), 34–39.

12 Flower, R.J., G.J. Blackwell: Anti-inflammatory steroids induce biosynthesis of a phospholipase A_2 inhibitor which prevents prostaglandin generation. *Nature* 278 (1979), 456–458.

13 Freitas, I.: Inflammation and photodynamic therapy. *J. Photochem. Photobiol. B: Biol* 8 (1991), 340–341.

14 Girotti, A.W.: Mechanisms of photosensitization. *Photochem. Photobiol.* 38 (1983), 745–751.

15 Goff, B.A., M. Bamberg, T. Hasan: Experimental photodynamic treatment of ovarian carcinoma cells with immunoconjugates. *Antibody Immunoconj. Radiopharm.* 51 (1992), 191–199.

16 Gomer, C.J.: Preclinical examination of first and second generation photosensitizers used in photodynamic therapy. *Photochem. Photobiol.* 54 (1991), 1093–1107.

17 Gomer, C.J., A. Ferrario, N. Hayashi, N. Rucker, B.C. Szirth, A.L. Murphree: Molecular, cellular and

tissue responses following photodynamic therapy. *Lasers Surg. Med.* 8 (1988), 450–463.

18 Gomer, C.J., N. Rucker, A. Ferrario, S. Wong: Properties and applications of photodynamic therapy. *Radit. Res.* 120 (1989), 1–18.

19 Gruner, S., H. Meffert, H. Volk, R. Grunow, S. John: The influence of hematoporphyrin derivative and visible light on murine skin graft survival, epidermal Langerhans cells and stimlation of the allogeneic mixed leucocyte reaction. *Scand. J. Immunol.* 32 (1985), 267–274.

20 Hall, A.K., H.R. Behrman: Prostaglandins: biosynthesis, metabolism, and mechanism of cellular action. In: *Prostaglandins*, (1982), pp. 1–38. Ed.: J.B. Lee, Elsevier, New York.

21 Haranaka, K.: Tumor necrosis factor: how to improve the antitumor activity and decrease accompanying side effects for therapeutic application. *J. Biol. Response Mod.* 7 (1988), 525–535.

22 Harber, L.C., D.R. Bickers: The porphyrias. In: *Photosensitivity Diseases. Principles of Diagnosis and Treatment*, (1989), pp. 241–287. Ed.: L.C. Harber, D.R. Bickers, B.C., Decker Inc., Philadelphia.

23 Hasan, T.: Photosensitizer delivery mediated by macromolečlar carrier systems. In: *Photodynamic Therapy: Basic Principles and Clinical Applications*, (1992), pp. 187–200. Eds.: B. Henderson, T. Dougherty, Marcel Dekker, New York.

24 Hawkins, C.W., D.R. Bickers, H. Mukhtar, C.A. Elmets: Cutaneous porphyrin photosensitization: murine ear swelling as a marker of the acute response. *J. Invest. Dermatol.* 86 (1986), 638–642.

25 Heltianu, C., M. Simionescu, N. Simionescu: Histamine receptors of the microvascular endothelium revealed *in situ* with a histamine-ferritin conjugate. Characteristic high-affinity binding sites on venules. *J. Cell Biol.* 93 (1982), 357–364.

26 Henderson, B.W.: Photodynamic therapy – coming of age. *Photodermatology* 6 (1989), 200–211.

27 Henderson, B.W., D.A. Bellnier: Tissue localization of photosensitizers and the mechanism of photodynamic tissue destruction. *CIBA Found. Symp.* 146 (1989), 112–125.

28 Henderson, B.W., J.M. Donovan: Release of prostaglandin E_2 from cells by photodynamic treatment *in vitro*. *Cancer Res.* 49 (1989), 6896–6900.

29 Henderson, B.W., T.J. Dougherty: How does photodynamic therapy work? *Photochem. Photobiol.* 55 (1992), 145–157.

30 Henderson, B.W., S.M. Waldow, T.S. Mang, T.S. Dougherty: Tumor destruction and kinetics of tumor cell death in two experimental mouse tumors following photodynamic therapy. *Cancer Res.* 45 (1985), 572–576.

31 Hong, S.C.L., L. Levine: Inhibition of arachidonic acid release from cells as the biochemical action of anti-inflammatory corticosteroids. *Proc. Natl Acad. Sci. USA* 73 (1976), 1730–1734.

32 Horton, E.W.: Prostanoids in health and disease. In: *Chemistry, Biochemistry and Pharmacological Activity of Prostanoids*, (1978), pp. 1–16. Eds.: S.M. Roberts, F. Scheinmann, Pergamon Press, Oxford.

33 Jori, G.: *In vivo* transport and pharmacokinetic behaviour of tumor photosensitizers. *CIBA Found. Symp.* 146 (1989), 78–86.

34 Korbelick, M., G. Krosl, D.J. Chaplin: Photofrin uptake by murine macrophages. *Cancer Res.* 51 (1991), 2251–2255.

35 Krutmann, J., M. Athar, I.U. Khan, D. Mendel, P.M. Guyre, H. Mukhtar, C.A. Elmets: Inhibition of the high affinity Fc-receptor (FcγRI) on human monocytes by porphyrin photosensitization is highly specific and mediated by the generation of superoxide radicals. *J. Biol. Chem.* 264 (1989), 11407–11413.

36 Labib, A., V. Lenaerts, F. Chouinard, J.-C. Leroux, R. Ouellet, J.E. van Lier: Biodegradable nanospheres containing phthalocyanines and naphthalocyanines for targeted photodynamic tumor therapy. *Pharmaceu. Res.* 8 (1991), 1027–1031.

37 Lim, H.W., I. Gigli: Role of complement in porphyrin-induced photosensitivity. *J. Invest. Dermatol.* 76 (1981), 4–9.

38 Lim, H.W., D. Parker, A.J. Marcus: Generation of eicosanoids from mast cells exposed to protoporphyrin and irradiation. *Clin. Res.* 34 (1986), 763.

39 Lynch, D.H., S. Haddad, C.J. Jolles, V.J. King, M.J. Ott, B. Robertson, R.C. Straight: Immunologic effects of peritoneal photodynamic treatment. In: *Proceedings SPIE – The International Society for Optical Engineering, Photodynamic Therapy: Mechanisms*, vol. 1065, (1989), pp. 55–65. Ed.: T.J. Dougherty, International Society for Optical Engineering, Photodynamic Therapy.

40 Manyak, M.J., A. Russo, P.D. Smith, E. Glastein: Photodynamic therapy. *J. Clin. Oncol.* 6 (1988), 390–391.

41 Menconi, M., P. Polgar: Radiation, lipid peroxidation and the role of oxygen radicals in eicosanoid metabolism. In: *Eicosanoids and Radiation*, (1988), pp. 119–131. Ed.: P. Polgar, Kluwer Academic Publishers, Boston.

42 Mew, D., V. Lum, C.K. Wat, G.H.N. Towers, R.J. Walter, W. Wright, M.W. Berns, J.G. Levy: Ability of specific monoclonal antibodies and conventional antisera conjugated to hematoporphyrin to label and kill selected cell lines subsequent to light activation. *Cancer Res.* 45 (1985), 4380–4386.

43 Moan, J., K. Berg: Photochemotherapy of cancer: experimental research. *Photochem. Photobiol.* 55 (1992), 931–948.

44 Musser, D.A., R.J. Fiel: Cutaneous photosensitizing and immunosuppressive effects of a series of tumor

localizing porphyrins. *Photochem. Photobiol.* 53 (1991), 119–123.

45 Nseyo U.O., R.K. Whalen, M.R. Duncan, B. Berman: Urinary cytokines following photodynamic therapy for bladder cancer. A preliminary report. *Urology* 36 (1990), 167–171.

46 Oleinick, N.L., A.R. Antunez, M.E. Clay, B.D. Rihter, M.E. Kenney: New phthalocyanine photosensitizers for photodynamic therapy. *Photochem. Photobiol.* 57 (1993), 242–247.

47 Oseroff, A.R., D. Ohuoha, T. Hasan, J.C. Bommer, M.L. Yarmush: Antibody-targeted photolysis: selective photodestruction of human T cell leukemia cells using monoclonal antibody-chlorin e6 conjugates. *Proc. Natl. Acad. Sci. USA* 83 (1986), 8744–8788.

48 Satein, M.B., S.N. Vogel, J.D. Sipe, P.A. Murphy, S.B. Migel, J.J. Oppenheim, D.L. Rosenstreich: The role of macrophages in the acute phase response: SAA inducer is closely related to lymphocyte activating factor and endogenous pyrogen. *Cell Immunol.* 63 (1981), 164–176.

49 Selman, S.H., M. Kreimer-Birnbaum, P.J. Goldblatt, R.W. Keck, S.L. Britton: Blood flow in transplantable bladder tumors treated with hematoporphyrin derivative and light. *Cancer Res.* 44 (1984), 1924–1927.

50 Star, W.M., H.P.A. Marijnissen, A.E. van den Berg-Blok, J.A.C. Versteeg, K.A.P. Fraken, H.D. Reinhold: Destruction of rat mammary tumor and normal tissue microcirculation by hematoporphyrin derivative photoradiation observed *in vitro* in sandwich observation chambers. *Cancer Res.* 46 (1986), 2532–2540.

51 Stenson, W.F., C.W. Parker: Prostaglandins and the immune response. In: *Prostaglandins*, (1982), pp. 39–89. Ed.: J.B. Lee, Elsevier, New York.

52 Tralau, C.J., A.R. Young, N.P.J. Walker, D.I. Vernon, A.J. MacRobert, S.B. Brown, S.G. Bown: Mouse skin photosensitivity with dihematoporphyrin ether (DHE) and aluminum sulfonated phthalocyanine (AlSPc): a comparative study. *Photochem. Photobiol.* 49 (1989), 305–312.

53 Wiegand-Steubing, R., S. Yeturu, A. Tuccillo, C.-H. Sun, M.W. Berns: Activation of macrophages by Photofrin II during photodynamic therapy. *J. Photochem. Photobiol. B: Biol.* 10 (1991), 133–145.

54 Yamamoto, N., S. Homma, J.K. Hoober, T.W. Sery, L.A. Donoso: Photodynamic immunopotentiation: *in vitro* macrophage activation by treatment with hematoporphyrin and light. *Photochem. Photobiol.* 49 (1989), 49S.

55 Zaidi, S.I.A., R. Agarwal, G. Eichler, B.D. Righter, M.E. Kenney, H. Mukhtar: Photodynamic effects of new silicon phthalocyanines: *in vitro* studies utilizing rat hepatic microsomes and human erythrocyte ghosts as model membrane sources. *Photochem. Photobiol.* 58 (1993), 204–210.

56 Zaidi, S.I.A., N.L. Oleinick, M.T. Zaim, H. Mukhtar.: Apoptosis during photodynamic therapy-induced ablation of RIF-1 induced tumors in C3H mice: electron microscopic, histopathologic and biochemical evidence. *Photochem. Photobiol.* 58 (1993), 771–776.

57 Zhou, C.: New Trends in photobiology: mechanisms of tumor necrosis induced by photodynamic therapy. *J. Photochem. Photobiol. B: Biol.* 3 (1989), 299–318.

58 Zurier, R.B.: Prostaglandins and inflammation. In: *Prostaglandins*, (1982), pp. 91–112. Ed.: J.B. Lee, Elsevier, New York.

20: Photoimmunology of PUVA and UVB therapy

Beatrix Volc-Platzer & Herbert Hönigsmann

Basic mechanisms

The skin — an immunoregulatory compartment

Efficient immune responses result from specific interactions between monocytes/macrophages capable of antigen presentation, B cells, T cells and certain immunoregulatory cytokines secreted by these cells. The immune system is anatomically compartmentalized within certain organs. It is now known that the non-lymphoid components of a particular microenvironment may contribute to the immune response generated within a particular organ system. Under various conditions, epithelial cells of the gut and keratinocytes (KC) of the epidermis synthesize and express immune response-associated antigens, in the murine as well as in the human situation (for review see [85]).

In 1978, Streilein [76] proposed the term 'skin-associated lymphoid tissue' (SALT), which has developed to provide immune protection of the skin against physical, chemical and microbial insults. Various discoveries and observations have supported this concept: identification of functional epidermotropic T-cell subsets [18], phenotypic and functional characterization of epidermal Langerhans cells (LC) (for review see [75]), the observation by Macher and Chase that T cells can recognize antigen within the cutaneous compartment [53] and the subsequent identification of unique regulatory circuits within the skin (for review see [77]), the discovery of the thy-1^+ dendritic T-cell population within murine skin [5, 82] equipped with the λ/δ T cell receptor (TCR) [38, 49], the observation that topically administered agents such as ultraviolet radiation (UVR) affect certain immune responses [6, 52, 55], the discovery of cytokines derived from KC which influence immune regulation, differentiation of T cells and cells of the monocyte/macrophage lineage and hematopoiesis [for review see 4, 51], and last but not least, the observation that upon appropriate stimulation KC express class II alloantigens [3, 85]. Evidence has accumulated within the last few years that KC are more than a physical barrier to the macroenvironment, but rather possess important immunological and secretory functions, i.e., expression of various ligands for recognition by T cells such as class II alloantigens [3, 85], intercellular adhesion molecule 1 (ICAM-1) [17] and BB-1 antigen [61]; in addition, KC of the upper layers contain tumor necrosis factor α (TNFα) [37], KC display high-affinity TNFα receptors [62], and, upon various stimuli, KC secrete TNFα [for review see 62] thereby initiating a cascade of pathophysiologic events. Moreover, interleukin-1 (IL-1) [1, 47] and granulocyte/macrophage colony stimulating factor (GM-CSF) [48] are produced by KC. By affecting any of these individual components UVR can influence and modify immune responses within the microenvironment of the skin. However, via the hematopoietic cells present in the skin which have access to the circulation and thus to other lymphoid organs on one hand, and via the release of cytokines into the systemic circulation on the other, topically applied agents such as UVR are capable of influencing and modifying systemic immune responses as well.

Immunosuppression by ultraviolet radiation

Ultraviolet radiation (UVR) has important local and systemic immunosuppressive effects, rather than upregulating cell-mediated immune responses. Immunosuppression by UVR has been recognized in the course of photobiologic investigations of the skin focusing on the carcinogenic properties of UVR. In 1974, Kripke reported that murine skin tumors induced by UVR were 'highly antigenic' and thus rejected even by syngeneic hosts [40]. Subsequent observations suggested that immunosuppressive mechanisms mediated predominantly by T suppressor lymphocytes [19] were responsible for this transplantation characteristic displayed by UVR-induced tumors [43].

Ultraviolet radiation exerts two major effects on the epidermal and dermal cellular and soluble constituents: (i) at the molecular level through interaction with the DNA which appears to be the main light-absorbing event; and (ii) at the biochemical level where immunologic changes are the result of complex interactions between the skin and components of the immune system. Both pathways appear to be inextricably related, in UV-induced carcinogenesis as well as in UV-induced immunosuppression [41]. Whereas these effects have been repeatedly investigated in experimental systems using UVB radiation (280–320 nm), the effects of UVA radiation (320–400 nm) have been less well explained. Although several 'receptors' for UV irradiation have been discussed, it has been shown by Kripke and coworkers that DNA is the major target of UV radiation in the generation of systemic immunosuppression [42]. Their studies using an *in situ* liposome-mediated delivery system of T4 endonuclease V provided proof that the primary molecular event mediating systemic immunosuppression is the formation of pyrimidine dimers [42].

UVR suppression of contact hypersensitivity and delayed-type hypersensitivity

Photoimmunologic effects of UVB radiation and PUVA have been studied extensively in animal models. Two major models have been investigated, one using high-dose (20–97 kJ/m^2), short-term UVB irradiation that severely alters the skin and suppresses the induction of contact hypersensitivity (CHS) and delayed-type hypersensitivity (DTH) at distant non-irradiated sites [65]. The low-dose UVR exposure has been reported without major skin changes and only local suppression of CHS induction [28, 57]. For immunologic events related to UVB phototherapy used for treatment and/or prophylaxis of skin diseases the low-dose model appears to be more relevant.

Topical PUVA in mice, given daily for seven days caused a marked decrease in ATPase$^+$ LC density in exposed skin sites [52], whereas psoralen or UVA alone did not exert such an effect. In contrast to PUVA-treated guinea-pigs [58], the depression in CHS responsiveness was observed when the site of induction had been PUVA treated whereas the site of elicitation had not been PUVA treated [52]. Higher doses of UVA could induce systemic suppression of CHS in mice [52]. The PUVA-induced systemic suppression of CHS exhibited several features similar to those induced by high-dose UVB [64, 65]. Similar mechanisms appear to be involved in UVB- and PUVA-induced suppression of CHS [10, 11]. Apparently both topical and systemic PUVA treatments cause depression of CHS. Whereas topical PUVA appears to cause Langerhans cell (LC) inactivation at the treated skin sites [11], suppression of CHS by systemic PUVA was reported to be indomethacin-sensitive, indicating that some prostaglandin (PG) mechanism is involved [10]. However, the results obtained by PUVA closely resemble local and systemic suppression of CHS caused by low- and high-dose UVB exposures respectively [52, 65, 80].

The LC is the primary epidermal antigen-presenting cell required for CD4$^+$ T-cell-mediated immune responses (for review see [75]). In addition to adhesion molecules and class II alloantigens which are required for the antigen-non-specific as well as the antigen-specific interactions with T cells [75], another molecule required for efficient antigen presentation has been identified on LC, i.e., B7/BB1 [79, 87] which interacts with the CD28 molecule on the CD4$^+$ T cell [39]. Tumor necrosis factor (TNF) as well as GM-CSF have been shown to support *in vitro* short-term culture of LC [74], and also to increase their antigen-presenting capacity [74].

In vivo exposure to UVB, UVA and PUVA is followed by changes of LC morphology and function [75], decreased LC density as determined by ATPase staining, and decreased class II antigen surface expression as visualized by immunomorphologic staining [75]. Although LC apparently lose the expression of various markers and may either leave the epidermis or die, a reasonable proportion of LC remains in the epidermis as evidenced by electron microscopy [75].

Photoimmunologic effects of PUVA and UVB in psoriasis and CTCL

Currently a wide variety of skin diseases is being treated with either PUVA, UVB or a combination of both (Table 20.1). In this chapter we attempt to give an overview on possible interactions between PUVA and UVB, and components of the immune system involved in pathogenetic mechanisms underlying disease states such as psoriasis and cutaneous T-cell lymphoma for which treatment with photo (chemo)therapy is already established.

Photoimmunologic mechanisms operative in psoriasis

The major therapeutic effect of PUVA on psoriasis is the antiproliferative effect based on the fact that psoralens (8-MOP, 5-MOP) intercalated between DNA base pairs absorb photons in the UVA range which then results in the formation of bifunctional adducts with interstrand crosslinks of the double helix [for review see 32]. The conjunction of psoralens with epidermal DNA results in the suppression of DNA synthesis and cell division [32]. However, crosslinking of DNA strands may not necessarily be the prerequisite of the beneficial

Table 20.1 Diseases to be treated with PUVA and/or UVB

Disease states	PUVA/UVB	Therapy(T)/Prevention(P)
Papulosquamous diseases		
Psoriasis	+/+	T
Palmoplantar pustulosis	+/+	T
Generalized lichen planus	+/+	T
Pityriasis rubra pilaris	+/−	T
Pityriasis lichenoides	+/−	T
Diseases with autoimmune background		
Cutanous graft-versus-host disease (cGvHD)	+/−	T
Vitiligo (poliosis)	+/−	T
Alopecia areata	+/−	T
Cutaneous neoplastic T-cell proliferative diseases		
Cutaneous T-cell lymphoma (mycosis fungoides, stages IA, IB)	+/−	T
Actinic reticuloid	+/−	T
Lymphomatoid papulosis	+/−	T
Others		
Atopic dermatitis/eczema	+/+	T
Generalized granuloma annulare	+/+	T
Urticaria pigmentosa	+/−	T
Polymorphous light eruption	+/−	P
Hydroa vacciniforme	+/−	P
Solar urticaria	+/−	P
Chronic actinic dermatitis	+/−	P(T)
Persistent light reaction	+/−	P
Erythropoietic protoporphyria	+/−	P

effect, for two reasons: (i) the same extent of remission of the disease can be achieved by UVA alone either from natural or from artificial light sources; and (ii) other diseases which are not associated with epidermal hyperproliferation also improve upon PUVA treatment. Moreover, UVB is also beneficial in certain forms of psoriasis, e.g. the guttate form, where DNA crosslinks are also not required. Thus PUVA may exert some other local effects on the cutaneous microenvironment and/or systemic effects on and through circulating blood cells.

Although the initiating events that lead to psoriasis are unknown, the fact that immunocompetent cells and proliferating KC are in close vicinity in psoriatic lesions suggests that self-perpetuating, immunologic processes are involved in the pathogenesis of the disease [83]. Cytokine up- and/or dysregulation may play a more dominant role rather then physical interactions between the cellular constituents of the cutaneous microenvironment. IL-1β, TNFα, TNFβ, NAP-1 (neutrophil attractant protein-1)/IL-8 and MCP-1 (monocyte attractant protein-1) have been found to be increasingly expressed by KC in psoriasis [8, 21, 23, 24]. TNFα as well as NAP-1/IL-8 have been localized to the upper layers of the epidermis [23, 37]. NAP-1/IL-8 is a potent chemoattractant for neutrophils [73] and may thus be responsible for the influx of neutrophils into the epidermis, eventually leading to the formation of subcorneal pustules. TNFα may be released upon environmental influences, e.g. mechanical trauma, UVR and microbial infection, thereby activating high-affinity TNFR [81], expressing KC of the lower epidermal layers as well as suprabasal LC, and thus initiating a cascade of events [62]. Indirect evidence exists that UVB radiation can prevent TNFα or -β-induced ICAM-1 expression by KC [46, 66] as well as by antigen-presenting monocytes [45]. Whereas this could in part be responsible for the therapeutic effect of phototherapy with UVB on guttate forms of psoriasis, no data exist with regard to PUVA treatment.

It has been shown that IL-1, TNFα and TNFβ induce MCP-1 production by fibroblasts and endothelial cells [24]. A similar induction mechanism for MCP-1 production by basal KC may be considered. By interfering with IL-1 and TNF release from KC it appears likely that MCP-1 production as a secondary phenomenon is also suppressed by UVB irradiation.

In addition, PUVA has been shown to exert dose-dependent effects resulting in changes of the function of lymphocytes [20], among them an impaired IL-2 production [67]. In psoriatic lesions, however, the overall density of intraepidermal mononuclear cells had been shown to decrease significantly [70]. Moreover, the T-cell population within the dermal inflammatory infiltrate was significantly diminished, and intraepidermal $CD1a^+$ dendritic cells disappeared [70]. ICAM-1 expression by papillary endothelial cells was observed to be unaffected by PUVA alone [70]. However, upon combination of PUVA with cyclosporine A (CSA), ICAM-1 expression by the papillary endothelium was significantly reduced [70].

The major effects of PUVA and UVB on psoriasis appear to be local ones, such as the primary antiproliferative effect of PUVA on KC on one hand, and the interference with cytokine production by KC on the other.

Since cytokine production is also effectively inhibited by UVB alone it appears that this effect can be mediated by UV radiation alone and is not dependent upon antiproliferative effects by PUVA.

Photoimmunological mechanisms in CTCL

PUVA treatment has been established as therapy for cutaneous T-cell lymphoma (CTCL) in the early stages IA and IB [22]. CTCL lesions are predominated by epidermotropic T cells being mainly of the $CD3^+/CD4^+/CD45RO^+$ T-helper memory cell phenotype [31]. In addition, $CD1a^+$ epidermal LC may be increased in these early stages of the disease [18]. Although effects on cytokine production [29] as described for psoriasis may also be operative in CTCL, the clinical remission in skin disorders associated with a dense T-cell infiltrate appear at least in part to be due to cytotoxic effects on the infiltrating T cells within the epidermis.

Another possibility is the influence on 'homing' receptors of epidermotropic malignant T cells. The cutaneous lymphocyte-associated antigen (CLA) [71] which is recognized by the monoclonal antibody HECA 452 (raised primarily against high endothelial cells) [15] has been demonstrated on the surface of skin-infiltrating lymphocytes in patients with patch/plaque CTCL [71]. In the murine model it has been demonstrated that a 6-day irradiation with UVB ($5000\,J/m^2$) leads to extensive loss of the murine counterpart of HECA, MECA [16], which is also expressed on high endothelial venules within lymph nodes and, like its human counterpart, is associated with the characteristic morphology of these specialized endothelial cells. Therefore it appears likely that the beneficial effect of PUVA and UVB on patch/plaque stage CTCLs appears to be largely mediated by downregulation of the expression of these homing molecules by UVR.

PUVA and UVB diminish the induction of allergic contact dermatitis and DTH [57, 58]. A major factor contributing to this phenomenon may be the reduction of ATPase$^+$, or class II-antigen$^+$, or CD1a$^+$ LC [52, 75]. Mechanisms comparable to CHS with 'antigenic persistence' [30] have been postulated for the development of CTCL lesions. Thus it appears possible that impairment of LC may result in the interruption of this type of cutaneous immune response.

Photoimmunologic effects in dermatoses with an autoimmune background

Systemic immunosuppressive effects have to be considered in the treatment of diseases which have an autoimmune background, such as vitiligo, atopic eczema, alopecia areata, lichen planus, chronic cutaneous graft-versus-host disease (GvHD) and some of the photodermatoses.

Patients with atopic eczema are generally treated similarly to patients receiving PUVA and UVB therapy for psoriasis [12]. At present it is unclear how PUVA or UVB contribute to the improvement of skin lesions. An alteration of lymphocyte function is postulated. In addition, an influence on the expression of high-affinity IgE receptors by LC in atopic dermatitis [7, 60, 78], comparable to the downregulation of class II antigens, may be postulated. New insights into the pathogenetic events may be gained with the use of high-dose UVA irradiation which has been reported to be beneficial in patients with acute atopic dermatitis lesions [44]. Whether the therapeutic effect is due to interaction with eosinophils, differential regulation of IgE-R and CD1 expression by epidermal LC, and/or an as yet unknown interaction with TH-2 cells [12] remains to be determined.

The use of phototherapy and photochemotherapy of vitiligo to restore pigmentation is already a historic one. In most cases photobiologic rather than photoimmunologic mechanisms are responsible for the therapeutic effect. However, vitiligo is often associated with autoimmune disorders. Since UVB as well as PUVA may induce a systemic T-suppressor cell population, one may postulate that melanocyte destruction could be prevented or interrupted [32].

Generalized lichen planus has been shown to respond to systemic PUVA treatment [68], although higher doses than for treatment of psoriasis may be required [32]. Whether this is due to the destruction of the epitheliotropic inflammatory mononuclear cells and/or interference with cytokines released remains to be clarified. The observation that oral lesions in chronic cutaneous GvHD following bone marrow transplantation (BMT) cleared after systemic PUVA treatment [84] suggests that systemic mechanisms may be involved as well.

Despite matching for histocompatibility antigens and extensive immunosuppressive regimens, chronic GvHD still occurs in approximately 50% of the recipients of allogeneic bone marrow [14]. Hymes *et al.* [33] reported on the successful treatment of a patient suffering from the lichenoid form of chronic cutaneous GvHD. Our own observations [84] as well as investigations by others support the use of PUVA in GvHD treatment [2, 34–36]. Even acute cutaneous GvHD has been reported to respond to PUVA treatment [2] (Reinauer *et al.*, manuscript submitted) Although experimental data to clarify the operative mechanism(s) do not exist, one may assume that a major effect is exerted on the release of TNFα. TNFα [72] as well as interferon gamma (IFN)-λ [63] appear to be crucial in the development of acute, and to a lesser extent, chronic GvHD. PUVA may interfere significantly with the production and/or release of these potent cytokines by UVR, whereas UVB does not seem to improve the disease. In animal models controversial data exist with regard to UVB radiation: an improvement has been reported by Glazier *et al.* [25] as well as the conversion from a lichenoid type to a sclerodermoid variant of cutaneous GvHD [9].

The most experimental use of UVR has been under investigation in transfusion/transplantation medicine for several years. Long-term survival of skin allografts has been observed after UVR of the recipient or the donor [26, 27, 56, 69]. Organ grafts have been prevented from rejection after UV-treated donor-specific blood transfusion [50]. Blood and blood products lose their capacity to induce allospecific reactions in the recipient after irradiation with UVB and UVC [15]. The most common mechanism operative in all these types of transplantation and transfusion appears to be the interaction with the dendritic cell, the most potent antigen-presenting cell circulating in the peripheral blood, which is also present in various organs. Downregulation of class II alloantigen expression by the dendritic cell population appears to be the most likely explanation of the phenomenon of reduced alloreactivity. However, accessory molecules such as BB-1/B7 are also required for the interaction with T-helper lymphocytes [87]. UVB radiation has been shown to alter the costimulatory activity of human dendritic cells [86], but at present it is not clear whether and to what extent accessory molecules on dendritic cells are downregulated. However, it appears from the data reported that the modulation by UVB is not sufficient to generate tolerizing signals [86].

Outlook

The number of disorders with cutaneous involvement being treated with one of the current UVR regimens is increasing steadily. The use of UVR may be more therapeutic or – as in certain photodermatoses – more prophylactic. Data from patients suffering from autoimmune-mediated diseases including those with GvHD, as well as experimental data from transplantation and transfusion settings, point to a broader applicability of UVR in the near future, adding another agent to the list of immunosuppressive and/or immunomodulating agents.

References

1 Ansel, J.C., T.A. Luger, D. Lowy, P. Perry, D.R. Roop, J.D. Mountz: The expression and modulation of IL-1a in murine keratinocytes. *J. Immunol.* 140 (1988), 2274–2278.

2 Atkinson, K., P. Weller, W. Ryman, J. Biggs: PUVA therapy for drug-resistant graft-versus-host disease. *Bone Marrow Transplant.* 1 (1986): 227–236.

3 Auböck, J., N. Romani, G. Grubauer, P. Fritsch: HLA-DR expression on keratinocytes is a common feature of diseased skin. *Br. J. Dermatol.* 114 (1986), 465–472.

4 Balkwill, F.R., F. Burke: The cytokine network. *Immunol. Today* 10 (1989), 299

5 Bergstresser, P.R., R.E. Tigelaar, J.H. Dees, J.W. Streilein: Thy-1 antigen-bearing dendritic cells populate murine epidermis. *J. Invest. Dermatol.* 81 (1983), 286–288.

6 Berman, B., D.S. France, G.P. Martinelli, A. Hass: Modulation of expression of epidermal Langerhans cell properties following *in situ* exposure to glucocorticosteroids. *J. Invest. Dermatol.* 80 (1983), 168–171.

7 Bieber, T., B. Dannenberg, J.C. Prinz, E.P. Rieber, W. Stolz, O. Braun-Falco: Occurrence of IgE-bearing epidermal Langerhans cells in atopic eczema: a study of the time course of the lesions and with regard to IgE serum level. *J. Invest. Dermatol.* 93 (1989), 215–219.

8 Camp, R.D.R., N.J. Fincham, J.S. Ross, K.B. Bacon, A.J.H. Gearing: Leukocyte chemoattractant cytokines of the epidermis. *J. Invest. Dermatol.* 95 (Suppl) (1990), 108S–110S.

9 Charley, M.R., J.N. Gilliam, R.D. Sontheimer: Ultraviolet B exposure converts murine lichenoid graft-vs-host disease (GvHSD) into sclerotic GvHSD (Abstract). *J. Invest. Dermatol.* 80 (1983), 328–329(A).

10 Chung, H.T., D.K. Burnham, B. Robertson, L.K. Roberts, R.A. Daynes: Involvement of prostaglandins in the immune alterations caused by exposure of mice to ultraviolet radiation. *J. Immunol.* 137 (1986), 2478–2484.

11 Chung, H.T., B.H. Cho, B.D. Jun, R.A. Daynes: The route of 8-methoxypsoralen administration dictates

the type of immunosuppression which occurs following PUVA therapy (Abstract). *J. Invest. Dermatol.* 88 (1987), 481(A).

12 Cooper, K.D.: Atopic dermatitis: recent trends in pathogenesis and therapy. *J. Invest. Dermatol.* 102 (1994), 128–137.

13 Deeg, H.J.: Modification of allointeractions after blood transfusion and marrow transplantation by ultraviolet light. *Sem. Hematol.* 29 (1992), 95–101.

14 Deeg, H.J., R. Storb: Graft-versus-host disease; pathophysiological and clinical aspects. *Ann. Rev. Med.* 35 (1984), 11–24.

15 Duijvestijn, A.M., E. Horst, S. Pals, B.N. Rouse, A.C. Steere, L.J. Picker, C.J.L.M. Meijer, E.C. Butcher: High endothelial differentiation in human lymphoid and inflammatory tissues defined by monoclonal antibody HECA-452. *Am. J. Pathol.* 130 (1988), 147–155.

16 Duijvestijn, A.M., M. Kerkhove, R.F. Bargatze, E.C. Butcher: Lymphoid tissue and inflammation-specific endothelial cell differentiation defined by monoclonal antibodies. *J. Immunol.* 138 (1987), 713–719.

17 Dustin, M.L., K.H. Singer, D.T. Tuck, T.A. Springer: Adhesion of T lymphoblasts to epidermal keratinocytes is regulated by interferon λ and is mediated by intercellular adhesion molecule 1 (ICAM-1). *J. Exp. Med.* 167 (1988), 1323–1340.

18 Edelson, R.L.: Cutaneous T cell lymphoma: mycosis fungoides, Sezary syndrome, and other variants. *J. Am. Acad. Dermatol.* 2 (1980), 89–106.

19 Fisher, M.S., M.L. Kripke: Suppressor T lymphocytes control the development of primary skin cancer in ultraviolet irradiated mice. *Science* 216 (1982), 1133–1134.

20 Friedman, P.S., S. Rogers: Photochemotherapy of psoriasis: DNA damage in blood lymphocytes. *J. Invest. Dermatol.* 74 (1980), 440–443.

21 Gearing, A.J.H., N.J. Fincham, C.R. Bird, M. Wadhwa, A. Meager, J.E. Cartwright, R.D.R. Camp: Cytokines in skin lesions of psoriasis. *Cytokine* 2 (1990), 68–75.

22 Gilchrest, B.A.: Oral methoxsalen photochemotherapy of mycosis fungoides. *Cancer* 38 (1976), 683–693.

23 Gillitzer, R., R. Berger, V. Mielke, C. Müller, K. Wolff, G. Stingl: Upper keratinocytes of psoriatic skin lesions express high levels of NAP-1/IL-8 mRNA *in situ*. *J. Invest. Dermatol.* 97 (1991), 73–79.

24 Gillitzer, R., K. Wolff, D. Tong, C. Müller, T. Yoshimura, A. Hartmann, G. Stingl, R. Berger: MCP-1 mRNA expression in basal keratinocytes of psoriatic lesions. *J. Invest. Dermatol.* 101 (1993), 127–131.

25 Glazier, A., W.L. Morison, C. Bucana, A.D. Hess, W.E. Beschorner, P.J. Tutschka: Suppression of epidermal graft-versus-host disease (GvHD) with ultraviolet B radiation (Abstract). *J. Invest. Dermatol.* 80 (1983), 328(A).

26 Granstein, R.D., L. Smith, J.A. Parrish: Prolongation of murine skin allograft survival by the systemic effects of 8-methoxypsoralen and long-wave ultraviolet radiation (PUVA). *J. Invest. Dermatol.* 88 (1987), 424–429.

27 Gruner, S., C. Riese, S. Schnitzler, H. Meffert, E. Karasek: Prolongation of the survival of skin grafts in mice by PUVA treatment. *Experientia* 40 (1984), 487–488.

28 Hanisko, J., R.P. Suskind: The effect of ultraviolet radiation on experimental cutaneous sensitization in guinea pigs. *J. Invest. Dermatol.* 40 (1963), 183–191.

29 Hansen, E.R., G.L. Vejlsgaard, S. Lisby, M. Heidenheim, O. Baadsgaard: Epidermal interleukin-1α functional activity and interleukin 8 immunoreactivity are increased in patients with cutaneous T cell lymphoma. *J. Invest. Dermatol.* 97 (1991), 818–823.

30 Heald, P., R. Edelson: The immunobiology of cutaneous T cell lymphoma. *J. Natl. Cancer Inst.* 83 (1991), 400–404.

31 Heald, P.W., S.-L. Yan, R.L. Edelson, R. Tigelaar, L.J. Picker: Skin-selective lymphocyte homing mechanisms in the pathogenesis of leukemic cutaneous T-cell lymphoma. *J. Invest. Dermatol.* 101 (1993), 222–226.

32 Hönigsmann H., T.B. Fitzpatrick, M.A. Pathak, K. Wolff: Oral photochemotherapy with psoralens and UVA (PUVA): principles and practice. In: *Dermatology in General Medicine*, fourth edition, (1993), pp. 1728–1754. Eds.: T.B. Fitzpatrick, A.Z. Eisen, K. Wolff, I.M. Freedberg, K.F. Austen, McGraw-Hill, New York.

33 Hymes, S.R., W.L. Morison, E.R. Farmer, L.L. Walters, P.J. Tutschka, G.W. Santos: Methoxsalen and ultraviolet A radiation in treatment of chronic cutaneous graft-versus-host reaction. *J. Am. Acad. Dermatol.* 12 (1985), 30–37.

34 Jampel, R.M., E.R. Farmer, G.B. Vogelsang, J. Wingard, G.W. Santos, W.L. Morison: PUVA therapy for chronic cutaneous graft-vs-host disease. *Arch Dermatol.* 127 (1991), 1673–1678.

35 Jaschonek, K., H. Einsele, H. Schmidt, H.J. Lenz, G. Ehninger, R. Dopfer, M. Steinert: 8-Methoxypsoralen and ultraviolet A therapy for cutaneous manifestations of graft-versus-host disease. *Lancet* i (1988), 599.

36 Kapoor, N., A.E. Pellegrini, E.A. Copelan, I. Cunningham, B.R. Avalos, J.L. Klein, P.J. Tutschka: Psoralen plus ultraviolet A (PUVA) in the treatment of chronic graft versus host disease: preliminary experience in standard treatment resistant patients. *Sem. Hematol.* 29 (1992), 108–112.

37 Kolde, G., K. Schulze-Osthoff, H. Meyer, J. Knop: Immunohistological and immunoelectronmicroscopic identification of TNF-α in normal human and murine epidermis. *Arch. Dermatol. Res.* 284 (1992), 154–158.

38 Koning, F., G. Stingl, W.M. Yokoyama, H. Yamada,

W.L. Maloy, E. Tschachler, E.M. Shevach, J.E. Coligan: Identification of a T3-associated λ/δ T cell receptor on Thy-1$^+$ dendritic epidermal cell lines. *Science* 236 (1987), 834–837.

39 Koulova L., E.A. Clark, G. Shu, B. Dupont: The CD28 ligand B7/BB1 provides costimulatory signal for alloactivation of CD4$^+$ T cells. *J. Exp. Med.* 173 (1991), 759–765.

40 Kripke, M.L.: Antigenicity of murine skin tumors induced by ultraviolet light. *J. Natl Cancer Inst.* 53 (1974), 1333–1336.

41 Kripke, M.L.: Immunological unresponsiveness induced by ultraviolet radiation. *Immunol. Rev.* 80 (1984), 87–102.

42 Kripke, M.L., P.A. Cox, L.G. Alas, D.B. Yarosh: Pyrimidine dimers in DNA initiate systemic immunosuppression in UV-irradiated mice. *Proc. Natl. Acad. Sci. USA* 89 (1992), 7516–7520.

43 Kripke, M.L., J.S. Lofgren, J. Beard, J.M. Jessup, M.S. Fisher: *In vivo* immune responses of mice during carcinogenesis by ultraviolet radiation. *J. Natl. Cancer Inst.* 59 (1977), 1227–1230.

44 Krutmann, J., W. Czech, T. Diepgen, R. Niedner, A. Kapp, E. Schöpf: High-dose UVA therapy in the treatment of patients with atopic dermatitis. *J. Am. Acad. Dermatol.* 26 (1992), 225–230.

45 Krutmann, J., I.U. Khan, R.S. Wallis, F. Zhang, E.A. Rich, J.J. Ellner, C.A. Elmets: Cell membrane is a major locus for ultraviolet B-induced alterations in accessory cells. *J. Clin. Invest.* 85 (1990), 1529–1536.

46 Krutmann, J., A. Köck, E. Schauer, F. Parlow, A. Möller, A. Kapp, E. Förster, E. Schöpf, T.A. Luger: Tumor necrosis factor β and ultraviolet radiation are potent regulators of human keratinocyte ICAM-1 expression. *J. Invest. Dermatol.* 95 (1990), 127–131.

47 Kupper, T.S., D.W. Ballard, A.O. Chua, J.S. McGuire, P.M. Fload, M.C. Horowitz, R. Langdon, L. Lightfoot, U. Gubler: Human keratinocytes contain mRNA indistinguishable from monocyte interleukin 1λ and βmRNA. Keratinocyte epidermal cell-derived thymocyte-activating factor is identical to interleukin 1. *J. Exp. Med.* 164 (1986), 2095–2100.

48 Kupper, T.S., F. Lee, D. Coleman, J. Chodakewitz, P. Flood, M. Horowitz: Keratinocyte derived T-cell growth factor (KTGF) is identical to granulocyte macrophage colony stimulating factor (GM-CSF). *J Invest. Dermatol.* 91 (1988), 185–188.

49 Kuziel, W.A., A. Takashima, M. Bonyhadi, P.R. Bergstresser, J.P. Allison, R.E. Tigelaar, P.W. Tucker: Regulation of T-cell receptor λ-chain mRNA expression in murine Thy-1$^+$ dendritic epidermal cells. *Nature* 328 (1987), 263–266.

50 Lau, H., K. Reemtsma, M.A. Hardy: Pancreatic islet allograft prolongation by donor-specific blood transfusions treated with ultraviolet irradiation. *Science* 221 (1983), 754–755.

51 Luger, T.A., T. Schwarz: Epidermal cell derived secretory regulins. In: *Epidermal Langerhans Cells,* (1991), pp. 217–257. Ed.: G. Schuler, CRC Press, Boca Raton, FL.

52 Lynch, D.R., M.F. Gurish, R.A. Daynes: Relationship between epidermal Langerhans cell density, ATPase activity and the induction of contact hypersensitivity. *J. Immunol.* 126 (1981), 1892–1897.

53 Macher, E., M.W. Chase: Studies on the sensitization of animals with simple chemical compounds. XII. The influence of excision of allergenic deposits on onset of delayed hypersensitivity and tolerance. *J. Exp. Med.* 129 (1969), 103–121.

54 Morison, W.L.: Effects of ultraviolet radiation on the immune system in humans. *Photochem. Photobiol.* 50 (1989), 515–524.

55 Morison, W.L., M.L. Kripke: Systemic suppression of contact hypersensitivity by ultraviolet B radiation or methoxsalen/ultraviolet A radiation in the guinea pig. *Cell Immunol.* 85 (1984), 270–277.

56 Morison, W.L., J.A. Parrish, M.E. Woehler, K.J. Bloch: The influence of PUVA and UVB radiation on skin graft survival in rabbits. *J. Invest. Dermatol.* 75 (1980), 331–333.

57 Morison, W.L., J.A. Parrish, M.E. Woehler, K.J. Bloch: The influence of ultraviolet radiation on allergic contact dermatitis in the guinea pig. I. UVB radiation. *Br. J. Dermatol.* 104 (1981), 161–164.

58 Morison, W.L., J.A. Parrish, M.E. Woehler, K.J. Bloch: The influence of ultraviolet radiation on allergic contact dermatitis in the guinea pig. II. Psoralen/UVA radiation. *Br. J. Dermatol.* 104 (1981), 165–168.

59 Morison, W.L., J.A. Parrish, K.J. Bloch, J.I. Krugler: *In vivo* effects of PUVA on lymphocyte function. *Br. J. Dermatol.* 104 (1981), 405–413.

60 Mudde, G.C., F.C. Van Reijsen, G.J. Boland, G.C. DeGast, P.L.B. Bruijnzeel, C.A.F.M. Bruijnzeel-Koomen: Allergen presentation by epidermal Langerhans cells from patients with atopic dermatitis is mediated by IgE. *Immunology* 69 (1990), 335–341.

61 Nickoloff, B.J., R.S. Mitra, K. Lee, L.A. Turka, J. Green, C. Thomson, Y. Shimizu: Discordant expression of CD28 ligands, BB-1 and B7 on keratinocytes *in vitro* and psoriatic cells *in vivo*. *Am. J. Pathol.* 142 (1993), 1029–1040.

62 Nickoloff, B.J., L.A. Turka: Keratinocytes: key immunocytes of the integument. *Am. J. Pathol.* 143 (1993), 325–331.

63 Niederwieser, D., J. Auböck, J. Troppmayer, M. Herold, G. Schuler, G. Borck, J. Lutz, P. Fritsch, C. Huber: IFN-λ mediated induction of MHC antigen expression in human keratinocytes and its influence on *in vitro* alloimmune responses. *J. Immunol.* 140 (1988), 2556–2564.

64 Noonan, F.P., E.C. DeFabo, M.L. Kripke: Suppression of contact hypersensitivity by UV radiation and its

relationship to UV-induced suppression of tumor-immunity. *Photochem. Photobiol.* 34 (1981), 683–689.

65 Noonan, F.P., M.L. Kripke, G.M. Pedersen, M.I. Greene: Suppression of contact hypersensitivity in mice by ultraviolet irradiation is associated with defective antigen presentation. *Immunology* 43 (1981), 527–533.

66 Norris, D.A., M.B. Lyons, M.H. Middleton, J.J. Yohn, M. Kashihara-Sawami: Ultraviolet radiation can either suppress or induce expression of intercellular adhesion molecule 1 (ICAM-1) on the surface of cultured human keratinocytes. *J. Invest Dermatol.* 95 (1990), 132–138.

67 Okamoto, H.: Alteration of lymphocyte functions by 8-methoxypsoralen and long-wave ultraviolet radiation. II. The effect of *in vivo* PUVA on IL-2 production. *J. Invest. Dermatol.* 89 (1987), 24–26.

68 Ortonne, J.P., J. Thivolet, C. Sannwald: Oral photochemotherapy in the treatment of lichen planus (LP). Clinical results, histological and ultrastructural observations. *Br. J. Dermatol.* 99 (1978), 77–88.

69 Osterwitz, H., D. Scholz, J. Kaden, M. Mebel: Prolongation of rat renal allograft survival by donor pretreatment with 8-methoxypsoralen and longwave ultraviolet irradiation of the graft (PUVA therapy). *Urol. Res.* 13 (1985), 95–98.

70 Petzelbauer, P., G. Stingl, K. Wolff, B. Volc-Platzer: Cyclosporin A suppresses ICAM-1 expression by papillary endothelium in healing psoriatic plaques. *J. Invest. Dermatol.* 96 (1991), 362–369.

71 Picker, L.J., S.A. Michie, L.S. Rott, E.C. Butcher: A unique phenotype of skin-associated lymphocytes in humans. *Am. J. Pathol.* 136 (1990), 1053–1068.

72 Piguet, P.F., G.E. Grau, B. Allet, P. Vassalli: Tumor necrosis factor is an effector of skin and gut lesions of the acute phase of graft-versus-host disease. *J. Exp. Med.* 166 (1987), 570–573.

73 Rot, A.: Endothelial cell binding of NAP-1/IL-8: role in neutrophil emigration. *Immunol. Today* 13 (1992), 291–294.

74 Schuler, G. (ed.): *Epidermal Langerhans Cells* (1991). CRC Press, Boca Raton, FL.

75 Stingl, G., E. Tschachler, V. Groh, K. Wolff, C. Hauser: The immune functions of epidermal cells. In: *Immune Mechanisms in Cutaneous Disease,* (1989), pp. 3–72. Ed.: Norris, D.A., Marcel Dekker, Inc., New York.

76 Streilein, J.W.: Lymphocyte traffic, T-cell malignancies and the skin. *J. Invest. Dermatol.* 71 (1978), 167–171.

77 Streilein, J.W.: Skin-associated lymphoid tissue. In: *Immune Mechanisms in Cutaneous Disease,* (1989), pp. 73–96. Ed.: D.A. Norris, Marcel Dekker, Inc., New York.

78 Sugiura, H., M. Uehara, T. Maeda: IgE-positive epidermal Langerhans cells in allergic contact dermatitis lesions provoked by DNCB in patients with atopic dermatitis. *Arch. Derm. Res.* 282 (1990), 295–299.

79 Symington, F.W., W. Brady, P.S. Linsley: Expression and function of B7 on human epidermal Langerhans cells. *J. Immunol.* 150 (1993), 1286–1295.

80 Toews, G.B., P.R. Bergstresser, J.W. Streilein: Epidermal Langerhans cell density determines whether contact hypersensitivity or unresponsiveness follows skin painting with DNFB. *J. Immunol.* 124 (1980), 445–453.

81 Trefzer, U., M. Brockhaus, H. Laetscher, F. Parlow, A. Kapp, E. Schöpf, J. Krutmann: 55-kD tumor necrosis factor receptor is expressed by human keratinocytes and plays a pivotal role in regulation of human keratinocyte ICAM-1 expression. *J. Invest. Dermatol.* 97 (1991), 911–916.

82 Tschachler, E., G. Schuler, J. Hutterer, H. Leibl, K. Wolff, G. Stingl: Expression of Thy-1 antigen by murine epidermal cells. *J. Invest. Dermatol.* 81 (1983), 282–285.

83 Valdimarsson, H., B.S. Baker, I. Jonsdottir, L. Fry: Psoriasis: a disease of abnormal keratinocyte proliferation induced by T lymphocytes. *Immunol. Today* 7 (1986), 256–259.

84 Volc-Platzer, B., H. Hönigsmann, W. Hinterberger, K. Wolff: Photochemotherapy improves chronic cutaneous graft-versus-host disease. *J. Am. Acad. Dermatol.* 23 (1990), 220–228.

85 Volc-Platzer, B., G. Stingl: HLA-DR synthesis and expression by human keratinocytes. In: *Experimental and Clinical Photoimmunology,* (1986), pp. 119–126. Eds.: R.A. Daynes, G. Krueger, CRC Press, Boca Raton, FL.

86 Young, J.W., J. Baggers, S.A. Soergel: High-dose UV-B radiation alters human dendritic cell costimulatory activity but does not allow dendritic cells to tolerize T-lymphocytes to alloantigens *in vitro*. *Blood* 81 (1993), 2987–2997.

87 Young, J.W., L. Koulova, S.A. Soergel, E.A. Clark, R.M. Steinman, B. Dupont: The B7/BB1 antigen provides one of several costimulatory signals for the activation of $CD4^+$ T lymphocytes by human blood dendritic cells *in vitro*. *J. Clin. Invest.* 90 (1992), 229–237.

21: Photoimmunobiology of blood transfusion, bone marrow and organ transplantation

H.-Joachim Deeg

Introduction

Immunological identity is expressed in the form of surface glycoproteins termed histocompatibility antigens. The major histocompatibility complex (MHC), in humans located on the short arm of chromosome 6, comprises genes encoding strong tissue antigens, termed HLA; corresponding antigens are called H-2 in the mouse, DLA in the dog, etc. In addition, so-called minor, non-MHC antigens encoded for by genes outside the MHC or even on different chromosomes serve as recognition signals. These minor antigens are presented as small peptides by MHC class I or II antigens. If allogeneic, i.e., genetically non-identical, organs, tissues or cells taken from one individual are transferred to another, an alloreaction occurs. T lymphocytes expressing the T-cell receptor and CD8 molecules preferentially recognize class I antigens, whereas CD4 positive T cells preferentially recognize class II antigens. If the host is immunocompetent a host-versus-graft (HvG) reaction, possibly followed by graft rejection, ensues. If, vice versa, the host is immunocompromised, immunocompetent donor-derived cells survive, recognize the host, and trigger a graft-versus-host (GvH) reaction which in turn may lead to the development of graft-versus-host disease (GvHD) [6].

Lindahl-Kiessling and Saefwenberg showed in 1971 that UV irradiation of allogeneic mononuclear cells eliminated their ability to stimulate in mixed leukocyte culture (MLC) *in vitro* [33]. Ultraviolet light across the entire UV spectrum, i.e., UVC (200–280 nm), UVB (280–320 nm) and UVA (320–400 nm) has been used in subsequent studies [14, 29]. While UVC and UVB are effective by themselves, UVA generally requires the presence of photosensitizing agents, such as 8-methoxypsoralen (8-MOP) to be active. Depending upon the UV wavelength and the target cells studied, the results may differ: best characterized are the effects of UV energy on DNA, including pyrimidine–dimer formation and protein–DNA crosslinks [46]. Ultraviolet light also induces cellular enzymes and cytokines, alters cell membrane characteristics, modifies cell surface proteins, and interferes with cell–cell interactions and cell homing. Ultraviolet light may directly affect the cell population studied or may operate indirectly via a secondary effect of the cells treated by UV light [2, 15]. Of interest is the further observation that UVB and UVA in conjunction with 8-MOP induce isomerization of *trans*-urocanic acid to its *cis* isomer [25, 38, 48]. *Cis*-urocanic acid, in turn, interferes with interleukin-1 secretion *in vitro*, inhibits HLA-DR expression, at least on epidermal cells, and suppresses stimulation of T lymphocytes by epidermal cells [48]. We now know that UV exposure also reduces MHC class II antigen expression on human and canine lymphocytes [20]; whether *cis*-urocanic acid is involved has not been determined. Cells so treated may still be able to trigger expression of interleukin-2 (IL-2) receptors in responding cells, but not the production of the cytokine required for receptor engagement (reviewed in [15]). Studies in rat models furthermore, show, that UV irradiation severely interferes with the homing of lymphocytes to lymph nodes and of dendritic cells (DC) to the spleen [42]. Both effects may interfere with the activation of an alloresponse.

In vitro models

DC are the most potent accessory cells in assay systems in several species [2, 3, 57]. T lymphocytes require DC for alloantigen and lectin responses [2, 3]. While DC maintain this function after gamma irradiation, UV exposure (200–300 nm) at very low doses completely abrogates this capacity. While normal or gamma-irradiated DC form dense clusters with mitogen-treated T lymphocytes, UV-irradiated DC fail to do so. If normal MLC supernatant or recombinant IL-2 is added, T lymphocyte responsiveness is restored, in agreement with the observation that UV-treated accessory cells induce IL-2 receptor expression, but fail to release the cytokine necessary for proliferative responses (reviewed in [2, 14]). Another reason for impaired cell interactions after UV exposure may be, as discussed above, the loss of MHC class II antigen expression. It is of note that the decline is very modest for HLA-DR but rather pronounced for -DP and -DQ [20]; HLA-DQ is expressed most prominently on DC [9].

The formation of clusters between DC and lymphocytes is Ca^{2+}-sensitive [55]. It is known that mobilization of Ca^{2+} and a rise in intracellular Ca^{2+} concentration ($[Ca^{2+}]i$) are pivotal to cell activation. Ultraviolet exposure increases $[Ca^{2+}]i$ within 2–3 h of irradiation. Dependent upon the UV (B or C) dose, $[Ca^{2+}]i$ can rise to 200–250% above baseline levels (130 nm). These UV-treated cells are then unable to be triggered by appropriate stimuli into additional Ca^{2+} mobilization. Gamma irradiation does not trigger a $[Ca^{2+}]i$ rise, and Ca^{2+} mobilization following subsequent addition of lectin is preserved. The functional impact of this UV-induced rise in $[Ca^{2+}]i$ on UV-mediated immune modulation is currently unclear [15, 55]. However, several investigators have shown that UV energy induces programmed cell death (apoptosis) in cell lines and fresh lymphocytes [5, 15, 35, 36]. Since low-dose UV irradiation induces gene transcription, e.g. metallothioneins [22] and c-myc [4], and causes shifts in Ca^{2+}, a cation necessary for DNA endonuclease activation, it was conceivable that UV irradiation led to cell elimination by way of apoptosis. Subsequent studies showed however that UV-induced DNA fragmentation was not consistently prevented by calcium chelators (EGDA) and, therefore, may not account completely for the observed effects. It is likely that a non-specific membrane effect also contributes to Ca^{2+} shifts and that eventual cell inactivation is dependent upon several mechanisms [5].

In vitro studies involving human bone marrow cells revealed a differential sensitivity to UVB light: UVB doses which completely inactivated mitogen responsive lymphocytes preserved the ability of hemopoietic precursors to form colonies *in vitro* [17]; this pattern was not observed with UVC (Fig. 21.1). Similar observations were made with murine spleen colony forming units [13], suggesting that it may be possible to use UV energy for marrow manipulation, as discussed below.

Models of transfusion-induced sensitization

Recipients transfused with unmodified blood components generally recognize donor alloantigens and destroy the cells on which these are expressed. The recognition process results in allosensitization of the recipient. With repeat exposure, a memory response may result in transfusion refractoriness or, in the case of transplantation, a graft rejection. This has been shown convincingly in preclinical transfusion studies, and clinically in patients with severe aplastic anemia treated with allogeneic marrow transplantation [2, 15, 31, 45, 52, 54, 58].

Allosensitization and marrow graft rejection

Normal dogs given 920 cGy of total body irradiation (TBI) and infused with marrow from a histocompatible (DLA-identical) sibling uniformly achieve engraftment. If, on the other hand, dogs are transfused pretransplant with whole blood from the intended marrow donor, they uniformly reject their graft [15, 58]. Rejection can be prevented by UV irradiation of the blood for approximately 1350 mJ/cm^2 at 200–300 nm, suggesting that UV light modifies transfused cells such that allorecognition, memory induction, or both, do

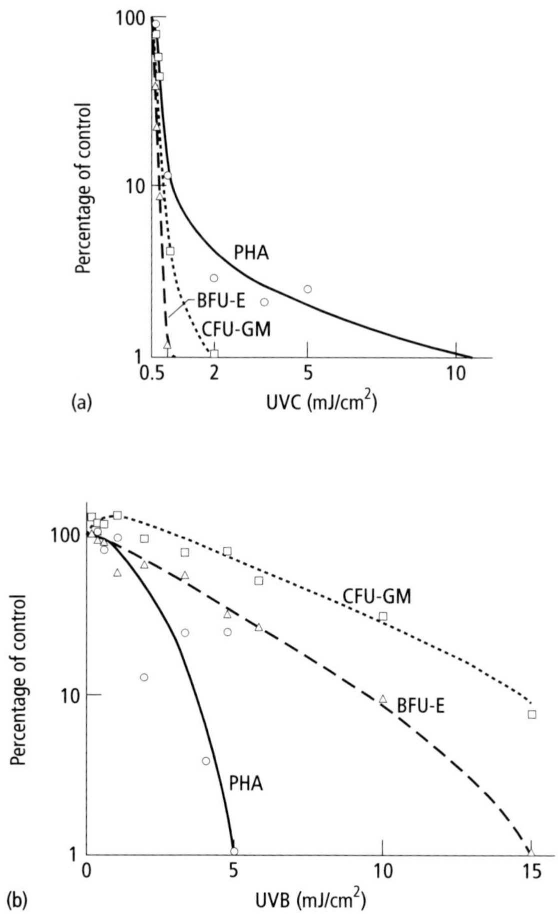

Fig. 21.1 Bone marrow mononuclear cells were UV irradiated and assayed for CFU-GM, BFU-E, and PHA-induced proliferation. Results are expressed relative to control (sham irradiated, 100%). (a) UVC, ultraviolet light (200–280 nm) with maximum energy at 254 nm. The means of ten separate experiments are shown; SD ranged from 3% to 34% for PHA responses and from 2% to 22% for CFU-GM, and BFU-E. (b) UVB, ultraviolet light (280–320 nm) with maximum energy at 302 nm. The means of ten experiments are shown; SD ranged from 15% to 35% for PHA responses and from 6% to 27% for CFU-GM and BFU-E. Reproduced with permission from [17] Deeg *et al.*, *Blood* 73 (1989), 369–371.

not occur. The addition of small numbers of normal DC ($12.5 \times 10^3/3$ transfusions) from the marrow donor to the UV-irradiated blood restores the ability of transfusion products to sensitize the recipient as determined by subsequent marrow graft rejection [16], an observation further supporting the role of DC in antigen presentation and alloreactivity [2, 3, 15].

Allosensitization and transfusion refractoriness

In another model, dogs transfused weekly with platelets from DLA-incompatible donors become refractory after an average of 3.1 transfusions; after 8 weeks 90% of the dogs are sensitized. If platelets are UV irradiated ($\geqslant 12\,mJ/cm^2$; 200–300 nm), 90% of the dogs remain transfusion-responsive. The beneficial effect appears to be dependent not only upon UV exposure but also upon a leukocyte-poor composition of the transfusion product [53]. Surprisingly, if after 8 weekly UVB-treated transfusions the transfusion of normal platelet concentrates is resumed, about two-thirds of the dogs remain transfusion responsive, not only to normal platelets from the original donor, but also from third party DLA-incompatible donors [54]. These observations suggest that a state of 'tolerance' is induced by the transfusion of UV-irradiated blood products [41].

Other investigators have utilized UVA in conjunction with 8-MOP (PUVA) to treat platelet concentrates both for sterilization and for the prevention of allosensitization [23].

Allosensitization and organ transplantation

Pretransplant transfusion has generally been thought to result in recipient sensitization and accelerated graft rejection. In the 1970s, studies by Opelz and Terasaki [43] indicated, however, that renal transplant recipients experienced improved graft survival following donor-specific blood transfusions. While this is still being debated, no such effect has been observed for other organs.

In the early 1980s, Lau and colleagues described a rat model in which donor-specific transfusions of UV-irradiated blood products not only failed to sensitize but actually resulted in improved pancreatic islet cell graft survival [31]. A similar effect was shown subsequently for other organs

[26]. Thus, as further discussed below, this approach may allow the induction of transplant tolerance.

Models of GvHD

Graft-versus-host disease is triggered by donor T lymphocytes. If UV energy inactivates T lymphocytes, UV irradiation should provide a means of GvHD prevention so long as other cell populations of relevance (platelets, erythrocytes, hemopoietic precursor cells) are left intact.

Transfusion-induced GvHD in dogs

Dogs conditioned with 920 cGy of TBI and transplanted with autologous marrow generally recover uneventfully; if they are transfused with viable DLA-incompatible leukocytes following autologous marrow infusion they develop GvHD, occasionally associated with marrow aplasia (transfusion-induced GvHD) [56]. In contrast, in a recent study, two of three dogs transfused with leukocytes irradiated with 20 mJ/cm^2 and none of three dogs given leukocytes irradiated with 1000 mJ/cm^2 of UV light (200–300 nm) developed GvHD [19]. *In vitro* studies with human leukocytes indicate that UV irradiation significantly reduces the precursor frequency of antirecipient cytotoxic precursor cells and may thereby prevent the development of transfusion-associated GvHD [62].

Transplant studies in rats

Lewis rats conditioned with 1050 cGy of TBI and infused with marrow and blood lymphocytes from histoincompatible ACI donors uniformly develop GvHD, which is generally fatal. If ACI cells to be transplanted are exposed to 70 mJ/cm^2 of UVB *in vitro*, the development of clinical GvHD is prevented. However, even 5 months posttransplant, ACI-derived chimera lymphocytes react strongly in MLC against host cells, suggesting that mechanisms which are not functional *in vitro* also contribute to the *in vivo* maintenance of tolerance [11, 47].

Transplant studies in mice

In a murine parent into F1 model, 5–15 mJ/cm^2 of UVB irradiation of donor spleen cells, transplanted along with normal marrow cells, completely prevents GvHD, while all mice given normal splenocytes develop GvHD. At the same time, recipients of syngeneic marrow exposed to 2.5–10 mJ/cm^2 of UVB show normal hemopoietic reconstitution [12]. B6D2 F1 recipients prepared with 1000 cGy of TBI and infused with B6 parental marrow cells and splenocytes, without the use of UVB irradiation all develop GvHD. UVB irradiation with 10 or 12.5 mJ/cm^2 prevents GvHD (Fig. 21.2), and chimeras show specific tolerance to skin grafts of the second parent strain [12].

Similar results have been obtained in another parent into F model (Balb/c donors and [Balb/c × CBA] recipients). In this strain combination, UV irradiation of marrow and spleen cells with 10 mJ/cm^2 results in graft failure of GvHD in a quarter of the mice. Approximately half of the remaining mice become complete donor chimeras, whereas all others have mixed or host type hemopoiesis [44]. With fully allogeneic transplants (Balb/c into CBA), GvHD developed in all animals.

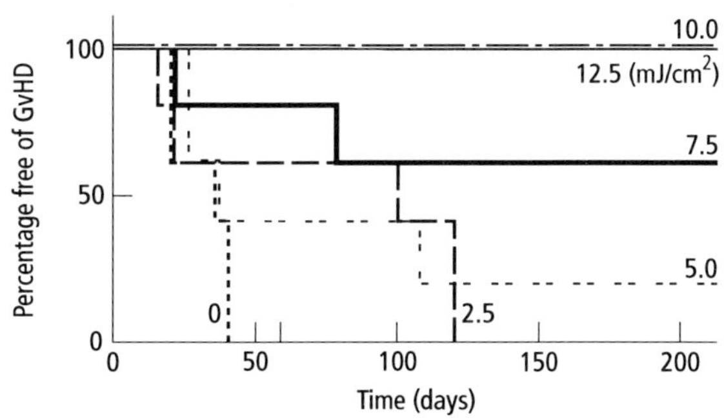

Fig. 21.2 Incidence of clinical GvHD after P → F_1 transplants. B6D2 F_1 recipient mice were given 1000 cGy of TBI and transplanted with 5 × 10^6 marrow and 10 × 10^6 spleen cells from C57BL/6 parental donors. Donor cells were either unmanipulated (0) or exposed to UVB (2.5 to 15 mJ/cm^2). Differences between groups were statistically significant by log rank trend test (p = 0.001). Reproduced with permission from [13] Cohn *et al.*, *Blood* 78 (1991), 3317–3322.

If cells were irradiated with 10 mJ/cm^2 of UVB, about 60% of recipients survived 80 days; half of these showed host-type hemopoiesis [44]. Thus it appears that Balb/c stem cells were damaged by UVB treatment, suggesting that UVB sensitivity may differ from strain to strain.

Other investigators have taken different approaches. Ullrich [60] injected C3H/HeN mice with 8-MOP (0.4 mg i.p.); 2 h later the spleen and marrow cells were harvested and exposed *in vitro* to 5–60 mJ/cm^2 of UVA radiation. These cells were transplanted into fully allogeneic Balb/c recipients irradiated with 900 cGy. Control mice transplanted with unirradiated spleen and marrow cells, or with cells from donor mice injected with 8-methoxypsoralen but not exposed to UVA, or cells only exposed to UVA (but not treated with 8-MOP) developed acute lethal GvHD. Mice transplanted with PUVA-treated cells survived significantly longer, 30% beyond 90–120 days posttransplant. About 85% of peripheral blood lymphocytes from long-term survivors were of donor H-2 type, indicating donor stem cell hemopoiesis. Spleen cells from long-term survivors were able to generate a proliferative response to third-party C57BL/6 cells, suggesting reconstitution of immune function.

Of interest is a comparison of the relative efficiency of UVB and PUVA in suppressing the induction of GvHD across fully allogeneic histocompatibility barriers. Although Pamphilon *et al.* [44] found UVB exposure to be more potent in suppressing the induction of GvHD, hemopoiesis by marrow cells thus treated was compromised. However, while the incidence of lethal GvHD was higher after transplantation with PUVA-treated cells, PUVA treatment appeared to damage stem cells to a lesser degree [60]. It is possible that these differences reflect different UV sensitivities of different strains, but they may also reflect an inherent difference in the biological response of stem cells to UVB and PUVA.

Finally, Gruner and colleagues determined that daily injections of *cis*-urocanic acid for 7 days after transplantation in [C57BL/6 × Balb/c] F_1 mice given spleen cells from Balb/c donors completely prevented GvHD in some and significantly delayed GvHD onset in the majority of mice [25].

Models of organ transplantation

Ultraviolet light has limited depth penetration and at first glance does not appear to hold much promise for the manipulation of solid organs. However, Lau and colleagues showed in a rat model (Lewis → ACI) that pancreatic islet cells could be UVB irradiated effectively and survived permanently, curing recipient animals of their diabetes mellitus [32]. Other investigators have shown that UV irradiation abrogates immunogenicity of other endocrine organ grafts, in particular of thyroid and parathyroid tissue [30]. Similar studies in cynomolgus monkeys and baboons have been less successful [26]. Some encouraging results have been reported with transplantation of UV-irradiated hepatocytes [28].

Some success has also been achieved with transplantation of UV-irradiated skin [12, 24, 25, 39]. Results depend upon donor/recipient combinations, the site from which the skin is taken and whether the skin is UV exposed from one or from both sides. The density of Langerhans cells in the skin and their inactivation by UV light are major determinants of graft outcome.

The cornea is another organ that can easily be UV irradiated while maintaining its viability. One study showed that UVB exposure in excess of 75 mJ/cm^2 yielded permanent graft survival in 33% of C57BL/6 mice given corneas from Balb/c donors [49].

More recent studies suggest that limitations due to non-permeability of organs for UV light can possibly be overcome. Several investigators have shown that PUVA treatment of experimental animals may reduce the immunogenicity of organs such as kidney or heart and result in prolonged graft survival in the recipient (reviewed in [25]). These observations are compatible with a mechanism involving intermediary effector molecules. As discussed already, studies by several investigators [21, 38, 48] suggest that one mechanism by which UV irradiation mediates its effect is by isomerization of urocanic acid from its *trans*- to its *cis*-configuration. In fact in one study, administration of *cis*-urocanic acid directly to the transplant recipient significantly prolonged first set skin graft survival and resulted in permanent acceptance of histoincompatible heterotopic heart grafts in 40% of rats so treated (Fig. 21.3) [25]. *Cis*-urocanic acid had no effect on second set skin grafts in mice [25].

Tolerance induction with UV-irradiated blood products

As discussed, dogs transfused with UVB/C-irradiated platelets do not become refractory, and even if after eight transfusions the transfusion of normal platelet concentrates is resumed, more than half of the dogs remain transfusion-responsive [54]. Some of these dogs fail to become refractory even to non-UV-irradiated platelets from different DLA-incompatible donors, suggesting that UV-irradiated transfusion products may be used to induce tolerance.

Conceivably, the production of a mediator molecule such as *cis*-urocanic acid is responsible

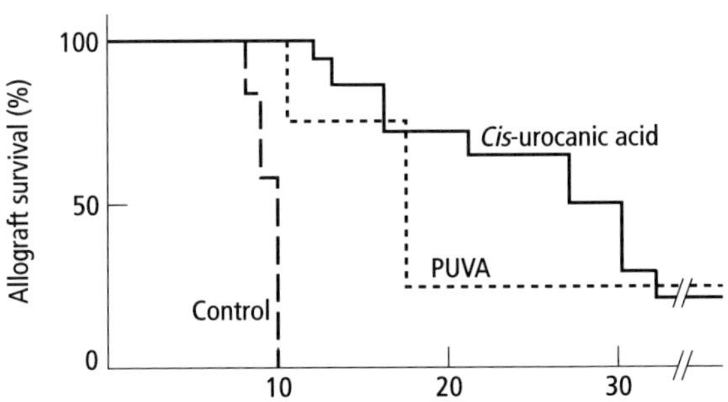

Fig. 21.3 Prolongation of rat heart allograft survival by treatment of the recipients with PUVA or *cis*-urocanic acid. BDIX rats were grafted heterotopically with a Sprague–Dawley heart graft and treated on seven consecutive days after transplantation with PUVA (injection of 1 mg/kg 8-methoxypsoralen and irradiation with 15 J/cm^2 UVA 1 h later at the shaved dorsal skin). Reproduced with permission from [25] Gruner *et al.*, *Sem. Hematol.* 29 (1992), 102–107.

for this effect. The administration of *cis*-urocanic acid has, indeed, been shown to have immunosuppressive properties in that it prevents hypersensitivity reactions to herpes simplex virus [50] and prolongs kidney graft survival [25].

Suppressive factors released by UV-irradiated keratinocytes have also been invoked as a cause of immunosuppression following UV exposure [15, 61]. Supernatants from UV-irradiated keratinocytes in mice mimic the effects of total body UV exposure and suppress the induction of hypersensitivity *in vivo*. This supernatant when injected into mice also suppresses the ability of their spleen cells to respond to alloantigen in MLC, apparently secondary to the appearance of splenic antigen-specific suppressor T cells ($CD3^+$, $CD4^-$, $CD8^+$). Treatment of keratinocytes with cycloheximide or treatment of supernatant with trypsin removed all suppressive activity, indicating that the active material was a protein. Activated keratinocytes release a wide spectrum of immunomodulatory substances, including some with suppressive activity, such as tumor necrosis factor alpha (TNFα), transforming growth factor beta (TGFβ) and IL-1 inhibitor [34]. It is currently unknown whether any of the known keratinocyte-derived cytokines are involved in immunosuppression following UV exposure. However, a role for TNFα in the suppression of contact hypersensitivity following UV irradiation has been described [64]. Whether this cytokine is also involved in the suppression of the immune response to alloantigen after UV exposure is not known.

Platelet transfusion studies in humans

Several technical problems have been noted with UV irradiation of transfusion products [45]. Most plastic materials utilized in transfusion medicine have been designed to prevent transmission of UV light, and only the recent development of plastic containers of different composition has allowed for high efficiency transmission of UV energy [1, 10]. Also, even though the intention is to inactivate leukocytes contained in the transfusion products, the aim is to preserve the function of red blood cells and platelets. Theoretically this must be assessed with each UV-irradiation device.

UV effect on platelets *in vitro*

UV irradiation in the UVC and UVB range at doses up to 1000 mJ/cm^2 does not damage platelets unacceptably [7, 45, 63], although there may be some changes in the ultrastructural appearance of platelets (for example, dilatation of the tubulo-canalicular system), increases in lactate dehydrogenase levels and pH. There may also be a concurrent decline in the expression of GPIb, a protein involved in platelet adhesion, but this does not affect platelet function. UVB irradiation in the range of 10 000 mJ/cm^2 results in a marked deterioration of hypotonic shock response after storage; the aggregation responses of those platelets are, nevertheless, preserved [45]. Alterations in serotonin, PF_4 and thromboxane-B2 after exposure to 8000–16 000 mJ/cm^2 of UV appear to be acceptable since platelets so treated are still functional *in vivo*.

In vivo function of UV-irradiated platelet concentrates

Studies on the recovery and half-life of autologous platelets in four healthy donors given platelets irradiated with 1500 mJ/cm^2 of UVB in a Stericell bag showed a mild reduction in survival (7 days versus 7.75 days in controls). Autologous platelet concentrates stored for up to 5 days and then irradiated with 2500 mJ/cm^2 of UVB in EVA containers (PL269) yielded normal *in vivo* recovery and survival [7]. Similarly, eight normal volunteers given platelets irradiated with 300 mJ/cm^2 of UVB in Stericell containers showed normal platelet survival. In one study UVB irradiation resulted in normal posttransfusion increments and corrected the template bleeding time [45]. Preliminary data from a prospective study indicate that patients transfused with platelet concentrates irradiated with 15 000 mJ/cm^2 of UVB and filtered red blood cells have a low probability of becoming immunized [45]. Additional studies, using EVA-MP or Teflon® containers for UVB irradiation [2, 7, 15] show that platelet concentrates so treated have normal *in vivo* function. It may be relevant that one study noted a lack of development of cytoxic antibodies in the recipients [15].

Treatment of established GvHD in humans

GvHD is a major complication of clinical marrow transplantation. Among patients who develop acute GvHD, only 20–30% achieve meaningful responses with available therapy, usually consisting of glucocorticoids, while 70–80% require additional therapy [37]. It is in this population of patients that UV light has been used, either in the form of photopheresis or PUVA [18, 27]. PUVA rather than UVB was chosen because of extensive clinical experience with this modality in patients with dermatological problems and with cutaneous T-cell lymphoma [59].

PUVA therapy

We have recently used PUVA in 18 patients who developed acute GvHD despite prophylaxis with methotrexate and cyclosporine, and who failed to respond completely to glucocorticoids given either alone or in conjunction with immunotoxin (Xomazyme CD5$^+$) or antithymocyte globulin [18]. Beginning at 4–7 weeks after transplantation these patients were given PUVA with UVA doses of 250–1000 mJ/cm^2 with dose increments as tolerated by the patient's skin in subsequent sessions. One to 32 treatments with PUVA were administered. Among 15 evaluable patients, 10 responded (Fig. 21.4) and five failed to do so. In several patients in whom acute GvHD had resolved completely, chronic GvHD developed subsequently, requiring additional therapy [15]. Kapoor and colleagues [27] recently reported a similar experience. PUVA can be used for the treatment not only of acute but also chronic GvHD.

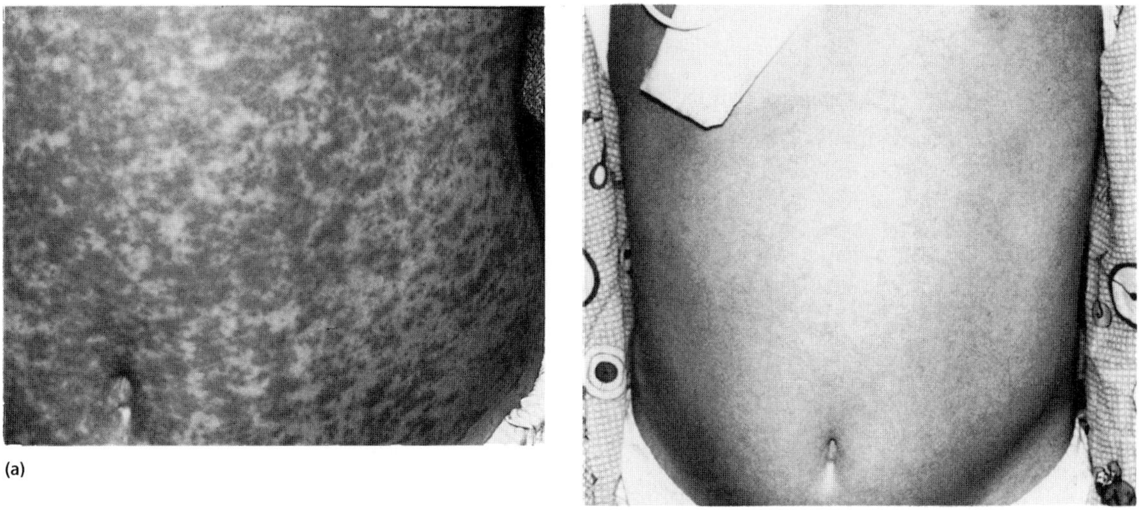

(a)

(b)

Fig. 21.4 Acute GvHD of the skin in a girl given a bone marrow transplant from an unrelated volunteer donor differing for one HLA antigen. Methotrexate and cyclosporine had been administered as GvHD prophylaxis. (a) Intense maculopapular rash 20 days postgrafting. The patient was treated with methylprednisone (daily) and antithymocyte globulin (for six doses given on alternate days). When her disease failed to respond, therapy with 8-methoxypsoralen and UVA (PUVA), three times a week, was initiated. (b) After 6 weeks of PUVA therapy (18 sessions) skin GvHD was basically resolved.

Photopheresis

Other investigators have used different approaches. Utilizing an apheresis device (THERAKOS) allowing for extracorporeal UVA irradiation, these investigators treated patients with chronic GvHD with extracorporeal UVA after *in vivo* sensitization with 8-MOP. In a small number of patients studied these investigators observed improvement of GvHD not only in the skin but also in the liver [8]. Additional studies on photopheresis are ongoing, including one on the use of photopheresis as adjuvant prophylaxis in high-risk transplant recipients (K. Atkinson, personal communication).

Clinical organ transplantation

Only very few studies on the usefulness of phototherapy in clinical organ transplantation have been reported. It is of interest, however, that Oesterwitz and colleagues reported a significant increase in 1 year graft survival and a lower probability of rejection episodes in patients given kidney grafts from PUVA-treated donors [40].

Conclusions

UV energy is currently being explored as a modality modifying immunity and immunogenicity. UVB or UVA in combination with photosensitizers are used in studies aimed at preventing allosensitization, at preventing GvHD, and at inducing tolerance in the transfusion recipient. While experimental studies are encouraging, the clinical usefulness remains to be determined.

Acknowledgments

We thank Bonnie Larson and Gloria Mairs for help with manuscript preparation. This work was supported in part by grants CA15704, CA18221, CA18029, CA31787, and HL36444 from the National Institutes of Health, DHHS.

References

1 Andreu, G., C. Boccaccio, C. Lecrubier, J. Fretault, J. Coursaget, J.P. Leguen, M. Oleggini, J.J. Fournel, M. Samama: Ultraviolet irradiation of platelet concentrates: feasibility in transfusion practice. *Transfusion* 30 (1990), 401.

2 Andreu, G., J.Y. Perrot, F. Pirenne, C. Boccaccio: The effect of UVB light on antigen presenting cells (APC). Implications for transfusion-induced sensitization. *Sem. Hematol.* 29 (1992), 122.

3 Aprile, J., H.J. Deeg: Ultraviolet irradiation of canine dendritic cells prevents mitogen-induced cluster formation and lymphocyte proliferation. *Transplantation* 42 (1986), 653.

4 Bazar, L.S., H.J. Deeg: Effect of ultraviolet B on c-myc, c-fos, IL-2, and IL-2R expression during T cell activation. *Exp. Hematol.* 18 (1990), 559.

5 Bazar, L.S., H.J. Deeg: Ultraviolet B-induced DNA fragmentation (apoptosis) in activated T-lymphocytes and Jurkat cells is augmented by inhibition of RNA or protein synthesis. *Exp. Hematol.* 20 (1992), 80.

6 Billingham, R.E.: The biology of graft-versus-host reactions. In: *The Harvey Lectures*, vol. 62, (1966), pp. 21–78. Academic Press, New York.

7 Boccaccio, C., T. Garcia, J. Klaren: Ultraviolet irradiation of platelet concentrates: *in vitro* and *in vivo* preclinical evaluation. *Transfusion Science* 2 (1990), 141.

8 Bolwell, B., A. Fishleder, A. Lichtin, A. Koo, C. Camisa, R. Green, B. Barna: Photopheresis in the treatment of chronic graft vs. host disease (Abstract). *Blood* 76 (suppl 1) (1990), 529.

9 Brooks, C.F., M. Moore: Differential MHC class II expression on human peripheral monocytes and dendritic cells. *Immunology* 63 (1988), 303.

10 Capon, S.M., R.A. Sacher, H.J. Deeg: Effective ultraviolet irradiation of platelet concentrates in teflon bags. *Transfusion* 30 (1990), 678.

11 Chabot, J.A., P. Pepino, T. Wasfie, M.D. Stegall, C. Marboe, M.A. Hardy: UVB pretreatment of rat bone marrow allografts. *Transplantation* 49 (1990), 886.

12 Claas, F.H.J., M. Rothert, I. Havinga, A.A. Schothorst, B.J. Vermeer, J.J. Van Rood: Influence of ultraviolet radiation treatment on the survival of heterotopic skin grafts in the mouse. *J. Invest. Dermatol.* 84 (1985), 31.

13 Cohn, M.L., R.A. Cahill, H.J. Deeg: Hemopoietic reconstitution and prevention of graft-versus-host disease in UVB irradiated haploidentical murine spleen and marrow cells. *Blood* 78 (1991), 3317–3322.

14 Deeg, H.J.: Ultraviolet irradiation in transplantation biology. Manipulation of immunity and immunogenicity. *Transplantation* 45 (1988), 845.

15 Deeg, H.J.: Modification of allointeractions after blood transfusion and marrow transplantation by ultraviolet light. *Sem. Hematol.* 29 (1992), 95.

16 Deeg, H.J., J. Aprile, R. Storb, T.C. Graham, R. Hackman, F.R. Appelbaum, F. Schuening: Functional dendritic cells are required for transfusion-induced sensitization in canine marrow graft recipients. Concise Report. *Blood* 71 (1988), 1138.

17 Deeg, H.J., L. Bazar, M. Sigaroudinia, M. Cottler-Fox: Ultraviolet B light inactivates bone marrow T lymphocytes but spares hematopoietic precursor cells. *Blood* 73 (1989), 369–371.

18 Deeg, H.J., K. Erickson, R. Storb, K.M. Sullivan: Photoinactivation of lymphohemopoietic cells: studies in transfusion medicine and bone marrow transplantation. *Blood Cells* 18 (1992), 151.

19 Deeg, H.J., T.C. Graham, L. Gerhard-Miller, F.R. Appelbaum, F. Schuening, R. Storb: Prevention of transfusion-induced graft-versus-host disease in dogs by ultraviolet irradiation. *Blood* 74 (1989), 2592.

20 Deeg, H.J., M. Sigaroudinia: Ultraviolet B-induced loss of HLA class II antigen expression on lymphocytes is dose, time and locus dependent. *Exp. Hematol.* 18 (1990), 916.

21 De Fabo, E.C., F.P. Noonan: Mechanism of immune suppression by ultraviolet irradiation *in vivo* I. Evidence for the existence of a unique photoreceptor in skin and its role in photoimmunology. *J. Exp. Med.* 157 (1983), 84.

22 Fornace, A.J. Jr, H. Schalch, I. Alamo Jr: Coordinate induction of metallothioneins I and II in rodent cells by UV irradiation. *Mol. Cell Biol.* 8 (1988), 4716.

23 Grana, N., K.J. Kao: Use of 8-methoxypsoralen and ultraviolet-A pretreated platelet concentrates to prevent alloimmunization against class I major histocompatibility antigens. *Blood* 77 (1991), 2530.

24 Granstein, R.D., L. Smith, J.A. Parrish: Prolongation of murine skin allograft survival by the systemic effects of 8-methoxypsoralen and long-wave ultraviolet radiation (PUVA). *J. Invest. Dermatol.* 88 (1987), 424.

25 Gruner, S., H. Oesterwitz, H. Stoppe, W. Henke, R. Eckert, N. Sönnichsen: Cis-urocanic acid as a mediator of ultraviolet-light-induced immunosuppression. *Sem. Hematol.* 29 (1992), 102–107.

26 Hardy, M.A., J. Chabot, G. Tannenbaum, H.T. Lau: Immunomodulation by ultraviolet irradiation. In: *Transplantation: Approaches to Graft Rejection. Progress in Clinical and Biological Research*, vol. 224, (1986), pp. 119–138. Ed.: H.T. Meryman, Alan R. Liss, Inc., New York.

27 Kapoor, N., A.E. Pelligrini, E.A. Copelan, I. Cunningham, B.R. Avalos, J.L. Klein, P.J. Tutschka: Psoralen plus ultraviolet A (PUVA) in the treatment of chronic graft versus host disease: preliminary experience in standard treatment resistant patients. *Sem. Hematol.* 29 (1992), 108.

28 Kawai, Y., J.B. Price, M.A. Hardy: Reversal of live failure in rats by ultraviolet-irradiated hepatocyte

transplantation. *Transplant Proc.* 19 (1987), 989.

29 Kripke, M.L.: Immunological unresponsiveness induced by ultraviolet radiation. *Immunol. Rev.* 80 (1984), 87.

30 Lafferty, K.J., S.K. Babcock, R.G. Gill: Prevention of rejection by treatment of the graft: an overview. In: *Transplantation: Approaches to Graft Rejection. Progress in Clinical and Biological Research*, vol. 224, (1986), pp. 87–117. Ed.: H.T. Meryman, Alan R. Liss, Inc., New York.

31 Lau, H., K. Reemtsma, M.A. Hardy: Pancreatic islet allograft prolongation by donor-specific blood transfusions treated with ultraviolet irradiation. *Science* 221 (1983), 754.

32 Lau, H., K. Reemtsma, M.A. Hardy: The use of direct ultraviolet irradiation and cyclosporine in facilitating indefinite pancreatic islet allograft acceptance. *Transplantation* 38 (1984), 566.

33 Lindahl-Kiessling, K., J. Saefwenberg: Inability of UV-irradiated lymphocytes to stimulate allogeneic cells in mixed lymphocyte culture. *International Archives of Allergy and Applied Immunology* 41 (1971), 670.

34 Luger, T.A., T.S. Schwartz: Epidermal cell derived cytokines. In: *Skin Immune System*, (1989), pp. 257–292. Ed.: J.D. Bos, CRC Press, Boca Raton, FL.

35 Marks, D.I., R.M. Fox: Mechanisms of photochemotherapy-induced apoptotic cell death in lymphoid cells. *Biochem. Cell Biol.* 69 (1991), 754.

36 Martin S.J., T.G. Cotter: Ultraviolet B irradiation of human leukaemia HL-60 cells *in vitro* induces apoptosis. *Int. J. Radiat. Biol.* 59 (1991), 1001.

37 Martin, P.J., G. Schoch, L. Fisher, V. Byers, C. Anasetti, F.R. Appelbaum, P.G. Beatty, K. Doney, G.B. McDonald, J.E. Sanders, K.M. Sullivan, R. Storb, E.D. Thomas, R.P. Witherspoon, P. Lomen, J. Hannigan, J.A. Hansen: A retrospective analysis of therapy for acute graft-versus-host disease: initial treatment. *Blood* 76 (1990), 1464.

38 Morrison, H., D. Avnir, T. Zarrella: Analysis of Z and E isomers of urocanic acid by high performance liquid chromatography. *J. Chromatogr.* 183 (1980), 83.

39 Oesterwitz, H., J. Kaden, D. Scholz, M. Mebel: Donor pretreatment with 8-methoxypsoralen (8-MOP) followed by ultraviolet irradiation (UVA) of the graft (PUVA therapy), a new approach to enhance rat renal allograft survival. *Transplant. Proc.* 18 (1986), 1297.

40 Oesterwitz, H., D. Scholz, J. Kaden, G. May, M. Mebel, R. Schirrow, W. Blank: Photochemical donor pretreatment in clinical kidney transplantation – preliminary report. *Urol. Res.* 15 (1987), 211.

41 Oluwole, S.F., J. Chabot, P. Pepino, K. Reemtsma, M.A. Hardy: Mechanisms of immunologic unresponsiveness induced by ultraviolet-irradiated donor-specific blood transfusions and peritransplant cyclosporine. *Transplantation* 46 (1988), 352.

42 Oluwole, S.F., K. Engelstad, C. De Rosa, T.S. Wang, R.A. Fawwaz, K. Reemtsma, M.A. Hardy: Migration patterns of dendritic cells in the rat: comparison of the effects of gamma and UV-B irradiation on the migration of dendritic cells and lymphocytes. *Cell. Immunol.* 133 (1991), 390.

43 Opelz, G., P.I. Terasaki: Dominant effect of transfusions on kidney graft survival. *Transplantation* 29 (1980), 153.

44 Pamphilon, D.H., A.A. Alnaqdy, V. Godwin, A.W. Preece, T.B. Wallington: Studies of allogeneic bone marrow and spleen cell transplantation in a murine model using ultraviolet-B light. *Blood* 77 (1991), 2072.

45 Pamphilon, D.H., E.L. Blundell: Ultraviolet-B irradiation of platelet transfusions: a strategy to reduce recipient allosensitization. *Sem. Hematol.* 29 (1992), 113.

46 Peak, M.J., J.G. Peak: Solar-ultraviolet induced damage to DNA. *Photodermatology* (1989), 1.

47 Pepino, P., M.A. Hardy, J.A. Chabot, C. Berger, C. Marboe, T. Wasfie: UVB irradiated allogeneic bone marrow transplantation in rats: prevention of graft versus host disease without immunosuppression. *Transplant. Proc.* 21 (1989), 2995.

48 Räsänen, L., C.T. Jansën, H. Hyöty, T. Reunala, H. Morrison: Cis-urocanic acid stereospecifically modulates human monocyte IL-1 production and surface HLA-DR antigen expression, T-cell IL-2 production and CD4/CD8 ratio. *Photodermatology* 6 (1989), 287.

49 Ray-Keil, L., J.W. Chandler: Reduction in the incidence of rejection of heterotopic murine corneal transplants by pretreatment with ultraviolet radiation. *Transplantation* 42 (1986), 403.

50 Ross, J.A., S.E.M. Howie, M. Norval, J. Maingay: Systemic administration of urocanic acid generates suppression of the delayed type hypersensitivity response to herpes simplex virus in a murine model of infection. *Photodermatology* 5 (1988), 9.

51 Schwartz, T., A. Urbanska, F. Gschnait, T.A. Luger: UV-irradiated epidermal cells produce a specific inhibitor of interleukin 1 activity. *J. Immunol.* 138 (1987), 1457.

52 Slichter, S.J.: Transfusion support in bone marrow transplantation. In: *Principles of Transfusion Medicine*, (1991), pp. 511–519. Eds.: E.C. Rossi, T.L. Simon, G.S. Moss, Williams & Wilkins, Baltimore, MD.

53 Slichter, S.J., M.A. Bean: Tolerance induction and prevention of platelet refractoriness in a dog transfusion model: effects of UV-B irradiation with or without WBC depletion (Abstract). *Blood* 78 (suppl 1) (1991), 389.

54 Slichter, S.J., H.J. Deeg, M.S. Kennedy: Prevention of platelet alloimmunization in dogs with systemic cyclosporine and by UV-irradiation or cyclosporine-loading of donor platelets. *Blood* 69 (1987), 414.

55 Spielberg, H., C.H. June, O.C. Blair, C. Nystrom-Rosander, N. Cereb, H.J. Deeg: Ultraviolet-irradiation of lymphocytes triggers increase in intracellular Ca^{2+} and prevents lectin-stimulated Ca^{2+} mobilization: evidence for UV and nifedipine sensitive Ca^{2+} channels. *Exp. Hematol.* 19 (1991), 742.

56 Spitzer, T.R.: Transfusion-induced graft-vs.-host disease. In: *Graft-vs.-Host Disease: Immunology, Pathophysiology, and Treatment*, (1990), pp. 539–555. Eds.: S.J. Burakoff, H.J. Deeg, J. Ferrara, K. Atkinson, Marcell Dekker, Inc., New York.

57 Steinman, R.M., M.C. Nussenzweig: Dendritic cells: features and functions. *Immunol. Rev.* 53 (1980), 127.

58 Storb, R., P.L. Weiden, H.J. Deeg, T.C. Graham, K. Atkinson, S.J. Slichter, E.D. Thomas: Rejection of marrow from DLA-identical canine littermates given transfusions before grafting: antigens involved are expressed on leukocytes and skin epithelial cells but not on platelets and red blood cells. *Blood* 54 (1979), 477.

59 Taylor, A., F.P. Gasparro: Extracorporeal photochemotherapy for cutaneous T-cell lymphoma and other diseases. *Sem. Hematol.* 29 (1992), 132.

60 Ullrich, S.E.: Photoactivation of T-cell function with psoralen and UVA radiation suppresses the induction of experimental murine graft-versus-host disease across major histocompatibility barriers. *Investigative Dermatol.* 96 (1991), 303.

61 Ullrich, S.E., B.W. McIntyre, J.M. Rivas: Suppression of the immune response to alloantigen by factors released from ultraviolet-irradiated keratinocytes. *J. Immunol.* 145 (1990), 489.

62 Van Prooijen, H.C., M.I. Aarts-Reimens, M.A. Grijzenhout, H. Van Weelden: Ultraviolet irradiation modulates MHC-alloreactive cytotoxic T-cell precursors involved in the onset of graft-versus-host disease. *Br. J. Haematol.* 8 (1992), 73.

63 Van Prooijen, H.C., K.M. Van Marwijk, H. Van Weelden, M.I. Aarts-Riemens, L. Borghuis, J.W. Akkerman: Evaluation of a new UVB source for irradiation of platelet concentrates. *Br. J. Haematol.* 75 (1990), 573.

64 Yoshikawa, T., J.W. Streilein: Tumor necrosis factor-alpha and ultraviolet light have similar effects on contact hypersensitivity in mice. *Reg. Immunol.* 3 (1990), 139.

22: Photoprotection from UVR-induced immunosuppression

Antony R. Young & Susan L. Walker

Sunscreens

Extensive research and development by the sunscreen industry in the past two decades has led to the development of highly effective sunscreens [41]. Their efficacy, assessed by the ability to inhibit human erythema 24 h after skin exposure to solar simulated radiation (SSR) is expressed as the sun protection factor (SPF). In practice, the minimal erythemal dose (MED) is determined with and without sunscreen application on the same person. The SPF = MED with sunscreen/ MED without sunscreen. This index of protection is specific for a given application density, the emission spectrum of the test source, and the action spectrum of the endpoint in question.

Action spectrum studies for human erythema show that UVB (280–315 nm) is three to four orders of magnitude more effective than UVA (315–400 nm) radiation [43]. Thus, despite the high (>95%) UVA content of solar UVR, it is solar UVB (295–315 nm) that is primarily the cause of sun-induced erythema. This means that sunscreens must contain effective UVB blocking agents. The vast majority of sunscreens contain UVB-absorbing molecules such as 2-ethylhexyl-4′-methoxycinnamate (2-EHMC), paraaminobenzoic acid (PABA) derivatives such as octyl dimethyl PABA (o-PABA) or camphor derivatives such as 1,7,7-trimethyl-3-(4-methylbenzylidine) bicyco [2.2.1]-2-hepatanone (3-(4-methylbenzylidine)-camphor). There has been an increasing trend, especially in the formulation of high SPF products, to include UVA-absorbing molecules such as 4-tert.butyl-4′-methoxydibenzoylmethane (4-TBMM) and 2-hydroxy-4-methoxybenzophenone (benzophenone-3) although the latter is a poor broadband UVA absorber. Micronized pigments such as titanium dioxide, which work by scattering rather than absorbing radiation, are also added to many formulations.

The SPF is an integrated index of protection against human erythema from a broadband SSR source. Protection from given wavelengths (or wavebands), which are dependent on the absorption spectrum of the overall formulation, can be determined by *in vitro* techniques and the distribution of wavelength protection, and can vary very considerably for a given SPF [9]. The consequence of this and the above considerations, as will be discussed later, is that a labeled SPF is an invalid index of protection for human erythema induced by non-SSR sources, and for endpoints, irrespective of inducing spectra, that have action spectra different from that of human erythema.

Use of effective high SPF sunscreens enables substantial changes of sun-exposure behavior especially in sun-sensitive skin types I and II; for example prolonged solar exposure periods without a sunburn. Thus, a person who might obtain an MED after 20 min exposure without a sunscreen might be able to stay in the sun for several hours without a sunburn if using a high SPF product. In addition, the use of a sunscreen, unless it is a neutral density filter (i.e., similar protection factors at each wavelength) changes the spectral distribution perceived by the skin. Thus efficient UVB protection alone allows exposure to higher doses of UVA than would otherwise be possible, the consequences of which are unknown.

Skin cancer

Epidemiologic studies have shown a relationship between skin cancer and solar exposure. The literature has recently been reviewed by an International Agency for Cancer Research (IARC) Working Group [29]. Non-melanoma cancers are related to cumulative exposure whereas melanomas (with the exception of lentigo maligna melanoma) appear to be related to patterns of exposure; for example exposure in childhood or intermittent high dose exposure. Sun-sensitive skin types I and II who burn readily and tan poorly if at all are much more skin cancer prone than skin types III and IV which tan well with less likelihood of burning.

Studies in the hairless albino mouse have shown that SSR results in non-melanoma skin cancers [78]. Action spectroscopy has shown that, as with human erythema [43] and mouse edema (a photobiological model for erythema) [4], UVB is about three to four orders of magnitude more carcinogenic than UVA per physical-unit dose [7]. In the absence of human action spectra data, it is reasonable to assume that chronic exposure to solar UVB is the prime cause of non-melanoma skin cancer in human skin.

It is widely accepted that, at both Northern and Southern latitudes, the incidence of malignant melanoma has increased dramatically over the past two decades [29]. Comparably reliable data for the assessment of incidence trends of non-melanoma cancers are less readily available because of often inadequate registration of tumors that are not generally life-threatening. However, such data that do exist also indicate an increase in the incidence of non-melanoma carcinomas [29]. Increases in skin cancer incidence are widely thought to be due to changes in sun-exposure behavior over the past few decades such as more outdoor activities, more vacations in sunny climates and exposure of more of the body surface. There is, at yet, no evidence to relate ozone layer depletion, which in theory would allow greater quantities of carcinogenic UVB radiation to reach the Earth's surface, to present skin cancer trends [47]. However, this could be an important factor in the future [30].

Little is known about the mechanisms of skin cancer. The exquisite skin cancer sensitivity of excision repair deficient xeroderma pigmentosum (XP) patients clearly indicates a critical role for DNA damage [35]. The elegant studies in the South American opossum *Monodelphis domestica*, in which the photoreactivation repair pathway of thymine dimers can be specifically induced by exposure to UVA radiation, show that DNA damage results in non-melanoma and melanoma skin cancers [38, 39]. More recently, it has been shown that postirradiation topical application of T4N5 endonuclease (which recognizes thymine dimers) reduces the incidence of skin cancer in mice [75]. Human skin cancers may arise from UVR-induced mutations to tumor suppressor genes such as p53 [3].

Mouse studies have shown that UVR-induced systemic immunosuppression plays a critical role in the pathogenesis of skin cancer. This is related to the generation of tumor antigen-specific T-suppressor cells which permit tolerance of normally highly antigenic tumors (Chapter 8). The UVR-induced inhibition of contact hypersensitivity (CHS), also mediated by the generation of antigen-specific T-suppressor cells, is widely used as a model of photocarcinogenesis (Chapter 8). It has been shown in mice that UVR-induced inhibition of contact and delayed-type hypersensitivity, and the generation of suppressor T cells, can be abrogated by the application of T4N5 endonuclease (specific for thymine dimers) which implicates DNA damage as the critical factor in UVR-induced immunosuppression [37, 72]. The role of UVR-induced immunosuppression in human skin cancer is unknown. The marked increase of skin cancer, especially on sun-exposed sites, in renal transplant patients maintained on immunosuppressive therapy suggests that immune surveillance plays an important role in skin tumor prevention [28, 44]. Susceptibility to UVR-induced immunosuppression, assessed by inhibition of CHS, may be genetically determined independent of skin type [67]. This trait has been reported in virtually all skin cancer patients in one study [76]. Thus, the present limited evidence suggests that UVR-induced immunosuppression plays a role in skin cancer.

Non-erythema photoprotection by sunscreens

Skin cancer

There is legitimate concern about current skin cancer trends and the possible future impact of ozone layer depletion. Public health campaigns advise sun avoidance and sunscreen use, usually with an SPF of at least 15. There are however no reliable data on the long-term benefits of human use of sunscreens; indeed it has been suggested, somewhat controversially, that there may be a positive correlation between malignant melanoma and sunscreen use [16]. The scientific justification for sunscreen use as a means of preventing skin cancer, and also photoaging, comes exclusively from mouse studies [14, 22, 73]. These have shown that, as might be intuitively expected, sunscreen application prior to UVR exposure reduces photocarcinogenesis and photoaging. However, the mouse studies published to date have failed to take into account how sunscreen use might alter sun-exposure behavior. In other words, the experimental design has been such that exposure time for untreated and treated animals has been the same. Clearly, this may not be the case with humans. The signficance of this should be considered in terms of the failure of time–dose reciprocity for skin cancer in mice [31]; it has been shown that a given daily UVR dose is more carcinogenic if delivered over a protracted period. Extrapolation to the human situation would suggest than an MED is more carcinogenic if delivered over several hours after sunscreen application compared with a shorter exposure without sunscreen use.

In order to prevent skin cancer a sunscreen should be effective against DNA damage. The data in this field are not extensive but studies have shown that sunscreen application inhibits UVR-induced DNA damage [15, 66, 69, 72]. Although, as discussed, the role of UVR-induced immunosuppression is not known, it would obviously be prudent if sunscreens were immunoprotective.

Experimental immunosuppression

The understanding and evaluation of immunoprotection is in its infancy. This section reviews the literature in chronological order, the results of which are summarized in Table 22.1. The first studies on the role of sunscreens in preventing UVR-induced damage to the immune system were reported in the early 1980s by Gurish *et al.* [21]. PABA, octyl dimethyl PABA, glycerol PABA and benzophenone-3 in single or combined formulations, with SPFs ranging from 5 to 15, failed to prevent UVR-induced susceptibility to implanted tumor tissue in C3Hf/HeN mice after 3 or 4 weeks irradiation (5 days/week) even though the sunscreens were effective in inhibiting histologically assessed epidermal damage during that period. However, PABA did prevent spleen lymphoid cell adoptive transfer of tumour susceptibility which was readily demonstrated in mice irradiated without the sunscreen. The same group [42] also reported that although PABA (5% w/w) protected C3Hf/HeN mice from UVB-induced epidermal Langerhans cell (ATPase +ve) depletion assessed at the end of repeat daily exposures (after an initial transient decrease in LC number), it failed to protect against inhibition of CHS.

Morison [49] reported that 5% PABA partially protected against UVB-induced inhibition of CHS in mice but offered no statistically significant protection against the induction of tumor susceptibility. Histological evaluation of the skin revealed that the sunscreen had offered good but not complete protection. Further studies by Morison *et al.* [50] showed that 5% PABA partially protected against UVB-induced inhibition of contact hypersensitivity and inflammation in guinea-pig skin exposed to sunlight.

Hersey *et al.* [23] exposed human volunteers for twelve consecutive days to about 1 MED of broadband UVB–UVA radiation. Comparisons were made between irradiated groups ($n = 11–12$) treated with a sunscreen of SPF 15 (8% octyl dimethyl PABA, 2% 2-hydroxy-4-methoxybenzophenone, 2% 4-tert.butyl-4′-methoxy-dibenzoylmethane) or its vehicle and a non-irradiated sunscreen-treated control. The sunscreen failed to prevent UVR-induced inhibition of natural killer

Table 22.1 Summary of studies on sunscreen immunoprotection

Reference	UVR source	Sunscreens (SPF)	Endpoint	Protection
Gurish *et al.* [21]	FS40	A PABA (12) B gly-PABA (13) C Benz-3 (5) D gly-PABA + Benz-3 (15) Commercial preparations	Mouse; C3HF/HeN	
			Inflammation	Yes
			Rejection of transplanted tumour fragments	No
			Induction of Ts cells	Yes
Lynch *et al.* [42]	FS40	PABA 5% (NS) Commercial preparation	Mouse; C3Hf/HeN	
			Inflammation	Partial
			CHS	No
			LC (ATPase$^+$) depletion	Partial
Morison [49]	FS40	PABA 5% (NS) Commercial preparation	Mouse; C3H/(MTV$^-$) HeNCR (MTV)	
			Inflammation	Partial
			CHS	Partial
			Mouse; BALB/c AnNCR	
			Rejection of transplanted tumour fragments	No
			Induction of T_s cells	No
Morison *et al.* [50]	Sun	PABA 5% (NS) Commercial preparation	Guinea-pig strain 2 (inbred)	
			Inflammation	Partial
			CHS	Partial
Hersey *et al.* [23]	Rellarium	o-PABA 8% + Benz-3 2% + 4-TBMM 2% (15) Commercial preparation	Human *in vivo*	
			NK numbers	Partial
			NK activity	No
			Decrease in T4/T8 cell ratio	No
			Increase in T8 cell numbers	
			DTH	No
			Ig production	No
Fisher *et al.* [13]	A SSR B FS36T12-UVB-VHO	A o-PABA + Benz-3 (6) B o-PABA + Benz-3 (15) Commercial preparations	Mouse; HRA/Skh	
			Inflammation	Yes
			CHS	No
Mommaas *et al.* [48]	Sylvania F20/T21	A 4-TBMM + 2-PBSA (10) B 4-IPDM + MB-cam + 2-PBSA (10) C 4-TBMM + MB-cam + 2-PBSA (6) Commercial preparations	Human *in vitro*	
			MLR	Partial B = C > A

PABA – para-aminobenzoic acid; o-PABA – octyl dimethyl PABA; gly-PABA – glycerol aminobenzoate; es-PABA – *N,N*-dimethyl-4-aminobenzoic acid-2-ethyl ester; 2-EHMC – 2-ethylhexyl-4′-methoxycinnamate; 4-TBMM – 4-tert.butyl-4′-methoxy-dibenzoylmethane; MB-cam – 3-(4-methylbenzlidine)-camphor; 2-PBSA – 2-phenylbenzimidazole-5-sulphonic acid; Benz-3 – benzophenone-3 (2-hydroxy-4-methoxybenzone); 4-IDPM – 4-isopropyldibenzoyl methane; NS – SPF not specified.

Table 22.1 *Continued*

Reference	UVR source	Sunscreens (SPF)	Endpoint	Protection
van Praag *et al.* [64]	Sylvania F75/85 W UV-21 (UVB) or in PUVA therapy with Sylvania FR90 Tα12 (UVA)	A 4-TBMM + MB-cam + 2-PBSA (6) B 4-TBMM + MB-cam + 2-PBSA + es-PABA (15) Commercial preparations	Human *in vivo* (psoriatic and normal) MECLR	 No
Reeve *et al.* [60]	Oliphant FL40SE (UVB)	A o-PABA 6.5% + Benz-3 3.0% (15) B 2-EHMC 7.5% + Benz-3 4.5% (15) Commercial preparations	Mouse; Skh: HR-1 Inflammation Rejection of transplanted tumour cells CHS	 Yes A No, B Yes A No, B Yes
Ho *et al.* [24]	1 × Oliphant FL40SE (UVB) + 6 × Sylvania F40 BL (UVA) filtered with cellulose acetate film	A o-PABA 6.5% + Benz-3 3.0% (15) Commercial preparation B o-PABA 6.5% in solvent (NS) C 2-EHMC 7.5% + Benz-3 4.5% (15) Commercial preparation D 2-EHMC 7.5% in solvent (NS)	Mouse; HRA:Skh-1, HRA:Skh-2 thy-1$^+$ depletion LC depletion CHS	 A, B Yes, C, D No Yes No
Wolf *et al.* [70]	FS40	A o-PABA 8.0% (4–6) B 2-EHMC 7.5% (4–6) C Benz-3 6% (2–4) In oil-in-water emulsion	Mouse; C3H/HeN (MTV$^-$) Inflammation Local CHS Systemic CHS	 Yes but decreases with > dose Yes Yes but decreases with > dose
Wolf *et al.* [72]	FS40	A o-PABA 8.0% (4–6) B 2-EHMC 7.5% (4–6) C Benz-3 6% (2–4) In oil-in-water emulsion	Mouse; C3H/HeN (MTV$^-$) Inflammation DNA damage (pyrimidine dimers) Systemic DTH	 A, B Yes, C partial Yes Partial
Wolf *et al.* [73]	FS 40	A o-PABA 8.0% (4–6) B 2-EHMC 7.5% (4–6) C Benz-3 6% (2–4)	Mouse; C3H/HeN (MTV$^-$) Inflammation Growth of injected melanoma cells	 A, B Yes, D partial No
Walker *et al.* [68]	2 × Wolff Helarium (UVB) 6 × Philips TL10 (UVA)	A 2-EHMC 9% in lotion (NS) B 2-EHMC 9% in ethanol (NS)	Mouse; Skh-1 Inflammation LC (Ia$^+$) depletion MECLR	 A partial; B yes Partial Yes Two applications of sunscreen

(NK) cell activity, delayed-type hypersensitivity (response to recall antigens), immunoglobulin production *in vitro* in pokeweed mitogen-stimulated B and T cells, a decrease in T4/T8 ratio and an increase in the percentage of $T8^+$ cells.

Fisher *et al.* [13] reported that two sunscreen preparations, containing octyl dimethyl PABA and benzophenone-3, with SPFs of 6 and 15 which although effective against inflammation afforded no immunoprotection from inhibition of CHS by SSR and UVB in inbred hairless albino mice (HRA/Skh).

Mommaas *et al.* [48] reported that sunscreens, but not their vehicles, partially inhibited that UVR-induced suppression of allogeneic activity of human peripheral blood mononuclear cells (mixed-lymphocyte reaction) in an *in vitro* assay. Products with comparable SPFs showed differing degrees of protection. In this assay the sunscreens were applied to quartz slides but this approach is not reliable for SPF determination which is better done *in vitro* using a substrate that approximates the microtopography of human skin [9].

van Praag *et al.* [65] found that sunscreens (with SPFs of 6 and 15) failed to prevent UVB-induced suppression of ability of human epidermal cells, from skin irradiated over a 4 week period, to stimulate in allogeneic response. Studies which were done on psoriatic patients undergoing whole-body UVB therapy and in which only a small part of the body was sunscreen-treated could not exclude a systemic effect. However, similar results were obtained in normal subjects who had local irradiation only.

Reeve *et al.* [60] compared two commercial SPF 15 sunscreens. One contained 7.5% 2-ethylhexyl-4′-methoxycinnamate and 4.5% benzophenone-3 and the other 6% octyl dimethyl PABA and 3% benzophenone-3. Vehicle groups were not studied. The cinnamate but not the PABA derivative product was effective in inhibiting UVR-induced systemic suppression of contact hypersensitivity and tumor susceptibility in inbred Skh: HR-1 hairless albino mice. However, both preparations were equally effective at presenting erythema and edema.

Comparisons of similar commercial preparations in hairless inbred HRA:Skh-1 (albino) and HRA:Skh-2 (pigmented) mice were made by Ho *et al.* [24]. Both preparations prevented the loss of Langerhans cells and thy-1^+ dendritic epidermal cells in both strains of mouse. The PABA derivative but not the cinnamate preparation protected the loss of thy-1^+ cells in both strains. There results were verified in experimental preparations of the two UVB sunscreens in a solvent without benzophenone-3. Neither commercial preparation prevented the inhibition of local CHS in the albino strain (pigmented strain not tested).

Wolf and colleagues [70] also compared 2-ethylhexyl-4′-methoxycinnamate (7.5%) with octyl dimethyl PABA (8%) and benzophenone-3 (6%) in three studies. The first [70] used UVR-induced local and systemic inhibition of CHS as well as edema in C3H/HeN (MTV) mice as endpoints. All sunscreens prevented edema and suppression of local CHS after exposure to two minimal edema doses (MOD). Protection against suppression of systemic CHS was observed between 4–8 MOD but showed considerable inter-experiment variation. In general, protection against edema was more consistent and was observed up to 32 MOD. The second study [72] examined the relationships between sunscreen protection of edema, immunosuppression, DNA damage and its repair in mice. The sunscreens prevented edema and thymine dimer formation but only partially protected against systemic suppression to DTH to *Candida albicans*. However, the postirradiation topical application of liposome-encapsulated dimer-specific repair enzymes almost completely prevented suppression of DTH, but had limited effect in preventing edema. The third study [73] showed that UVR enhanced the growth of melanoma cells injected into the irradiated ears of mice. Application of the sunscreens prevented edema and histological damage but offerred no protection against the UVR-enhanced growth of melanoma cells.

Walker *et al.* [68] have studied the protective properties of single and double applications of 2-ethylhexyl-4′-methoxycinnamate at 9% in two vehicles. Unlike other workers, these authors used a long exposure period to UVB plus UVA to simulate a day's sunbathing. The endpoints in the Skh-1 mouse were edema, LC (assessed by Ia staining) and the functional ability of epidermal cells irradiated *in vitro* to stimulate an allogeneic response by mixed epidermal cell lymphocyte response (MECLR). Sunscreen application which completely protected against edema afforded good but not total protection against LC depletion. These endpoints showed a highly significant linear correlation with each other at different UVR doses, and with and without sunscreen application. MECLR could not readily be correlated with LC number. Suppression of immune function could be completely prevented if the sunscreen was reapplied halfway through the irradiation period. These authors noted a lack of correlation between protection from edema and immune function; the former being more readily inhibited.

Why do sunscreens fail to afford immunoprotection?

The above literature review of experimental immunoprotection shows no overall coherent pattern. There is a striking lack of correlation in studies in which protection of immunocompetent epidermal cells has been compared with immune function. However, all the LC studies have shown that these cells can be protected, to varying degrees, but differences have been shown between the protective abilities of 2-ethylhexyl-4′-methoxycinnamate in the comparative protection of LC and thy-1$^+$ dendritic cells [24], although a PABA derivative protected both cell types. One study has shown a clear quantitative relationship between protection of LC and inflammation [68]. Thus, the data suggest overall that sunscreen protection against inflammation is indicative of protection against LC depletion, although protection indices may be different. Studies have shown a lack of correlation between LC numbers *in situ* and immune function [5, 42, 54] which is consistent with the sunscreen studies which generally show a clear, but varying, dissociation between protection of inflammation (and by implication LC depletion where this has not been done) and immune function. Where immunoprotection has been observed it is generally less good than protection from inflammation, either by measurement or by comparison with the product label SPF. This differential is even greater than it might seem as the use of UVB sources (as in many of the experiments) rather than SSR would result in higher SPFs for erythema than the product label [11].

Edema in the mouse may be justified as a photobiologic model for human erythema as their action spectra are similar [4]. It is difficult to escape the conclusion from the mouse studies that there must be fundamental mechanistic differences between UVR-induced immunosuppression and inflammation. At the most fundamental level this could represent differences in chromophores and hence action spectra. If so, the SPF might have little significance as an index of immunoprotection because, it must be stressed, this value is dependent not only on application density but also the action spectrum of the endpoint and the absorption spectrum of the sunscreen as well as the emission spectrum used for its determination. It is worth noting that the action spectrum for human erythema shows differences of three orders of magnitude over the 300–320 nm range [43] so that differences in action spectra in this range could result in substantial differences in protection indices for other endpoints. Comparisons with inflammation are difficult because there are few wavelength dependence studies for endpoints of immune damage [6, 10, 18]. There is evidence for spectral differences between inflammation and immunosuppression [17] and that different immunosuppressive factors have different wavelength dependencies [32].

Human and mouse stratum corneum is rich in urocanic acid (UCA) which undergoes a *trans* to *cis* isomerization after UVR exposure [55]. There is a considerable body of evidence from several workers which implicates *trans*-urocanic acid as a chromophore, and the *cis* isomer as a first-stage mediator for immunosuppression in the mouse [57]. The original suggestion that UCA was a chromophore for immunosuppression resulted from the similarity of the action spectrum for CHS and the absorption spectrum of UCA [6]. Recently it has been reported that the action spectrum for UCA photoisomerization is rather different from the absorption spectrum of UCA and action spectrum of CHS, with higher than expected sensitivity in the UVA region [18]. A recent study has shown that sunscreens are very effective at affording protection against UCA photoisomerization, in humans, after a single dose of UVR [35]. The protection efficiency for UCA was found to exceed the sunscreen SPF by a factor of about ten. However, sunscreen protection against UCA photoisomerization was substantially reduced if five successive daily exposures were used as *cis*-UCA was shown to accumulate in the stratum corneum. Many of the reviewed sunscreen studies used multiple UVR exposures to test the immunoprotection afforded by sunscreens [13, 21, 23, 24, 42, 48, 60, 65]. UCA photoisomerization was also shown to be independent of skin type, indicating that UCA photoisomerization and erythema are independent processes. Therefore a role for UCA could readily account for a dissociation between sunscreen protection of erythema and immune function. Two recent mouse studies have provided very convincing evidence that skin DNA is also a chromophore and that thymine dimers are lesions of considerable significance [37, 72]. Photochemical modification of a chromphore is but the first step in a photobiological response and it is possible that similar biological responses may be mediated by different photochemical mechanisms. Clearly it is important to protect the stratum corneum as UCA photoisomerization plays a role in immunosuppression. On the other hand, an important role for DNA as an immunochromophore would require protection of the viable epidermis and such protection might well correlate with the inflammatory response which may also share DNA as a chromophore [74]. However, mechanistic differences may occur even with a common chromophore and could for example relate to the number and type of specific lesions. In this respect the work of Wolf *et al.* [72] is interesting as it shows differences in inflammation and immunosuppression for a given level of thymine dimers. Studies to date on the role of UVR-induced DNA damage in skin photobiology have focused almost entirely on thymine dimers although it is increasingly appreciated that other lesions, such as the pyrimidine-(6–4)pyrimidone photoproduct and its photoisomer the Dewar pyrimidone, may be important [46]. Different photobiological endpoints may be DNA lesion specific in both quality and quantity, and show dependence on differences in relative DNA repair, which in human skin *in vivo* saturates at low UVR doses [27]. Saturation is also seen in mouse skin *in vivo* [34]. Mouse and *in vitro* studies indicate that 6–4 photoproducts are preferentially repaired in comparison with thymine dimers [45]. Modification of lesion-specific repair could influence biological outcome. The addition of PABA alone, or PABA after UVC exposure, to mouse embryo fibroblasts *in vitro*, can induce DNA repair synthesis [40], the significance of which is not known.

Molecular damage may be sustained by both direct absorption of UVR by the chromophore and by energy transfer reactions after UVR absorption by another chromophore (i.e., sensitizer) which may be endogenous or exogenous. Most sunscreens work by absorbing UVR energy but little is known about its fate, although it has been shown that PABA can sensitize the formation of thymine dimers in DNA [64] and lethality of mammalian cells *in vitro* [58], it has been speculated that such a sensitization process could account for the apparent failures of immunoprotection by PABA [13, 60, 70, 72]. Thymine dimer sensitization has also been reported for 2,2′-dihyroxy-4,4′ dimethoxybenzophenone sulfonic acid [2]. Thus a molecule may act as a filter or a sensitizer depending on its concentration and location. Sensitized formation of thymine dimers is not possible with 2-ethylhexyl-4′-methoxycinnamate or the UVA sunscreen 4-tert.butyl-4′-methoxydibenzoylmethane [19], which are widely used in European broad spectrum products, and so cannot explain the failures of the sunscreens to immunoprotect. Energy transfer reactions between sunscreens and UCA do not seem to have been studied. Sensitized *trans* to *cis* isomerization of UA has been reported [54] and is theoretically possible with sunscreens as the triplet energy level of UCA, at 55 kcal/mol [54], is lower than those of 2-ethylhexyl-4′-methoxycinnamate (57 kcal/

mol), 4-tert.butyl-4′-methoxy-dibenzoylmethane (59.5 kcal/mol) [19] and PABA (75 kcal/mol) [58]. Molecular damage to DNA may also be sustained by photoaddition reactions. For example, the photobinding of UA to DNA *in vitro* has been reported [12, 53]. Little is known about sunscreen/DNA photobinding, although it has been reported that PABA has such properties [64] as well as being able to form photoproducts with thymine and thymidine [63].

To date only one study has compared sunscreen protection of inflammation and immune function at different UVR doses [70], the results of which suggested differences in dose–response curves. The studies of Walker *et al.* [68] in mice show that LC depletion is much more responsive to increasing UVR dose than the MECLR. Difference of dose–response would result in different protection factors for different endpoints even though they shared a common chromophore.

It is well known that the SPF of a given concentration of sunscreen molecules is very vehicle dependent [33]. Walker *et al.* have shown vehicle differences in protection from mouse edema and immunologic endpoints [68]. It may be necessary to study pharmacological aspects. Hydration of the stratum corneum by sunscreen vehicles could enhance the transfer of *cis* UCA to the viable epidermis. Sunscreen molecules could modify the behavior of each other or immunoactive mediators or interfere with DNA repair. For example, sunscreens may modify the transepidermal penetration of 5-methoxypsoralen [1]. Cinnamate sunscreens may also quench triplet formation from other chromophores [51, 52]. Clearly, apart from simple optical screening, sunscreens have the theoretical ability to undergo many photochemical reactions with molecules of biological significance. In addition, sunscreens and their vehicles may modify photobiological responses. Explanations for lack of immunoprotection and the diverse results from different research groups may not be possible until we gain a greater understanding of the interactions of sunscreens and the skin.

Skin infections and photodermatoses

The use of sunscreens for the management of skin infections and photodermatoses that are suspected of having a photoimmunological component is an obvious strategy. There is experimental and clinical evidence to relate activation of herpes simplex virus (HSV) to solar exposure [61, 62]. UVB-induced labial human HSV infection was very successfully prevented with a sunscreen in a double-blind, placebo (vehicle) -controlled crossover study [61]. The sunscreen, with an SPF of 15, contained 2.8% glyceryl PABA, 3.3% octyl dimethyl PABA and 5.6% benzophenone-3.

Immunological factors are suspected in photodermatoses such as polymorphic light eruption [56] and chronic actinic dermatitis (actinic reticuloid) (Chapter 14) in which an inflammatory response is provoked by exposure to UVR, conceivably by the formation of a photoantigen. The action spectra for these conditions show considerable interperson variation but frequently extend well into the UVA region. Successful use of broad spectrum sunscreens to prevent these conditions has been reported [8, 20, 26].

Photochemoprotection

Psoralen photochemotherapy (PUVA) is used in the treatment of some photodermatoses which may have an immunological basis [25]. The mechanisms of PUVA are not well understood, but assuming that an inflammatory response may be initiated by the formation of a photoantigen, PUVA may be effective (at least in part) by acting as a photochemoprotective agent which induces an 'endogenous sunscreen' and thereby inhibits the formation of the photoantigen. Recent human skin studies *in vivo* have shown that tanning and stratum corneum thickening induced by a 5-methoxypsoralen (5-MOP)-containing UVB sunscreen affords superior protection from subsequent SSR-induced DNA damage (thymine dimers) than comparable changes induced by SSR alone [59]. Such photochemoprotection was particularly striking in sun-sensitive skin types I and II [77]. There have to date been no studies of the possible immunoprotective role of a psoralen-induced tan.

Conclusions

A review of the literature shows that sunscreens effective in preventing erythema are either ineffective or less effective at preventing UVR-induced changes of immune function despite generally good levels of protection against Langerhans cell damage. The reasons for these discrepancies are not known and cannot be linked to a particular type of sunscreen molecule. It is perhaps worth stressing that SPF is determined from a single exposure but most of the sunscreen studies on immune function are based on repeat exposure studies which correspond better with the 'real-life' situation.

Despite the current lack of knowledge about the role of UVR-induced immunosuppression in human skin cancer, there is a clear public health concern about the use of effective erythema-preventing sunscreens which fail to afford comparable immunoprotection, especially when used to prolong solar exposure time. A better understanding of sunscreen photoimmunology should enable the development and testing of products that are immunoprotective.

References

1 Averbeck, D., S. Averbeck, L. Dubertret, A.R. Young, P. Morliere: Genotoxicity of bergapten and bergamot oil in *Saccharomyces cerevisiae*. *J. Photochem. Photobiol. B: Biol.* 7 (1990), 209–229.

2 Bolton, K., B.S. Martincigh, L.F. Salter: The potential carcinogenic effect of Uvinul DS49 – a common UV absorber used in cosmetics. *J. Photochem. Photobiol. A: Chem.* 63 (1992), 241–248.

3 Brash, D.E., J.A. Rudolph, J.A. Simon, A. Lin, G.J. McKenna, H.P. Baden, A.J. Halperin, J. Ponten: A role for sunlight in skin cancer: UV-induced p53 mutations in squamous cell carcinoma. *Proc. Natl. Acad. Sci. USA* 88 (1991), 10124–10128.

4 Cole, C.A., R.E. Davies, P.D. Forbes, L.C. D'Aloisio: Comparison of action spectra for acute cutaneous responses to ultraviolet radiation: man and the albino hairless mouse. *Photochem. Photobiol.* 37 (1983), 623–631.

5 Cooper, K.D., P. Fox, G. Neises, S.I. Katz: Effects of ultraviolet radiation on human epidermal cell alloantigen presentation: initial depression of Langerhans cell dependent function is followed by the appearance of $T6^-Dr^+$ cells that enhance epidermal alloantigen presentation. *J. Immunol.* 134 (1985), 129–137.

6 DeFabo, E.C., F.P. Noonan: Mechanism of immune suppression by ultraviolet irradiation *in vivo*. I. Evidence for the existence of a unique photoreceptor in skin and its role in photoimmunology. *J. Exp. Med.* 157 (1983), 84–98.

7 de Gruijl, F.R., J.C. van der Leun: Action spectra for photocarcinogenesis. In: *Biological Responses to Ultraviolet A Radiation*, (1992), pp. 91–97. Ed.: F. Urbach, Valdenmar Publishing Company, Overland Park, KS.

8 Diffey, B.L., P.M. Farr: Sunscreen protection against UVB, UVA and blue light: an *in vivo* and *in vitro* comparison. *Br J. Dermatol.* 124 (1991), 258–263.

9 Diffey, B.L., J. Robson: A new substrate to measure sunscreen protection factors throughout the ultraviolet spectrum. *J. Soc. Cosmet. Chem.* 40 (1989), 127–133.

10 Elmets, C.A., M. J. LeVine, D.R. Bickers: Action spectrum studies for induction of immunologic unresponsiveness to dinitrofluorobenzene following *in vivo* low dose ultraviolet radiation. *Photochem. Photobiol.* 42 (1985), 391–397.

11 Farr, P.M., B.L. Diffey: How reliable are sunscreen protection factors? *Br. J. Dermatol.* 112 (1985), 113–118.

12 Farrow, S.J., T. Mohammad, W. Baird, H. Morrison: Photochemical co-valent binding of urocanic acid to polynucleic acids. *Chem. Biol. Interact.* 75 (1990), 105–118.

13 Fisher, M.S., J.M. Menter, I. Willis: Ultraviolet radiation-induced suppression of contact hypersensitivity in relation to padimate O and oxybenzone. *J. Invest. Dermatol.* 92 (1989), 337–341.

14 Forbes, P.D., R.E. Davies, C.P. Sambuco, F. Urbach: Inhibition of ultraviolet radiation-induced skin tumors in hairless mice by topical application of 2-ethylhexyl p-methoxycinnamate. *J. Toxicol. Cut. Ocular Toxicol.* 8 (1989), 209–226.

15 Freeman, S.E., R.D. Ley, K.D. Ley: Sunscreen protection against UV-induced pyrimidine dimers in DNA of human skin *in situ*. *Photodermatology* 5 (1988), 243–247.

16 Garland, C.F., F.C. Garland, E.D. Gorham: Could sunscreens increase melanoma risk? *Amer. J. Pub. Health.* 82 (1992), 614–615.

17 Gibbs, N.K., M. Norval, N.J. Traynor, J.C. Crosby, G. Lowe, B.E. Johnson: Comparative potency of broad-band and narrow-band phototherapy sources to induce edema, sunburn cells and urocanic acid photoisomerisation in hairless mouse skin. *Photochem. Photobiol.* 58 (1993), 643–647.

18 Gibbs, N.K., M. Norval, N.J. Traynor, M. Wolf, B.E. Johnson, J. Crosby: Action spectra for the trans to cis photoisomerization of urocanic acid *in vitro* and in mouse skin. *Photochem. Photobiol.* 57 (1993), 584–590.

19 Gonzenbach, H., T.J. Hill, T.G. Truscott: The triplet energy levels of UVA and UVB sunscreens. *J. Photochem. Photobiol. B: Biol.* 16 (1992), 377–379.

20 Greaves, K., A.J. Cripps, D.J. Cripps: Actinic reticuloid: action spectra and UVA protection factor sunscreens. *Clin. Exp. Dermatol.* 17 (1992), 94–98.

21 Gurish, M.F., L.K. Roberts, G.G. Krueger, R.A. Daynes: The effect of various sunscreen agents on skin damage and the induction of tumour susceptibility in mice subjected to ultraviolet irradiation. *J. Invest. Dermatol.* 76 (1981), 246–251.

22 Harrison, J.A., S.L. Walker, S.R. Plastow, M.D. Batt, J.L.M. Hawk, A.R. Young: Sunscreens with low sun protection factor inhibit ultraviolet B and A photoaging in the skin of the hairless albino mouse. *Photodermatol. Photoimmunol. Photomed.* 8 (1991), 12–20.

23 Hersey, P., M. MacDonald, C. Burns, S. Schibeci, H. Matthews, F.J. Wilkinson: Analysis of the effect of a sunscreen agent on the suppression of natural killer cell activity induced in human subjects by radiation from solarium lamps. *J. Invest. Dermatol.* 88 (1987), 271–276.

24 Ho, K.K.-L., G.M. Halliday, R.St.C. Barnetson: Sunscreens protect epidermal Langerhans cells and Thy-1^+ cells but not local contact sensitisation from the effects of ultraviolet light. *J. Invest. Dermatol.* 98 (1992), 720–724.

25 Hönigsmann, H.: Psoralen photochemotherapy – mechanisms, drugs, toxicity. In: *Therapeutic Photomedicine,* (1986), pp. 52–66. Eds.: H. Hönigsmann, G. Stingle, Karger, Basel.

26 Hönigsmann, H.: Wirkung einer therapeutischen Lichtschutzcreme auf die Phototestreaktion bei polymorpher Lichtdermatose. *Z. Hautkrank.* 64 (1989), 1065–1068.

27 Hönigsmann, H., W. Brenner, A. Tanew, B. Ortel: UV-induced unscheduled DNA synthesis in human skin: dose response, correlation with erythema, time course and split dose exposure *in vivo*. *J. Photochem. Photobiol. B: Biol.* 1 (1987), 33–43.

28 Hoxtell, E.O., J.S. Mandel, S.S. Murray, L.M. Schuman, R.W. Goltz: Incidence of skin cancer after renal transplantation. *Arch. Dermatol.* 113 (1977), 436–438.

29 Studies of cancer in humans. In: *IARC Monographs of the Evaluation of Carcinogenic Risks to Humans. Solar and Ultraviolet Radiation,* vol. 55, (1992), pp. 73–135. IARC, Lyon.

30 Kelfkens, G., F.R. de Gruijl, J.C. van der Leun: Ozone depletion and increase in annual carcinogenic ultraviolet dose. *Photochem. Photobiol.* 52 (1990), 819–823.

31 Kelfkens, G., H. van Weelden, F.R. de Gruijl, J.C. van der Leun: The influence of dose rate on ultraviolet tumorigenesis. *J. Photochem. Photobiol. B: Biol.* 10 (1991), 41–50.

32 Kim, T-Y., M.L. Kripke, S.E. Ullrich: Immunosuppression by factors released from UV-irradiated epidermal cells: selective effects on the generation of contact and delayed hypersensitivity after exposure to UVA or UVB radiation. *J. Invest. Dermatol.* 94 (1990), 26–32.

33 Klein, K.: Formulating sunscreen products. In: *Sunscreens; Development, Evaluation and Regulatory Aspects,* (1990), pp. 235–266. Eds.: N.J. Lowe, N.A. Shaath, Marcel Dekker Inc., New York.

34 Kodama, K., T. Ishikawa, S. Takayama: Dose response, wavelength dependence, and time course of ultraviolet radiation-induced unscheduled DNA synthesis in mouse skin *in vivo*. *Cancer Res.* 44 (1984), 2150–2154.

35 Kraemer, H.H.: Xeroderma pigmentosum. A prototype disease of environmental–genetic interaction. *Arch. Dermatol.* 116 (1980), 541–542.

36 Krein, P.M., D. Moyal: Sunscreens with broad-spectrum absorption decrease the *trans* to *cis* photoisomerisation of urocanic acid in the human stratum corneum after multiple UV light exposures. *Photochem. Photobiol.* 60 (1994), 280–287.

37 Kripke, M.L., P.A. Cox, D.B. Yarosh: Pyrimidine dimers in DNA initiate systemic immunosuppression in UV-irradiated mice. *Proc. Natl. Acad. Sci. USA* 89 (1992), 7516–7520.

38 Ley, R.D., L.A. Applegate, R.J. Fry, A.B. Sanchez: Photoreactivation of ultraviolet radiation-induced skin and eye tumors of *Monodelphis domestica*. *Cancer Res.* 51 (1991), 6539–6542.

39 Ley, R.D., L.A. Applegate, R.S. Padilla, T.D. Stuart: Ultraviolet radiation – induced malignant melanoma in *Monodelphis domestica*. *Photochem. Photobiol.* 50 (1989), 1–5.

40 Long, S.D., J.B. Little: Sunscreen agents induce DNA repair activity in mouse embryo fibroblasts. *J. Environ. Pathol. Toxicol. Oncol.* 5 (1984), 193–200.

41 Lowe, N.J., N.A. Shaath (Eds.): *Sunscreens; Development, Evaluation and Regulatory Aspects,* (1990). Marcel Dekker Inc., New York.

42 Lynch, D.H., M.F. Gurish, R.A. Daynes: Relationship between epidermal Langerhans cell density ATPase activity and the induction of contact hypersensitivity. *J. Immunol.* 126 (1981), 1892–1897.

43 McKinlay, A.F., B.L. Diffey: A reference action spectrum for ultraviolet induced erythema in human skin. *CIE Journal* 6 (1987), 17–22.

44 McLelland, J., A. Rees, G. Williams, T. Chu: The incidence of immunosuppression-related skin disease in long term transplant patients. *Transplantation* 46

(1988), 871–874.
45 Mitchell, D.L., J.E. Cleaver, J.H. Epstein: Repair of pyrimidine(6-4)pyrimidone photoproducts in mouse skin. *J. Invest. Dermatol.* 95 (1990), 55–59.
46 Mitchell, D.L., D. Karentz: The induction and repair of DNA photodamage in the environment. In: *Environmental UV Photobiology*, (1993), pp. 345–371. Eds.: A.R. Young, L.O. Bjørn, J. Moan, W. Nultsch, Plenum Press, New York.
47 Moan, J. A. Dahlback: Ultraviolet radiation and skin cancer: epidemiological data from Scandinavia. In: *Environmental UV Photobiology*, (1993), pp. 255–290. Eds.: A.R. Young, L.O. Bjørn, J. Moan, W. Nultsch, Plenum Press, New York.
48 Mommaas, A.M., M.C.G. van Praag, J.N.B. Bavinck, C. Out-Luiting, B.J. Vermeer, F.H.J. Class: Analysis of the protective effect of topical sunscreens on the UVB-radiation-induced suppression of the mixed-lymphocyte reaction: *J. Invest. Dermatol.* 95 (1990), 313–316.
49 Morison, W.L.: The effect of a sunscreen containing para-aminobenzoic acid on the systemic immunologic alterations induced in mice by exposure to UVB radiation. *J. Invest. Dermatol.* 83 (1984), 405–408.
50 Morison, W.L., R.A. Pike, M.L. Kripke: Effect of sunlight and its component wavebands on contact hypersensitivity in mice and guinea pigs. *Photodermatology* 2 (1985), 195–204.
51 Morlière, P., O. Avice, T. Sa E. Melo, L. Dubertret, M. Giraud, R. Santus: A study of the photochemical properties of some cinnamate sunscreens by steady state and laser flash photolysis. *J. Photochem. Photobiol.* 36 (1982), 395–399.
52 Morlière, P., G. Huppe, D. Averbeck, A.R. Young, R. Santus, L. Dubertret: *In vitro* photostability and photosensitizing properties of bergamot oil. Effects of a cinnamate sunscreen. *J. Photochem. Photobiol. B: Biol.* 7 (1990), 199–208.
53 Morrison, H.: Photochemistry and Photobiology of urocanic acid. *Photodermatology* 2 (1985), 158–165.
54 Morrison, H., D. Avnir, C. Bernasconi, G. Fagan: Z/E photoisomerization of urocanic acid. *Photochem. Photobiol.* 32 (1980), 711–714.
55 Noonan, F.P., C. Bucana, D.N. Sauder, E.C. DeFabo: Mechanism of systemic immune suppression by UV irradiation *in vivo*. II. The UV effects on number and morphology of epidermal Langerhans cells and the UV-induced suppression of contact hypersensitivity have different wavelength dependencies. *J. Immunol.* 132 (1984), 2408–2416.
56 Norris, P.G., J.N. Barker, M.H. Allen, K.M. Leiferman, D.M. MacDonald, D.O. Haskard, J.L.M. Hawk: Adhesion molecule expression in polymorphic light eruption. *J. Invest. Dermatol.* 99 (1992), 504–508.
57 Norval, M., T.J. Simpson, J.A. Ross: Urocanic acid and immunosuppression. *Photochem. Photobiol.* 50 (1989), 267–275.
58 Osgood, P.J., S.H. Moss, D.J.G: The sensitization of near ultraviolet radiation killing of mammalian cells by the sunscreen agent para-aminobenzoic acid. *J. Invest. Dermatol.* 79 (1982), 354–357.
59 Potten, C.S., C.A. Chadwick, A.J. Cohen, O. Nikaido, T. Matsunaga, N.W. Schipper, A.R. Young: DNA damage in UV-irradiated human skin *in vivo*: automated direct measurement by image analysis (thymine dimers) compared with indirect measurement (unscheduled DNA synthesis) and protection by 5-methoxypsoralen. *Int. J. Radiat. Biol.* 63 (1993), 313–324.
60 Reeve, V.E., M. Bosnic, C. Boehm-Wilcox, R.D. Ley: Differential protection by two sunscreens from UV radiation-induced immunosuppression. *J. Invest. Dermatol.* 97 (1991), 624–628.
61 Rooney, J.F., Y. Bryson, M.L. Mannix, M. Dillon, C.R. Wohlenberg, S. Banks, C.J. Wallington, A.L. Notkins, S.E. Straus: Prevention of ultraviolet-light-induced herpes labialis by sunscreen. *Lancet* 338 (1991), 1419–1422.
62 Rüdlinger, R., M. Norval: Herpes simplex virus infections: new concepts in an old disease. *Dermatologica* 178 (1989), 1.
63 Shaw, A.A., L.A. Wainschel, M.D. Shetlar: Photoaddition of *p*-aminobenzoic acid to thymine and thymdine. *Photochem. Photobiol.* 55 (1992), 657–663.
64 Sutherland, J.C., K.P. Griffin: *p*-aminobenzoic acid can sensitize the formation of pyrimidine dimers in DNA: direct chemical evidence. *Photochem. Photobiol.* 40 (1984), 391–394.
65 van Pragg, M.C.G., C. Out-Luyting, F.H.J. Claas, B.-J. Vermeer, A.M. Mommaas: Effect of topical sunscreens on the UV-radiation-induced suppression of the alloactivating capacity in human skin *in vivo*. *J. Invest. Dermatol.* 97 (1991), 629–633.
66 van Praag, M.C.G., L. Roza, B.W. Boom, C. Out-Luijting, J.B.A. Bergen Henegouwen, B.J. Vermeer, A.M. Mommass: Determination of the photoprotective efficacy of a topical sunscreen against UVB-induced DNA damage in human epidermis. *J. Photochem. Photobiol. B: Biol.* 19 (1993), 129–134.
67 Vermeer, M., G.J. Schmieder, T. Yoshikawa, J.-W. van den Berg, M.S. Metzman, J.R. Taylor, J.W. Streilein: Effects of ultraviolet B light on cutaneous immune responses of humans with deeply pigmented skin. *J. Invest. Dermatol.* 97 (1991), 729–734.
68 Walker, S.L., J. Morris, A.C. Chu, A.R. Young: Relationship between the ability of sunscreens containing 2-ethylhexyl-4'-methoxycinnamate to protect against UVR-induced inflammation, depletion of epidermal Langerhans (Ia^+) cells and suppression of alloactivating capacity of murine skin *in vivo*. *J. Photochem. Photobiol. B: Biol.* 22 (1994), 29–36.
69 Walter, J.F., P.R. DeQuoy: The hairless mouse as a

model for evaluating sunscreens. *Arch. Dermatol.* 116 (1980), 419–421.

70 Wolf, P., C.K. Donawho, M.L. Kripke: Analysis of the protective effect of different sunscreens on ultraviolet radiation-induced local and systemic suppression of contact hypersensitivity and inflammatory responses in mice. *J. Invest. Dermatol.* 100 (1993), 254–259.

71 Wolf, P., C.K. Donawho, M.L. Kripke: Effect of sunscreens on UV radiation-induced enhancement of melanoma growth in mice. *J. Natl Cancer Inst.* 86 (1994), 99–105.

72 Wolf, P., D.B. Yarosh, M.L. Kripke: Effects of sunscreens and a DNA excision repair enzyme on ultraviolet radiation-induced inflammation, immune suppression, and cyclobutane dimer formation in mice. *J. Invest. Dermatol.* 101 (1993), 523–527.

73 Wulf, H.C., T. Poulsen, H. Brodthagen, K. Hou-Jensen: Sunscreens for delay of ultraviolet induction of skin tumors. *J. Am. Acad. Dermatol.* 7 (1982), 194–202.

74 Yarosh, D.B.: The role of DNA damage and UV-induced cytokines in skin cancer. *J. Photochem. Photobiol. B: Biol.* 16 (1992), 91–94.

75 Yarosh, D., L.G. Alas, V. Yee, A. Oberyszyn, J.T. Kibitel, D. Mitchell: Pyrimidine dimer removal enhanced by DNA repair liposomes reduces the incidence of UV skin cancer in mice. *Cancer Res.* 52 (1992), 4227–4231.

76 Yoshikawa, T., V. Rae, W. Bruins-Slot, J.-W. van den Berg, J.R. Taylor, J.W. Streilein: Susceptibility to effects of UVB radiation on induction of contact hypersensitivity as a risk factor for skin cancer in man. *J. Invest. Dermatol.* 95 (1990), 530–536.

77 Young, A.R., C.S. Potten, C.A. Chadwick, G.M. Murphy, A.J. Cohen: Photoprotection and 5-MOP photochemoprotection from UVR-induced DNA damage in humans: the role of skin type. *J. Invest. Dermatol.* 97 (1991), 942–948.

78 Young, A.R., S.L. Walker, J.S. Kinley, S.R. Plastow, D. Averbeck, P. Morliere, L. Dubertret: Phototumorigenesis studies of 5-methoxypsoralen in beramot oil: evaluation and modification of risk of human use in an albino mouse skin model. *J. Photochem. Photobiol. B: Biol.* 7 (1990), 231–250.

Index

Page references in *italic* refer to figures; those in **bold** refer to tables